Eurosensors XII

Sensors Series
Series Editor: **B E Jones**

Solid State Gas Sensors
Edited by P T Moseley and B C Tofield

Techniques and Mechanisms in Gas Sensing
Edited by P T Moseley, J O W Norris and D E Williams

Hall Effect Devices
R S Popović

Sensors: Technology, Systems and Applications
Edited by K T V Grattan

Thin Film Resistive Sensors
Edited by P Ciureanu and S Middelhoek

Biosensors: Microelectrochemical Devices
M Lambrechts and W Sansen

Sensors VI: Technology, Systems and Applications
Edited by K T V Grattan and A T Augousti

Sensors and their Applications VII
Edited by A T Augousti

Sensors and their Applications VIII
Edited by A T Augousti and N M White

Advances in Actuators
Edited by A P Dorey and J H Moore

Automotive Sensors
M H Westbrook and J D Turner

Intelligent Sensor Systems, Revised Edition
J E Brignell and N M White

Ultrasonic Sensors
R C Asher

Eurosensors XII

**Proceedings of the 12th European Conference
on Solid-State Transducers and
the 9th UK Conference on
Sensors and their Applications,
Southampton, UK, 13–16 September 1998**

Volume 2

Edited by

N M White

University of Southampton

Institute of Physics Publishing
Bristol and Philadelphia

British Library Cataloguing-in-Publication Data

A catalogue record for this book is available from the British Library

ISBN 0 7503 0536 3 (Two Volume Set)

ISBN 0 7503 0595 9 (Volume 1)

ISBN 0 7503 0596 7 (Volume 2)

Library of Congress Cataloging-in-Publication Data are available

Series Editor: **Professor B E Jones**, Brunel University

Published by Institute of Physics Publishing, wholly owned by The Institute of Physics, London
Institute of Physics Publishing, Dirac House, Temple Back, Bristol BS1 6BE, UK
US Office: Institute of Physics Publishing, The Public Ledger Building, Suite 1035, 150 South Independence Mall West, Philadelphia, PA 19106, USA

Printed in the UK by J W Arrowsmith Ltd, Bristol

Contents

Acoustics and Ultrasonics   67

VOLUME 2

Flow

Novel integrated gas flow sensor based on porous silicon technology

G. Kaltsas and A. G. Nassiopoulou

Institute of Microelectronics, NCSR Demokritos,
P.O. Box 60228, 15310 Aghia Paraskevi Attikis, Athens, GREECE
Tel: (301) 6542783, Fax: (301) 6511723, e-mail: A.Nassiopoulou@imel.demokritos.gr

Abstract. We report on a novel C-MOS compatible silicon gas flow sensor with thermal isolation assured by a thick porous silicon layer. The good thermal isolation, combined with a small device size (of the order of 1.1mm x 1.5mm) lead to very fast response with a time constant as small as 1.5ms. The principle of operation is based on heat transfer from a polysilicon resistor to the gas, while two Al/polysilicon thermopiles integrated on both sides of the resistor measure the induced temperature differences. The heater and the hot contacts of the thermopiles are thermally isolated. The sensor was tested by measuring nitrogen flows from 0 to 0.4 m/sec. First results showed very good sensor characteristics: sensitivity per heating power 9.8 mV/(m/sec)W, responsivity 1.4 V/W, noise equivalent power and minimum detectable velocity 7 10^{-9} W/Hz$^{1/2}$ and 4.1 10^{-3} m/sec respectively.

1. Introduction

Gas flow sensors are currently used in many applications as for example in natural gas supply, in automotive and environmental applications, in biomedical instrumentation, in air conditioning and refinement of chemicals. Integrated silicon flow sensors have the advantage of being mass fabricated and may be combined with the control and readout electronics on the same chip. In order to overcome problems related to slow response, low signal, high power consumption and the necessity to use a second temperature reference chip, the use of bulk micromachining for adequate thermal isolation is necessary [1,2]. This type of sensors is usually in the form of micro-bridges, which makes them fragile, they don't allow much design freedom and they also demand double side lithography and long etching times. Additionally most of them use KOH for silicon etching which is not C-MOS compatible or EDP which is toxic. As an alternative solution, this paper describes a novel integrated flow sensor which uses porous silicon for thermal isolation [3]. Preliminary results seem to be very promising.

2. Sensor design and fabrication

The flow sensor consists of two series of Al/p-type polysilicon thermocouples forming two thermopiles, placed on both sides of a p-type polysilicon heater on a thick porous silicon fixture consisted of an aluminium tube with dimensions: 4mm x 10mm x 210mm. This fixture can take up to 3 packaged flow sensors at the same time. Care was taken to position the packaged sensors outside the hydrodynamic entry length. The input and output signals were controlled by a PC using the LABVIEW 4.0 package.

The devices used for evaluation consisted of two pairs of thermopiles of 12 thermocouples each, with 10 μm width strips. The distance from the hot contact to the heater was 40 μm and the width of the porous silicon layer was 240 μm. Testing of other sensor geometries are in progress.

Preliminary experiments showed very good sensor characteristics. Polysilicon resistance was found to increase almost linearly with temperature with a slope of 2.84 Ω/°C. This behavior is in agreement with previous results, in the literature [8]. The heater resistance was found to depend almost linearly on the input heating power with a rate of 4.6 Ω/mW, so the heater temperature increase per input power is 1.67 °C/mW. This rate of temperature increase is an indication of very effective thermal isolation by the porous silicon layer.

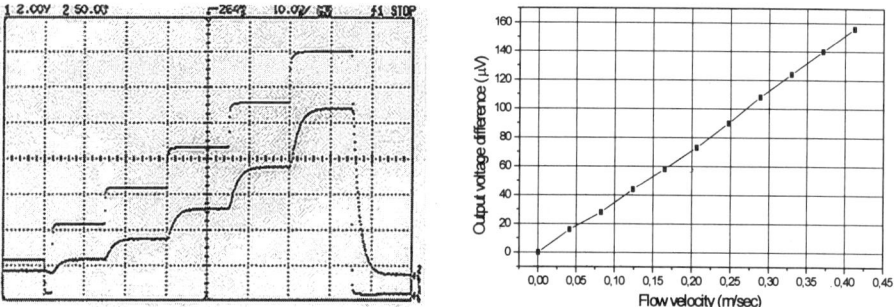

Figure 2: Oscillographs of the thermopile response to six successive heat pulses from 1 mA to 6 mA with a 1 mA step. The width of each step is 15 ms.

Figure 3: Thermopile output voltage difference as a function of flow velocity.

The total signal is obtained as the sum of the signals from each thermopile. The sum increases linearly with power, with a slope of 0.85 V/W. Fig. 2 shows an oscillograph of the thermopile response to six successive heat pulses from 1mA to 6 mA with a step of 1mA. Pulse width was 15 ms and the raise and fall times were 1.5 ms and 1.7 ms respectively. These time constants are very small corresponding to very fast response, which makes the sensor suitable for many interesting applications, as is for example the measurement of the fluctuations in turbulent flow.

For the flow measurements we operate our sensor at constant current of 4mA which leads to an input power of 41 mW. With the previous mentioned rate of change, the temperature of the polysilicon heater was 68 °C above ambient. The signal from each thermopile was 29 mV, which gives a responsivity value for the total signal from the two thermopiles equal to:

$$R = \frac{V_{th}}{P_{in}} = 1.4 \; V/W$$

The noise equivalent power (NEP) of the sensor is defined as the required heating power when the signal-to-noise ratio equals to one:

$$NEP = V_s / R$$

where V_s is the thermopile noise voltage, which is basically Johnson noise, originating from the ohmic resistance of the thermopiles and it can be expressed as:

$$V_s^2 = 4KTR\Delta f$$

layer. The thickness of the porous silicon layer is about 60 µm. The thermal conductivity of porous silicon is 100 times less than that of silicon [4] which makes it an excellent thermal isolation layer [5]. The hot contacts of the thermopiles are near the polysilicon heater and their cold parts are outside the porous silicon area and they are used as reference for the ambient temperature.

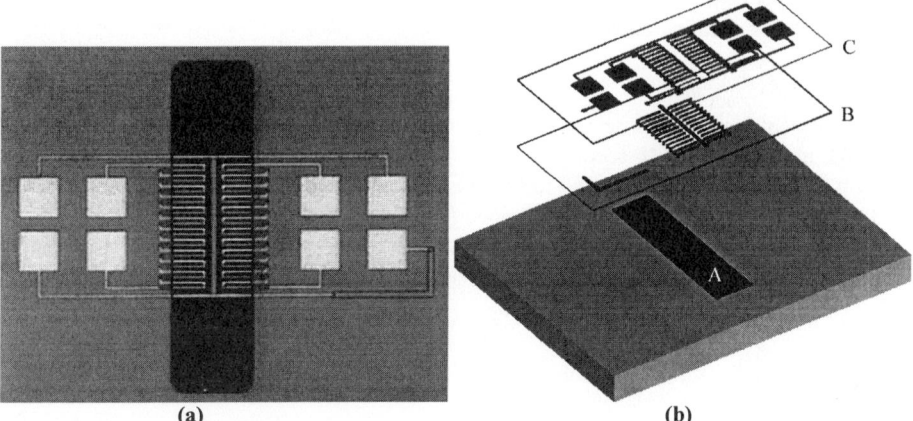

Figure 1: (a) Top view of the flow sensor.
(b) The main parts of the flow sensor. A) Porous silicon isolation layer B) Polysilicon part of thermopiles and heater C) Al part of thermopiles and metal pads.

An image of the sensor is shown in fig. 1(a), while (b) shows the different steps in its fabrication. A second polysilicon resistor is integrated on bulk silicon which serves to stabilise the electrical power of the heater [1]. In order to determine the optimum geometry of the flow sensor different device designs were fabricated with differences in the distance between the heater and the thermopiles, the number of the thermocouples and the porous layer thickness. The polysilicon resistance width was 20 µm in all cases. The sensor was fabricated on a (100), 1-5 Ωcm, p-type wafer in a four mask process. 100 nm SiO_2 was first grown and a 100 nm polysilicon layer was deposited on top by LPCVD. This SiO_2/polysilicon bilayer acts as mask for the selective porous silicon formation [6]. After porous silicon formation a bilayer of TEOS oxide and polysilicon was deposited and the polysilicon layer was implanted by boron and patterned in order to form the heater, the compensation resistance and the first part of the thermopiles. Polysilicon resistivity was 4.6 10^{-3} Ωcm in order to provide both high Seebeck coefficient for the thermopiles [7] and very low temperature coefficient of resistance [8] for the heater. The final step was the aluminum deposition and patterning in order to form the second part of the thermopiles and the metal pads. The overall chip dimensions were: 1.1mm x 1.5 mm.

3. Experimental results and discussion

The experimental set-up for sensor testing consisted of a gas tube, the mass flow controllers, the sensor fixture and the electronics. All the flow measurements were made with pure nitrogen flow at room temperature. Flows in the range of 0 to 1000 sccm were used, which correspond to a flow velocity of 0 to 0.4 m/sec for our measurement set-up. The hydraulic diameter was Dh=5.7 10^{-3} m [9], which for a flow velocity of 0.4 m/sec corresponds to a Reynolds number Re=147, well below the value for turbulent flow (Re=2300). The sensor

The calculated noise was Vs=10^{-8} V and the NEP was 7 10^{-9} W/Hz$^{1/2}$. The signal from the upstream thermopile decreases while this from the downstream thermopile increases with flow. This is consistent with theory [10] and with other results published in the literature [11]. Fig. 3 shows the thermopile difference vs. flow velocity. The signal can be determined as:

$$S = \frac{V_{diff}}{U} = 0.4 \, mV \Big/ \left(m \big/ \mathrm{sec}\right)$$

and the sensitivity per heating power as:

$$S_n = \frac{S}{P_{in}} = 9.8 \, mV \Big/ \left(m \big/ \mathrm{sec}\right) W$$

For our experimental set-up the determined minimum detectable flow velocity was: 4.1 10^{-3} m/sec, which corresponds to 10 sccm mass flow. We believe that the upper limit of the full scale region is well above 0.4 m/sec and further experiments are in progress in this direction.

The flow sensor performance, as obtained from our preliminary testing experiments is very satisfactory. These performances, combined with the advantages of the applied isolation (planar and C-MOS compatible [12]), open new possibilities for future applications.

4. Conclusions

A monolithic flow sensor was fabricated using porous silicon as thermal isolation layer. The sensor was tested at flows from 0 to 0.4 m/sec and showed sensitivity per heating power equal to 9.8 mV/(m/sec)W. The responsivity and the noise equivalent power was 1.4 V/W and 7 10^{-9} W/Hz$^{1/2}$ respectively. The overall sensor dimensions were 1.1mm x 1.5mm. The small sensor size, in addition to the very good thermal isolation, provided by porous silicon, gives to the sensor very fast thermal response with a time constant of 1.5 ms. The above sensor design can be easily combined with C-MOS on-chip interface circuitry, which opens many important possibilities towards smart sensor fabrication.

References

[1]   Moser D and Baltes H 1993 *Sensors and Actuators A* **37-38** 33-37

[2]   Oudheusden B W Van, Herwaarden A W Van 1990 *Sensors and Actuators* **A21-A23** 425-430

[3]   Nassiopoulou G and Kaltsas G *Patent application*, priority No PCT/GR97/00040

[4]   Lang W 1997 *Properties of Porous Silicon* (INSPEC, United Kingdom)

[5]   Tabata O 1986 *IEEE Trans. on Electr. Devices* **ED-33**(3) 361-365

[6]   Kaltsas G, Nassiopoulos A G 1998 *Sensors and Actuators: A*, **65/2-3** 175-179

[7]   Herwaarden A W Van and Sarro P M 1986 *Sensors and Actuators* **10** 321-346

[8]   Obermeir E, Kopystynski P, Niebl R 1986 *IEEE Solid-State Sensors Workshop*

[9]   Incoropera F P, Witt D P De 1990 *Fundamentals of heat and mass transfer* (John Wiley & Sons, New York, Third Edition)

[10]   Schlichting H 1979 *Boundary Layer Theory* (McGraw-Hill, New York)

[11]   Mayer F, Paul O and Baltes H 1995 *Transducers '95 - Eurosensors IX*

[12]   Imai K, Unno H 1984 *IEEE Trans. on Electr. Devices* **ED-31**(3) 297-302

Flow
Paper presented at Eurosensors XII, 13–16 September 1998
© 1998 IOP Publishing Ltd

MOVING GAS OUTLETS FOR THE EVALUATION OF FAST GAS SENSORS

P. Tobias, A. Lloyd Spetz, P. Mårtensson, A. Baranzahi, A. Göras*, and I. Lundström

Swedish Sensor Centre, S-SENCE, and Applied Physics, Linköping University, S-581 83 Linköping, Sweden.
Mecel AB, Box 73, S-662 00 Åmål, Sweden.

Abstract. It is shown that platinum thin-insulator-silicon carbide Schottky diodes operated at about 600°C are fast enough to monitor the air fuel ratio in the individual cylinders in the exhaust from a petrol engine. These chemical sensors have time constants of the order of 1 ms. We describe a simple laboratory technique, which can be used to change the gas composition at a chemical sensor within milliseconds. It is based on mechanically oscillating gas outlets placed close to the sensor surface. The properties of and possibilities with such "moving gas outlets" are described.

1. Introduction

Chemical sensors with fast gas responses are of interest for the monitoring and control of combustion engines. Time constants smaller than about 10-20 ms will make monitoring of individual cylinders in a normal automobile engine possible with only one sensor in the manifold. It has e.g. been shown that films of materials like strontium titanate operated around 900°C are fast enough to follow the combustion in individual cylinders of a petrol engine [1,2].

We have recently demonstrated that platinum - (thin insulator) - silicon carbide Schottky diodes can be used to monitor the air fuel ratio in the exhaust from individual cylinders by mounting one such sensor at the branching point of the manifold [3]. Fig. 1 shows the response of the Pt-SiC diode to changes in the air fuel ratio caused by one cylinder running fat during a number of cycles and the other three cylinders of this 4-cylinder engine, running almost stoichiometric. The response to a rich mixture is a decreased voltage across the diode. The curves for the different cylinders are plotted so that their relative position corresponds to the crank shaft position at the ignition of the particular cylinder. The curves represent from top to bottom cylinder 1, 3, 4 and 2, i.e. the ignition order of the engine. The engine speed was 2400 rpm and the nominal length of an air/fuel pulse from one cylinder was therefore 12.5 msec. The length of the response pulse depends, however, on how rich the cylinder was, the richer the longer the response pulse. A fast enough chemical sensor gives therefore well resolved transient responses, related to individual cylinders and provide also new diagnostic information e.g. from the length of the response pulse. It is therefore of large interest to study such fast sensors to optimize their performance and to understand how they operate. It is, however, difficult to provide gas pulses in the laboratory, which mimic the exhaust pulses of a car engine. This severely limits the possibility to make more detailed investigations of the transient properties of fast gas sensors. We have tried to overcome this problem by using a simple device that we term a "moving gas outlet", which enables a fast modulation of the gas composition outside a sensor surface. The use of such a device has previously been briefly described in a short communication [4]. In this contribution we describe the possibilities and problems with a moving gas outlet in some more detail. This is done in connection with Pt-SiC Schottky diodes.

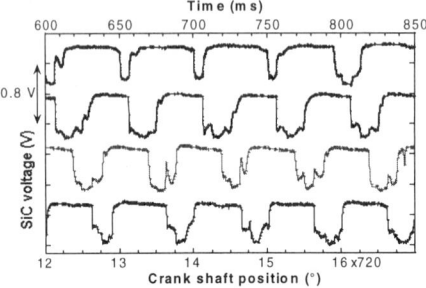

Fig. 1. Response of a Pt-SiC-sensor placed in the manifold of a 4-cylinder petrol engine run at 2400 rpm. The curves show the responses to a fat air fuel mixture (decreasing voltage across the device) in one cylinder at a time. From top to bottom: cylinder1, 3, 4 and 2. Sensor temperature ~700°C. (see the text for further information).

2. Experimental details

Schottky diodes with gate metals of 1 nm Ta or 10 nm $TaSi_x$ and 100 nm platinum were processed as described elsewhere [5]. The SiC surface was ozone cleaned before the deposition of the metals which gives a native oxide, see Fig. 2a. Hydrogen and hydrogen containing gases dissociate on the catalytic metal surface. Hydrogen atoms diffuse within microseconds through the metal and create an electrically polarized layer at the metal insulator interface. The polarized layer causes a shift in the current voltage, IV, curve of the Schottky diode along the voltage axis, see Fig. 2b.

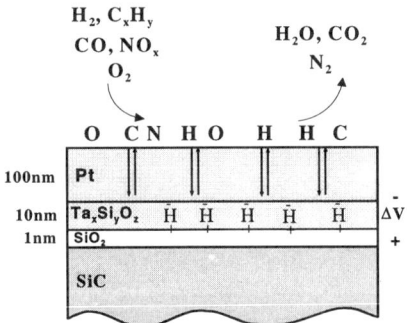

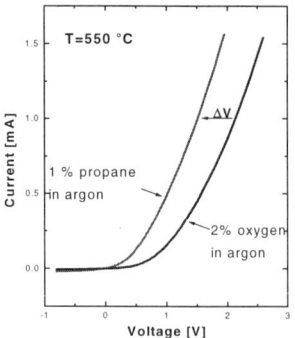

Fig. 2. (a) Schematic drawing of a platinum thin-insulator-silicon carbide Schottky diode and an illustration of the detection mechanism.

Fig. 2 (b) Current voltage characteristics of the device in (a). The voltage across the diode necessary to obtain a given current is smaller in a reducing atmosphere. The response, ΔV, is defined in the drawing.

We developed a gas modulation system consisting of a moving gas outlets as schematically shown in Fig. 3 and the insert of Fig. 4b. The first system had two outlets moving back and forth at a short distance from the sensor surface to rapidly modulate the atmosphere at the sensor (Fig. 4b). This set up avoids dead volumes, although other problems may occur such as the leakage of laboratory air from the sides and that it may be difficult to exactly determine the composition of the gas mixture hitting the sensor surface. An obvious extension of the system in 4b is to use more than two outlets allowing more complicated exposure schemes and also very short test gas pulses by making e.g. the middle outlet very small in Fig. 3. The scheme in Fig. 3 could also minimize leakage of ambient from the sides.

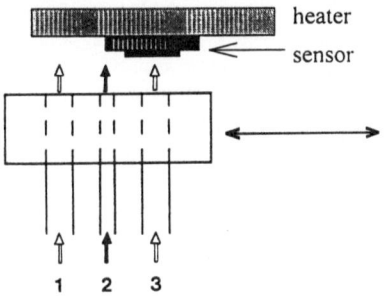

Fig. 3. Schematics of a moving gas outlet with three outlets illustrating a possible extension (to three or more outlets) of the device shown in the insert of Fig. 4 (b) used in the experiments so far

3. Results and discussion

Several parameters will influence the shape of the gas pulses hitting the sensor surface such as the frequency of the motion of the gas outlets, the size of the outlets, their separation and distance from the sensor surface, as well as the amplitude of the motion of the outlets. Furthermore the sensor response may be recorded either as a change in current at a constant voltage, or as a change in voltage at a constant current. Examples are given in Fig. 4 using both possibilities and also different modes of applying the gas outlets. Fig. 4 shows the oscilloscope trace from the current change through a diode subjected to 3% butane in argon as the test gas and 5% O_2 in argon as the reference gas at flows of 60 and 200 ml/min respectively. The response to the test gas is upwards (50 μA/div.) and the time scale is 10 ms per division (frequency 15 Hz, test gas pulse length ~30 ms). In this case, the amplitude of oscillations of the gas outlet was small.

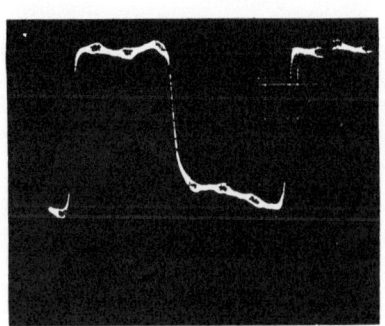

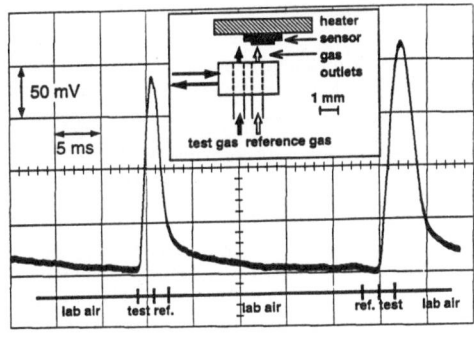

The sensor surface was always above or between the outlets. Furthermore the experimental set up was placed in an atmosphere of argon. In Fig. 4b as illustrated in the insert the amplitude of the oscillations was much larger, and furthermore, the ambient was laboratory air. In this case the voltage necessary to keep a constant current (of 1 mA) was used as the sensor signal. ΔV is positive upwards in the drawing. The frequency of oscillation was 20 Hz. In this case a clear asymmetry is observed in the response to the test gas (1% propane in argon), both in amplitude and speed of response, depending on the fact that in one case reference gas (2% oxygen in argon) proceeds the sample pulse and in the other case laboratory air. (Note that the drawing shows only half of a total cycle). Flow rate 200 ml/min and test gas pulse length ~2 ms. Many details have, however, still to be studied in order to optimize the moving gas outlets. One such detail is related to the number of outlets which of course can be more than two. Already the case with three outlets as illustrated in Fig. 3 gives a number of degrees of freedom. It is now possible, by making the middle outlet very small, to obtain very short test gas pulses. Furthermore, interesting experiments can be made by exposing the sensing surface either to a test gas containing a mixture of two types of molecules or expose the surface to the molecules in different sequences. In such and similar exposing schemes the influence of eventual interactions between molecules in the response of the sensor can be elucidated. Such experiments are presently taking place in our laboratory. One possible difficulty with the moving gas outlets is that e.g. test gas molecules may remain for a long time in the region close to the sensor surface increasing the background response. The background signal increases then with the frequency of the moving outlets. Careful design and proper choice of flow rates, size and position of the outlets will minimize such effects.

4. Conclusions

We believe, that the experimental technique demonstrated here might be very helpful in the characterization of fast chemical sensors. The results indicate that test gas pulses are fast enough to elucidate time constants of the order of 1 msec with the simple device proposed. Furthermore it is obvious that care must be taken in the arrangements of the moving gas outlet and the choice of amplitude for the movement of the outlets.

Acknowledgements

Our research on high temperature chemical sensors based on silicon carbide is supported by a grant from the Swedish Research Council for Engineering Sciences. S-SENCE, Swedish Sensor Center, is financed by Swedish industry, Swedish National Board for Industrial and Technical Development, and Linköping University. The exhaust measurements were performed at Mecel AB in Åmål, Sweden.

References

[1] U Lampe, J Gerblinger and H Meixner, *Sensors and Actuators B*, 7 (1992) 787-791
[2] H Meixner, U Lampe, J Gerblinger, and M Fleischer, *Fresnius J. Anal. Chem.* 348 (1994) 536-541
[3] A Baranzahi, A Lloyd Spetz, P Tobias, I Lundström, P Mårtensson, M Glavmo, A Göras, J Nytomt, P Salomonsson, and H Larsson, *SAE Technical Paper Series 972940, Combustion and Emisson Formation in SI Engines (SP-1300)* (1998) 231-240
[4] P Tobias, A Baranzahi, A Lloyd Spetz, O Kordina, E Janzén, and I Lundström, IEEE Electron. Device Letters, 18 (1997) 287-289
[5] A Lloyd Spetz, A Baranzahi, P Tobias, and I Lundström, Phys. Stat. Sol. (a) 162 (1997) 493-511

Caloric liquid flow sensor for lowest flow rates

J.M. Köhler, V. Baier, T. Schulz, U. Dillner
Institute of Physical High Technology e.V., P.O.Box 100239, D-07742 Jena, Germany
Th. Gehring, S. Howitz
GeSiM mbH, Bautzner Landstr. 45, D-01454 Großerkmannsdorf, Germany

Abstract. The precise determination of flow rates is an essential precondition for the application of micro fluid devices in clinical diagnostics, in quantitative product measurement in miniaturised combinatorial chemistry, and in miniaturised high throughput screening. Based on the well established thermal measurement principle, a sensor concept has been realised, which combines very low transport rates in micro channels with the extremely low parasitic heat capacity of a free-standing thin film transducer. This way, flow rates in the nl/min region can be measured precisely. The sensor is a 3-chip module. It can easily be connected with other micro fluid modules as reactors, pumps and chemical sensors, and is, therefore, well suited for freely configurable micro fluid arrangements. It detects the motion of a liquid column caused by the release of single droplets (500 pl) from a Si chip pump in the closed channel.

1. Introduction

Many efforts have been done during the last two decades in order to minimise volumes for analytical procedures. Most of these activities have been focussed in the field of chemical sensing. But the precise handling of small amounts of liquids has remained an essential demand beside the chemical detection. The development of active micro fluid devices like pumps and valves has opened the way to the application of micro system technology in chemistry, molecular biology, and biochemistry. Miniaturised devices are under development for biological analysis, medical diagnostics, combinatorial chemistry and high throughput screening. Piezoelectrically driven micropumps based on silicon chip technology are very efficient tools for dispensing small droplets. However, the size of generated droplets is hard to control. There is a lack of small devices, which allow to detect low volume flow rates or to count single droplets in the nanolitre and the picolitre region. Therefore, it should be tested, if a microcaloric chip device could be able to measure lowest flow rates and detect transport of liquid in micro channels connected with the release of smallest droplets.

2. Sensor concept

The measurement of a mass flow by the change of a thermal signal is used in many different classical and chip devices. Devices with a symmetrical arrangement of a heater and one or more pairs of detectors proved to be particular suited in the detection of small flows in gases and liquids [1-6]. In these arrangements, thin film transducers on free standing membranes offer a very important advantage due to the low parasitic heat capacity of the tiny volume of the measurement system and the low heat transport rates in the membrane plane.

The sensor reported here is based on the inclusion of a free-standing thin film transducer stack, which is in contact with the streaming liquid from both faces, in the centre of a micro fluid channel. This way, the exchange of heat between liquid and transducer through the walls of the fluid channel should be reduced. The central volume part of the streaming liquid determines the heat exchange between liquid and transducer. In comparison to the gas flow sensors, the relative parasitic heat capacity of the transducer is clearly reduced because of the fact, that the heat capacity of liquid is much larger then the heat capacity of gases. This way, the heat exchange conditions are nearly completely controlled by the properties of the liquid and the flow rate itself. Hence, the sensor should not only provide a high sensitivity but it should also exhibit a very low time constant. Therefore, it should become possible to detect fast changes in the rate of small liquid flows.

Each of the three pairs of detector resistors is fabricated with a different distance to the central heater (cf. Fig. 1, b). Thus, it becomes possible to use the sensor for very different flow rates. The detectors situated in the immediate neighbourhood of the heater are suited for the detection of smallest deformation in the temperature profile on the right and the left of the heater in the case of smallest flow rates. The outer detectors are suited best for flat and strongly asymmetrical temperature profiles in the case of high flow rates.

3. Sensor preparation

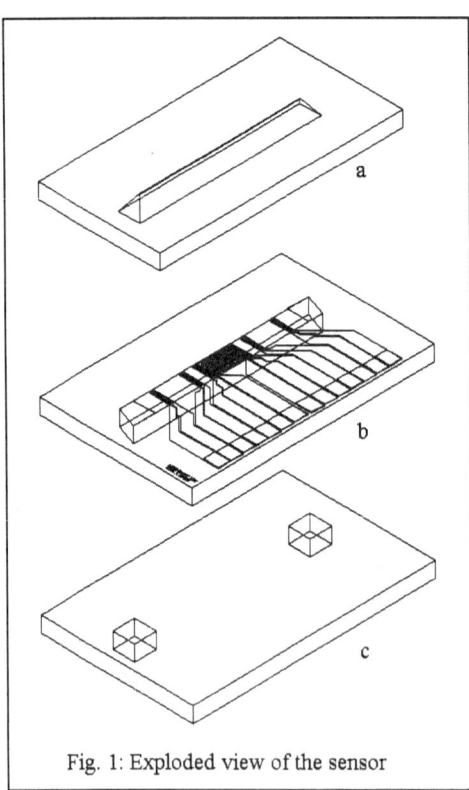

Fig. 1: Exploded view of the sensor

The three sensor chips (cf. Fig. 1) were fabricated by thin film deposition and microlithography. Silicon (100) was used for the upper covering chip (Fig. 1, a). The fluid channel was wet etched anisotropically by KOH leading to a V-groove shape. The transducer chip (Fig. 1, b) was made by silicon (110) in order to prepare perpendicular side walls of the fluid channel. The channel was formed by back side etching. Silicon nitride was used as mask layer for this anisotropical wet etching process. The front side was coated by a silicon oxinitride layer, which was stress compensated. This layer remained as a free standing membrane of about $1\mu m$ thickness after wet etching of the silicon. To form the heater and the detector resistors on this free-standing layer, a platinum film with an adhesive layer of Ti was deposited and patterned by photolithography and sputter etching in an argon plasma. Aluminum bond pads were fabricated by high rate sputter deposition and subsequent wet etching. The resistors for detection are smaller and have a higher electrical resistance than the heater, which is formed by a larger meander. The free-standing membrane also supplies the orifices for the liquid exchange between both part of fluid channel, which are formed by reactive ion etching of the membrane system in an SF_6 plasma. Moreover, CVD of SiO_2 is necessary for passivation of the heater and detector films against the fluids. Finally, the third chip (Fig. 1, c) is made of silicon (100) with the inlet and outlet channels fabricated by wet etching. The sensor was completed by glue bonding of the three chips, making the electrical contacts by ultrasonic bonding and attaching the fluid connectors.

4. Measurement

The measurements were carried out by using a bridge circuit with the two corresponding detector resistors of the sensor in one side of the bridge. Therefore only the differential signal which is related to the changing fluid flow is registered. The influence of temperature variations related to the incoming fluid on the result of measurement is minimised by the bridge arrangement.

5. Test

At first, the sensor was integrated in a steady (pulse-free) micro fluid arrangement with a micro ejector pump. The flow rate was stepwise increased from 1 µl/min up to 10 µl/min. As shown in Fig. 2, the onset of the sensor signal is very clearly observed even at the lowest flow rate.

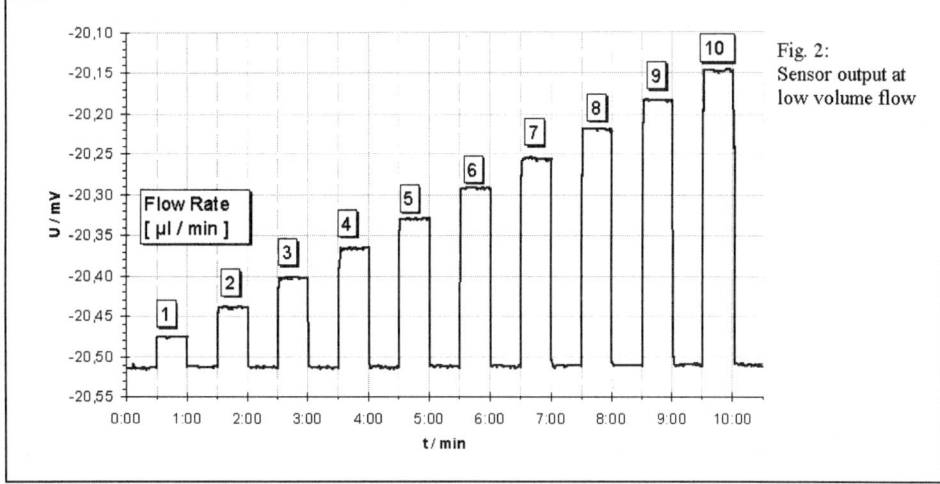

Fig. 2: Sensor output at low volume flow

The difference in the output signal between no flow and a flow of 1 µl/min is about 40 µV, which can easily be processed in the electronic amplifier. The signal to noise ratio was estimated to be better than 10 in case of 1 µl/min flow. The measurement over one decade of flow rates reflects the linearity of the signal.

At second, the sensor was tested in order to detect the movement of the liquid column inside the closed fluid channel system during the formation of a single droplet at the outlet of a micropump. Series of pulses for the generation of single droplets with a frequency of 0.5 up to 15 Hz were applied for testing the dynamic behaviour of the sensor. The upper frequency was restricted by the maximum conversion speed of the analog-digital converter of 30 samples per second.

Fig. 3 shows the signal for 5 single droplets with a volume of approximately 1 nl and a delay of 2 seconds. The signals of the 5 pulses are completely separated in time. The time constant, which results from the superposition of the real liquid movement and the thermal

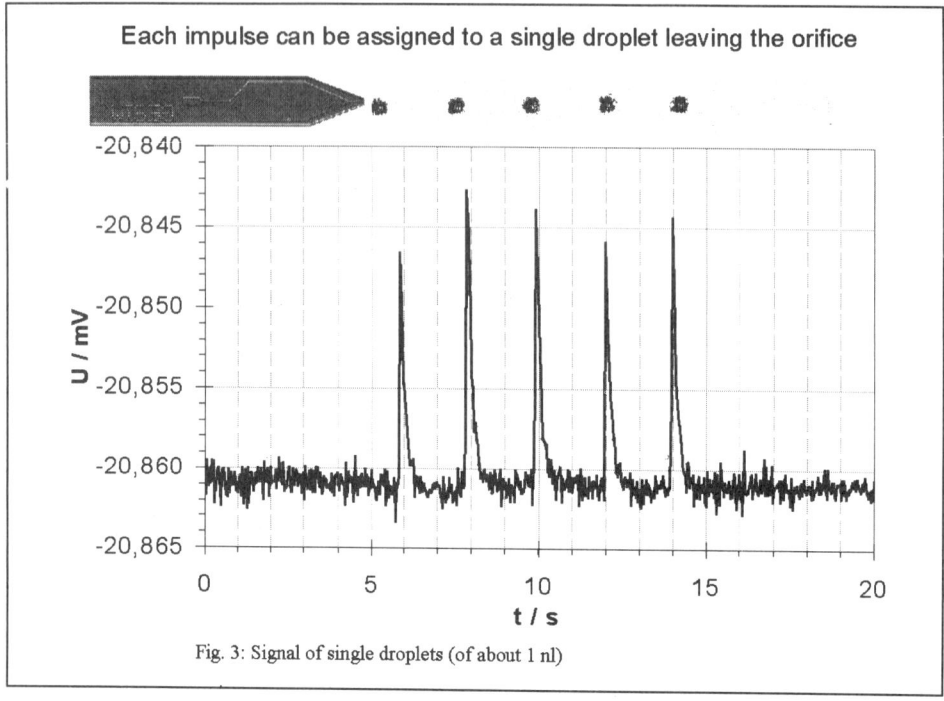

Fig. 3: Signal of single droplets (of about 1 nl)

transducer system is in the range of 250 ms. The signal maximum amounts to about 15 μV, which is sufficient for a convenient electronic processing.

6. Conclusion

A very sensitive and fast working flow sensor for liquid was manufactured using the caloric principle connected with a transducer system placed on a free-standing thin film membrane. Thus, a strong reduction of parasitic heat capacities of the transducer system was successfully realised. Micro system technology including metallic thin film transducers supplied an arrangement suitable for fast detection of liquid motion in micro channels.

References

[1] Lammerink S J, Tas N R, Elwenspoek M and Fluitman H J 1993 *Sensors and Actuators A* 37-38 45-50
[2] Van der Wiel A J, Hoogerwerf A C and de Rooij N F 1993 *Tech. Digest, 7th Int. Conf. Solid-State Sensors and Actuators, Yokohama* 800-3
[3] Mayer F, Paul O and Baltes H 1995 *Tech. Digest, 8th Int. Conf. Solid-State Sensors and Actuators, Stockholm* Vol. 1, 528-31
[4] Nguyen N T and Kiehnscherf R 1995 *Sensors and Actuators A* 49 17-20
[5] Qiu L, Hein S, Obermaier E and Schubert A 1996 *Sensors and Actuators A* 54 547-51
[6] Dillner U, Kessler E, Poser S, Baier V and Müller J 1997 *Sensors and Actuators A* 60 1-4
[7] GeSiM mbH: Flowsensor Data Sheet 1997

Ferrofluids on micromechanics: preliminary results

R. Perez-Castillejos, J. A. Plaza, J. Esteve, P. Losantos, M.C. Acero, C. Cané, and F. Serra-Mestres.

Centro Nacional de Microelectrónica (CNM) CSIC
Campus UAB 08193 Bellaterra Spain
Telephone: +34 3 580.26.25, Fax: 580.14.96
E-mail: raquel@cnm.es

Abstract

An introduction to ferrofluids in MEMS applications is presented. Ferrofluids are fluids with magnetic properties. By applying a magnetic field, the balance of forces within the ferrofluid is varied so that the magnetic fluid can move or even apply pressure. These capabilities can be used in the field of the microsystems. In this work we have obtained the typical values of pressure which can be expected from a ferrofluid. By a piezoresistive pressure sensor a pressure of 0.05 bars has been obtained.

Introduction

In the last years, new principles of actuation for micromechanics have been searched. Although using magnetic forces has always been an interesting solution, its final realisation has to face the same problem. Typically, Miniaturised permanent magnets [1] and ferro-magnetic layers have been used in microelectronics in order to integrate a material that was able to answer to an applied magnetic field. This kind of deposited layers is mainly bidimensional and, as the magnetic forces depend on the volume, the generated magnetic forces are too small. Ferrofluids provide a solution to this problem by offering an absolute adaptability to any geometry. Moreover, these fluids can be easily moved in a microsystem as channels, and cavities, for instance.

The use of magnetic fluids in combination with microsystem technology will define new concepts and applications of integrated microsystems that can be incorporated into new products as for example, biosensors and magnetic microactuators. Until now, ferrofluids have never been used in combination with microsystems technologies. In the approach of the present work, we will study the viability of using ferrofluids in microsystems by a first and simple experiment. For the first experiment, we have studied the typical pressure that a ferrofluid can generate over a silicon membrane.

Ferrofluids

Magnetic fluids or ferrofluids are stable suspensions of single domain magnetic nanoparticles in a carrier liquid [2]. The magnetic particles are coated with a molecular layer of a dispersant which is compatible with both the carrier liquid and the magnetic particles. Thermal agitation keeps the particles suspended because of Brownian motion, and coatings act as an elastic cushion preventing the particles from sticking to each other. Well-designed ferrofluids do not settle out and preserve their magnetic properties over long periods of time.

Size of the particles makes the difference between the two types of fluids which depend on the applied magnetic field: ferrofluids and magneto-rheological fluids. Magneto-rheological fluids are stable suspensions of magnetically polarisable micron sized particles suspended in a carrier fluid. These fluids vary their viscosity with the applied magnetic field and they can solidify in the presence of a sufficient magnetic field. In comparison, colloidal ferrofluids (typically, a suspension of magnetite particles of approximately 100Å in diameter) retain liquid flowability even in the most intense applied magnetic fields. Ferrofluids use to be described as magnetically *soft* material, which means that the magnetism vector follows the applied field without hysteresis and a small applied field is required to produce saturation. Ferrofluids can also be characterised as isotropic and they exhibit *superparamagnetic* behaviour [2] which is similar to paramagnetic behaviour except that the magnetisation in low to moderate fields is much larger.

The strength of magnetisation, M_S, of a ferrofluid determines how much pressure it can produce in a given magnetic field [3]: the higher the M_S, the greater the magnetic pressure exerted by the fluid. Higher saturations of magnetisation are achieved by increasing the volume concentration of solids in suspension when the fluid is manufactured, therefore viscosity requirements will define the upper limit of magnetisation of the ferrofluid. Summarising, ferrofluids behave as an electrically non-conducting homogeneous continuum which trends to move towards the region where the magnetic field is stronger. The colloidal ferrofluid must be synthesised, for it is not found in nature. Since the first ferrofluid was synthesised [4] many uses have been proposed and realised: sealing for several industrial processes, loudspeaker applications, inertia dampers, angular position sensors or inkjet printing. Ferrofluids have also been used as a biologically compatible material for biomedicine and for magnetic separation in biotechnology.

Experimental set-up

In order to test the magnitude of the pressure that can be achieved using ferrofluids, a pressure sensor was used. As it was previously introduced, a ferrofluid responds to a non-uniform magnetic field by collecting in regions where the field is more intense. To reproduce this behaviour was our intention when designing the experimental set-up (fig. 1).

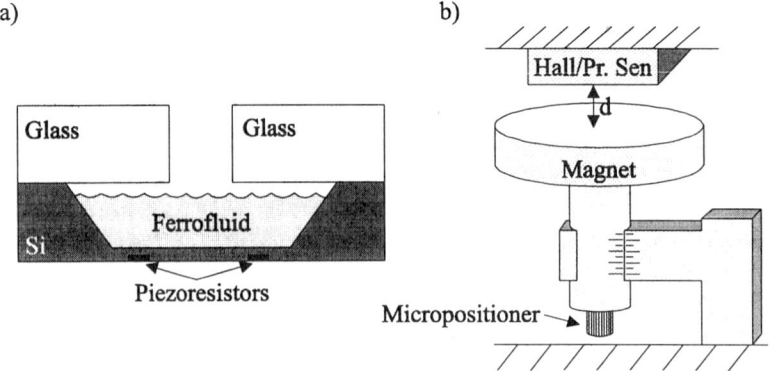

Fig.1: a) Pressure sensor filled in with ferrofluid, b) experimental set-up.

The used pressure sensor was based in the piezoresistive effect given by a Wheatstone bridge. After testing, the sensibility of this sensor was found to be 600mV/bar [5]. The area of the membrane is 1500x1500µm² and the thickness is 15µm. The cavity inside this pressure sensor was filled in with ferrofluid (fig. 1.a). We chose a hydrocarbon based ferrofluid (WHJS1) purchased by Liquid Research Ltd. The magnetic saturation of this magnetic fluid is 350G.

The gradient magnetic field was supplied by a magnet which was placed over a micropositioner (fig. 1.b). In this way the strength of the applied field could be precisely controlled by the distance between the sensor and the magnet. A calibrated linear Hall effect sensor (USM 503) was used to measure the field flux density (B) generated by the magnet as a function of the distance (d).

Results

Firstly, the Hall sensor was placed over the magnet to determine the level of magnetic field versus the position of the magnet (d). The magnetic field versus the distance from the magnet is shown in figure 2.a.

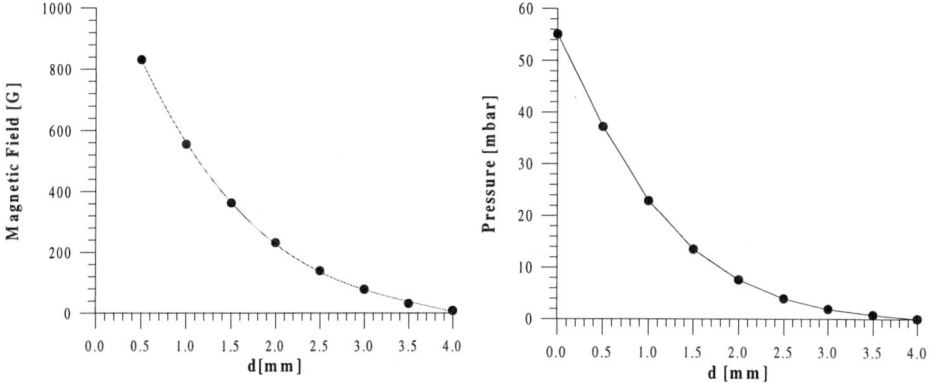

Fig.2: a) Magnetic field and b) magnetic pressure versus the distance to the magnet.

After measuring the magnetic field, the Hall sensor was replaced by the pressure sensor to check whether any change could be induced by the magnetic field over an empty pressure sensor. No differences were found between the sensor's characteristics with or without an applied magnetic field.

Then, the pressure sensor was filled in with ferrofluid. When the magnet was moved towards the sensor, so that the magnetic field increased, the magnetic pressure relative to the ferrofluid bent the membrane. The magnetic pressure versus the distance of the magnet to the sensor is plotted in figure 2.b. The maximum pressure achieved was 0.05bar, but of course, this value depends on the quantity of ferrofluid filling the cavity. The high level of pressure that can be achieved with this device suggests the high interest that the use of ferrofluids has as the actuator, e.g. for a micropump. Another interesting value is the maximum displacement induced by this pressure. The membrane has been simulated by the FEM using ANSYS 5.3. The results

772

obtained by a non-lineal solution are shown in figure 3. The deformation of the membrane is shown in figure 3.a and the maximum displacement of the membrane versus the magnetic pressure, in figure 3.b. More than 6μm displacements were obtained for the maximum magnetic pressure.

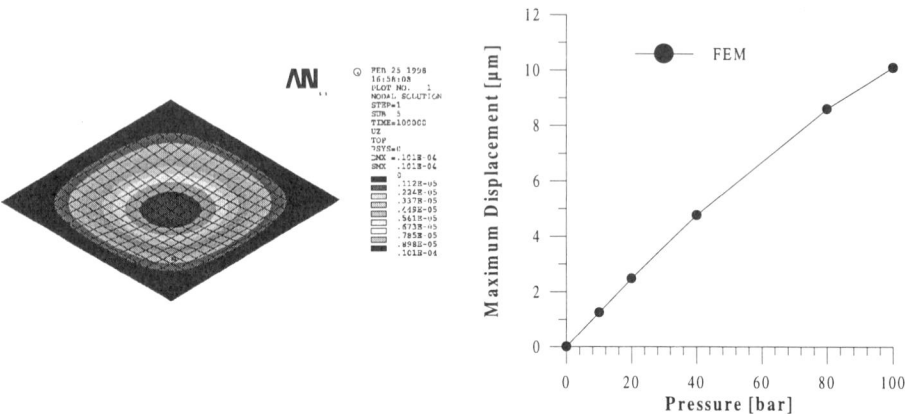

Fig. 3: a) Deformation of the membrane, b) maximum displacement of the membrane versus the magnetic pressure.

Conclusions

Ferrofluids have been proved to be useful for the microsystems applications. In this work our purpose was to start a basic research activity in order to evaluate the potentiality of ferrofluids. Our experiments with a pressure sensor show that a level of pressure in the range of decades of mbar can be achieved with applied magnetic fields in the range of hundreds of Gauss. Results of this preliminary experiment show a high potential in combining both ferrofluids and microsystems, therefore further research is under development for new applications.

Acknowledgements

This work has been supported by the Brite Euram III Program (BRPR-CT97-0598). We are especially grateful to Liquids Research Ltd. in Dublin, Ireland, for their procurement of the ferrofluids used in this study.

References

[1]   W. Benecke, Scaling behaviour of Micro-Actuators, Actuator 94, Bremen, June 1994.

[2]   Rosensweig, R.E., Ferrohydrodynamics, 1985, Cambridge University Press, New York.

[3]   Miller, C.W., Magnetic fluids: magnetic forces and pumping mechanisms, NTIS Final Report AD/A-006323, September 1973, The Sibbley School of Mechanical and Aerospace Engineering, Cornell University, Ithaca, New York.

[4]   Rosensweig, R.E., Magnetic Fluids, 1966, International Science and technology, Vol. 55, pg. 48-56.

[5]   Ángel Merlos, Joaquín Santander, Francesca Campabadal, Desarrollo de un sensor de presión, CNM's Internal Report, December 1997.

Fluidic Oscillator for Flowmeter

Miroslav Husák

Czech Technical University in Prague, Faculty of Electrical Engineering,
Department of Microelectronics, ,
Technicka 2, Prague 6, Czech Republic
Tel.: +420-2+2435 2267, Fax.: +420-2-2431 0792, e-mail::husak@feld.cvut.cz,

Summary

The article describes design and realisation of a flowmeter structure. The principle of a flowmeter sensor operation is based on the mechanism of a fluidic oscillator. It uses fluidic paths in feedback. Oscillator samples have been made using different technologies and materials. Micromechanical shaping, etching and sticking as basic operations have been used. Further there is described design, construction and measured results of the developed flowmeter with a fluidic oscillator for measurement of fluid flow. Measured fluid flow is displayed on a digital or analog measuring device calibrated in values of immediate flow amount or flow velocity.

Introduction

Flowmeters with fluidic oscillators enable to measure flow of liquids and gases without necessity to re-adjust for various substances. The core of the fluidic oscillator is made of a fluidic bistable amplifier, completed with a feedback. This feedback is created by connecting output and control channels (see Figure. 1). In the feedback circuit, oscillations of fluid emerge. In injection jets the pressure energy of the fluid is changed to kinetic energy. In collectors the fluid flows in the direction of opening enlargement and kinetic

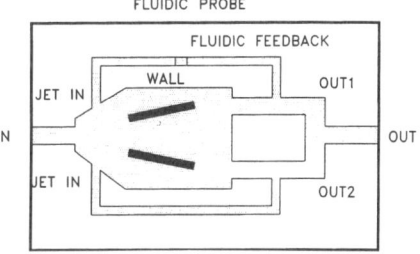

Figure 1: Principle of the fluidic oscillator

energy of captured fluid changes into pressure energy there. Effect of Coandon phenomenon causes adhesion of liquid/gas flow from the input jet IN to one or another holding wall. It is possible to turn over the flow from one wall to another one by impulses led to injection jets JET IN. Connection of output fluid paths OUT1 and OUT2 with injection jets creates feedback that constitutes an astable flip-flop system with turn-over of the fluid flow alternately to one or another wall. In the feedback the signal is delayed - shifted in phase. This phenomenon arises because the accummulation chamber is filled through the resistance (narrowing path). Further there is utilized fluid inertia that causes spreading of the signal in the form of a pressure wave. In the channels it is manifested by pulsating fluid. The signal has an impulse character. In this way a fluidic system is created that is similar in its behaviour to an electric astable flip-flop circuit with pulse shaped output signal. With respect to the analogy with the electric astable system it is called „fluidic oscillator". The output flow from the fluidic oscillator is marked OUT and it is composed of the outputs OUT1 and OUT2. Frequency of these oscillations is dependent only on fluid flow velocity and diameter of feeding pipes. It does not depend on properties of the flowing fluid in a certain range of Reynold's numbers [1]. Sensor systems with fluidic oscillators can be used for measurement of liquid and gas flow with temperatures ranging from -20 to +120 °C.

The conversion characteristic is a linear function for big Reynold's numbers. The output signal is electrical or pneumatical with impulse or analog shape. The fluidic oscillator does not contain any moving parts, it is suitable for gases and liquids with different properties without necessity of readjustment (air, water), its operation is independent on temperature in a wide range.

Construction And Production Technology of Fluidic Oscillator

For development of this sensor probe, various materials and technologies have been used. First samples have been made by precise microshaping of polymethylmetaacrylate. Individual middle reliefs of the probe have been made by precise microshaping. Then the outer walls have been sticked precisely to individual reliefs. Various relief shapes with different dimensions have been designed and realized. Afterwards simulation has

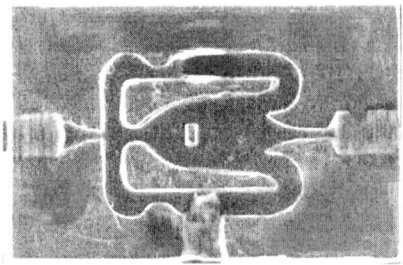

been performed and subsequent measurements for optimal setup of mutual position of individual relief parts have been done so that optimal parameters could be obtained. One of these parameters is reaching maximum possible oscillation amplitude at minimum possible volume of higher harmonics. The ratio height/width of input jet IN has been chosen 3/1. The angle between holding wall and oscillator axis is 11° or 12°. On Figure 2 photography of one realized structure of fluidic oscillators is shown.

Figure 2 Samples of realized structures of fluidic oscillators

Conclusion

The fluidic oscillator has been completed with an electronic connection for impulse processing from the sensor.. A block diagram of the whole sensor system is on the Figure 3. Several samples of sensor systems with fluidic oscillators with heated resistance temperature sensors have been made and measured. The differencies between samples have been in size and layout of the inner relief. For individual systems with fluidic oscillators, the oscillation frequency has been measured by a counter. The frequency has been measured in dependence on flow volume Q at different fluid temperatures in the range from 5 to 35 °C, namely both at increasing and decreasing fluid flow volume. Fluid temperature has no influence on measured values in the given range.

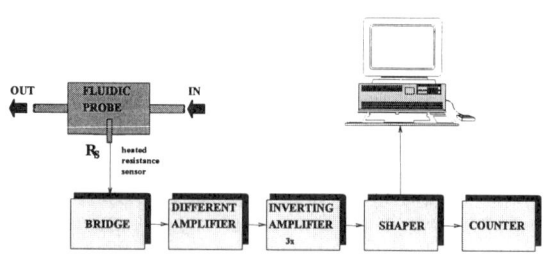

Figure 3: A Block diagram of the whole sensor system

Flow
Paper presented at Eurosensors XII, 13–16 September 1998
© *1998 IOP Publishing Ltd*

A thermal flow sensor for fluids with dynamic velocity, temperature or conductivity profiles

Dan S. Popescu*, Philippe Lerch**, Philippe Renaud**, Camelia Dunare***, Mircea Modreanu***, Dan Dascalu***

*University "Politehnica" Bucharest, Department of Electronics, Physical Electronics Centre, Iuliu Maniu 1-3, Bucharest, Romania
*Swiss Federal Institute of Technology Lausanne, Institute of Microsystems, CH-1015 Lausanne, Switzerland
***National Institute for R@D in Microtechnology, Bucharest, Romania, CP38-160

Abstract
A new type of integrated thermal flow sensor containing two simetrical measuring channels is presented. The structure permit optimisation of flow sensing in applications with transient velocity, temperature or conductivity profiles in the fluid and has improved sensitivities for small temperatures differences above ambient, necessary for medical applications. Analysis is made for fluid flow and heat transfer using analitical and FEA technics. Technological implementation optimise the compensation of the internal stresses in the suspended structures and passivation of the sensor structure for work in aggressive media. Electronic interfaces for processing are derived for development of a fully integrated variant of the sensor.

1. Introduction

This sensor is an anemometer type thermal flow sensor. The main problem in such structures is the dependence of the sensor parameters also of the fluid temperature [1-2]. In our design a second identical channel is added at the structure. This give a reference related with the fluid temperature and permits a differential type processing with improved performances. The temperature of the heater in the first channel can be kept at a constant temperature difference from the fluid also for variable temperature profiles of the fluid.

Top view and a cross-section on the sensor are schematically shown in Fig.1.

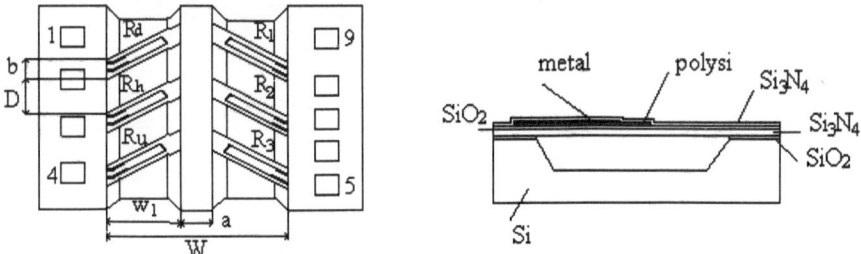

Fig.1 Schematic top view and cross-section of the sensor

All the resistors are from polysilicon deposited on beams from Si_3N_4 and oxide layers and coated with a BPSG or with a Si3N4 protection layer.

2. Flow sensor structures

The silicon part of the sensor is presented in Fig.2. The sensor can be closed by bonding with a cap of Pyrex containing two cavities with the same depth as the two silicon channels, for applications in microfluidics or can be used only the silicon structure in other applications as in respiratory control.

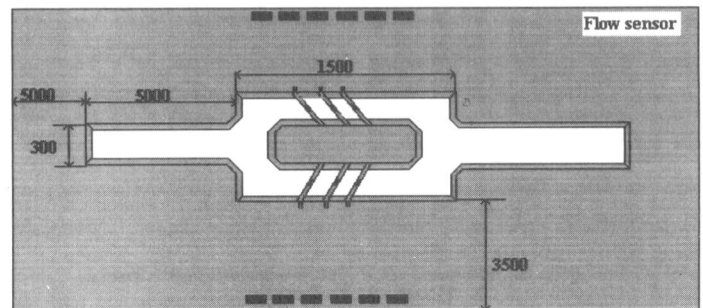

Fig.2 The silicon structure of the sensor

3. Analysis and modelling

First a 2D finite element analysis in the upper part of height l_z of the vertical cross-section have been made. In Fig.3 temperature distributions are plotted for: $h=20\mu m$, $b= 30\mu m$, $D=60\mu m$ and for a difference of temperature of $10^{o}C$ between the heater and the fluid. The results are very dependent of the fluid thermal diffusivity coefficient, $\qquad \alpha=K/\rho c$ $\qquad\qquad\qquad$ (1) K-the thermal conductivity, c-the heat capacity of the fluid.

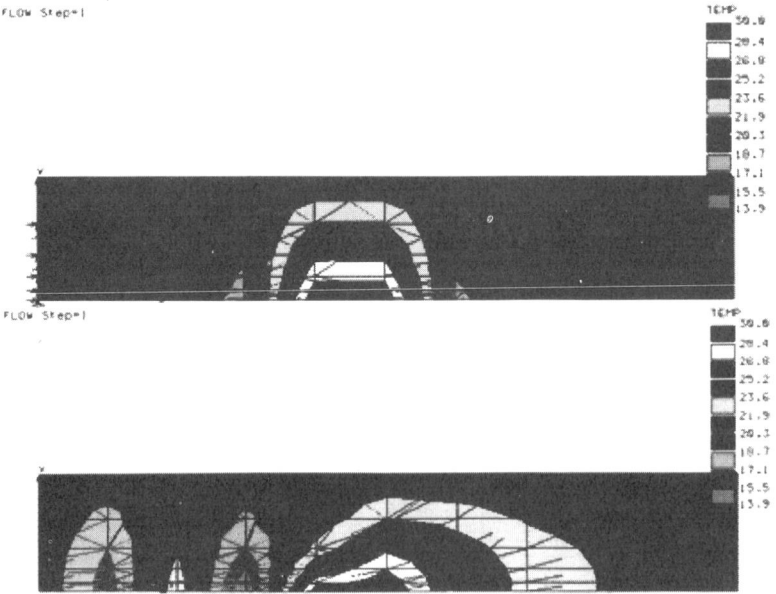

Fig.3. Temperature distribution for $2l_z=20\mu m$, V=60mm/min and
a) K=1.4e11mg $\mu m/s^3\,^{o}C$, b) K=1.4e9 mg $\mu m/s^3\,^{o}C$

As is shown in Fig.3.a, due to the small height of the channel the sensor response for the propanol flow is very low. In Fig. 3.b a difference of temperature of almost 2°C is obtained in the same channel if the thermal diffusivity coefficient decrease with two order of magnitude.

In order to increase the sensitivity for low velocities when the coefficient α is not very low, the height of the channel must be increased. For a 200μm height channel a temperature difference of about 2°C result for propanol. The change of the thermal conductivity has a high influence. This can be used in microseparation systems but a compromise with the small dead volume is necessary for the channel depth.

Sensor sensitivity is given by the relation, [1]:

$$S = (T_h - T_{fl}) \cdot \frac{x_m - b}{\alpha} \cdot \exp{-(x_m - b)/l_z} \quad \text{where } x_m = x_u = x_d \tag{2}$$

Heat transfer from the fluid to the measuring resistors

With m, mass of the suspended resistors, $T_r(t)$ and T, resistors and fluid temperatures, result, [3]:

$$T_r(t) = T_r(0) + (T_{fluid} - T_r(0)) \cdot \left(1 - e^{-\frac{\bar{h} \cdot A}{m \cdot c} \cdot t}\right) \quad \text{for an input step temperature} \tag{3}$$

where: $\tau = \dfrac{mc}{\bar{h} \cdot A}$, the time constant and $\omega_o = \dfrac{2\pi}{\tau}$, the cut-off frequency

$$T_r(t) = \frac{\Delta T_{fluid}}{2 \cdot \sqrt{1 + (\omega/\omega_o)^2}} \cdot \sin(\omega t - \Phi) + \bar{T} \quad \text{for a sinusoidal input temperature} \tag{4}$$

The heat transfer coefficient can be calculated from the Nusselt number:

$$\overline{Nu} = \frac{\bar{h} \cdot \Delta x}{K} = 0.33 \cdot Re_x^{1/2} \cdot Pr^{1/3} \quad \text{with} \quad Re = \frac{u_m \cdot L}{\mu/\rho}, \quad Pr = \mu \cdot c / K \tag{5}$$

Case study

For: $b=30$μm, $x_m=l_z=100$μm, $A=30$μm*30 μm, $L=1000$μm, fluid propanol: $K_{fl}=0.14$W/m °k, $\rho_{fl}=785$kg/m³, $\mu=2.4*10^{-3}$Ns/m², polysilicon parameters: $\rho=2300$kg/m³, $\alpha=1.1*10^{-3}/°C$ for phosphorus doped polysilicon, velocity $u_m=10$mm/min
$T_h-T_{fl}=10$°C, $U=5$V-the supply voltage of the measuring Wheastone bridge,
result: $\Delta T=T_d-T_u=0.63$°C, $\Delta U = (\alpha \cdot \Delta T/2) \cdot U = 1.73$mV, Re=$54*10^3$, Pr=49, $\overline{Nu}=2.77*10^{-2}$
$\bar{h}=1292$ W/m²°C, $\tau=0.6$ms, $\omega_o=16$kHz

4. Technology experiments

The minimum obtainable dimensions, width and thickness, for the supporting beams are dependent of the stress induced in the layers. Experiments are made for the following variants: sandwiches of LPCVD Si3N4 and thermal SiO2 for total stress compensation, beams of LPCVD silicon rich nitride, and beams of SiON.

Passivation oxide layers are experimented for resistance in KOH during etching. Results are in table1. The best results are for BPSG layers.

Table1

Tip oxide	Etching rate (Å/min)		
	Non-densified	Densified* 1000 °C	Densified** 425 °C
SiO2	250	175	200
PSG3%P	300	205	255
BSG10%B	135	95	100
BPSG4%P,4%B	115	92	97,5

*Densification at 1000°C, 35 min in wet O2. **Densification at 425 °C, 18h in H2O vapours.

778

1. Processing interface

An electronic interface variant is presented in Fig.4. The temperature difference between the heater and the fluid is dynamically kept constant.

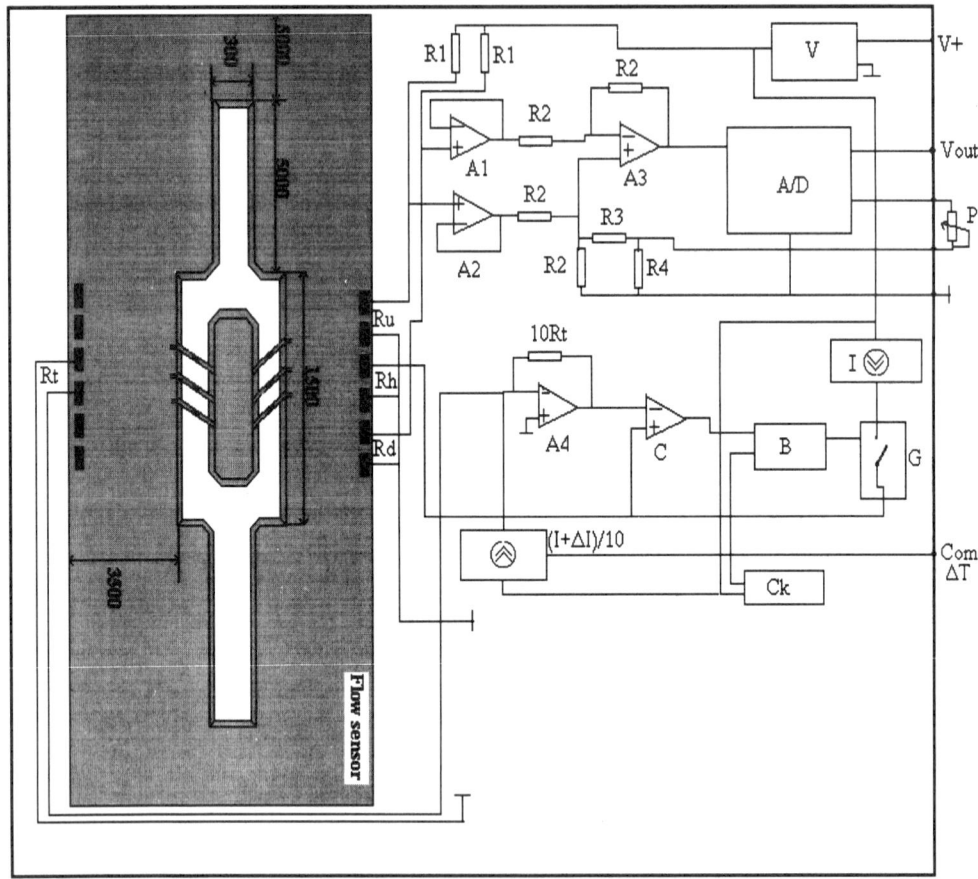

Fig.4 An electronic interface for the sensor

5. Conclusion

A high performance thermal flow sensor can be integrated for use in applications with dynamic profiles for the fluid temperature and velocity.

Acknowledgements

This work was supported by the Swiss National Foundation under the project no. 7RUPJO48726

References

[1] Theo S.J. Lammerink, N. Tas, Miko Elwenspoek and Jan Fluitman, "Micro-liquid flow sensor". Sesors and Actuators A, 37-38 (1993), 45-50.
[2] Neda T., Nakamura K., Takumi T., "A polysilicon flow sensor for gas flow meters", Proc. of Transducers'95, pp.548-551.
[3] Dan S. Popescu, Bogdan Mihalea, Philippe Lerch, Philippe Renaud, Dan Dascalu, "Design and modelling for a thermal flow sensor", Proc. of CAS'97 Conf., Oct.'97, Sinaia, Romania, pp. 511-515.

Flow
Paper presented at Eurosensors XII, 13–16 September 1998
© *1998 IOP Publishing Ltd*

Calibration of a novel airflow transducer for use in pneumotachography

E F J Coolen and M J West

Coventry University, School of Engineering, Priory Street, Coventry, CV1 5FB, UK
Tel.: +44-1203-838870, Fax: +44-1203-838949
email: cey145@coventry.ac.uk, m.j.west@coventry.ac.uk

Abstract. This paper describes the calibration of a highly sensitive flow transducer, exhibiting a linear pressure-flow relationship, as used in a novel type of pneumotachograph. After an introduction of the pneumotachograph and its operating principles the calibration procedures and results are presented. The device offers a suitable sensor for the wide range of flow-rates encountered in human breathing.

1. Introduction

A novel pneumotachograph is currently under development which exploits the properties of Wells turbine used in many wave energy projects during the eighties and the nineties. One such project resulted in the Clam wave energy converter as proposed by the SEA wave energy group at Coventry (Lanchester) Polytechnic [1]. In small scale model tests of the Clam the need arose to develop a linear damper which emulated the characteristics of Wells turbine. Experiments had shown that Wells turbine maintains its linear pressure-flow relationship even when the blades are in a stall condition, suggesting that the profile of the blades is not significant in ensuring the linear pressure-flow relationship [2]. The linear damper used in the small scale model of the Clam consisted of a slotted disk with unprofiled blades driven by a dc motor. The damper disk produced very little aerodynamic torque due to its thin flat blades and hence could be driven by a small motor with suitable speed control.

The damper disk results obtained from its use in the Clam [2] suggested its ideal suitability for applications requiring a very sensitive flow-rate transducer in oscillating airflows. One such application area is pneumotachography with typical flow-rates for humans in the range of 0.1 l/s for neonates during sleep up to 16 l/s for a young athlete.

A damper disk used inside a pneumotachograph has resulted in a highly sensitive flow transducer exhibiting a linear pressure-flow relationship when the disk is driven by a small permanent magnet dc motor at a constant angular velocity [3], for which a suitable motor speed controller has been developed [4]. The disk is directly attached to the motor shaft and both are located inside a horizontal breathing pipe of appropriate diameter, figure 1. The disk is operated at rotational speeds of up to 10,000 rpm and with flow-rates of up to 16 l/s operation of the damper stays within the region of a linear pressure-flow relationship [2]. A breathing-induced pressure drop across the damper disk is picked up by pressure tappings in the pipe wall on both sides of the disk and measured using a pressure transducer of the strain-gauge in bridge-configuration type. After A/D conversion of the signal very little data processing is needed to extract the original breathing flow-rate information. A data analysis package written in the LabView graphical programming language allows the airflow data to be manipulated in order to display key parameters and charts as required in pneumotachography [4].

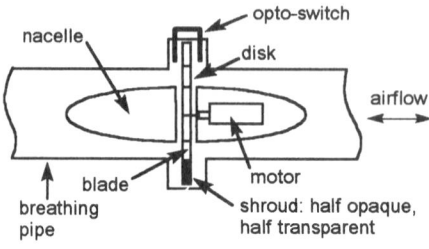

Figure 1: The pneumotachograph rig

2. Calibration procedures

The damper disk rotating at a constant angular velocity converts a breathing-induced airflow into a linearly proportional pressure drop across the disk. Preliminary evidence has indicated that the relationship between these two parameters is given by the equation [2, 5]

$$D_f = \frac{Q}{\Delta P} = K \frac{d_t}{\rho \omega Sa^n}$$

(1)

where D_f is the damping factor, Q is the flow-rate, ΔP is the pressure drop across the disk, d_t is the blade tip diameter, ρ is the air density, ω is the disk angular velocity, Sa is the blade solidity, and $K = 0.185$ and $n = 1.65$ are empirically determined constants [2]. The original airflow can then be easily determined from the pressure drop by the data analysis package employing equation 1.

The calibration experiments were all done for unidirectional flows, as it had been discovered in tests by the wave energy group at Coventry (Lanchester) Polytechnic and at Queen's University that the pressure drop across Wells turbine is not dependent on airflow direction and hence the results obtained for unidirectional flow tests are applicable to oscillating airflow conditions [2, 6].

For the purpose of calibration additional piping of approximately 4 pipe diameters length with a bellmouth was attached at the pneumotachograph entry to smooth the incoming airflow. A fan attached to the pneumotachograph exit was employed to pull variable amounts of air through the rig. Four pipe diameters downstream from the bellmouth four holes in the pipe wall were used to determine static pressure from which the actual flow-rate Q_{act} can then be calculated as

$$Q_{act} = KA \sqrt{\frac{2(P_0 - P_s)}{\rho}}$$

(2)

where A is the cross-sectional area of the pipe, P_0 is the atmospheric pressure, and P_s is the static pressure at the holes in the pipe wall. The loss factor K was determined from separate Pitot traverse experiments to obtain the velocity profile and hence the flow-rate. Traverse results were taken at three points around the circumference of the pipe wall, all at the same distance from the pipe entry and located slightly further downstream from the static pressure points. The value for K was found to be consistent at 0.89 over the entire flow-range of interest.

It was decided to use ten flow-rates and for each flow-rate pressure drop readings were taken for ten different damper disk velocities. The flow-rates were in the range of 3.36 to 18.08 l/s and the damper disk speed was varied from 2142 to 10,020 rpm. The damper disk tested had six blades, each with a chord of 10 mm, a blade tip diameter of 64 mm and a hub diameter of 45 mm. The blade tip diameter matches that of the breathing pipe and the hub diameter corresponds to the diameter of the nacelle housing the dc motor as in figure 1. This yielded a blade solidity of 0.346.

3. Calibration results

Plotting the pressure drop readings versus flow-rate at different damper disk speeds N resulted in the graph in figure 2. Although the two curves for 2142 and 3168 rpm exhibit non-linearity, the linear pressure-flow relationship is clearly demonstrated for higher rotational speeds of the damper disk (5340 rpm and above in figure 2).

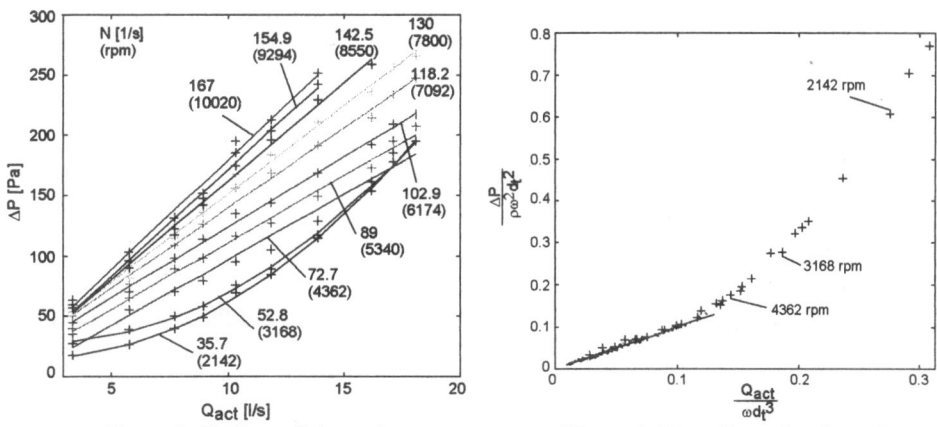

Figure 2: Damper disk results   Figure 3: Non-dimensional results

The dependency on N can be removed by dimensional analysis techniques resulting in the graph in figure 3. As expected all results lie on the same curve, which shows a linear pressure-flow relationship at the lower end of the abscissa (high N). The three data sets spanning the remainder of the abscissa, displaying a square relationship, are those for 2142 rpm and 3168 rpm. Thus figure 3 again demonstrates that the useful range, within which the flow transducer is to be operated, is for $N \geq 5340$ rpm.

Figure 3 still exhibits a dependency on solidity which when removed yields the plot in figure 4. Comparing the results in figure 4 to performance data for Wells turbines tested in the UK during the eighties (figure 5 [2]) demonstrates excellent correlation. Although this would prove equation 1 in terms of the dependency of damping factor on solidity, a further five damper disks, varying in number of blades and chord, were designed [5] and are currently undergoing calibration following the procedure set out in section 2.

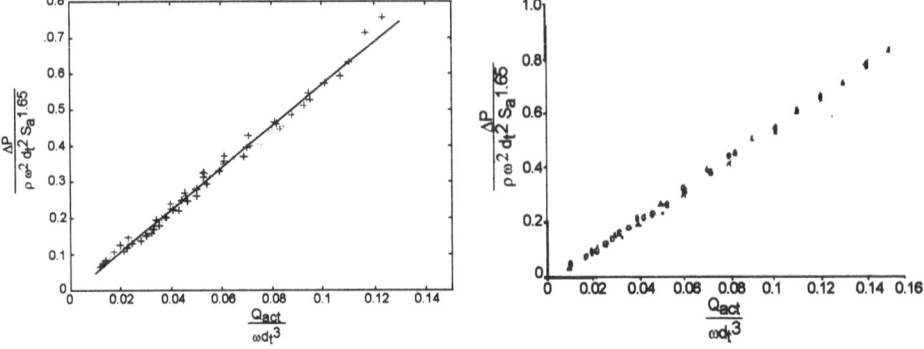

Figure 4: Results independent of solidity   Figure 5: Wells turbine results

4. Conclusions

A highly sensitive flow transducer employing a sectored damper disk has been investigated with respect to its useful range within which a linear pressure-flow relationship exists. It has been shown that the transducer exhibits a linear pressure-flow relationship for rotational speeds of 5340 rpm and above, irrespective of its blade solidity. From these results it can be concluded that the transducer is suitable to the field of pneumotachography and its associated range of flow-rates, requiring very little processing of the pressure measurements taken from the transducer to obtain the flow-rate. Further calibration results involving the use of damper disks with different solidities are expected to reinforce the conclusions derived from the results presented in this paper.

References

[1]   Bellamy N.W. and Peatfield A.M., *Design and performance of the circular SEA Clam wave energy converter*, 3rd Int. Symposium on Wave, Tidal, OTEC, and Small Scale Hydro Energy, Brighton, May 1986, 239-248

[2]   White P.R.S. et al, *The design, instrumentation and use of linear orifices to simulate the performance of Wells turbines*, 4th Int. Conf. on Systems Engineering, Coventry, Sept. 1985

[3]   West M.J. and Coolen E.F.J., *A transducer for lung function monitoring*, Tenth Int. Conf. on Systems Engineering, Coventry, Sept. 1994, 1391-1398

[4]   Coolen E.F.J., West M.J. and Lewis C.P., *The use of LabView and Matlab in the development of a novel pneumotachograph*, Colloq. "The use of systems analysis and modelling tools: experiences and applications", IEE, London, March 1998

[5]   Coolen E.F.J. and West M.J., *Further development of a new form of pneumotachograph*, 12th Int. Conf. on Systems Science, Wroclaw, Poland, Sept. 1995, 351-358

[6]   Raghunathan S., *The aerodynamic aspects of Wells turbine*, 3rd Int. Symposium on Wave, Tidal, OTEC, and Small Scale Hydro Energy, Brighton, May 1986, 261-279

An ionic liquid flow sensor

A Kruusing, S Leppävuori and A Uusimäki

Microelectronics and Material Physics Laboratories and EMPART research group of
Infotech Oulu, University of Oulu, P.O.Box 444, FIN-90570 Oulu, Finland

Abstract. A new liquid flow sensor suitable for miniature systems is proposed. The
sensor's operation relies on ion-drag effects of the flow which result in an increase of
liquid resistance between the electrodes at the periphery of the flow channel. The
sensor is simple and has little influence on the liquid or the flow. In this paper the
physical processes in this sensor are explained together with qualitative theory.
Experimental data gained from sensing the water flow in a 1 mm inner diameter tube
are also presented.

1. Introduction

Miniature liquid flow sensors are necessary in microdispensers, flow injection analysis, drug
delivery systems, micro chemical plants, fuel delivery systems for micro/nano spacecraft and
for evaluation and surveillance of micropumps and fluid systems. Micro liquid sensors
should be simple, reliable, have small dimensions, consume little energy and should not
interfere with the fluid flow or alter the fluid properties.

A review of recent advances in micro liquid sensor technology can be found in [1-3].
The most widely used are thermal dilution sensors and thermal transit time (time of flight)
sensors. These devices include a heating element and one or two temperature sensors placed
in the flow. Other micro liquid flow sensors rely on pressure drop measurement across a flow
restriction or on force measurement on an element placed in the flow stream.

In this paper we report the attempt to apply to a liquid sensor the ion drag principle
[4] which is well known in air flow sensors. This results in a very simple flow sensor for
liquids where ions are inherently present.

2. Theory

Let us consider an arrangement of two infinite electrical insulator plates in the xy-plane
between which liquid flows in the x-direction. Suppose the flow to be invariant of z (Fig. 1).

Suppose that at $x = 0$ two infinite long thin linear conductors are placed along the z-
axis the at edges of the channel. Suppose the liquid to be a weak ionic conductor and the
fluid flow regime to be laminar.

If a voltage is applied across the conductors, an electrical field builds up in the liquid, carrying the ions from one electrode to another so that the macroscopic current or conductance between the electrodes can be detected. Because the current is carried by ions, and the flow influences the movement of the ions, the current (conductance) becomes sensitive to flow.

For a positive ion we can write (see also [5])

$$\vec{v} = \mu \vec{E} + \vec{v}_l \tag{1}$$

where $\vec{v}$ is the fluid velocity, μ is the ions mobility, $\vec{E}$ the electrical field strength, and $\vec{v}_l$ the fluid velocity at the ion's site.

For a laminar flow the velocity profile is parabolic,

$$v_l(y) = \frac{3Q^*(d^2 - 4y^2)}{2d^3}, \tag{2}$$

where d is the distance between the plates and Q^* is the volume flow rate for unit width of the plates [6].

Using the known formula for the electric field strength between long conductors [7] we can write for the x and y components of the ion's velocity in the flow and the electric field

$$v_x = \frac{C^*U}{2\pi\varepsilon}\left\{\mu x + \frac{b}{d^3}\left[y^2 - \left(\frac{d}{2}\right)^2\right]Q^*\right\} \times \left[\frac{1}{x^2 + (y - d/2)^2} - \frac{1}{x^2 + (y + d/2)^2}\right], \tag{3}$$

$$v_y = \frac{C^*U}{2\pi\varepsilon}\left[\frac{\mu(y - d/2)}{x^2 + (y - d/2)^2} - \frac{\mu(y + d/2)}{x^2 + (y + d/2)^2}\right], \tag{4}$$

where C^* is the capacity per unit length of the electrodes and U is the voltage between them.

In Fig. 1 the ions' trajectories in a channel are depicted for the cases of a standing and a moving fluid. Note that at the edges of the channel the ions' motion is dictated by the electric field and in the middle of the channel by the flow field.

The points at which $v_x = 0$ are determined by the formula

$$x + \frac{6Q^*}{\mu d^3}\left[y^2 - \left(\frac{d}{2}\right)^2\right] = 0 \tag{5}$$

and lie on parabola (dashed line in Fig. 1).

The ion motion equations (3), (4) determine the current (conductivity/resistivity) between the electrodes, but do not have a simple analytical solution. A simple estimate for the resistivity change due to the flow measured between the electrodes may be calculated in the following way.

Consider first the situation where the liquid is standing. The time taken for an ion starting at an electrode to reach the middle plane is

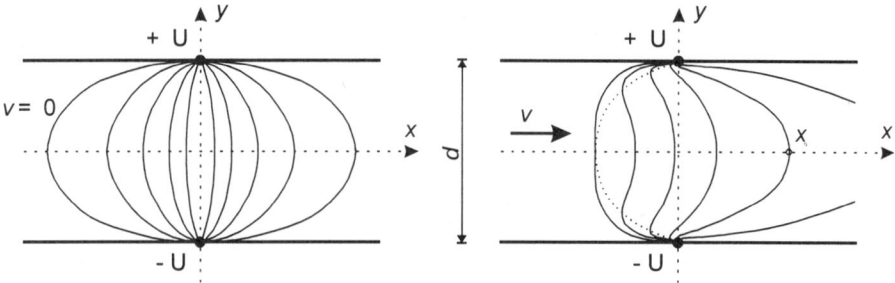

Figure 1. Ion trajectories under electric field of two linear conductors and laminar flow

$$\Delta t = \frac{d^2}{2\mu U}.$$ (6)

Now suppose that in the flow the same ion reaches the middle plane in the same time and that its declination from the plane of the electrodes is

$$x_0 = \bar{v}\Delta t = \frac{d^2 \bar{v}}{\mu U} = \frac{dQ^*}{\mu U},$$ (7)

where $\bar{v}$ is the average fluid velocity in the channel.

Suppose that the resistivity measured between the electrodes is proportional to the fluid velocity at x_0 (considering x_0 to represent the middle point of the current beam. The lower the ions' velocity, the smaller the current, because the current is the charge transferred in unit time).

From (4) and (7) we obtain

$$\frac{R}{R_0} = 1 + \left(\frac{Q^*}{\mu U}\right)^2.$$ (8)

We see that the resistivity of the liquid between the electrodes depends quadratically on the flow rate. The resistivity change is larger if the mobilities of the ions, and the electric field strength, is smaller because then the fluid carries the ions away to a greater extent.

3. Experiment

The feasibility of the proposed sensor was proved by resistivity measurements between two nickel coated wire electrodes in a tube of inner diameter 6,5 mm and filled with distilled water. The distance between the wires was 3,5 mm. The resistivity was measured at 1 μA DC. The results sre shown in Fig. 2. The sensor responds to the flow, but the static resistivity is not stable due to the elecrode processes. A recent observation is that the scattering of data may be suppressed by insulator coated electrodes and high frequency excitation.

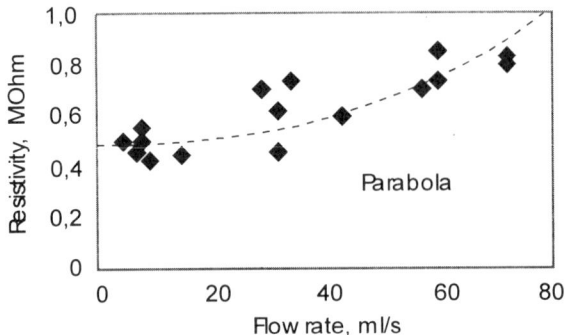

Figure 2. Resistivity change of moving liquid

4. Conclusions

A new liquid flow sensor scheme has been proposed. The sensor is simple, small and has an integrated response over the entire flow field in the channel. It does not heat the fluid and does not obstruct the flow. Its drawbacks are the necessity of ionic conduction in the liquid, sensitivity to the properties of the liquid and possible instability due to the electrode condition. The instability may be suppressed using selected electrode materials, differential and four point detection techniques.

Acknowledgements

This work was supported by a Nordic Council of Ministers research grant to Arvi Kruusing to Oulu University in spring semester 1998.

References

[1]   Gravesen P, Branebjerg J and Jensen O S 1993 *J. Micromech. Microeng.* **3** 168-182

[2]   Shoji S and Esashi M 1994 *J. Micromech. Microeng.* **4** 157-171

[3]   Jerman H and Terry S 1997 In book: Ray-Choudhury, Ed. *Handbook on Microlithography, Micromachining, and Microfabrication* vol. 2 (Bellingham: SPIE Optical Engineering Press) p 379-433

[4]   Malaczynski G W and Schroeder T 1992 *IEEE Trans. Ind. Appl.* **28** 304-309

[5]   Fleury V, Chazalviel J-N and Rosso M 1992 *Physical Review Letters* **68** 2492-2495

[6]   White F M 1988 *Fluid Mechanics* (New York: McGraw-Hill)

[7]   Elliott R S 1966 *Electromagnetics* (New York: McGraw-Hill)

Noninvasive flowmeter

Jiří Jakovenko

Czech Technical University in Prague, Faculty of Electrical Engineering,
Department of Microelectronics,
Technicka 2, Prague 6, Czech Republic
Tel.: +420-2+2435 2211, Fax.: +420-2-2431 0792, e-mail:jakovenk@feld.cvut.cz,

Abstrakt. For noninvasive monitoring of small flow rate quantitis (into 10 l per hour), a flowmeter has been designed. Its principle activities is measurement of warming liquid running through the tube wall. Flowmeter works with flow detection in range 0,1 - 10 l/hour. with high sensitivity for small flows. The system contains circuits for processing and signal analysis in a compact implementation together with the sensitive part of a sensor. Output data are transfered across standard serial interface RS-232 to personal computer. Advantage of noninvasive flow monitoring is absence of mechanical sensor section, which increases reliability, and enables measuring of polluted and aggressive liquids.

Introduction

One way of monitoring mass flows by noninvasive way is electrocaloric method, which use bouh conductive and convective heat transfer. On fig.1 there is showed the principle of the calorimeteric probe, which analysing warming of flow liquid. Medium is led through good

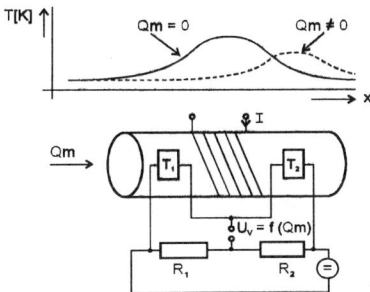

Fig.1 The principle of calorimetric probe

thermal conducting metal tube, on which heat winding is placed and up both parties toward it there are placed symmetricaly two resistive temperature sensors which measure temperatures T1 and T2. Sensors are connected to a bridge. The whole arrangement has thermal insulating. If liquid doesn't flow, give out owing to free convection and liquid conductivity to steady warming-up, which span on both sides of heating winding and bridge is balanced. Liquid flow affects however, symmetrical temperature distribution, so that between measuring points T1 and T2 thermal difference arises, which breaks balance of the bridge and then the output voltage is function of flow.

Probe design

Design and construction of proper measure probe has been realized first on the basis further explicited requirements:
- *independence on environmental influence (temperature, pressure)*
- *independence on liquid temperature*

- *fastest time response (response time is changing according to flow magnitude, longest is close to lowest values of flow)*
- *non flow detection*
- *prevention of liquid overheating at flow absence*

Between heating element and measuring thermal sensor there is necessary to place thermal insulating material, for increase of sensitivity, which prevents diffusion of heat through the wall of the tubule from heating to the thermal sensors.

To reach better sensitivity close to higher flows and overheating liquid prevention at absence of flow is desirable to regulate power of heating winding, that is to keep constant temperatures difference of entering liquid and temperature of the tube wall. It means upkeeping constant temperatures difference Tc - Tv. Thereby we can also prevent temperature influence of entering liquid.

The designed system comprises proper sensitive parts (a heating element and thermal sensors which measure liquid warming), electronic circuits and a PC, which compiles and displays measured values. The centre of electronic circuits is microprocessor, which drives function of analog signal circuits from thermal sensors and generates driving signals for A/D converter. From obtained data the microprocessor computes length of pulse-width modulation pulse PWM, which adjusts electric power heating of tube probe. Further it provides data transmission through serial interface RS-232 into the PC, which proceses data. Appliance is designed as a compact functional unit, to which they are connected supply of measuring liquid and serial interface for PC communication. Block diagram of electronics is on fig.2. Block T/u is temperatures - voltage converter, (semiconductor resistive temperature sensor, in this case). Tv mean temperature of liquid input, Tc - tube wall temperatures, T1 means liquid temperature close before warming-up and T2 close behind heating.

Fig.3b) shows dependence of dimensionless number Cq on the flow. Number Cq, which is flow function, indicate voltage magnitude $\dot{u}_q$ translated by A/D converter to 12 - bitwise number and it's given by equation:

$$C_Q = \frac{2^n}{U_{ref}} u_Q \quad [-]$$

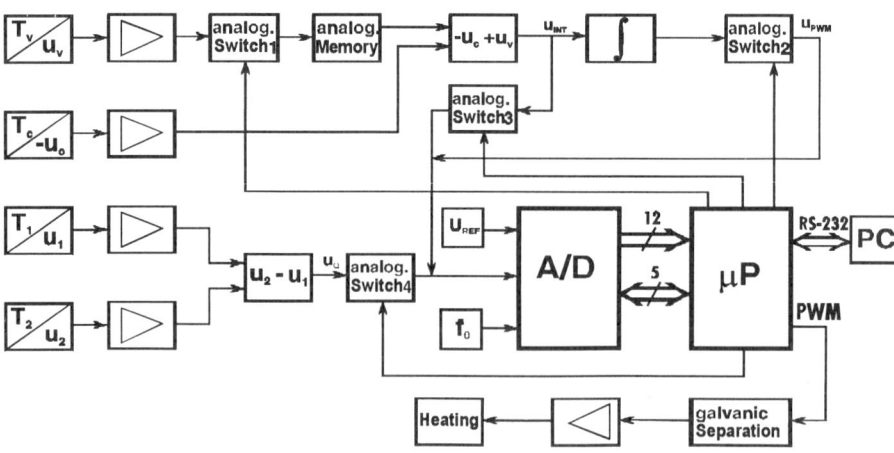

Fig.2 Electronics block diagram

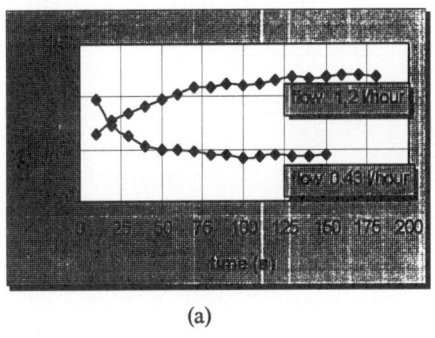 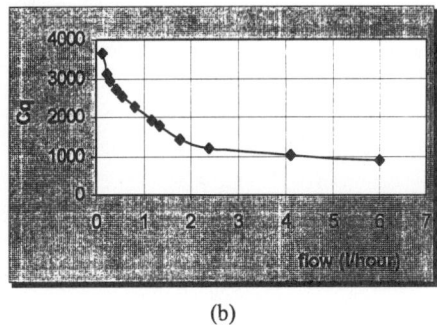

(a)   (b)

Fig. 3   (a) response on unit flow step
          (b) dependence of numbers Cq on flow

These two quantity are storaged in computer calibration table .
In graph on fig. 3a) there is showed time response on unit step, (from zero flow value to 1,2 l/hour and 0,43 l/hour) As showed the waveforms doesn't rise from one's point, which is caused by indefinable phenomenas at flow absence.

Conclusion

Thanks to connection to the standard interface it is possible to add such facilities to measurement network of complex technological process. From measuring system parameters long response time close to small flow rates results. By dimension minimisation of active probe part (it is. distance of thermal pick-up difference $\Delta T12$) it is possible to improve this parameter participally. Further system improvement of accuracy has been reached by realization of thermal regressive loop of heating tube wall. In this case there comes up traffic delay between heating winding and temperature sensors. It results in small change of temperatures drift by tube wall and also fluctuation of thermal difference $\Delta T12$. Metering isn't dependent upon temperature of liquid and system work with flow rate 0.1 - 3 l/hour. with very good sensitivity (this range is possible to regulate slightly by changing of heating) and 10% accuracy, that however has been reached just for a short-time. Worse reproducibility has been caused by worse thermal insulation of whole system.
For measuring of calibration waveform system has been completed by autocalibration unit, which is capable adjust calibration flow waveform in needed range. Autocalibration runs quite automatically and is drived by microprocessor.

References

[1] Oudheuden, B.W. - Bruijn, J.M. - Hoogeboom, P. J. - Beaufort, d. - Huijsing, J. H. :*Integrated Senzor for Non-invasive Monitoring of Flow in Pipes.* Sensor and Actuators, 18, 1989, pp. 259 - 267
[2] Lydersen, A. L. :*Fluid Flow and Heat Transfer.* John Wiley, Chicester 1979
[3] Knudsen, J. G.- Katz, D. L. :*Fluid Dynamics and Heat Transfer.* McGraw-Hill Company, New York 1958

Biosensors II

Biosensors II
Paper presented at Eurosensors XII, 13–16 September 1998
© *1998 IOP Publishing Ltd*

Functionnalization of Si/SiO$_2$ substrates with homooligonucleotides for a DNA biosensor

J.P. Cloarec*†, J.R. Martin*, C. Polychronakos‡, I. Lawrence†, M.F. Lawrence†, E. Souteyrand*

* IFOS-PCI, Ecole Centrale de Lyon, BP 163, 69131 Ecully Cedex, France

† Concordia University, Department of Chemistry and Biochemistry, 1455 de Maisonneuve West, Montreal, PQ, Canada, H3G 1M8

‡ Montreal Children's Hospital, 2300 Tupper Street, Montreal, PQ, Canada, H3H 1P3

Abstract. Si/SiO2 substrates are functionnalized with single stranded homooligonucleotides in order to implement a DNA biosensor based on field effect measurement. Two methods of immobilization of homooligonucleotides on Si/SiO$_2$ substrates, using an aminosilane and a glycidoxysilane, are described. Both methods are characterized by radiolabelling and electrochemical impedance measurements on Electrolyte / Dielectric / Semiconductor structures. The protocol using glycidoxysilane allows a strong and high density immobilization of oligonucleotides.

1 Introduction

The detection of homooligonucleotides hybridization by field effect, using a Si/SiO$_2$ substrate functionnalized with single stranded oligonucleotides, has been demonstrated in a previous publication [1]. Such a method allows direct and in situ detection of specific oligonucleotidic sequences, without using any enzymatic or electrochemical reaction, nor any labeled molecule. Although electric measurements are fairly mastered, a main problem encountered with such a device is the difficulty of obtaining reproducible surface of immobilized oligonucleotides, with an appropriate density and a robust grafting. We present investigations concerning the functionnalization of silica with oligonucleotides.

2 Materials and methods

2.1 Immobilization of oligonucleotides on Si/SiO$_2$ substrates

The Si/SiO$_2$ substrates are 1 cm^2 <100> doped silicon chips, covered with a layer of 100 Å thickness of silicon dioxide. The substrates are first hydroxylated using sulfochromic acid, and derivatized with a silane bearing a terminal moiety on which modified oligonucleotides can react to achieve covalent coupling. The functionnalized substrate (figure 1) can then be monitored by impedance measurements for the detection of complementary strands. Two methods of immobilization of oligonucleotides on silica have been investigated.

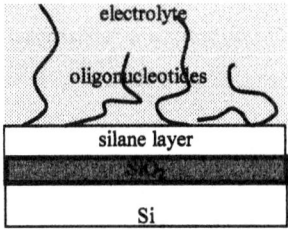

figure 1. Si/SiO$_2$ substrate with immobilized oligonucleotides

The first method, called APTS / bromine, consists in immersing the hydroxylated substrate in an aqueous solution of APTS (AminoPropylTriethoxySilane [2]), followed by coating of brominated [3] oligonucleotides (oligodT$_{20}$) on the aminosilane. The second method, called GPTS / amine, consists in heating the hydroxylated substrate to remove surface water, followed by immersing the substrate in a mixture of xylene and GPTS (GlycidoxyPropylTriethoxysilane) in anhydrous conditions, in presence of a catalyst (diisopropylethylamine) [4]. OligodT$_{20}$ bearing an aminolinker at the 5' end are then left to react on the surface in water under alkaline pH [5].

2.2 Radiolabelling

In order to quantify the immobilization, radiolabeled oligonucleotides are coated on the silane layer, and detected by phosphorimaging. The radioactivity on the chip is measured before and after strong washings using detergent solution. These washings are aimed at removing physisorbed molecules.

2.3 Electrochemical impedance measurements

The Si/SiO$_2$ substrates are used as working electrodes in a classical three electrodes potentiostatic set-up, and are characterized by impedance measurements at each step of their implementation (hydroxylated silica, silanization, oligonucleotide coupling). The thickness of the silica layer is large enough to prevent transfer of electrons between the electrolyte and the semiconductor, and is used as a blocking layer. The set-up allows measurements versus DC voltage of in-phase and out-of-phase impedance of the silicon / functionnalized silica / electrolyte structure.

3 Resultats and discussion

3.1 Radiolabelling

The figure 2 indicates the amount of radiolabeled oligodT immobilized on the substrate, before and after washings, for the methods GPTS/amine, APTS/bromine, and for a silica substrate without silane (SiOH). No meaningful difference is observed after washings between APTS derivatized substrates, and a silica substrate without silane.

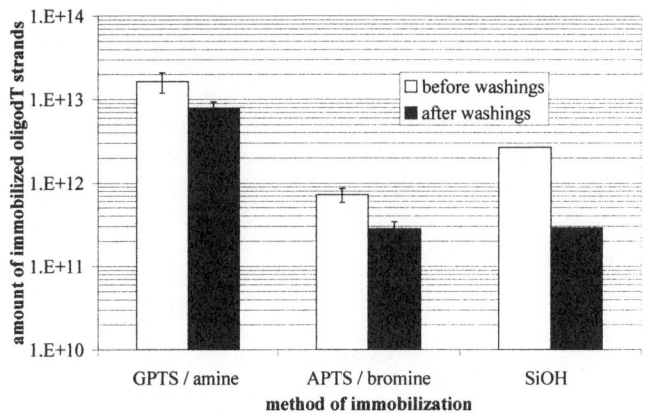

Figure 2. immobilization of radiolabeled oligodT on silane.

In our reaction conditions, covalent binding of oligonucleotide on the surface is thus not achieved. On the contrary the GPTS / amine method allows binding of high amounts of oligodT, corresponding to 10^{12} to 10^{13} strands per cm^2. The oligodT are not removed by the SDS washings, which proves robust binding of the silane on the surface, and strong attachment of the oligonucleotides to the silane layer.

3.2 Electrochemical impedance measurements

Each step of the immobilization process is monitored by following the evolution of the out-of-phase impedance curve. The shifts along the voltage axis are related to the variation of electric charges at the dielectric / electrolyte interface, which are reflected, by field effect, by the change in the flat band potential of the semiconductor.

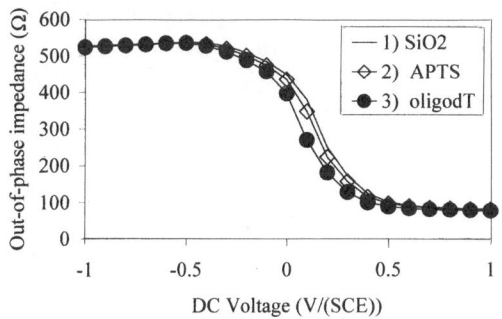

Figure 3.

Out-of-phase impedance curve characterizing APTS / bromine method.

The substrate is p type Si/SiO$_2$. The impedance is measured under a frequency of 100 kHz.

The figure 3 shows evolution of out-of-phase impedance of a Si/SiO$_2$ substrate functionnalized with APTS / bromine method. A small shift of about -20 mV of the curve along the potential axis occurs for the silanization step (curve 2), indicating a small variation of the net negative charge at the dielectric / electrolyte interface. The intensity and the direction of this shift is not reproducible, indicating that the silanization process is not controled. Coating of oligodT (curve 3) causes a shift of -30 to -120 mV. Although the amplitude of the shift is not reproducible, this shift is always oriented towards the negative potentials, indicating an increase of the net negative charge on the dielectric surface. This is consistent with the electrical properties of oligonucleotides, which are negatively charged molecules.

Figure 4 shows evolution of out-of-phase impedance of a Si/SiO₂ substrate functionnalized with GPTS / amine method. It exhibits a very reproducible shift of +250mV for the silanization step (curve 2), indicating a strong decrease of the net negative charge at the dielectric / electrolyte interface. This is consistent with GPTS condensation on the surface silanols of the silica, which are negatively charged in the measurement buffer. The coating of oligodT causes a reproducible shift of –300 mV (curve 3), indicating a strong increase of net negative charge on the dielectric surface.

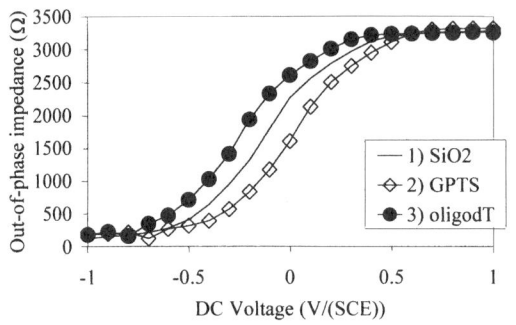

Figure 4.

Out-of-phase impedance curve characterizing GPTS / amine method.
The substrate is n type Si/SiO₂. The impedance is measured under a frequency of 20 kHz.

Radiolabelling and electrochemical impedance measurements give consistent results, showing that the GPTS / amine method is more reliable and controlled than the APTS / bromine method. The low yield of immobilization for the APTS / bromine method is firstly caused by the conditions of silanization, provoking APTS polymerization in solution and adsorption on the surface, rather than creating covalent binding with the silica. Interactions between the primary amine moiety and the silica surface silanols are also expected to decrease the yield of oligonucleotide coupling.

The conditions of silanization are better controlled for the GPTS / amine method. The use of anhydrous reagents and atmosphere, the control by heating of adsorbed water on the silica surface, and the presence of a catalyst (amine) allow covalent binding of the silane on the silica. Though reproducibility has not been reached, the coupling reaction of oligonucleotides on the GPTS yields 10^{12} to 10^{13} strands to be strongly bound to the surface.

4 Conclusion

Correlation between electrochemical impedance measurements and radiolabeling shows the GPTS / amine method to yield dense and robust immobilization of oligonucleotides on the Si/SiO₂ substrates. This method is expected to give reproducible surfaces for direct detection of specific oligonucleotidic sequences and to allow correlation between electric signal and amount of hybridized strands.

5 References

[1] Souteyrand, E, Cloarec, JP, Martin, JR, Wilson, C, Lawrence, I, Mikkelsen, S, Lawrence, MF 1997 *Journal of Physical Chemistry* **101** 2980-2985.

[2] Caravajal G S, Leyden D E, Quinting G R, Maciel G E 1988 *Anal. Chem.* **60** 1776-1786

[3] Keller G H, Cumming C U, Huang D P, Manak M M, Ting R 1988 *Anal. Biochem.* **170** 441-450

[4] Maskos, U, Southern, E M 1992 *Nucleic Acids Research* **20** 1679-1684

[5] Lamture *et al.* 1994 *Nucleic Acids Research* **22** 2121-2125

Biosensors II
Paper presented at Eurosensors XII, 13–16 September 1998
© *1998 IOP Publishing Ltd*

Sensitive electrochemical detection of antigens using gold electrodes functionalized with antibody moieties

S. AMEUR[1,2], C. MARTELET[1], J. M. CHOVELON[1], H BEN OUADA[2], N. JAFFREZIC-RENAULT[1], D. BARBIER[3].

[1] IFoS, UMR CNRS 5621, Ecole Centrale Lyon, BP 163, 69131 Ecully, France.
[2] LPI Faculté des Sciences de Monastir, Monastir 5000, Tunisie.
[3] LPM, INSA Bat 502, 6ème etage, 20 Av. A. Einstein, 69621 Villeurbanne, France.

ABSTRACT. This paper deals with direct electrochemical detection of antigens using a new method of functionalization of gold electrodes. The antibody was treated in order to form SH groups allowing direct fixation onto gold substrates with coverage factor close to $\theta = 0.9$. An interesting sensitivity for antigen of about few pg/ml can be reached.

1.Introduction

For direct detection without labeling of immune complexes, it has been proposed several techniques based on optical (ellipsometry [1], surface plasmon resonnance [2]), piezoelectric [3] or electrochemical (immunochemical field effect transistor [4], impedimetric immunosensor [5]) devices.

Recently immunosensors using gold as a substrate has been proposed [5-7]. Gold is often functionnalized by alkanethiols and other thiol derivatives, then antibodies (or antigens) were allowed to be fixed at the modified gold electrode by different immobilization processes.

In this work electrochemical detection on gold electrodes using a new method of functionalization has been investigated. Gold electrodes were covered with reduced antibodies (containing free sulfhydryl groups). An immunochemical reaction was carried out on the electrode surface and changes of the electrode impedance were evaluated.

2. Experimental

The 2-mercaptoethylamine (2-MEA) was from Aldrich. The goat anti-rabbit IgG, the rabbit IgG and phosphate buffered saline tabletes were purchased from Sigma. The polyoxyethylen sorbitan monolaurate ("Tween 20") was from Merck.

Dissolving 1 tablet of phosphate buffered saline in 200ml of distilled water gave 10mM phosphate buffer, 2.7 mM potassium chloride, 137 mM sodium chloride, pH 7.4 at 25°C. In some cases 0.2 % (v/v) of "Tween 20" surfactant was added in solution in order to reduce the non specific interactions when washing electrodes containing antibodies.

2.1. Sensor, electrode design, immobilization protocol and electrical measurements

Gold disk electrodes (diameter 1.0 mm) sealed in a Mecaprex resin were polished first with emery paper then followed by polishing with a 6, 3 and 1μm diamond pastes respectively. After ultrasonicating with water and absolute ethanol bath for 15min, gold electrodes were thoroughly rinsed with desionised water and dried in a stream of pure nitrogen.

Electrodes were dipped for 12h in a solution of reduced anti-rabbit antibodies (as it will be described later). After thoroughly rinsing with PBS Tween , electrodes were inserted in a flow injection system. The flow rate was 1ml.min^{-1}, corresponding to a linear speed of 20 cm.min^{-1} inside the measurement cell [8].

By proper selection of the immobilization process and a knowledge of the antibody structure, the molecule can be oriented on the support so its bivalent binding capacity for antigen can be realized fully. The disulfides in the hinge region that hold the heavy chains together can be reduced selectively with 2-mercaptoethylamine (2-MEA) to form two half IgG molecules, each containing an antigen binding site and free SH moieties [9]. Immobilization of resulting sulfhydryls using any of the activation process designed to couple these groups will provide covalent attachement away from the antigen binding site.

The binding formed between the sulfur atom and gold is very strong [10]. It would therefore appear appealing to develop an immunosensor with reduced antibodies immobilized, through their free sulfhydryls, on gold substrates.

A protocol describing the formation of reduced antibodies and their immobilization onto iodoacetyl-activated support materials was described in [11]. We choose to stop at the second step of this protocol since we assume that only components containing a free sulfhydryl group can be coupled to the surface. Since, in such a mixture only 2-MEA, reduced antibodies or a mixture of them will be present at the surface of gold substrates.

The apparatus used for impedance measurements was composed of an impedance analyser EG&G instruments (Princeton Applied research Model 6310) connected to a computer. Impedance and cyclic voltammetry were measured using respectively the M398 and M270 commercial programs. A $[Fe(CN)_6]^{3-/4-}$ redox couple was used since it represents an electrochemically reversible, well defined, one-electron redox reaction with moderate normal potential. The open circuit potential was monitored against the reference electrode after obtention of reproducible voltammograms. The potential of the working electrode, in impedance measurements, was then held at this value.

3.Resultats and discussions

3.1.Characterization of layers by cyclic voltammetry

Figure 1 shows typical cyclic voltammograms obtained with bare gold electrode, 2-MEA coated electrode and with reduced antibodies. In the last case, although a small current due to ferrocyanide oxidation appears with large over potentials, most of the current observed at the modified electrode was capacitive. However for 2-MEA coated electrodes electron transfer process was not blocked showing a partially covered electrode.

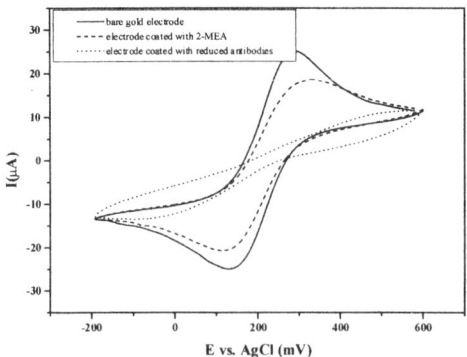

Figure 1: Cyclic voltammograms for: bare and different coatted gold electrodes. Electrolyte solution PBS containing the$[Fe(CN)_6]^{3-/4-}$ redox couple. Scan rate, 100mV.s^{-1}

3.2. Characterization of layers by impedance spectroscopy

Since impedance spectra of the gold and modified gold electrodes were recorded under the same conditions, the value of R_{ct}, may be assumed to be affected only by the effective surface area. A surface coverage factor, θ, was estimated from: $(1-\theta)=R_{ct}(Au)/ R_{ct}(Au\text{-modified}$ electrode) [13]. For two different electrodes, θ, was found respectively, 0.92 and 0.87 showing a good reproducibility. Concerning 2-MEA the coverage factor was also estimated using the same procedure, the θ value does not exceeded 0.57.

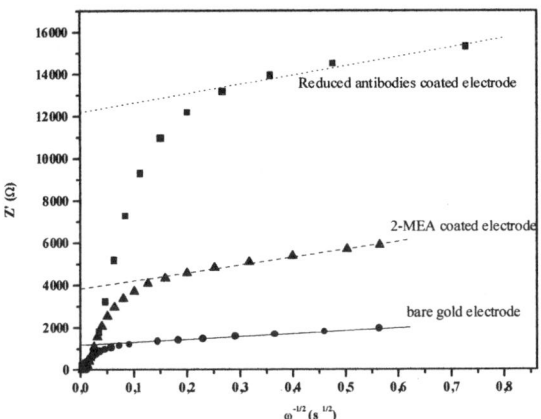

Figure 2: Determination of the coverage factor for different coated electrodes.

In the presence of redox-active species in solution, an impedance response of the electrode-solution system is conveniently expressed in terms of Randles' equivalent circuit.Figure 3a shows complex impedance plots obtained for bare and modified gold electrodes in the presence of the $[Fe(CN)_6]^{3-/4-}$ redox couple. In the case of reduced antibodies modified gold electrode the electrochemical impedance changes drastically. As the faradic reaction of a redox pair becomes increasingly hindered, the apparent charge transfer resistance increases.

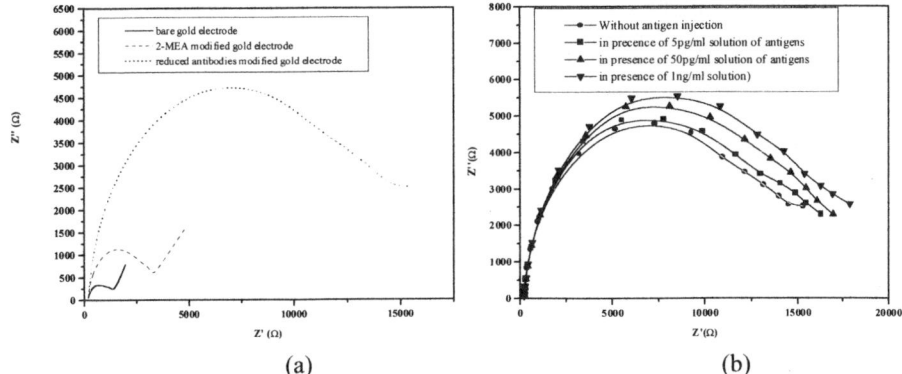

(a)   (b)

Figure 3: (a) : Nyquist plot of different coated electrodes (b) influence of antigen concentration

Antigen injection was investigated and shown in curve 3b. A very interesting variation in the impedance diagram was observed. Thus, antibodies seems retain their activity, sensitivity and allow a good detection limit of about 5pg/ml.

4. Conclusion

The feasibility of immunosensing using reduced antibodies modified gold electrodes was demonstrated. The layer formed on gold surface covers well the surface. Such coverage rate proves the presence of active antibodies in the surface of gold electrodes. Using electrochemical impedance spectroscopy on reduced antibodies allow antigen detection as low as 5 pg/ml.changes when we are in presence of antigens in solution showing that antibodies retain their activity towards antigens.

References:
[1]   Jin G, Jansson R, Lundström I and Arwin H 1995 Transdurs '95 Eurosensors IX 509-512.
[2]   Millot M. C, Martin F, Bousquet D, Sébille B and Lévy Y. 1995 Sensors and Actuators B 29 268-273.
[3]   Cohen Y, Levi S, Rubin S and Willner I 1996 J. Electroanal. Chem 417 65-75.
[4]   Rickert M. S, Zhang Y, Hesketh P. J, Maclay G. J Gendel S. M and Stetter 1995 Biosensors & Bioelectronics 10 675-681.
[5]   Berggren C and Johansson G1997; Anal. Chem 69(18); 3651-3657.
[6]   Knichel M, Heiduschka P, Beck W, Jung G and Göpel W 1995 Sensors and Actuators B 28 85-94.
[7]   Rickert J, Göpel W, Beck W, Jung G and Heiduschka P 1996 Biosensors & Bioelectronics 11 757-768.
[8]   Ameur S, Maupas H, Martelet C, Renault N. J, Ben OUada H, Cosnier S and Labbe P1995 Materials Science and Engineering c 5 111-119.
[9]   Hermanson G. T., Mallia A. K. and Smith P. K. 1992 Immobilization Affinity Ligand Techniques (USA: Academic Press).
[10]   Gebbert A.; Alvarez-Icaza M.; Stöcklein W.and Schmid R. D. 1992 Anal. Chem., 64, 997-1003
[11]   Blank M and Vodyanoy I 1994 Biomembrane Electrochemistry (Washington: ACS)

Potentiometric detection of trace level of mercury using biosensor with Ir/IrO$_2$ internal electrode

T Krawczyński vel Krawczyk, M Moszczyńska, M Trojanowicz

Department of Chemistry, The University of Warsaw, 02-093 Warsaw, Poland

Abstract: A potentiometric biosensor was developed with urease entrapped in PVC film, which can be reversibly used for determination of mercury with detection limit 50 nM and practically the same sensitivity to monovalent and divalent mercury species.

1. Introduction

Among various methods of mercury determination especially valuable are simple and fast procedures used as screening tests for industrial process water or foodstuffs to indicate the presence of different forms of this species. Heavy metals are well known to inhibit the activity of enzymes and application of this phenomenon to the determination of these hazardous toxic elements offers several advantages. First of all the detectors used in the method of enzyme inhibitor determination can be very sensitive because the reduction of enzyme activity by single inhibitor molecule can be large due to amplification effect. Besides enzymes are often specific for the inhibitor and in many cases the inhibition effect of the pollutant under investigation is related to its biological toxicity. So, the inhibition of enzymatic activity by mercury may offer a good choice as a simple and sensitive screen test.

In several papers published so far on the use of inhibition of urease for the detection of trace level of heavy metal ions urease was used in dissolved form in the solution [1-6] or immobilised in flow-through reactor of the measuring set-up [7-10]. The aim this work was to develop a biosensor which could be easier applied in field screening tests based on inhibition of urease.

2. Experimental

Urease was immobilised physically by multiple immersion of thermally grown Ir/IrO$_2$ electrode in a stirred urease suspension in THF containing dissolved PVC. After each immersion electrode was dried in a stream of argon for 10 min.

The operation of Ir/IrO$_2$ electrode was checked by recording the potential changes due to pH changes caused by addition of solid sodium hydroxide to 0.1 M phosphoric acid (both from POCh, Gliwice, Poland). The stepwise pH changes in the range from 2 to 11 were monitored by glass electrode from Orion (USA) and the reference electrode used was from SCE from Label (Poland). The calibrations of urea biosensors were carried out by addition of adequate portions of 1 M urea solution (from 5 to 800 μl) with Eppendorf micropipettes to 100 ml of diluted TRIS - NaCl buffer. Inhibition measurements were done as following: urea biosensor was placed in stirred diluted buffer solution and potential was recorded until its change were smaller than 0.1 mV/min. Next urea solution was added to obtain the concentration of urea 1 mM (100 μl of 1 M urea to 100 ml of buffer) and again potential was recorded. Next the biosensor was transferred to solution of Hg(II) ions to inhibit the enzyme during a strictly defined time period (from 10 to 60 min.) and again the electrode was transferred to buffer solution and the signal height after addition of urea to obtain 1 mM concentration was recorded as before inhibition process. The inhibition percent was calculated from the decrease of the signal using calibration curve for urea biosensor. Another way was the addition of 100 μl of 1 M urea to 100 ml of TRIS/NaCl buffer containing mercury ions at different concentration and measurement of the rate of initial changes of potential after inhibition period (15 min.). The rate of potential change was plotted vs. concentration of Hg(II) ions.

3. Results and discussion

3.1. Biosensor with urease entrapped in PVC layer

Biosensor with urease entrapped in PVC layer at the surface of Ir/IrO_2 pH sensitive electrode exhibits satisfactory reproducibility of signals for given concentration of urea, as well as good long-term stability and sensitivity. The long term stability was the best when the biosensor was stored in TRIS buffer solution containing EDTA and thioacetamide.

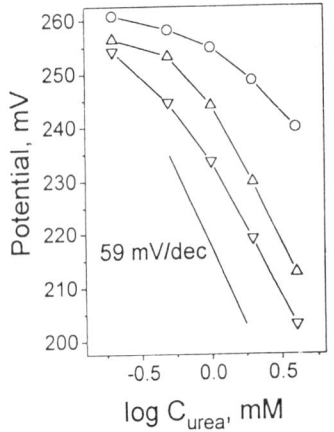

Fig.1. Influence of number of enzymatic layers on calibration plot of Ir/IrO_2-PVC/urease sensor. o - 1 layer, Δ - 2 layers, ∇ - 3 layers.

The base drawback of this sensor was its relatively long equilibration time. This time was shorter for higher concentration of urea, however for small concentration (0.05-0.1 mM) the equilibrium potential value was attained after at least 5-7 min. and the time of equilibration was increased twice for one-week-old biosensor. Especially in TRIS buffer without added urea this time was extremely long and could even be 1 to 1.5 hour using drift of 0.1 mV/min as a criterion of obtaining stable potential value. This time was only insignificantly shorter when the sensor was immersed in TRIS buffer solution containing small concentration (50 μM) of urea as well as it was not shorter very much when the temperature of the measurement was increased up to 40°C. The last operation additionally caused significant reduction of enzyme activity.

It was noticed that the increase of number of deposited layers (up to 10) increased the linear range of calibration plot for low concentration of urea. The typical calibration plot for potentiometric biosensor with urease immobilised in PVC presented in Fig.1 is S-shaped, although one can notice quasi-linear part from 0.2 to 4 mM. The slope of the linear part of calibration plot was close to the Nernstian value although in the very narrow linear concentration range (from 0.5 to 2 mM) it could be even super-Nernstian (up to 72-74 mV/dec.).

When stored in regeneration solution in refrigerator between series of measurements the decrease of slope of the calibration graph was about 20 % after one month while for dry stored sensor the slope decrease was about 40 % after one day storage.

3.2. Inhibition of activity of Ir/IrO$_2$-urease/PVC biosensor by mercury

For inhibition measurements based on the decrease of biosensor signal for urea the definite concentration 1 mM of the substrate was chosen. For this concentration the signal is high enough to observe its decrease after inhibition. Besides this concentration value is located in the middle of linear calibration range, so, the calculation of inhibition % can be based on linear regression of the calibration graph using very simple expression:

$$\%_{inh} = (1 - 10^{[\Delta E(inh) - \Delta E(1)]/S}) \times 100\%$$

where: $\Delta E(inh)$ - signal for 1 mM urea for inhibited biosensor in the presence of Hg(II) ions;
$\Delta E(1)$ - signal for 1 mM urea for non-inhibited (regenerated) biosensor;
S - the slope of the calibration graph.

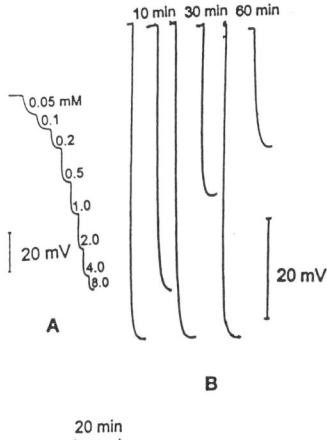

Fig.2. Response to urea for the sensor (A) and signals for 1 mM urea (B) before and after inhibition with 5 µM Hg(NO$_3$)$_2$ during various times. After each inhibition sensor was soaked for 10 min. in regeneration solution.

It was found that for freshly prepared biosensor inhibition % increases up to 30 min. inhibition time and for longer time the increase is significantly smaller. The course of this dependence is different for biosensor stored in regeneration solution where significant increase is obtained up to 10 min. and for longer time this increase is smaller. These results indicate that the most advantageous of inhibition lies in the range between 10 and 30 minutes. The recording of signals for biosensor uninhibited and inhibited for different time id presented in Fig.2. After inhibition procedure during definite time the biosensor was soaked in regeneration solution until stable potential value was obtained but not less than 10 minutes.

The dependence of inhibition % calculated as above on the concentration of mercury ions is better linear in the narrow range from 1 to 5 µM Hg(II) than for stored in regeneration solution for which the dependence of inhibition % for higher concentration is less expressed which suggests the limited possibility of determination of mercury above 2 μM concentration.

Better results, *i.e.* wider determination range was obtained for measurement of initial rate of potential change after addition of 1 mM urea. For biosensor inhibited during 15 minutes in solution of Hg(II) of different concentration the change of potential was measured during first 5-7 min. after addition of substrate in the linear region of the dependence of potential change in time. The results of such measurements are presented in Fig.3, where obtained range of calibration was from 0.5 to 10 μM Hg(II).

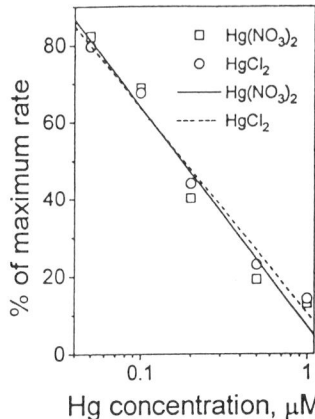

Fig.3. Dependence of initial rate of potential changes of Ir/IrO_2-PVC/urease biosensor after addition of 1 mM urea to the TRIS/NaCl buffer after 15 min. inhibition at different concentration levels of Hg(II) ions.

4. References

[1]   Torren E C, Burger Jr. and F J 1968 *Mikrochim.Acta* [Wien] 1049

[2]   Winquist F, Lundström I and Danielsson B 1988 *Anal. Lett.* **21** 1801

[3]   Wittekind E, Werner M, Reinicke A, Herbert A and Hansen P 1996 *Environ. Technol.* **17** 597

[4]   Liu D, Yin A, Chen K, Ge K, Nie L and Yao S 1995 *Anal. Lett.* **28** 1323

[5]   Narinesingh D, Mungal R and Ngo T T 1994 *Anal. Chim. Acta* **292** 185

[6]   Prell-Swaid A and Schwedt G 1994 *Acta Hydrochim. Hydrobiol.* **22** 70

[7]   Andres R T and Narayanaswamy R 1995 Analyst **120** 1549

[8]   Ögren L and Johansson G 1978 Anal. Chim. Acta **96** 1

[9]   Mattiasson B, Danielsson B, Hermansson C and Mosbach K 1978 *FEBS Lett.* **85** 203

[10] Bryce DW, Fernández-Romero J M and Luque de Castro M D 1994 *Anal. Lett.* **27** 867

Biosensors II
Paper presented at Eurosensors XII, 13–16 September 1998
ⓒ *1998 IOP Publishing Ltd*

Molecular Dimensionality, Surface Chemistry and 3D Enzyme Structure Studies for Formation of Efficient Interfacial Connections between Transducers and Enzymes

T. D. Gibson * and D. Steele.

Enzyme Biotechnology Group, School of Biochemistry and Molecular Biology, Irene Manton Building, University of Leeds, Leeds , LS2 9JT, UK.
Telephone +44 - 113 - 233 - 1940: *Fax* +44 - 113 - 233 - 2593: *E-mail* t.d.gibson@leeds.ac.uk

Abstract. The fabrication of enzyme based biosensors usually involves several different steps before the final device is assembled. During the fabrication process the biological activity of the enzyme needs to be preserved and after production it is preferable to have a long shelf-life. In this report we report the necessity of careful investigation of the chemistry and structure of the transducer materials and the 3D structure of the enzyme. Optimisation of the immobilisation procedures used and promotion of a correct interfacial bond between the transducer and the enzyme is crucial to biosensor performance. The enzyme acetylcholinesterase will be used as the example in the abstract, with other enzymes being discussed in detail in the full presentation.

1. Introduction and background.

One of the major problems in producing biosensors is encountered during the fabrication step, where the actual amount of active protein deposited is often very small. Also preserving the stability of the protein employed (e.g. enzymes) is an extremely important goal. Generally a minimum requirement of 1 year's shelf stability is required, with 2 years being fully acceptable. Where enzymes are immobilised onto surfaces the type of chemistry and the support used generally affects the activity observed [1, 2]. It has been observed that enzymes which are simply adsorbed on carbon surfaces often lose significant amounts of activity at low concentrations. Sensors produced from horseradish peroxidase adsorbed onto carbon black powder (50mg) gave no electrochemical response to peroxide when the enzyme concentration was below $8mg.ml^{-1}$ indicating the enzyme was completely inactivated. At slightly higher concentrations ($10 mg.ml^{-1}$) an electrochemical response was recorded, indicating that the enzyme retained some activity (unpublished results).

With the enzyme acetylcholinesterase the immobilisation method had significant effects on the amperometric response of the biosensors produced as well as the long term shelf stability [2, 3] This report is directed at the mechanisms of enzyme stabilisation and the creation of efficient molecular interfaces for the production of biosensors, using a combination of molecular structure evaluation and experimental results.

2. Materials and methods

2.1 Production and Stabilisation of Acetylcholinesterase Biosensors.

The production of the acetylcholinesterase biosensors was carried out by the method reported in Rippeth et al, 1997. Several different stabilisers were added to the matrix including dextran sulphate, DEAE-dextran, lactitol, trehalose and the positively charged co-polymer Gafquat 755N.

2.2 Molecular Structure Determination and Electrostatic Surface Mapping.

Structural information was obtained from the Brookhaven databank (acetylcholinesterase, PDB code 2ACE) and molecular images with electrostatic surface potentials and the hard sphere representations were produced using the program GRASP (graphical representation and analysis of surface properties). Only residues with whole charges are shown coloured, with atoms coloured blue carrying positive charge, those coloured red are negative. This is indicated by the presence of + and - symbols in the b/w figures in this abstract.

3. Results and Discussion

It can be clearly seen that the structure of the enzyme acetylcholinesterase is relatively large, with a distinct negatively charged domain around the active site, figure 1.

Figure 1. The Electrostatic Surface Potential of Acetylcholinesterase.

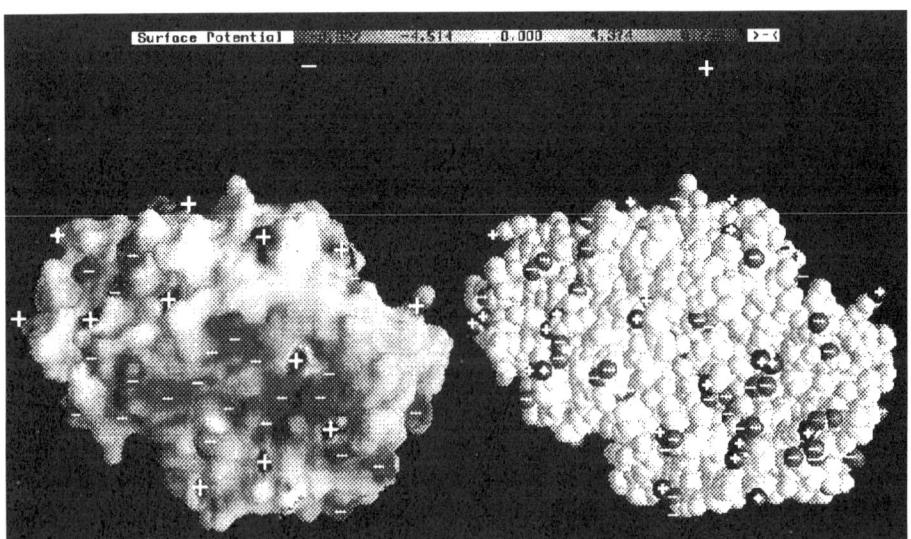

Negative residues are shown marked with - signs and positive residues with + signs.

The enzyme also has a defined electrostatic boundary with the majority of the negative charge distributed on the face of the enzyme where the active site is situated. The reverse face of the enzyme is mainly positive in nature. The substrate is positively charged and it has been postulated that the negative domain around the active site plays a part in directing the acetylcholine into the site itself, thereby enhancing the enzyme activity [4].

In the construction of the biosensor devices, the use of the bifunctional reagent glutaraldehyde is thought to be rather non-selective in promoting crosslinking between the enzyme molecules, in that any amino group which is available for chemical reaction is likely to be a target for the reaction to occur. The net result of glutaraldehyde crosslinking is the production of an insoluble layer of enzyme on the transducer surface which is not chemically attached to the surface of the carbon and which is likely to act as a diffusional barrier to substances being electrochemically detected at the transducer surface, e.g. the thiocholine produced by the enzyme reaction. Addition of stabilisers into this system will only tend to increase the thickness of the insoluble layer, albeit the enzyme stability is enhanced somewhat.

Direct activation of the transducer surface using a carbodiimide reported by Rippeth et al [2] will tend to produce a more open matrix, with the enzyme immobilised by covalent interaction on the surface of the carbon. The higher amperometric responses observed using this immobilisation method indicate that the diffusional constraints are less compared to the glutaraldehyde immobilised biosensors. The responses can be seen in table 1.

Table 1. Amperometric Response of Acetylcholinesterase Biosensors.

Immobilisation Method - Stabiliser Combination	Current Response Day 0 (nA)	Current Response Day 16 (nA)
Glutaraldehyde - Buffer Alone	48	12
Glutaraldehyde - Trehalose	111	50
Glutaraldehyde - Lactitol	160	123
Glutaraldehyde - Gafquat and Lactitol	49	40
Carbodiimide - Buffer Alone	275	177 (day 27)
Carbodiimide - Trehalose	532	469
Carbodiimide - Lactitol	574	510
Carbodiimide - Gafquat and Lactitol	335	290
Carbodiimide - Dextran Sulphate and Lactitol	947	1057

The effects of the stabilisers when added indicate multiple effects, notably that the positive charged polymers tend to reduce the amperometric signal which is likely to be due to i) interaction of the positive polymer with the negative face of the enzyme to create a diffusional barrier and ii) substrate partitioning of the positive substrate away from the active site by the positive matrix so formed. Addition of negative polymers are likely to interact with the positive face of the enzyme and produce a large increase in signal response which is probably mainly due to substrate partitioning. Figure 2 overleaf shows the enzyme structure and a diagrammatic carbon surface at approximately the same molecular scale, to give an indication of the interfacial connection and the distances involved. It is thought that the interactions leading to covalent immobilisation will tend to be on the positive face of the molecules as the transducer surface is more negative due to unreacted carboxylic acid groups and the preponderance of quinones on the carbon surface. The work carried out using acetylcholinesterase as the model enzyme system is being extended into other amperometric

808

biosensor systems using the combination of enzyme structure determination and selective chemistry for immobilisation and stabilisation of the enzyme - transducer interfaces.

The newly published crystal structure of horseradish peroxidase [5] has already begun to be examined and the EU funded project DIAMONDS is utilising such molecular information to produce electronic interfaces for new types of biosensors for future use.

Figure 2. Molecular Dimensionality of the Enzyme Structure of Acetylcholinesterase and the Carbon Surface of the Transducer.

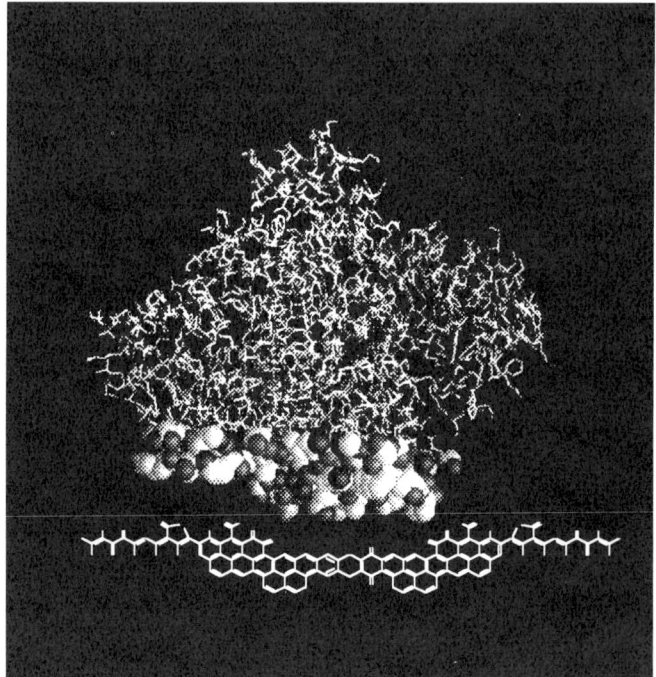

4. Acknowledgements

The authors acknowledge the experimental help of Mr John Rippeth and Dr Matthew Bates of GEM Ltd. Also DGXII, Biotechnology Programme of the EU for the support of the structural biology and electronic interface project DIAMONDS.

5. References

1)Appleton B, Gibson T D and Woodward J R. 1997 *Sensors and Actuators B.* **43** 65 - 69.

2) Rippeth J J, Gibson T D, Hart J P, Hartley I C and Nelson G. 1997 *Analyst* **122** 1425 - 1429.

3) Rippeth J J, Gibson T D, Hart J P and Nelson G. 1997 *Proceedings of the American Chemical Society. Spring Meeting. San Francisco* **76** 608- 609.

4) Ripoll D R, Faerman C H, Axelsen P H, Silman I and Sussman J L. 1993 *Proc.Nat.Acad.Sci. USA.* **90** 5128 - 5132.

5) Gajhede M, Schuller D J, Henriksen A, Smith A T and Poulos T L. 1997 Nature Structural Biology **4** 1032 - 1038.

Biosensors II
Paper presented at Eurosensors XII, 13–16 September 1998
© *1998 IOP Publishing Ltd*

Sensitive detection of pesticide using ENFET with enzymes immobilized by cross-linking and entrapment method

K. WAN[1], J.M. CHOVELON[1], N. JAFFREZIC-RENAULT[1], A.P. SOLDATKIN[2]

[1] IFoS, UMR CNRS 5621, Ecole Centrale Lyon, BP 163, 69131 Ecully Cedex (France)
[2] Institute of Molecular Biology and Genetics, Academy of Sciences of the Ukrainian, SSR, 150, Zabolotny St., Kiev 252143, Ukraine

Abstract. Trichlorfon as a common pesticide is detected using a BuChE-FET sensor which is prepared with cross-linked BSA-glutaraldehyde membrane or photocrosslinkable PVA/SbQ (poly(vinyl alcohol) containing styrylpyridinium) membrane. On the basis of these BuChE-FET sensors, a biochemical test for trichlorfon in liquid was developed by enzyme inihibition. PVA/SbQ membrane shows better characteristics compared to BSA membrane : larger linear dynamic range in kinetic mode, better reproducibility, and above all, better stability in storage. Nevertheless, the both membranes present the same detection limit for trichlorfon (10^{-6} M).

1. Introduction:

Since pesticides are among the most toxic products of the chemical industry, a strong demand for disposable, low cost and simple monitoring devices exists. In this context, biosensors seem to be a promising tool. In this case, the scheme for the detection of the pesticides is associated with their ability to inhibit one enzyme such as the acetyl- or butyrylcholinesterase (AcChE or BuChE). Normally AcChE and BuChE hydrolyze acetylchloride and butyryl chloride, while acid and choline species are formed. In the case of electrochemical biosensors, products of these reactions are usually detected with amperometric [1,2], potentiometric [3,4], and conductimetric [5] electrodes. Concerning potentiometric detection, Ion Sensitive Field Effect Transistors (ISFET), represent one interesting alternative, although a few works have been devoted to them.

The basic requirements for a reliable biosensing probe are its sensitivity, linearity, response time, reproducibility and long lifetime. These parameters are usually controlled by immobilization procedures and the sensitivity of the basic electrochemical sensors. Up to now, works dealing with ISFET have shown either a low reproducibility [4] and/or a short lifetime [4,5], probably due, among other things, to the immobilization procedure used.

In this work, the performances of the ISFETs using two different immobilization procedures have been compared. In the first case, a cross-linking procedure with BSA-glutaraldehyde is used, while in the second one, enzymes are immobilized by entrapment in polyvinylalcohol bearing styrylpyridinium groups.

2. Experimental:

Preparation of the two different membranes:

BSA membrane: a mixture of 5mg BSA, 5mg BuChE (EC 3.1.1.8., 10-20 U mg[-1], Pseudocholinesterase, from horse serum, Sigma), 10µl of glycerol in 90µl of phosphate buffer (1mM, pH 8.0) was prepared. 0.5µl of this mixture was deposited on the sensitive area of a FET. Then the sensor chips were placed in a saturated glutaraldehyde vapor for 30 min.

PVA/SbQ membrane: a mixture of 45 mg PVA/SbQ (SPP-H-13, Toyo Gosei Kogyo Co. LTD, Japan), 5 mg BuChE in 50mg phophate buffer (1mM, pH 8.0) was prepared. 0.5 µl of this mixture was deposited on the sensitive area of a FET. Then, the sensor chips were exposed under UV light for 25min.

Concerning the REFET for both membranes, the same procedures were used but the BuChE was replaced by 5mg of the polymer used for constituting the membrane .

For both enzymatic membranes, the best sensitivities were obtained for pH comprises between 8-8.5 and at a buffer concentration (phosphate buffer) of 1 mM.

Design of biosensors.

The used ISFETs were produced at the Microdevices Research Institute, Ukraine [6]. The sensor chip contained two identical pH-sensitive FETs micromachined on the same p-Si wafer.

3. Results and discussion:

Biosensors calibration

The biosensor calibration curves were presented following two basic methods : first, when steady-state sensor response was registered (Fig 1a and 2a) and secondly, when kinetic response, i.e. the maximum slope of the response curve versus time was measured (Fig. 1b and 2b).

Fig. 1a and 2a, show that the dynamic ranges for both membranes are very similar. On opposite, the dynamic linear ranges in kinetic mode (Fig. 1b and 2b) differ strongly: 0.2-1 mM for BSA membrane and 0.2-5.8 mM for PVA/SbQ membrane. In the same way, the apparent Km values calculated in kinetic mode are different (2mM for BSA membrane and about 3.8mM for PVA/SbQ membrane). This result can be explained by considering the different structures of the enzymatic membrane: in one case the enzyme is cross-linked with BSA while in the second case, the enzyme is entrapped in the photopolymer PVA/SbQ. As a resulting effect, the enzyme is more free in PVA/SbQ membrane than in BSA membrane. It is noteworthy that here these values are comparable between themselves, because the same experimental conditions were used. It is known [7] that values of Km also depend on the ionic strength and pH of the buffer etc...

Stability of the sensor

The lifetime of enzymatic membranes constitutes a limiting factor for biosensor application. In this context a storage stability for both membranes has been investigated. After a storage of 4 months in dry and dark at 4°C, 80% of the inital enzymatic activity for PVA/SbQ membrane was maintained while it remains only 15% of the initial enzymatic activity after 3 weeks in buffer solution at 4°C with BSA membrane. It is noteworthy that results presented here concern only the optimal storage conditions for each membrane.

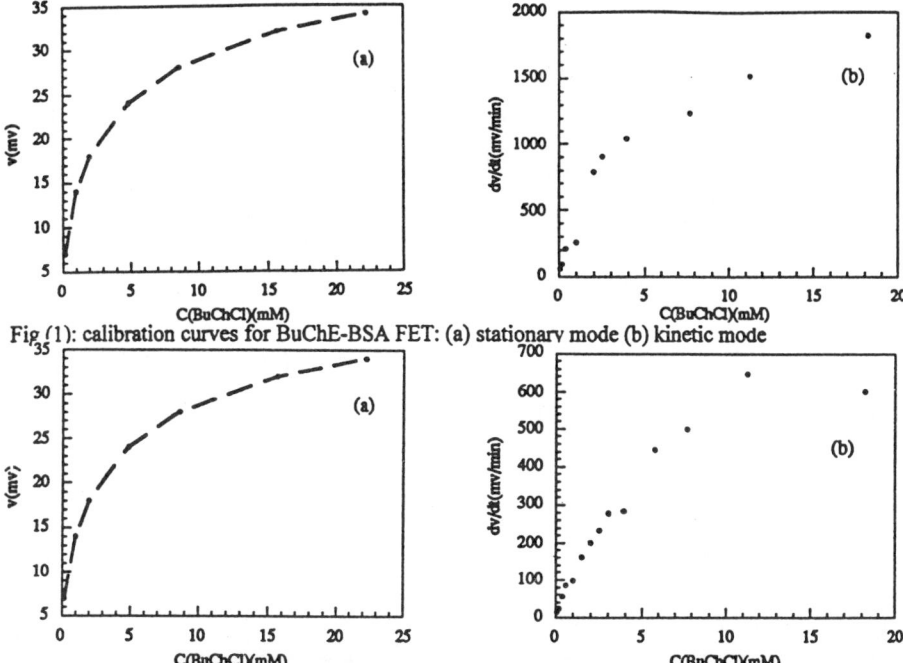

Fig (1): calibration curves for BuChE-BSA FET: (a) stationary mode (b) kinetic mode

Fig(2): calibration curves for BuChE-PVA/SbQ FET: (a) stationary mode (b) kinetic mode

Storage stability in the different enzymatic membranes

In addition, for these two enzymatic membranes, the number of assays plays an important role. For PVA/SbQ membrane, 80% of its initial activity is maintained after 40 cycles of assays while the BSA one losses 60% of its initial activity after the first assay and during the 20 following assays, keeps 40 % of its initial activity

Determination of trichlorfon with BuChE FET sensors:

Fig. 3 and 4 show that the degree of enzyme inhibition depends on the trichlorfon concentration and incubation time. In addition, these figures show a range of detected concentration comprises between 10^{-3}M and 10^{-6}M for both membranes which corresponds to the values found in the literature [6]. On the other hand, since Fig 3a and 4a show that the incubation time is an important factor, 30 min for BuChE inhibition time will be chosen for further experiments.

4. Conclusion:

In this paper, two different procedures have been used for immobilised enzymes on FET transducer for detecting pesticides . In a first one, enzymes were entrapped in a photocross-linkable PVA/SbQ membrane, while in the second one, enzymes were cross-linked with BSA-glutaraldehyde membrane. Although the both membranes present the same detection limit for trichlorfon (10^{-6}M), PVA/SbQ membrane show a better reproducibility and above all, a longer lifetime.

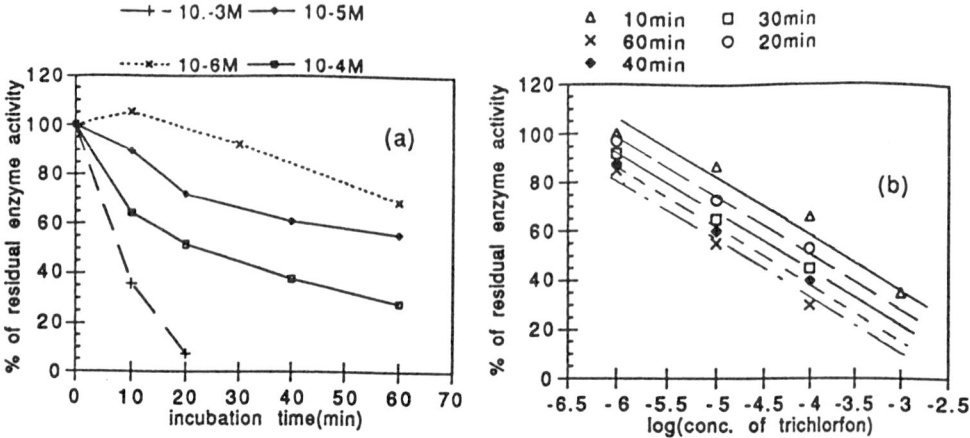

Fig(3) % of residual enzyme activity of BuChE-BSA FET sensor as a funtion of incubation time (a) or trichlorfon concentration (b)

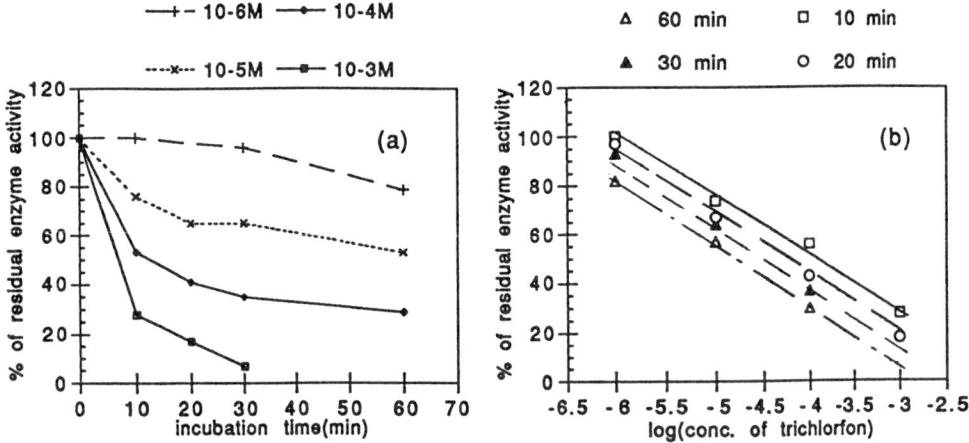

Fig(4) % of residual enzyme activity of BuChE-PVA/SbQ FET sensor as funtion of incubation time (a) or trichlorfon concentration (b)

References

[1] P.Skladal, M.Fiala, J.Krejci, Intern. J. Environ.Anal. Chem.65 (1996) 139

[2] N. Mionetto, J.L. Marty, I. Karube, Biosensors & Bioelectronic 9 (1994) 463

[3] L. Campanella, C. Colapicchioni, G. Favero, M.P. Sammartino, M. Tomasseti, Sensors and Actuators, B33 (1996) 25

[4] C. Dumschat, H. Müller, K. stein, G.schwedt, Analytica, Chimica Acta, 252 (1991) 7

[5] S. V. Dzydevich, A.A. Shul'ga, A.P. Soldatkin, A.M. Nyamsi Hendji, N. Jaffrezic-Renault, C. Martelet, Electroanalysis 6 (1994) 752

[6] A.M. Nyamisi Hendji, N. Jaffrezic-Renault, C. Martelet, P. Cléchet, A.A. Shul'ga, L.I. Netchiporouk, A.P. Soldatkin, W.B. Woldarski, Analytica Chimica Acta, 281/1 (1993) 3

[7] M. Hanss and A. Rey, Biochim. Biophys; Acta, 227 (1971) 618

Biosensors II
Paper presented at Eurosensors XII, 13–16 September 1998
© *1998 IOP Publishing Ltd*

Rapid differentiation of microbial cultures in liquid media using an electronic nose

S T Heron and T D Gibson

School of Biochemistry and Molecular Biology, University of Leeds, Leeds, LS2 9JT, UK

Abstract. The rapid differentiation of bacteria grown in a liquid media was achieved using an electronic nose containing a conducting polymer sensor array. The differential ability of the electronic nose was compared between the headspace of micro-organisms grown in nutrient broth and on nutrient agar plates. Initial results suggest that microbial differentiation is not improved by using liquid media headspace.

1. Introduction

Research into the usefulness of electronic noses in the detection and identification of micro-organisms, using various types of sensor and analysis techniques are presently under investigation. Work carried out on the microbial headspace samples taken from above microbial cultures grown on nutrient agar plates [1] and from bacterial cultures grown in nutrient medium [2] have shown results which are very encouraging. The development of an electronic nose which could detect and identify micro-organisms from the volatiles released during growth, could enable the use of a new rapid method for microbial classification.

2. Materials and Methods

The Electronic Nose was developed in the Department of Biochemistry and Molecular Biology at Leeds University and has been commercialised by Bloodhound Sensors Ltd. The Instrument used to supply the data presented in this poster was a prototype Bloodhound® BH114 Sensory Array System. Data handling and manipulation other than the Bloodhound control software was carried out in Microsoft Excel. The sensor array used consisted of 16 different conducting polymer sensors. The prototype BH114 is a dynamic flow detection system (flow rate 150ml min^{-1}). Exposure to an odour produces a response from each sensor in the array. Four parameters are extracted from the response profile of each sensor which is then used for data analysis. This is shown in diagram 1.

Headspace sampling was taken from above pure cultures of micro-organisms growing on nutrient agar plates and from broth cultures (all media supplied by Oxoid Ltd) sealed inside 'Teglar' sample bags after a 24 hour growth period and a 1 hour volatile equilibrium period.

814

Diagram 1   Parameters measured for each sensor response

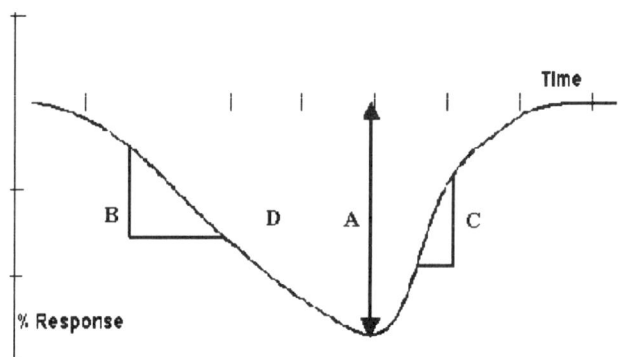

A=Divergence, B=Absorption slope, C=Desorption slope and D=Area.

3.   Data analysis

3.1   Principal Component Analysis (PCA) and Discriminant Analysis (DA)

Principal component analysis [3] and discriminant analysis are both multivariate analysis techniques. They were used to reduce large groups of data in which there was a large number of interrelated variables and produce a two-dimensional graphical representation of the data. Therefore large data sets could be easily analysed to observe any similarity or differentiation between the microbial samples. PCA and DA was carried out using a Microsoft Excel add-in package Xlstat. Selected raw data was analysed using the techniques to accentuate any differences between the samples.

3.2   Back propagation - neural networks

Back propagation also known as backward error propagation or backprop, is a very popular network - learning algorithm. It is described as an adaptive system that minimises an error signal by using gradient descent [4]. The response data collected by the Bloodhound software to each micro-organism were used to train the neural network. After the network had been trained, another set of unknown (to the neural network) microbial response data was then used to see if the network could discriminate between the different bacterial profiles and correctly classify the samples.
Some initial results have been analysed in collaboration with Dr Paul Corcoran, Sensing and Control Research Group - University of Derby, using a neural network approach with a multilayered perceptron back propagation network..

4. Results

The discriminant analysis technique depicted as two-dimensional graphical representations show that sampling microbial volatiles from the headspace of nutrient broth cultures does not increase the discrimination power of the electronic nose when compared to the samples taken from the headspace of nutrient agar plate cultures. Principal Component Analysis carried out on three different micro-organisms show that the electronic nose is clearly discriminating between the headspace samples from above nutrient agar plates, whereas similar results taken from liquid cultures are less clear.

4.1   Discriminant Analysis (DA)

DA of 8 test micro-organisms grown on nutrient agar

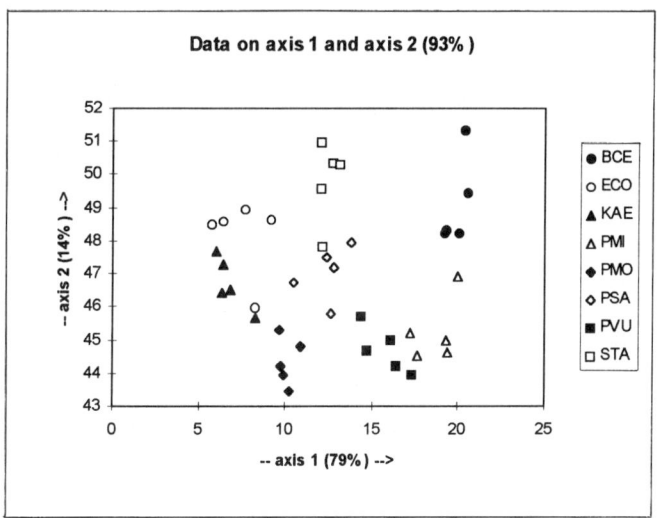

DA of 8 test micro-organisms grown in nutrient broth

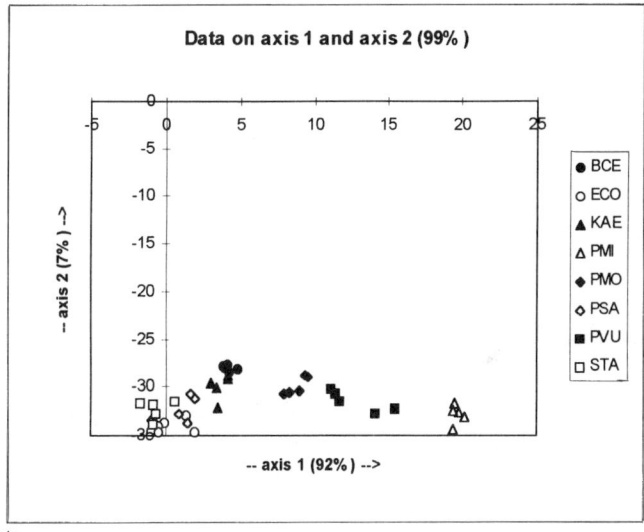

Key to the DA plots
Bacillus cereus - BCE, *Escherichia coli* - ECO, *Klebsiella aerogenes* - KAE, *Proteus mirabilis* - PMI, *Proteus morganni* - PMO, *Proteus vulgaris* - PVU, *Pseudomonas aeruginosa* - PSA and *Staphylococcus aureus* - STA.

It can be noticed that the discriminant analysis technique gave a larger spread of data for the plate cultures than for the broths. This is likely to be due to the effect of water vapour on the sensor response, which may be reduced by using sensors with low sensitivity to water vapour [5].

816

4.2   Principal Component Analysis (PCA) plot of three test micro-organisms grown on nutrient agar plates *Escherichia coli* (ECO), *Staphylococcus aureus* (STA) and *Pseudomonas aeruginosa* (PS1)

Data on axis 1 and axis 2 (60%)

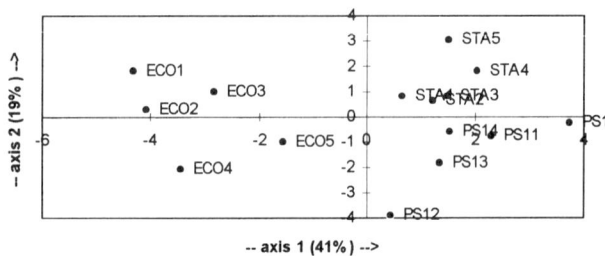

-- axis 1 (41%) -->

4.3   Example of neural network classification of microbial headspace samples taken from above nutrient broth culture and nutrient agar plates.

	Percentage Correct Classification %	
Micro-organism (strain)	Nutrient Agar Broth	Nutrient Agar Plate
Bacillus cereus 603	64	42
Escherichia coli 10418	89	100
Klebsiella aerogenes 719	45	62
Proteus mirabilis 715	56	62
Proteus morganni 868	67	12
Proteus vulgaris 195	100	56
Pseudomonas aeruginosa 10662	78	88
Staphylococcus aureus 6571	35	53

The results obtained with the neural network are rather inconclusive in that both types of growth conditions (plate or broth cultures) show a similar percentage of correct classification in all of the test micro-organisms.

5.   Conclusion

Initial results would seem to indicate that microbial differentiation using an electronic nose is not improved if the micro-organisms are grown in a broth culture, with the indication that better discrimination occurs from samples grown on nutrient agar plates. Bearing in mind that the sensors employed in this study were conductive polymers having some response to water vapour, further work will be necessary to establish if the use of water insensitive materials will improve the discrimination power of the technique. It is likely that improved sampling techniques combined with a different series sensors will give a much higher level of discrimination in liquid cultures, this hypothesis is being investigated in some detail at the present time.

References

[1]   Gibson T D, Prosser O P, Hulbert J N, Marshall R W, Corcoran P, Lowery P, Ruck-Keene E A and Heron S 1997 *Sensors and Actuators B* **44** 413-422
[2]   Gardner J W, Craven M, Dow C and Hines E L 1998 *Meas. Sci. Technol.* **9** 120-127
[3]   Jolliffe I T 1986 *Principal Component Analysis* (Springer-Verlag New York Inc.)
[4]   Anderson J A 1995 *An Introduction to Neural Networks* (Massachusetts Institute of Technology)
[5]   Clements J, Boden N, Gibson T D, Chandler R C, Hulbert J N and Ruck-Keene E A *1997 Proceedings of Eurosensors XI Warsaw* Vol. **2** 889-892

CONTROL OF A MYOGLOBIN LEVEL IN SOLUTION BY THE BIOAFFINIC SENSOR BASED ON THE PHOTOLUMINESCENCE OF POROUS SILICON

Valentyna M. Starodub, Leonid L. Fedorenko[1], Nickolaj F. Starodub*

A.V.Palladin's Institute of Biochemistry, National Academy of Sciences, 9 Leontovicha Str., 252030 Kiev, Ukraine, Fax: 380-44-2296365; E-mail: prof@progress.FreeNet.Kiev.UA
[1]Institute of Semiconductor Physics, National Academy of Sciences, Kiev, Ukraine

Summary

The results about work testing of early developed bioaffinic sensor based on the photoluminescence of porous silicon are presented. It is shown that this biosensor is suitable for control of a myoglobin level in solution. The sensitivity of such a sensor is comparable with the standard ELISA method but the overall time of analysis accomplished by the sensor is much shorter. Data about operational stability of bioaffinic sensor showed that transducer together with the sensitive layer may be used for one measurement only.

1. Introduction.

For express diagnostic and environmental monitoring it is necessary to have methods which could allow to fulfill analysis in real time regime. Biosensors with the direct registration of specific biological interactions excite great interest. Early [3,4] we informed about creation of sensor based on the porous silicon photoluminescence. It is a principally new approach in biosensor technology. It provides simple and fast registration of antigen-antibody interaction. Sensitivity and operational stability of such an immune sensor for myoglobin detection have been studied. Besides, registration of other biospecific interactions (for example, enzyme labeled immune components) is analysed too. Obtained results are presented in this report.

2. Materials and methods.

Intensity of photoluminescence of porous silicon at the influence of He-Ne laser beams was measured. The wave length for observation of the visible photoluminescence of porous silicon was about 650 nm. The time of photoluminescence decay of the porous silicon was described by "stretch" exponent. Specific antibodies were immobilized on the porous silicon surface by spontaneous sorption. At first the surface of the porous silicon crystals was cleaned by consecutive alcohol and distilled water washing. For immobilization of antibodies on the surface of the crystals (1x1 mm) they were immersed into solutions (100 µg/ml) of these substances for 1 h at the room temperature. Then crystals were washed by tris-HCl buffer (pH 7.3) containing 140 mmol/l sodium chloride. Human myoglobin was used as antigen and mouse monoclonal IgG to myoglobin as specific antibodies. We used antibodies of two different types: the first class was obtained to one site of myoglobin and the second had specificity to another site of the antigen. Moreover, last type of monoclonal antibodies was labelled by β-glucose oxidase. As a model of enzymatic reaction of β-glucose oxidase with glucose was used.

---

* Author to whom correspondence should be sent

3. Results and discussion.

Early [4] we demonstrated that typical spectrums of the photoluminescence of porous silicon at the contact with distilled water, or 10 mmol/l tris-HCl buffer (pH 7.3) solution, or 140 mmol/l buffered sodium chloride solution, or solution of IgG, or solution of anti-IgG prepared in 10 mmol/l tris-HCl buffer (pH 7.3) containing sodium chloride, remained almost immutable during no less than 2-2,5 hours. At the same time after previous immobilization of antibodies or antigens on the porous silicon surface and their consecutive contact with the solution of corresponding antigens or antibodies a great reduction of photoluminescence intensity was observed. In this experiments we found that during 120-150 min spontaneous degradation of photoluminescence of porous silicon was no more than 10%. Moreover, during first 60 min spontaneous degradation was not observed (Table 1). If monoclonal antibodies to human myoglobin were immobilized on the porous silicon surface the level of photoluminescence did not change. It decreased if the surface, treated in the way mentioned above, was immersed in the solution of myoglobin.

Table 1.

The level of degradation of porous silicon photoluminescence at different time of its incubation in buffer solution.

Incubation time, min	5	60	90	120	150	270
Level of photoluminescence,%	100	98	95	93	90	74

Decrease in the intensity of photoluminescence was registered in 5 min after the beginning of measurement and it achieved maximal level after 30-35 min. In the case of use of myoglobin solution at the concentration of 1 μg/ml after 16-18 min the level of photoluminescence decreased for 50% (Table 2). According to existing ideas about the nature of the visible photoluminescence of porous silicon, extinguishing of the photoluminescence can be explained by a dehydrogenization of porous silicon surface after specific immune complex formation. Hydrogen is released from silicon bonds and subsequently captured by the immune complex. In turn, torn Si bonds are known to intensify nonradioactive channel of recombination [1,2,5].

Table 2.

Change of the level of the photoluminescence of porous silicon at the immune complex formation at the different time of its incubation in myoglobin solution (1 μg/ml).

Incubation time, min	5	10	15	20	25	30	35
Level of photoluminescence,%	97	80	65	40	20	10	8

Dependence of the photoluminescence level on the concentration of myoglobin in solution is presented on the Table 3. It is shown that intensity of chemiluminescence did not change at the myoglobin concentration less than 1.0 ng/ml. The level of detected concentration of myoglobin is in the range from 0.01 to 10 µg/ml.

Table 3.

Change of the level of porous silicon photoluminescence in the case of immune complex formation at the different concentrations of myoglobin (15 min incubation).

Concentration of myoglobin, µg/ml	0.001	0.01	0.1	1	10
Level of photoluminescence,%	100	85	65	30	15

It is necessary to note that immune complex, which was formed by immobilized monoclonal antibodies, myoglobin and soluble monoclonal antibodies labelled by β-glucose oxidase, caused more essential decrease in photoluminescence level in comparison with the case when labelled monoclonal antibodies were absent in the solution (Table 4). Moreover, we want to pay attention to the fact that presence of substrate for β-glucose oxidase in the solution to be analysed promoted decrease of the level of photoluminescence. It may be caused by realization of different mechanisms of influence of enzymatic reaction on porous silicon photoluminescence. One of them may be connected with the generation of oxygen radicals during enzymatic reaction.

Table 4.

Change of the level of the photoluminescence of porous silicon at the immune complex formation at different time of its incubation in solution of myoglobin and anti-myoglobin labelled by β-glucose oxidase (at equal concentrations of 1 µg/ml) without substrate for enzyme and at its presence.

Incubation time, min /Level of photoluminescence,%							
	5	10	15	20	25	30	35
Without substrate	85	72	49	26	11	6	5
In the presence of substrate	80	68	44	20	8	4	3

The operational stability of the immune sensor based on the porous silicon photoluminescence was studied in the following way. After every cycle of immune complex formation and measurement of photoluminescence intensity the surface was treated by 0.1 N HCl or acetate buffer with pH 2.2. It was found that intensity of photoluminescence dramatically decreased after the first cycle of porous silicon usage (Table 5). Maybe, it is a result of either porous silicon degradation at the condition of the destruction of immune complex or incomplete destruction of this complex. The first assumption is more real than the second one.

Table 5.

Operational stability of the signal of immune sensor *

Number of measurements/Level of photoluminescence changes, %		
1	2	3
100	52	10

*) Explanation in the text.

4. Conclusion.

For the first time the immune sensor based on the photoluminescence of porous silicon for the measurement of myoglobin concentration in the solution was created. Some working and operational characteristics of this sensor were studied.

Acknowledgments

This work was supported by Ukrainian Ministry of Sciences and Technology (grants N 5.4 / 255).

References

1. Canham, L. 1990. Silicon quantum wire array fabrication by electrochemical dissociation of wafers. 1990. *Appl. Phys. Lett.* 57: 1046-1048.
2. Canham, L.T., Leong, M.Y., Cox, T.T., Beale, M.T.J. and Marsh, K.J. 1993. Efficient visible photoluminescence and electroluminescence from highly porous silicon. *21st ICPS on the physics of the semic.*32 Beijing, China: 1423-1430.
3. Starodub, N.F., Fedorenko, L.L., Starodub, V.M., and Dikij, S.P. 1998. Silicium crystal photoluminescence as transducer for biosensors. In: *Advanced Electronic Technologies and Systems Based on Low-Dimensional Quantum Devices,* edited by M.Balkanski and N. Andreev. Kluver Academic Publishers: 91-92.
4. Starodub, N.F., Fedorenko, L.L., Starodub, V.M., Dikij, S.P. and Svechnikov, S.V. 1996. Use of the silicon crystals photoluminescence to control immune complex formation. *Sensors and Actuators.* 35-36 (B): 44-47.
5. Vial, J., Billet, S., Bsiesy, A., Fishman, G., Gaspard, F., Herino, R., Legion, M., Madeare, R., Michalcesku, T., Miller, F., and Romestaine, F. 1993. Bright visible light emission from electro-oxidised porous silicon. *Physica B.* 185: 593-602.

Biosensors II
Paper presented at Eurosensors XII, 13–16 September 1998
© 1998 IOP Publishing Ltd

Complex Study of Gas-responsive Properties of Oligo- and Polyphthalocyanines

S.A. Krutovertsev*, A.E. Tarasova*, S.I. Sorokin*, L.S. Krutovertseva*,
A.V. Zorin*, A.I. Sherle**, and V.V. Promislova**.

*Joint Stock Company "PRACTIC-NC", Zelenograd, Moscow, 103460 Russia
**Institute of Chemical Physics, RAS, ul. Kosygina 4, Moscow, 117977 Russia

Abstract. More than 150 various oligo- and polyphthalocyanines, both metal-free and metal-containing, were synthesized. The methods of oligo- and polyphthalocyanines films producing by vacuum thermal sputtering and that of oligophthalocyanines by precipitation from solution were developed. The electric, sorption and gas-responsive properties of the films produced were investigated. More than 30 samples were proved to be of adequate properties for using them as the basis of gas sensors.

1. Introduction

The phthalocyanines (Pc) are known to modify their electrophysical properties, e.g. electric conductivity, dielectric permittivity, etc., when affected by some active gases, which is the basis for utilization of those as gas sensors [1,2]. The major advantages of such sensors as comparable to, for instance, SnO_2-based semiconductor sensors, are higher sensitivity to such compounds as NO_2, lower operating temperature of the sensitive element and, consequently, lower heating input, which is particularly important for designing of portable instruments, and in most cases higher selectivity. Usually, two types of the gas sensors on the Pc basis are used. Those are resistive and piezosorption sensors. Besides, the Pc, containing Pb, Fe, Al, Cu, Co, and Ru, are used as sorbents [1-3].

At the same time, great amount of polyphthalocyanines is known to be of valuable physical and physico-chemical properties, which exceed sometimes the original Pc characteristics [4]. The polyphthalocyanines could be thought of as rather advanced materials for sensitive layers of the gas sensors, which is, however, not widely reflected in the literature . The reason for the latter might be the fact that polyphthalocyanines obtained by usual methods are nonmelting and insoluble compounds that are poorly reproducible by their structure and properties and contain uncontrolled admixtures. The producing of materials on their base requires separate long-term investigations.

An attempt of complex investigation of electric and gas-responsive properties of a series of synthesized metal-free and metal-containing oligo- and polyphthalocyanines (OPc and PPc, respectively) has been done.

2. Experimental

Metal-free OPc and PPc as well as those containing Cu, Fe, Mn, Co, and Ni were examined. The samples were synthesized by polycyclotetramerization of tetranitrile of pyromellitic acid in bulk and in polar solvents at 180-300 °C for 5-30 hours in the presence of 0-5 mol% carbamide. The synthesis and some properties of the compounds obtained were detailed in [5, 6].

A special test chip (Fig. 1) on an alumina or sitall 5×15 mm base was devised for the study of PPc electric and sensor properties. Two comb-like electrodes, made of nickel with vanadium or chromium sublayer, were located on one side of the base; the opposite side contained nickel film heater for the measurements at the temperature below 250 °C or that of nichrome for higher temperatures. The distance between the neighbor electrode tines was 20 micrometers, the metal layer width was 0.2-0.3 micrometers. The active side with the comb-like electrodes and the heater at the opposite side was 4×4 mm in size.

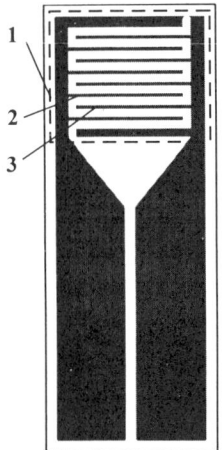 

Fig.1 *Design variant of test structure for sensor's investigation:*
1-sorbent;
2,3-interdigital electrodes:
4-heater.

The surface of the test chip was treated with the sample powders by precipitation from solution for soluble OPc and/or by vacuum thermal evaporation. The latter was carried out using UVN-71P3 equipment at an after-pressure not higher than $4×10^{-6}$ Pa and temperature from 600 to 1200 ^{0}C; the distance between evaporator and the base was 100 mm. Simultaneously, both test chips and quartz resonator of the temperature independent AT-section with 8-9 Mhz fundamental frequency were sputtered. The samples of soluble PPc were applied to the test chip surface by centrifuging or immersing from solution into appropriate solvent, most frequently DMFA.

Electric properties of the films were studied upon direct as well as alternating current at locked frequency 1 kHz and 1 Mhz. Adsorption properties of the films were examined by piezoquartz microbalans method [7]. A testing unit was used for investigation of the gas sensitive properties of the films as well as RD-230 Riken Keyki (Japan) devise for gas blending. The gas sensitive films were studied in relation to the model gases - O_2, H_2, CO, NO_2, NH_3, H_2S, as well as water and alcohol vapor. A sample "two pressures" generator of humid gas "Rodnik-2" was applied for analysis of environmental humidity effect on the films properties.

3. Results

Over 150 various OPc and PPc macromolecules differed by size, structure as well as metal type, concentration and its location were synthesized. The samples were dark-colored powder-like compounds, some of which, i.e. OPc, were acetone- and dimethylformamide-soluble. All the samples were H_2SO_4-soluble and could be precipitated from this solution into water.

Thermogravimetry analysis was carried out using "Derivatograf OD-102" at 6 grad/min heating rate and 100 mg sample, according to the data all the polymers were heat resistant in air up to 300-400 ^{0}C. Further heating resulted in loss of the sample, the rate of which depended on the structure of the compounds obtained. The Ppc heat resistance in vacuum was higher than that in air. Indeed, dependent on the method of synthesis, the sample losses when heated at 350 ^{0}C for 15 hours in vacuum were 3-20 %, whereas 10-80 % in air.

The width and color of the films produced, the latter ranging from almost colorless and transparent to dark blue, green-blue and brown, differed with the original both OPc and PPc and the sputtering manner. Their electrophysical characteristics differed considerably as well.

Indeed, the films resistance varied from Ohm units to GOhm; in most cases, i.e. for the films with the resistance lower 200 MOhm, it was of ohmic nature and frequency independent.

For the most rich-colored films, produced by sputtering, the resistance was found to diminish with the temperature increase in the range 20-250 $^\circ$C, Fig. 2. The same relationship was observed for the film samples developed from solution, while, as a rule, those films were more high-ohmic.

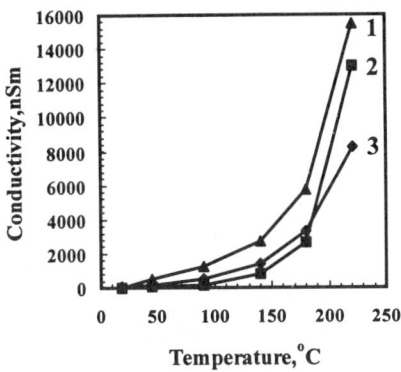

Fig. 2 *Typical temperature dependence of conductivity for different polyphthalocyanines:*
1- Fe polyphthalocyanine,
2- Ni/Cu polyphthalocyanine,
3- Cu polyphthalocyanine.

When treated continuously (18 hours) in conc. H_2SO_4 the film substance dissolved and could be precipitated from the acid solution into water. The data of electronic absorption spectra attested to the presence of azaporphin macrocycles in structure of the film compound.

The analysis of the gas-responsive properties of the films revealed more than 30 samples investigated to be of adequate sensitivity in relation to some model gases. In particular, sensitivity toward concentration of O_2, NO_2, H_2S, NH_3 and humid gases was found for a number of materials. The data on metrological characteristics of the experimental structures are shown in Tab. 1. Typical relationships of PPc film resistance versus concentration of ammonium gas at different temperatures are shown on Fig. 3.

Table 1. *Metrological characteristics of test structures.*

Types of the sensors developed	Sensitivity threshold, ppb	Range of measurements, ppb	Time constant, s	Operating temperature of the unit, $^\circ$C
Sensor of O_2 concentration	10^4	10^4- 10^9	10-60*	100-200*
Sensor of NO_2 concentration	0,5	$1 - 10^4$	10-30*	150-200*
Sensor of H_2S concentration	0,5	$1-10^4$	10-30*	100-180
Sensor of NH_3 concentration	10^3	$10^3 - 10^6$	1- 15*	-20 - +60

* In dependents of measurements conditions

824

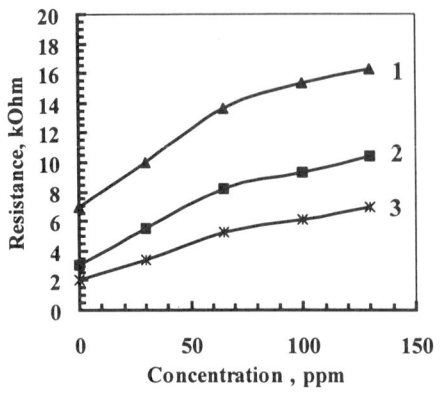

Fig.3 *Dependence of CuPPc resistance on ammonia (NH₃) concentration for different temperatures:*
1-140 ºC ;
2-190 ºC;
3-240 ºC.

4. Conclusions.

The investigation of electrophysical, adsorption and sensor properties of the films based on the synthesized both Opc and PPc ascertains that some of the materials due to their properties can be applied for developing of the sensors for detecting of active gas ingredients in air.

Specifically, some of the samples yield a unique sensitivity, about 0.5 ppb, to NO_2 and H_2S. A number of materials is highly sensitive to O_2, NH_3 and humidity in gases. At the same time, unlike the known semiconductive sensors on the basis of inorganic sorbents, SnO_2, ZnO and so on, the sensors we developed are in operation at considerably lower temperatures, 100--200 ^0C, which enables decreasing heating input and their utilization for portable equipment.

The sensor design and microelectronics technology have been worked out based on the present research. The test sensors for the control of O_2, NO_2, H_2S, NH_3 and humidity of gases concentration have been devised.

Acknowledgments.

This work was supported by International Science and Technology Center (ISTC), project number 082.

References.

[1]. Jones T.A., Bott.B.. and Thorpe S.C. Fast response metal phthalocyanine-based gas sensors, *Sensors and Actuators*, 17 (1989), pp. 467-474.

[2] Hamann C., Mrwa A., Müller M., Göpel W. and Rager M., Lead phthalocyanine thin films for NO_2 sensors, *Sensors and Actuators B, 4* (1991), pp. 73-78.

[3] Di Natale C., Repole G., Davide F., D'Amico A., Fogletti V., Paolesse R., Spagnoli M., Tagliatesta P., Boschi T., Investigation of the sensing properties of conductive solid-state metalloporphyrins, *Proceedings of Eurosensors X The 10th Europian Conference on Solid-State Transducers. Leuven, Belgium, 8-11 September 1996,* v.2, pp. 661-663.

4]. Berlin. A.A. and Sherle. A.I., Synthesis and Properties of Polymer Azoporphines.*Usp. Khim.,* 1979, vol. 48, no. 11, p. 2037.

[5]. Sherle. A.I., Promislova. V.V., Shapiro. N.I., Epstein. V.R., and Berlin. A.A., Synthesis of Soluble Polyazoporphines *Visokomol. Soedin.,* 1980, vol. A22, no. 6, p. 258.

[6]. Epstein. V.R. and Sherle. A.I., Oligophthalocyanines and Polymers on Their Basis: Synthesis, Structure and Thermal Stability *Visokomol. Soedin,* 1990, vol. A32, no. 8, p. 1655.

[7] Hlavay J. and Guilbault G.G. Applications of the piezoelectric crystal detector in analytical chemistry, *Anal.* Chem., 49 (1977) 1980.

Biosensors II
Paper presented at Eurosensors XII, 13–16 September 1998
© *1998 IOP Publishing Ltd*

High Speed Separations of Phenols on a Glass Device

Martin Arundell, Andreas Manz

Zeneca/Smithkline Beecham Centre for Analytical Sciences, Imperial College of Science, Technology & Medicine, London, SW7 2AY, United Kingdom

A micromachined capillary electrophoresis system has been fabricated on a glass chip for the separation and detection of phenols. The five phenols, viz., phenol, 2-chlorophenol, 2,4-dichlorophenol, 2,4,6-trichlorophenol and pentachlorophenol were separated within 150 seconds with a 22mM Tris-borate EDTA buffer containing 1mM fluorescein. Detection was by indirect fluorescence using a conventional light source.

I Introduction

Within the last decade the requirement for on-site continuous monitoring has increased as a result of the introduction of stringent Environmental Quality Standards (EQS's) concerning the level of pollutants in aquatic systems. This puts pressure on the water industry and the regulators to carry out frequent analyses for a large number of pollutants. With a monitor *in-situ* a constant check on all the priority pollutants can be maintained. By incorporating the analysis of environmental pollutants into μ -Tas (miniaturised - Total Analysis System) technology we aim to develop an on-line monitor for pollutants found in surface or waste water.

II Experimental

I Layout of Microfabricated Glass Device

The glass microstructures were fabricated and bonded by the Alberta Microelectronic Centre, Canada. The design of the glass device was centred around three key components of an environmental monitor which include the extraction of potential organic pollutants and the separation and detection of the extracted material.

The dimensions of the glass plate is 2 by 4 inches. The extractor channel (Figure1) is 2.5cm long, 1mm in width and etched to a depth of 200μm. The injection channels (Figure 2) and separation channel (Figure 3) are 200μm and 40μm wide respectively and etched to a depth of 10μm.

II   Instrumentation

Basis of the instrument was a Leica BMIL microscope. The microchip is placed on a chipholder, fabricated in the engineering workshop at Imperial College, which in turn sits on an object holder. The high voltage system for capillary electrophoresis is manually switched. Two high voltage power supplies (HCN 7E 12500 and HCN 7E 35000, F.U.G., Rosenheim, Germany) are used to switch between injection and separation. For the high voltage switching two relays are used (Magnecraft/Struthers Dunn W102HVX-3, Kempston Electrical, Rushden). The relays are controlled via two manual plain rocker switches supplied by a 24v dc 45W ac adapter from RS Components Ltd.

III   Indirect laser Detection

The light source, from a Leica quicksilver 50w high pressure lamp, was passed through a filter cube with an excitation range of 420-490nm. Background fluorescence emitted from the fluorophore was collected with a 30x microscope objective and passed through a 515nm long pass filter. The fluorescent image was focused onto a photomultiplier tube (Hamamatsu) which was connected to a current amplifier (Hamamatsu).

I Layout of Microstructure showing Fluid Flow in Microextractor and CE Injection and Separation

Extraction

Figure 1 shows direction of fluid flow through the microextractor. The large arrows highlight the main flow through the chip with the smaller arrows indicating the small leakage of water to one of the injection waste channels.

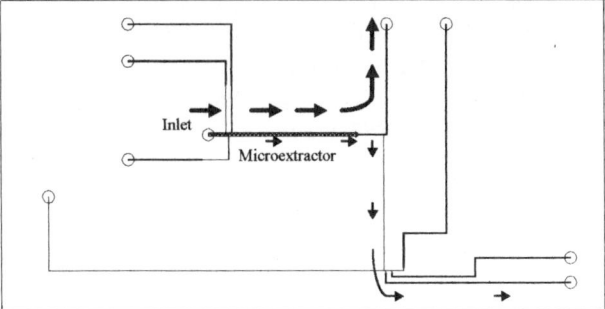

Injection

Figure 2 shows electromigration through the injection channel and out through one of the waste channels depending on the injection plug volume to be utilised (1) 0.1nl or (2) 0.5nl .

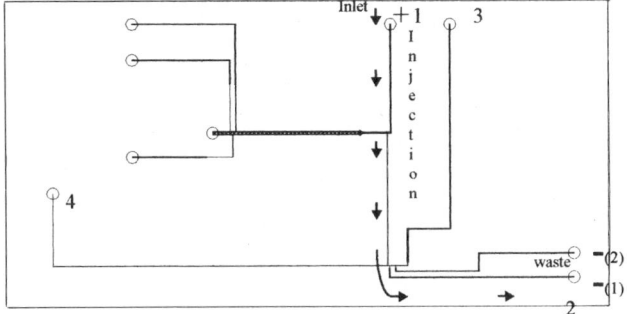

Separation and Detection

Figure 3 shows direction of the buffer solution from the inlet reservoir as it sweeps the sample plug (indicated by the asterix) along the separation channel and towards the waste reservoir.

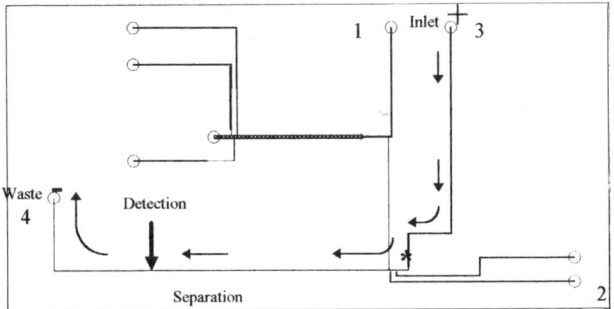

III Phenol Analysis

Five phenols were made up in a TBE buffer at a concentration of 100ppm. Applying a voltage between 1 and 2 (Figure 2) +3.75 kV and ground between the two injection reservoirs caused the sample solution to be pumped electroosmotically into the 0.1nL section. A separation voltage between 3 and 4 (Figure 3) of 1.375kV and ground was then applied between the two separation reservoirs (Figure 30). The five phenols, viz., phenol, 2-chlorophenol, 2,4-dichlorophenol, 2,4,6-trichlorophenol and pentachlorophenol were then detected by indirect fluorescence detection.

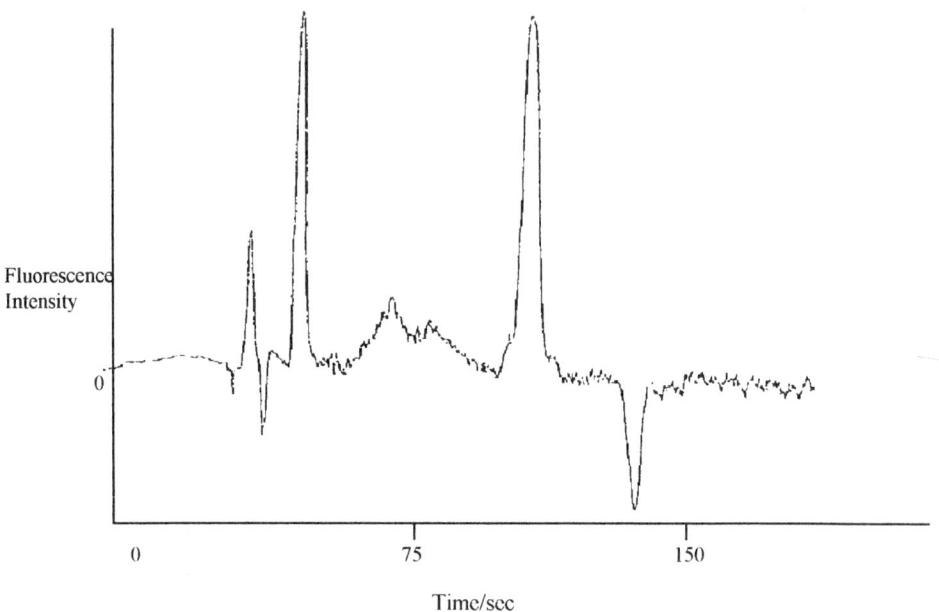

IV Conclusion

CZE on a glass chip results in high speed separations of phenols. Further work is required in optimising the indirect fluorescent detection of phenols and interfacing extraction on chip with the separation.

Biosensors II
Paper presented at Eurosensors XII, 13–16 September 1998
© *1998 IOP Publishing Ltd*

Mathematical Simulation of Enzyme Biosensors with Multilayer Charged Membranes

V.Rossokhaty, N.Rossokhataja

National University of Ukraine, P.O.Box 493, 253222 Kiev-222, Ukraine,
e-mail: ross@nv.kiev.ua

Abstract. Mathematical model of the enzyme biosensor with multilayer charged membrane is developed. The charged layer of membrane is supposed to be penetrable for any particles and formed by built-in uniformly distributed charge. The Michaelis-Menten theory is used for description of the reaction velocity. The model is reduced to one-dimensional initial boundary value problem for the system of second-order partial differential equations describing diffusion-drift transport of reaction components and products in membrane and Poisson equation for electrostatic potential. The discrete model is constructed. The results of numerical experiment are in good qualitative fit with results of physical one. Created package can be easily adopted for membranes with any number of layers.

1. Introduction

Biosensors combining the basic principals of microelectronics, chemistry and biotechnology attract considerable attention in such fields as medicine, environmental monitoring, pharmaceutics and food quality control. Now the most important problem for manufacture is to develop the inexpensive, commercially available, reliable devices with high operation performances. The latter can be achieved in particular by employment of multilayer charged enzyme membranes [1]. The prediction of response of such type of sensor requires a detailed analysis of reaction-diffusion-drift phenomena occurring in the multilayer membrane. It is complicated by necessity of taking into account interaction between charged species of test solution and charge encapsulated in some layer(s) of membrane.

As a rule, authors of many of known works [2,3] on biosensor modelling consider single-layer non-charged membranes and neglect influence of Donnan potential on species transport, difference between diffusion coefficients of species etc. In the case of electrically neutral systems such simplifications are legitimate. But in the cases when charged particle movement is controlled by not only diffusion but by drift in an electric field as well such simplifications become incorrect and do not fit to model species behavior in multylayer charged membranes. An aim of the present work is to develop a mathematical model of an enzyme biosensor with multilayer charged membrane, to develop and realise a numerical method and to carry out the numerical experiment that will predict the transient performance parameters of these devices and will allow optimisation of its performance.

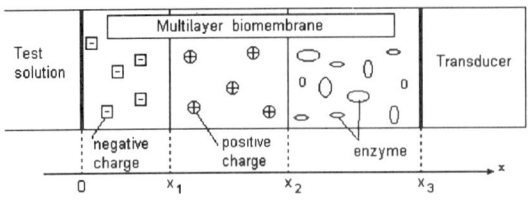

Fig.1. Diagram of the biosensor model

2. Model

For the definiteness but without loss of generality, we consider biosensor for express measurement of glucose concentration in human blood because this problem is of great importance for contemporary medicine. The model of enzyme biosensor with multilayer charged membrane can be represented as it is shown in Fig.1. Solid-state transducer linearly transforms the value of protons concentration near membrane-transducer interface at x_3 into electrical signal. Multilayer membrane (3-layer in the present model) contains at least one enzyme layer with uniformly distributed immobilized enzyme molecules of a given concentration and other layers involving (if necessary) encapsulated charge of constant concentration. Test solution is considered to be infinite source of five types of species: glucose, oxygen, protons, buffer and its conjugate base with concentrations C_G, C_O, C_H, C_W and C_{WH} respectively. The solution-membrane interface and the interfaces between layers into membrane have some "resistance" for species penetration $R_s^{j,j+1} \geq 0$, $s=G, O, H, W, WH$, where $j=0, 1, 2$ indicates the layer (0 - test solution). The transient distributions of species within the membrane can be described by following system of partial differential equations:

$$\frac{\partial}{\partial x}\left(\varepsilon \frac{\partial \varphi}{\partial x}\right) = -\frac{1}{\varepsilon_0}e\sum z_s C_s + Q, \tag{1}$$

$$\frac{\partial C_s}{\partial t} = -\frac{\partial}{\partial x}J_s - U_s, \tag{2}$$

where $\quad J_s = C_s \dfrac{e z_s D_s}{kT}\dfrac{\partial \varphi}{\partial x} - D_s \dfrac{\partial C_s}{\partial x};\qquad \varepsilon, Q, D_s = \begin{cases} \varepsilon_1, Q_1, D_s^1, & x \in [0, x_1); \\ \varepsilon_2, Q_2, D_s^2, & x \in (x_1, x_2); \\ \varepsilon_3, Q_3, D_s^3, & x \in (x_2, x_3]. \end{cases}$

Here e is the charge of proton; φ is electrostatic potential; k is Boltsman constant, ε_0, ε are absolute and relative dielectric constants; Q is the density of charge encapsulated in the layer; J_s, D_s, C_s, z_s are flux, diffusion coefficient, concentration and charge (in units of e) of species (for $s=G, O, WH$ $z_s=0$; $z_H=+1$; $z_W=-1$); U_s is the generation-recombination rate of species; T is absolute temperature.

For $s=G,O$ the recombination rates are assumed to obey the Michaelis-Menten kinetics

$$U_G = \gamma U_O = V_{max} \frac{C_G C_O}{K_G^M C_O + K_O^M C_G + C_G C_O}, \tag{3}$$

where K_s^M are the Michaelis constant; V_{max} - maximum rate of enzymatic reaction depended in particular, on enzyme concentration; γ reflects stoehiometry of a reaction (in our case $\gamma=0.5$).

Since the species H (which is the product of enzymatic reaction between glucose and oxygen) also undergoes an association-dissociation reaction with conjugate base of buffer W, where R_H, R_W and R_{WH} can be written as follows:

$$U_H = -U_G - k_b\left(K_W C_{WH} - C_W C_H\right), \quad U_W = -U_{WH} = -k_b\left(K_W C_{WH} - C_W C_H\right), \quad (4)$$

where K_W is reaction constant and k_b is binding rate of W and H species.

For the interfaces the points x_ℓ ($\ell=1,2$) one can write

$$\varphi\left(x_\ell - 0, t\right) = \varphi\left(x_\ell + 0, t\right); \quad \varepsilon\left(x_\ell - 0\right)\frac{\partial\varphi\left(x_\ell - 0, t\right)}{\partial x} = \varepsilon\left(x_\ell + 0\right)\frac{\partial\varphi\left(x_\ell + 0, t\right)}{\partial x}; \quad (5)$$

$$J_s\left(x_\ell - 0, t\right) = J_s\left(x_\ell + 0, t\right), \quad C_s\left(x_\ell - 0, t\right) = C_s\left(x_\ell + 0, t\right) - R_s^{\ell, \ell+1}\frac{\partial C_s\left(x_\ell + 0, t\right)}{\partial x}. \quad (6)$$

The boundary conditions are evaluated in correspondence to following assumptions: i) where are no fluxes through the membrane-transducer boundary and ii) the biosensor chip is surrounded by test solution which is at zero potential. Under these assumption, the right-hand-side boundary condition for potential is defined by a structure of a transducer because potential is distributed along the transducer correspondingly to electrophysical properties of the layers which the transducer consists from. For example, conductometric transducer, as a rule, consists from only one insulating layer of thickness d_{ins} and one can write the boundary condition for potential as follows:

$$\varphi\left(0, t\right) = 0, \quad \varphi\left(x_3, t\right) = -\frac{\varepsilon_3}{\varepsilon_{ins}} d_{ins}\frac{\partial\varphi\left(x_3, t\right)}{\partial x}; \quad (7)$$

$$C_s\left(0, t\right) = C_{s0}^B + \left(C_s^B - C_{s0}^B\right)\left(1 - e^{-t/\tau}\right) + R_s^{0,1}\frac{\partial C_s\left(0, t\right)}{\partial x}, \quad J_s = (x_3, t) = 0, \quad (8)$$

$$C_s\left(x, 0\right) = C_{s0}^B. \quad (9)$$

where ε_T and d_T are relative dielectric constant and thickness of transducer material; C_{s0}^B, C_s^B are species concentrations in test solution at $t=0$ and $t>0$ respectively; τ is the time constant (τ is assumed to be much less then any of characteristic times of processes in membrane).

Numerical Experiment and Results

For discretization of the nonlinear problems (1)-(9) we have applied the technique of method of lines with the use of the finite difference method with respect to the space variable, x. Because the concentration profiles change most rapidly near $x=0$ and especially at the beginning of intervals corresponding to enzyme layers, irregular grid which nodes are chosen automatically in dependence on physical data of the problem were used.

To carry out the numerical experiment the developed model of biosensor was adopted to the experimental 2-layer membrane system described in [1], with the inner layer being enzyme one and the outer layer containing the uniformly distributed negative or positive charge. Thickness of enzyme membrane is 20μm. Thickness of the additional membrane as well as concentration of encapsulated charge in it varied in a wide ranges. Since there is very poor information about values of such parameters of membrane in buffer such as dielectric constants, diffusion coefficient for proton and buffer components W and WH they were admitted to be equal to those of water.

As a result of computer simulation the transient and position dependencies of all species concentrations and potential, time dependencies of output signal and calibration

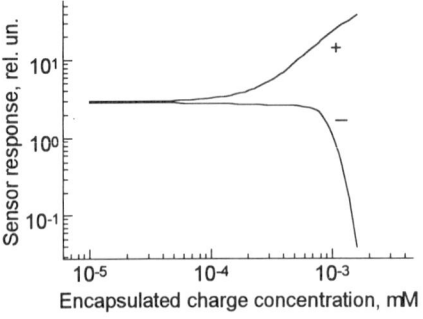

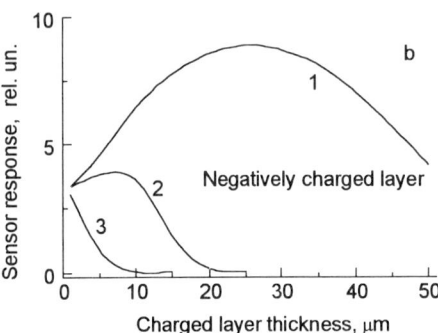

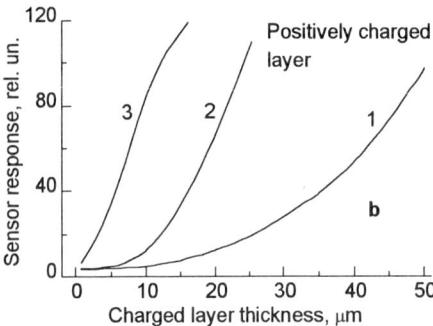

curves of glucose sensor were obtained. As it can be seen from Fig.2a, one can significantly control sensor sensitivity by varying concentration of charge encapsulated in outer layer of the membrane. In practice it is easier to vary a thickness of this layer [1]. However, in such a case it is very hard to predict a shift in a sensor response even qualitatively since behaviour of species is complicated due to additional diffusion barrier produced by the outer layer. Numerical experiment shows that dependencies of sensor sensitivity on charged layer thickness can be non-monotone with substantial difference between maximal and minimal values (Fig.2b). This gives opportunity for optimisation performances of enzyme potentiometric biosensors by varying combinations of physical and biochemical parameters of the membrane layers.

Fig.2. Dependence of the glucose biosensor output on encapsulated charge concentration (a) and thickness of the charged outer membrane (b, c).

a: charged layer thickness is 1 μm;
b, c: concentration of charge encapsulated in the outer layer is 10^{-6} mM (1), 10^{-5} mM (2) and 10^{-4} mM (3).

References

[1] Soldatkin A, El'skaya A, Shul'ga A, et all. Glucose-sensitive field-effect transistor with additional Nafion membrane. 1993 *Analytica Chimica Acta.* **283** 695-701.

[2] Ruckenstein E and Varanasi S Acid-generating immobilized enzymic reactions in porous media-activity control via augmentation of proton diffusion by weak acids. 1984 *Chem.Eng.Sci.* **39** 1185-1200.

[3] Ruckenstein E, Ogundiran S O and Varanasi S An Algebraic Equation for the Steady-State Response of Enzyme-pH Electrodes and Field-Effect Transistors. 1988 *Biosensors,* **3** 269-295

[4] Caras S, Janata J, Saupe D and Schmitt K pH-Based Enzyme Potentiometric Sensors. 1985 *Anal.Chem.* **57** 1917-1920.

Biosensors II
Paper presented at Eurosensors XII, 13–16 September 1998
© *1998 IOP Publishing Ltd*

Medical pressure sensors on the basis of silicon microcrystals and SOI layers

A.Druzhinin, E.Lavitska and I.Maryamova

Lviv Polytechnical University. Kotlarevsky street 1, Lviv, 290013, Ukraine

Abstract. Piezoresistive pressure sensors on the basis of semiconductor microcrystals and laser recrystallized SOI layers have been developed. Design examples of the sensors and some devices for medical applications are described.

1. Introduction

In the development of piezoresistive pressure sensors for medical applications both semiconductor silicon microcrystals and polysilicon-on-insulator layers were used as an element basis.

Semiconductor microcrystals (SM) grown in form of silicon whiskers due to their perfect crystalline structure have the high mechanical strength. They are miniature and may be used directly as sensitive elements of the mechanical sensors [1].

Polysilicon layers on insulating substrates (SOI structures) represent a perspective material for microelectronics. Microzone laser recrystallization of polysilicon layers provides significant increase of the gauge factor that is important for the development of miniature high-sensitive mechanical sensors [2].

2. Intracranial and intrabone pressure sensors

On the basis of SM miniature pressure sensors have been developed for the ranges from 0-300 H_2O to 0-500 mm Hg to operate in the temperature range 10-40°C. The diaphragm transforms the pressure measured to the displacement and transmits it to the intermediate transformer - universal strain unit. The universal strain unit with mounted on it SW piezoresistors forms a design basis of the sensor. It consists of the covar ring 9 mm in diameter punched together with the cantilever beam $3\times0.1\times1$ mm^3, two silicon strain gauges mounted on the beam and the metal-glass lead unit. Height of the strain unit is 0.7 mm. Sensor's application defines size of the diaphragm and housing design.

The intracranial pressure sensor has a plane droplet shape. It was designed, in particular, for the diagnostics of cerebral traumas in neurosurgery. Sensor's package is fabricated of stainless steel or plastic. Its shape avoids damage of the ambient tissues. Total height of the sensor is less than 2.5 mm while mass is less than 3 g. The measured pressure range is (0-4) $\times10^4$ Pa (0-300 mm Hg). The designed sensor is able to measure intracranial pressure directly in different sections of the brain with further pressure registration in analog or digital form. Output signal of the sensor is 40-60 mV for 10 mA excitation current. To provide high resolution and coordination with the registrating device the sensor is combined with the amplifier. Special commutation provides data polling from few pressure sensors.

Modification of the above sensor is the sensor for intrabone pressure measurement. This sensor is equipped by the special device providing connection between the underdiaphragm chamber with the intrabone cavity. The measured pressure range is 0-300 mm H_2O with pressure resolution of 5 mm H_2O. The device is supplied with a scale indicator.

3. Pressure sensor for gastroenterological investigations

The probe for gastroenterological investigations contains two intracavitary pressure sensors combined with the pH-meter. This probe was designed for the diagnostics of stomach and intestinal diseases. It provides the measurement of the intracavitary pressure that indirectly reflects a motor activity of organs investigated.

The pressure sensor for gastroenterological investigations contains the rectangular polymeric diaphragm 4×10 mm that is connected with the cantilever beam of the strain unit by the rod. Draft design of the sensor is shown in Fig.1 where proportion between the sensor's height and diameter has been deformed to provide the better clarity. In real design the encapsulated strain unit itself forms the sensor housing 1.2 mm 1.2 mm in height. The sensor is mounted in the capsule fabricated of stainless steel. The capsule is connected with the pipe of the probe.

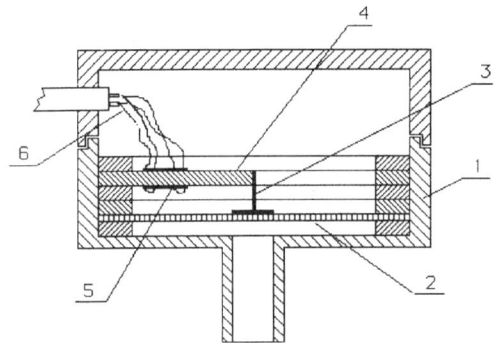

Fig.1. Sensor for gastro-
enterological investigations
(draft design):
1 – housing;
2 – diaphragm;
3 – rod;
4 – cantilever beam;
5 – strain gauges;
6 – contact wires.

Table 1.

Technical performances of the pressure sensor for gastroenterological

Parameter	Value
Measured pressure range, mm H_2O	0 - 300
Full-scale output signal without amplification at 2 V supply, mV	60
Size, mm^3	$5 \times 10 \times 1.2$
Mass, g	< 1
Whole probe diameter, mm	7
Resolution, mm H_2O	5

During investigations the first pressure sensor is introduced in the stomach and the second one is placed in the *Duodenum* zone. Main performances of the sensors are presented in Table 1.

Sensors output is registrated by the multichannel recorder.

4. Device for the functional diagnostics of the hearing tube

On the basis of the developed pressure-rarefaction sensor for the range 0...±500 mm H_2O the device to determine equipressure function of the hearing tube at perforated eardrum has been developed. Draft design of the device is shown in Fig.2 where SM pressure sensor (1) forms a basis of the device. It provides the measurement and registration of the differential pressure. Its sensitivity is 0.15 mV/(mm H2O), output nonlinearity is less than 1.5%. Pneumatic cylinder (2) and water manometer (4) are used for creation the excess pressure or rarefaction ±500 mm H_2O (49 kPa) and its control.

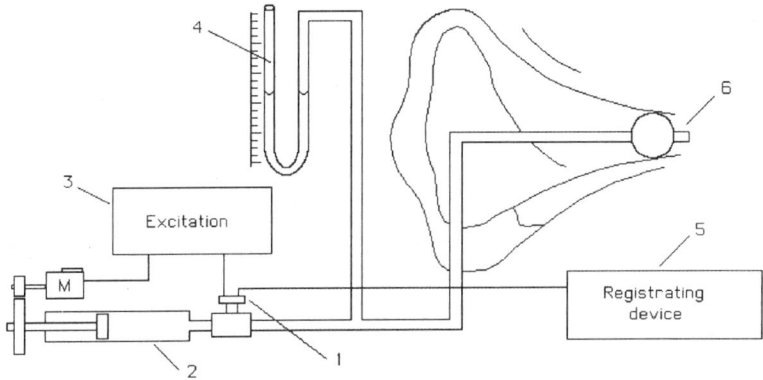

Fig.2. Device for the functional diagnostics of the hearing tube: 1 - SM pressure sensor, 2 - pneumatic cylinder, 3 - power unit, 4 - water manometer, 5 - registrator.

An operation of the device includes the following procedure. Extra air pressure is pumped into the ear. Through the perforated eardrum it goes to the drum cavity and then to the hearing tube. After the pressure pumping/exhausting into/out of the ear the patient is proposed to make 20 swallowing movements. For the normal functioning hearing tube the pressure step by step declines to zero that is registrated by corresponding device (5). When the patient makes swallowing movements opening the hearing tube the excess pressure comes into the gullet. Residual pressure in the ear defines a dysfunction level of the hearing tube.

The developed sensors were successfully tested in the clinics of the Lviv Medical University.

5. Microelectronic pressure sensors for arterial pressure measurement

The batch technology of microelectronic pressure sensors with polysilicon piezoresistors has been developed. Selection of the initial material and optimization of its processing, in particular, laser recrystallization of poly-Si layers [2,3], make it possible to obtain SOI-structures for the development of medical pressure sensor.

The simplified design of the pressure sensor's chip (Fig.3) includes 2 mm × 2 mm diaphragm fabricated by anisotropic etching of the Si substrate in KOH water solution. The pressure range was adjusted by the diaphragm's thickness. Taking into account anisotropic nature of the etching process the diaphragm edges and scribing strips were oriented in [110] direction. The longitudinal axes of poly-Si strain gauges connected into a fully active Wheatstone bridge were aligned in [110] direction corresponding to the laser scanning direction; poly-Si resistor was used for thermal compensation.

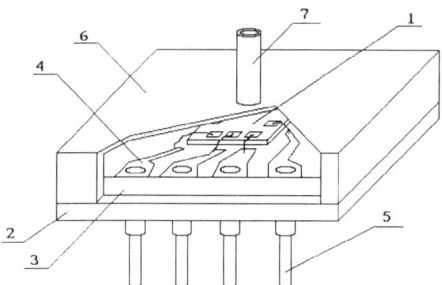

Fig.3. Simplified design of microelectronic pressure sensor for medical application: 1 - SOI chip, 2 - package; 3 - glass base; 4 - Al contact regions, 5 - pins, 6 - cap, 7 - pressure tube.

Operating pressure range of the sensor is 0-300 mm Hg (0-0.4 bar), operating temperature range 20-60°C, its sensitivity at 20°C is about 10 mV/(V×bar), temperature coefficient of sensor output is less than 0.04 %F.S./deg.

The sensor was applied in the device for the arterial pressure measurement.

Conclusions

The investigations fulfilled showed silicon whiskers to be a prospective material as concerns the development of high-sensitive pressure sensors for medical applications. But such kind of sensors might be produced only by small series, while the developed SOI sensors might be prodused commercially. The batch technology of SOI pressure sensors has been developed as a joint project of the MST group from the Lviv Polytechnical University and "Rodon" joint-stock company (Ivano-Frankivsk). Experimental high-sensitive sensors were successfully tested in devices for the arterial pressure measurement.

References

[1] Voronin V., Maryamova I., Zaganyach Y. et al. 1992 *Sensors and Actuators* **30A** 27-33
[2] Voronin V.A., Druzhinin A.A., Maryamova I.I. et al. 1992 *Sensors and Actuators* **30A** 143-147
[3] A.Druzhinin, E.Lavitska, I.Maryamova, V.Voronin 1997 *Sensors and Actuators* **61A** 400-404

Biosensors II
Paper presented at Eurosensors XII, 13–16 September 1998
© *1998 IOP Publishing Ltd*

OPTIMIZATION METHODS OF ENZYME INTEGRATION WITH TRANSDUCERS FOR ANALYSIS OF IRREVERSIBLE INHIBITORS

N.F.Starodub, W.Torbicz[*], D.Pijanowska[*], V.M.Starodub, M.I.Kanjuk, M.Dawgul[*]

Institute of Biochemistry, National Academy of Sciences,
9 Leontovicha Str., 252030 Kiev, Ukraine
[*]Institute of Biocybernetics & Biomedical Engineering, Polish Academy of Sciences,
ul. Ks. Trojdena 4, 02-109 Warsaw, Poland

ABSTRACT
Manufacturing and application of biosensors cause some problems, the main of which are: 1) optimization of enzyme immobilization process to preserve molecular enzyme structures and maximal value of enzyme activity and 2) development of approaches to accomplish a repeated analysis of substances which are irreversible inhibitors. In this report both problems are analysed. For stabilization of the structure of glucose oxydase, urease and cholinesterases during their immobilization several treatments were examined. Optimization of these processes allows to preserve about 70-80% of initial enzyme activity. A repeated analysis of phosphororganic pesticides and heavy metal ions, which are irreversible inhibitors requires application of special approaches.

1. Introduction

To immobilize biological molecules and to preserve high level of their activity during this process, a number methods are used [1-5]. Among them the following methods can be distinguished: 1) direct immobilization with or without utilisation of bi-functional reagents, 2) direct immobilization as indicated in point 1 but after preliminary treatment of the transducer surface by chemical agents and 3) application of special membranes with or without inclusion of bi-functional reagents. As a rule, investigators endeavour not to use strong chemical reagents and to keep up low temperature as well as to work with the substances which have high initial specific activity and to immobilize enzyme in redundancy. Often, all these precautions are insufficient and it is necessary to apply special approaches, especially, if it is necessary to obtain a biosensor for durable, continuos and repeated analysis. A lot of enzymes are very labile in usual conditions and sensitive to some inhibitors since they have an unstable structure. To prevent their full or partial inactivation at the immobilization process it is necessary to utilise special chemical substances (substrates, chelates, thiols etc.). Additionally, a number of analytes (heavy metal ions, phosphororganic pesticides) are irreversible inhibitors for enzymes which are used as sensitive components in enzymatic sensors, so arise the problem, how the repeated analysis can be optimized. In the paper we discuss the influence of different technological procedures on the preservation of activities of the following enzymes as: glucose oxydase, urease and cholinesterases during their immobilization. Some useful methods for performing of repeated analysis of heavy metal ions and phosphororganic pesticides with enzymatic sensors are given.

2. Experimental

2.1. Material

Urease, acetyl- (AChE), butyrylcholine (BChE) esterase, acetylcholine chloride (AChCl), butyrylcholine chloride (BChCl), GOD, bovine serum albumin (BSA),PAM-2 iodide, EDTA, DTT saccharose and glycerol were obtained from Sigma ; chlorpyrifos from the Industrial Institute of Organic Chemistry, Poland; nitro-cellulose (NC) sheets (hybond-N-type) from Amersham. Salts and other reagents were of the analytical grade.

2.2. Sensor transducer

pH-sensitive field effect transistors (ISFET), made in the Institute of Biocybernetics and Biomedical Engineering, Polish Academy of Sciences, were used as the transducers.

2.3. Fabrication of the enzymatic layer

Before deposition of the enzymatic layer the ISFET gates were cleaned with concentrated sulphuric acid and chromic acid mixture for 2 min, washed with distilled water and dried at 75 °C for 1 hr. To obtain enzymatic biosensors - EnFETs, enzymes were immobilized on the ISFET gate surfaces by several procedures [5]. In the first one, enzymes were deposited on the gate surface in polymerized GA. The mixture of enzyme-water solution, GA and BSA was deposited on the gate surface and the sensor was left in refrigerator at humid atmosphere and optionally an extra polymerization was performed in GA vapour. According to the second procedure, the gate surface was preliminary treated (30 min) by GA (25%). Then, the surface was washed by distilled water and the covalently bonded enzymatic layer was deposited by drop coating of water enzyme (1%) solution. Non-attached enzyme molecules were washed out from the sensor surface by vigorously stirred 10 mM tris-HCl buffer solution (pH 7.3). The third procedure was accomplished by the enzyme entrapment in alginate gel. The protocol was as follows: sodium alginate solution (1%) was mixed with the above mentioned enzyme solution in ratio 1:1. The mixture was deposited on the gate surface and then calcium chloride solution (1%) was applied on the top of it. The sensor was left in the refrigerator at humid atmosphere for 30 min. Formed membrane was washed by calcium chloride (10 mM) buffer at pH 7.3 to remove free enzyme molecules from the membrane surface. We also prepared the special enzymatic strip (NC-strip) biosensor by the following procedure [4]. NC strips were soaked in enzymatic solution (5 ml at the concentration of 10 mg/ml in the 50 mM, pH 7.3 tris-HCl buffer, containing 140 mM sodium chloride). Then they were washed with the buffer for 30 min and dried at room temperature.

2.5. Determination of enzyme activity and reactivation of enzymatic membrane

To determine the residual activity of immobilized enzymes they were immobilized on the silicon plates of dimensions of 4x4 mm with the methods described in the previous section. These plates with enzymatic membranes were washed several times by 50 mM, pH 7.3 tris-HCl buffer, containing 140 mM sodium chloride and then they were immersed in special mixture for determination of residual activity by photometric methods [1]. For estimation of enzyme inhibition the enzymatic membranes with the immobilized AChE, BChE and urease were treated in the same way but the solution contained inhibitors (pesticides, heavy metal ions). The reactivation procedure of ChEs membranes was realised by immersing them in 6 – 10 mM solution of PAM-2 iodide for 1 h or over night. In the case of urease membranes EDTA and DTT were used for their reactivation

3. Results and discussion

3.1. Optimization of enzymatic membrane formation by GA polymerization

The detailed investigation of the influence of the enzyme immobilization process parameters on the residual enzyme activity were performed using GOD. Enzyme was immobilized by polymerization of GA and optionally this process was followed by GA vapour conditioning. It was stated that for the membrane solution containing GA (0,5%), GOD (1%) and BSA 0,5% and 1%. The maximum of

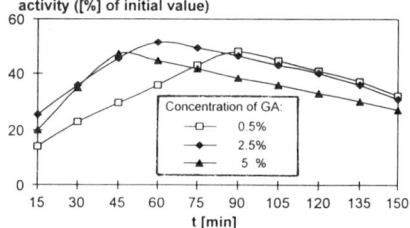

Fig. 1. Time dependence for residual activity of GOD for different concentration or GA at the BSA concentration 1%.

residual activity is achieved for 90 min of GA polimerization. For the increased GA concentration (from 0.5 to 5.0%) the polimerization time to achieve the maximum of the enzyme activity in membrane reduces (Fig.1).

Increasing of GA concentration over 5% dramatically decreases the residual enzyme activity in membrane. So, our experiments demonstrate that the optimal concentration of GA solution and time polimerization are 2.5% and 60 min respectively. The next experimental works were devoted to investigation of dependence of the residual activity of the enzyme in membrane on the enzyme/BSA ratio in the solution used for immobilization of GOD and urease. It was stated that the maximum of the enzyme activity in membrane can be achieved for the enzyme/BSA ratio equal to 1:1 for both enzymes.

Urease in comparison to GOD is more sensitive for inactivation during immobilization process and the activity of this enzyme is decreased more intensively while membrane is preserved. This was the reason for examination of several approaches of stabilization of urease and other enzymes activity in the immobilization process. At first we studied the possibility of protection of enzyme active centres by substrates or their analogues which for urease are: thiourea and formamide. All these substances were added to the polymerization medium at the concentration which was 4-5 time higher than the enzyme amount. It was shown that the membrane formation was disturbed in the presence of urea in the polymerization medium. Maybe, it is due to the local pH changes as a result of catalytic depletion of substrate. As it is known, the rate of enzymatic depletion of formamide is considerably lower than that for urea. Application of formamide in polymerization medium results in increasing of residual activity of urease in membrane about 25-30%. Slowly depleted formamide protects active centre of urese in presence of GA. At the same time, tiourea as non-hydrolyzed substrate creates such bonds with enzyme molecules which cause destabilization of their structure and active centres of enzyme become accessible for reactive groups of GA. It is necessary to note that in case of glucose presence in polymerization medium, the residual activity of GOD doesn't change its value but the process of membrane formation is slightly disturbed. Disturbances in membrane formation without considerable changes of the residual enzyme activity were also stated for ChEs membranes if substrates had been used during immobilization process.

It is known that saccharose and glycerol can be effectively used for stabilization of protein structure and preservation of native enzyme conformation. We have shown that addition of saccharose to the polymerization medium in the range of concentrations of 0.5 to 2.5% increases the residual enzyme (GOD, urease, AChE) activity (Table 1).

Table 1. Residual activity of enzymes in % of the initial value for stabilization of enzyme performed with saccharose

Enzyme	Concentration of saccharose [%]					
	0,5	1,5	2,0	2,5	3,0	3,5
GOD	60	64	68	72	67	3,5
Urease	48	51	56	62	58	52
AChE	56	60	64	68	61	59

Increasing of the concentrations over this range (up to 5%) impedes formation of the membrane and decreases its activity. Adding of glycerol of the concentration from 1 to 20% results in the increasing of enzyme activity (Table 2). The maximal value of GOD, urease and AChE activities appear if saccharose (2.5%) and glycerol (5%) are added commonly to the membrane solution.

Table 2. Residual activity of enzymes in % of the initial value for stabilization of enzyme performed with glycerol

Enzyme	Concentration of glycerol [%] at (2.5% of saccharose)				
	1	3	5	10	20
GOD	74	78	82	69	63
Urease	62	65	70	58	50
AChE	69	73	78	64	52

3.2. Repeated use of enzymatic sensors of irreversible inhibitors

Urease, AChE and BChE can be used in sensors of heavy metal ions and phosphororganic pesticides, which are irreversible inhibitors for these enzymes. Due to that, the enzymatic membranes should be replaced or reactivated for repeated use of these sensors. Since urease has thiol group in active centre, so this enzyme can be reactivated by special chelates or thiol containing substance (e g.: EDTA or DTT respectively) after inhibition by heavy metal ions. Phosphororganic pesticides are able to phosphorylate OH groups of serine in active site of any ChEs resulting in dramatic decreasing of enzyme activity. In this case, for reactivation of enzymatic membrane PAM-2 iodide is often used. Unfortunately, it is not possible to achieve the initial enzyme activity. So we tried to develop a simple approach for replacement of enzymatic membranes. It was shown that alginate gel and NC-strips are very suitable materials for creation of replaceable enzymatic membranes. The response time of enzymatic sensor based on the alginate gel membrane is of the same order to that of the biosensor the membrane with polymerized GA. The response time of the sensor based on the NC-strip is much longer. The standard deviation of the output signals the enzymatic sensors based on alginate gel and NC-strip membranes for series of measurements and for different membrane castings did not exceed 10%.

3.3 Optimization of conditions for enzymatic sensors preservation

There are two principal methods of storing of enzymatic sensors: 1) keeping them in dry state and 2) immersion in special solution. We optimized conditions of preservation of enzymatic sensors using urease and ChEs sensors. Enzymatic membranes of these sensors lose their activity very quickly in dry state at room temperature. After 1 month of preservation in such conditions the urease sensors retained only 25% of initial level of response. For preservation of enzymatic sensors in solution it is needed to prevent bacterial degradation of enzymatic membranes. For this purpose we utilised sodium azide. It was shown that in the presence of this substance of the concentration of 0.1%, the activity of urease membrane decreased approximately to 30%. But this activity was restored to the initial level after careful membrane washing in 3-5 mM tris- HCl buffer, pH 7.4. The best preservation of the enzymatic membrane activity and the sensor responses were observed for the solution of 20mM buffer,0.1% sodium azide and 1 mM EDTA, pH 7.4.

4. Conclusions

It is possible to choose the composition of membrane solution and time of GA polimerization to receive the optimal values of enzyme activities for GOD, urease and AChE immobilized in membranes. Alginate gel and NC-strips are suitable materials for creation of replaceable enzymatic membranes.

Acknowledgements

This work was supported by Commission of the European Communities (grant N PL965131), Ukrainian Ministry of Sciences and Technology (grants N 6.4 / 194) and Committee of Scientific Research, Poland (grant 8T10C000109)

References

[1] N.F.Starodub, L.N.Chustochka, A.V.Lazarenko, O.A.Bubrjak, A.G.Terent'ev, A.V.El'skaya: Integration of biological material into electrochemical biosensors, J. Anal. Chem. (Russ.) 45, (1990) pp1432-1440.

[2] M. Nakako, Y. Hanazato, M. Maeda and S. Shiono: Neutral lipid enzyme electrode based on ion-sensitive field effect transistors, Analyt. Chim. Acta, 185, (1986).pp.179-185.

[3] N.F.Starodub, I.M. Samodumova, V.M.Starodub, Usage of organosilanes for integration of enzymes and immunocomponents with electrochemical and optical transducers", Sensors and Actuators, B 24-25, (1995), pp.173-176.

[4] O.A. Bubrjak, L.N. Chustochka, A.P. Soldatkin, N.F. Starodub, Investigation of immobilization and properties of urease with the purpose of creation of biosensors based on semiconductor structures, Ukr. Biochem Zhurn., 64, 1992, pp. 66-71.

[5] C. Tranh-Minh, Immobilized enzyme probes for determining inhibitors, in Ion-Selective Electrode Rev. 7, (1985), pp. 41-75.

Biosensors II
Paper presented at Eurosensors XII, 13–16 September 1998
© 1998 IOP Publishing Ltd

Volatile Organic Compounds Mixtures Quantification with TGS Sensor Array

A Szczurek[1], P M Szecówka[2], B W Licznerski[2]

(1) Institute of Environment Protection Engineering, Wroclaw University of Technology, Wybrzeze St. Wyspianskiego 27, 50-370 Wroclaw, Poland
(2) Institute of Microsystem Technology, Wroclaw University of Technology
Wybrzeze St. Wyspianskiego 27, 50-370 Wroclaw, Poland
email: szecowka@ite.ite.pwr.wroc.pl

Abstract. A concept of intelligent sensor system for quantification of volatile organic compounds appearing in 2-component mixtures is presented. The system uses sensor array containing 4 TGS 800 series sensor array and artificial neural network processing their responses. Characterisation of sensors was made with in-house developed installation. Facing strong sensors sensitivity for humidity, appropriate samples were collected to provide sufficient pattern set for neural network training. System developed for butanol-xylene mixtures analysis provides accuracy of ±10 %.

1. Introduction

Volatile organic compounds are classified among the basic and most common pollutants. Because of toxic, explosive, cancerogenic, mutagenic and other human-aggressive properties, their permanent monitoring becomes significant task in environmental protection. Traditional analytical methods (including e.g. spectrophotometry, chromatography), providing very good accuracy and reliability of measurements, are not sufficient in most of practical applications requiring continuous work in real time, often far away from the laboratories with smart instruments.

Thus there is strong need for low-cost and handy instruments capable of measurements of specified sets of compounds with reasonable accuracy. For some time several scientific groups have tried to apply gas sensor arrays with smart data processing techniques to resolve relevant problems (e.g. in electronic noses [1]). Although relatively seldom there were presented systems providing quantitative analysis of mixed compounds [2,3,4]. And many real applications require strict information about concentration of selected pollutants rather than qualitative type detectors or classifiers. The paper presents system consisting of a few gas sensors connected with medium size neural network, providing quantification of butanol and xylene appearing simultaneously in atmosphere with changing humidity.

2. Sensors characterisation

During the experiment special installation was developed, providing simultaneous characterisation of a few gas sensors in atmospheres of alcohols and aromatic compounds mixtures, with different humidity levels. Vapour samples were prepared by injection of

842

appropriate liquid into the coil and then blowing it with zero air (humidified if necessary) into tedlar sample bag. High purity zero air generator (made by Horriba) was used. The samples obtained this way were finally pumped into the glass vessel (volume of 0.5 l) containing sensor head with wires. Thus the atmosphere over the sensors changed dynamically, with constant flow (2l/min) of the mixture. For all investigated samples the moment of sensors responses collection was fixed to 4 min from the start of mixture flow [5].

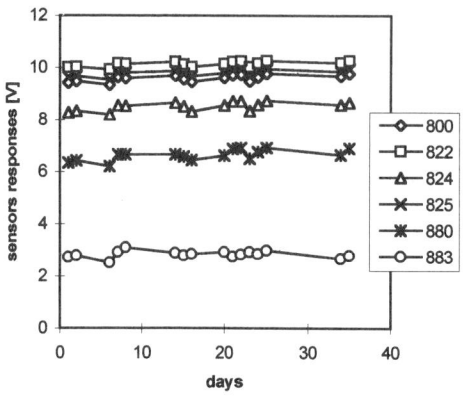

Fig. 1 TGS sensors stability; calibration with 300 mg/m³ of ethanol in dry air

Sensor head contained 6 sensors, made by Taguchi Gas Sensors Inc., including TGS 800, TGS 822, TGS 825, TGS 824, TGS 880 and TGS 883. Voltages on reference resistor, for 12V circuit supply, were taken as sensors responses. The sensors were operated in appropriate temperatures by applying constant 5V supply to heaters. It should be noted that they have been kept in high temperatures for about 420 days - all the time of several following experiments. In that period none strict trend in responses fluctuations was observed, although there were day-to-day differences of stochastic kind Figure 1 presents part of daily calibration measurements, taken during 35 days of investigation described here). Characterisation of sensors was made with mixtures containing xylene (0-300 mg/m³) and butanol (0-240 mg/m³) in 4 levels of humidity (dry, 4 g/m³, 10 g/m³ and 100% rh from saturator). Total amount of 148 samples was collected. Figure 2 presents part of the results, for clarity including only samples with pure compounds in one humidity. Analysis of the data brought following conclusions:

• TGS 824 provides the best discrimination between the two compounds
• TGS 883 reveals good sensitivity for humidity and poor sensitivity for both compounds
• all the sensors are sensitive for humidity

Thus 4 sensors were selected for construction of the system: TGS 824, TGS 883 for mentioned reasons and TGS 822, TGS 880 to provide a kind of bias.

3. Neural processing

Artificial neural networks, especially feedforward structures with sigmoid transfer function in neurones, trained with BP algorithm or other, became recently popular tool for processing of data arising from gas sensor arrays. Although their capabilities for approximation of nonlinear multi-variable functions were strictly proved much earlier, e.g. [6,7].

In-house developed software tool was applied for construction of several variants of neural networks providing translation of selected 4 sensors responses to continuous values determining concentrations of butanol and xylene respectively. The best results were obtained for the network containing 4 input units, two hidden layers with 30 and 45 neurones respectively and 2 output units. Simple summation of weighted signals and sigmoid transfer function were applied in all the neurones excluding ones from input (dummy) layer. 109 patterns finely distributed in 3-dimensional space of compounds concentrations and scaled were used for network training and whole data set for the tests of evaluated structures. Learning strategy was based on error backpropagation (BP) algorithm with special tracking procedure, providing automatic long time computer processing, with saving good results and overtraining protection. Strong momentum ratio (0.8) and relatively small learning ratio (0.001 decreasing to 0.0001) were applied. Evaluation of the network providing absolutely best result took over 3 millions of iterations. Such long time was found as a kind of optimum by mentioned tracking procedure, although there were also structures of similar quality obtained after e.g 180 thousands of iterations (with higher learning ratio applied).

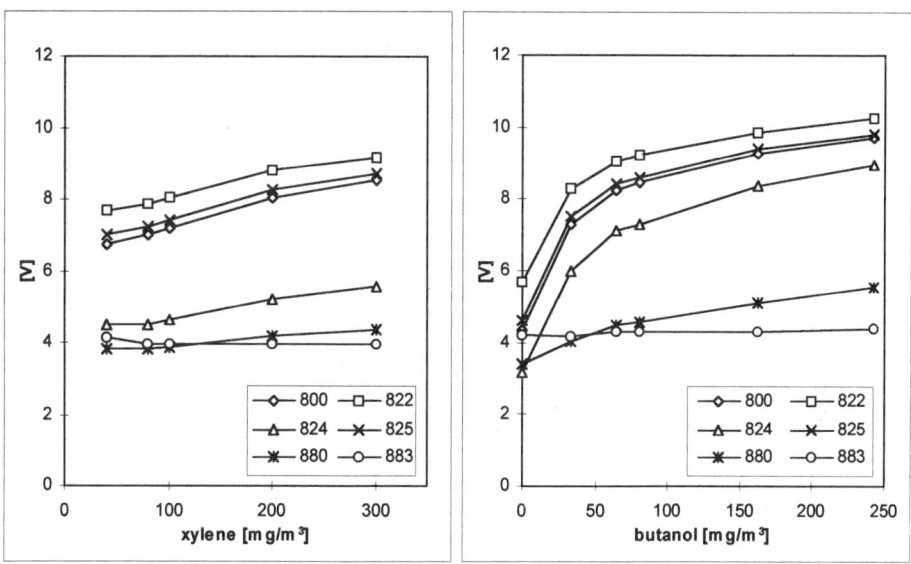

Fig. 2 Sensor array responses for pure xylene and butanol vapours; constant humidity 4 g/m^3

4. Results

The results of test on whole available data set are presented on Figure 3. Horizontal axes denote real concentrations of each compound in the mixtures, while vertical axes stand for two output units of the neural network (responses are scaled to physical range). Thus dotted lines denote fields of ideal neural network responses. For most of the samples ±10% of range accuracy is reached (there were 2 exceptions with errors of 11% and 12%). None significant difference in quality of responses for „known" and „unknown" samples was observed, what enables optimistic conclusion, that in real life conditions the system would provide similar performance. Although it should be noted that if any „unknown" compound, influencing the sensors, appeared in the atmosphere over the sensor array it would disturb system performance.

Complexity of neural processing algorithm applied enables implementation on simple microcontroller (e.g. [8]), thus eventual portable realisation of the whole system is possible, after resolving problem of sensor head physical construction and finding sample inlet method.

844

5. Conclusions

A concept of simple butanol-xylene mixture analyser was proposed. Reasonable accuracy of compounds quantification, independence on atmosphere humidity and partially on sensors fluctuations make it potential solution for selected compounds emission monitoring problem.

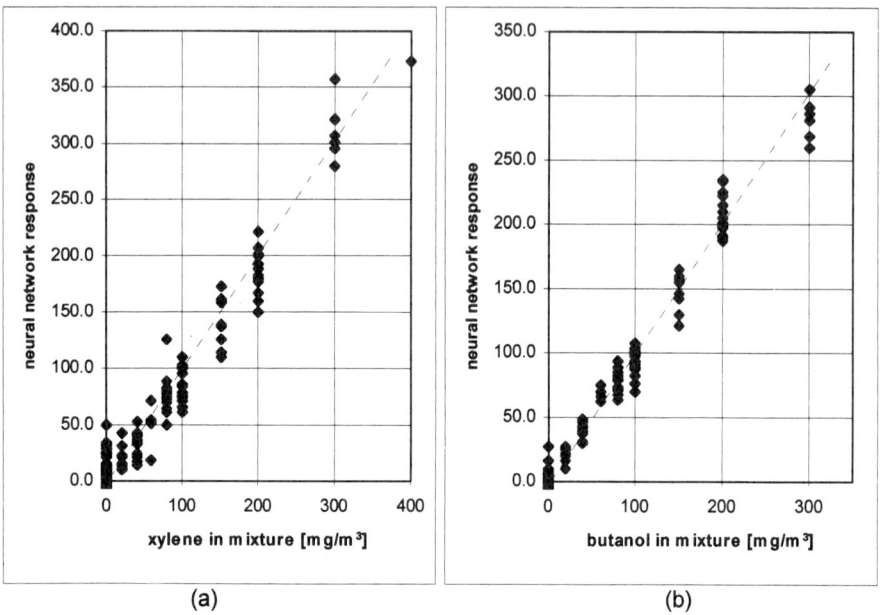

(a)   (b)

Fig 3 Neural network responses for butanol and xylene mixtures in different humidities for
(a) xylene and (b) butanol dedicated output.

Acknowledgements
Presented research was financed by KBN, project „Badania wlasne, zlecenie nr 342 223"

References

[1] Di Natale C, Macagnano A, Mantini A, Davide F, D'Amico A, Paolesse R, Boschi T, Faccio M and Ferri G 1997 *Proc ISIE'97 Guimaraes Portugal* SS122-27
[2] Gutierez F J, Ares L, Robla J, Horillo M C, Sayago I, Getino J and Garcia C 1997 *Proc ISIE'97 Guimaraes Portugal* SS113-15
[3] Baltes H, Koll A and Lange D, 1997 *Proc ISIE'97 Guimaraes, Portugal* SS152-7
[4] Huyberechts G, Szecówka P M, Roggen J and Licznerski B W 1997 *Sensors and Actuators B* **45** 123-130
[5] Szczurek A, Licznerski B W, Rusek S, Szecówka P M 1997 *Proc. ISSE'97 Szklarska Poręba Poland*, 242-247
[6] Funahashi K 1989 *Neural Networks* **2** 183-192
[7] Kornik K, Stinchcombe M, White H 1989 *Neural Networks* **2** 359-366
[8] Janiczek J, Stepien S, Licznerski B W, Szecówka P M, Huyberechts G, 1997 *Proc. of the Third Conference Neural Networks and Their Applications, Kule-Czestochowa, Poland* 570-575

Biosensors II
Paper presented at Eurosensors XII, 13–16 September 1998
© *1998 IOP Publishing Ltd*

Comparison of the determination of affinity constants with surface plasmon resonance and quartz crystal microbalance

C Kößlinger*, E Uttenthaler*, T Abel*, S Hauck*, S Drost*

*Fraunhofer-Institute for Solid State Technology, München, Germany

Abstract. A comparison between quartz crystal microbalance (QCM) and surface plasmon resonance (SPR) as devices for the determination of affinity constants is presented. The association and dissociation rate constants k_d and k_a respectively were determined by non-linear curve fitting of the time dependent measurement signals during association and dissociation processes. We used the virus-protein vp73 from the african swine fever (ASF) virus as analyte and immobilised an anti-ASF-antibody. The SPR measurements yielded $7.9 \cdot 10^{-4}$ s^{-1} for k_d and $5.6 \cdot 10^3$ (Ms)$^{-1}$ for k_a. The QCM results were $3.1 \cdot 10^{-4}$ s^{-1} for k_d and $1.1 \cdot 10^4$ (Ms)$^{-1}$ for k_a. k_d and k_a differ for about a factor of 2 for the QCM and SPR respectively.

1. Introduction

In the field of biosensors, Quartz Crystal Microbalance (QCM) and Surface Plasmon Resonance (SPR) have become known independently as methods, which are suitable for the direct on-line detection of affinity reactions [1,2]. This property is an essential condition for the determination of dissociation and association rate constants (k_d and k_a respectively).

The resonance signal of the QCM changes due to the binding of antigen and antibodies. The time dependent behaviour of the signal can be described by the following mathematical models.

Association:
$$\Delta F = -\frac{k_a \cdot C \cdot \Delta F_{max}}{k_a \cdot C + k_d} \cdot \left(1 - e^{-(k_a \cdot C + k_d)(t - t_0)}\right) \qquad \text{(eq. 1)}$$

Dissociation:
$$\Delta F = -\Delta F_0 \cdot e^{-k_d \cdot (t - t_0)} \qquad \text{(eq. 2)}$$

A long reaction time for association ($t \to \infty$) results in an equilibrium state (Langmuir Isotherm).

Langmuir:
$$\Delta F = -\Delta F_{max} \cdot \frac{C}{C + K} \qquad \text{(eq. 3)}$$

ΔF is the change of the sensor signal, C is analyte concentration, t is the time and ΔF_{max} is the signal change of a completely covered surface, ΔF_0 and t_0 are fitting boundary values and K is the equilibrium dissociation constant which is composed by k_d / k_a. The resonance signal of the SPR is analogous except of the minus sign.

In this paper, we compare the dissociation and association rate constants as well as the equilibrium dissociation constant K of the reaction between the virus protein vp73 of the African Swine Fever (ASF) virus and a monoclonal antibody determined by SPR and QCM.

2. Materials and Methods

2.1 Quartz Crystal Microbalance

The whole system consisted of the quartz crystal mounted in a flow-through cell, the oscillator, a frequency counter, a pump, an injection valve and a computer for the system control and data processing [3].

The quartz crystals were AT-cut with 8 mm diameter and a thickness of 0.084 mm, corresponding to a resonance frequency of 20 MHz. The gold surface was activated with 0.01 M dithiobis-succinimidyl-propionate (DSP) solved in DMSO for 10 minutes. Anti-ASF-antibodies antibodies (1mg/ml) were incubated for 12 hours at room temperature. As carrier buffer a 140 mM phosphate buffered saline (PBS), adjusted to pH 7.5, was used.

Only the measurements with vp73 were done with flow injection analysis technique. The sample volume was 100 µl. The flow rate was 30 µl/min. Regeneration was performed with the regeneration buffer boric acid/KCl-NaOH (pH 12.3). Data processing was done with the data analysis software Origin 4.0.

2.2 SPR

We used the BiaCore 2000 of the Company BIACORE AB for the SPR measurements. The main components of that device are the sensor-chip, the optical system and the fluid-system. The sensor-chip consists of glass plate coated with gold. A carboxymethyl-dextranhydrogel overlays the gold. A flat and clean goldsurface has a capacity for a monolayer coating with antibodies of about 1 - 5 ng / mm^2. By accumulation in the porous matrix a capacity of up to 50 ng / mm^2 is possible.

HBS buffer (10 mM Hepes pH 7.4, 150 mM NaCl, 3.4 mM EDTA, 0.005 % Tween 20) was used as carrier liquid. To prepare the dextran coating for amino coupling the hydrogel was activated with 100 mM N-hydroxysuccinimide (NHS) und 400 mM N-ethyl-N´-(dimethyl-aminopropyl)carbodiimide (EDC) in equal amounts. The activated dextran layer was coated with anti-ASF-antibodies (500 µg/ml) of the company INGENASA. The free active groups were saturated with 1M ethanolaminhydrochloride pH 8.5. The regeneration buffer was Glycin pH 2.5.

All steps (activation, saturation, immobilisation, measurement and regeneration) were carried out under flow condition. The flow rate was 10 µl/min. The injection-volume for the activation with NHS/EDC was 35 µl. Saturation, immobilisation, regeneration and measurement needed 50 µl injection volume each. The temperature was adjusted to 25 °C. Data processing was done with the BIA Evaluation Software.

Different concentrations of vp73 obtained from INGENASA were injected in the QCM and SPR device. vp73 is a virus protein from the core of the african swine fever virus with 73 kD molecular weight.

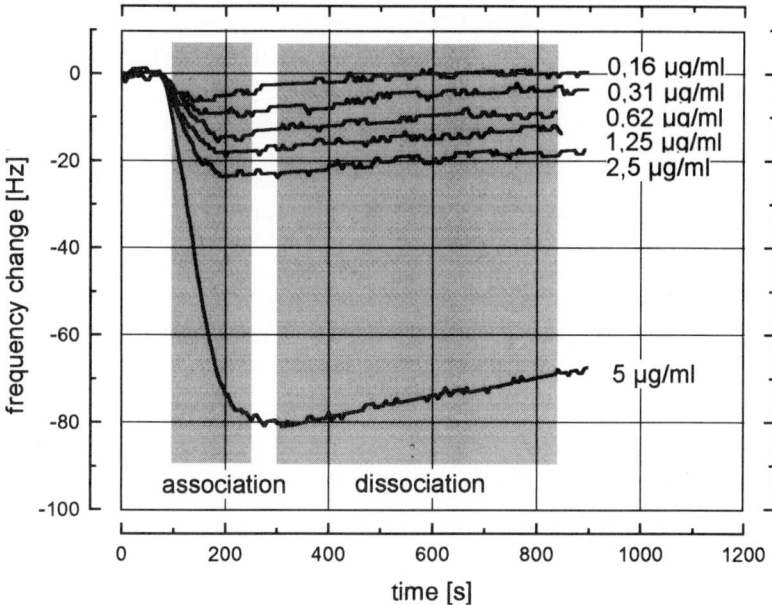

Figure 1: The change of the resonance frequency of the QCM due to different concentrations of vp73.

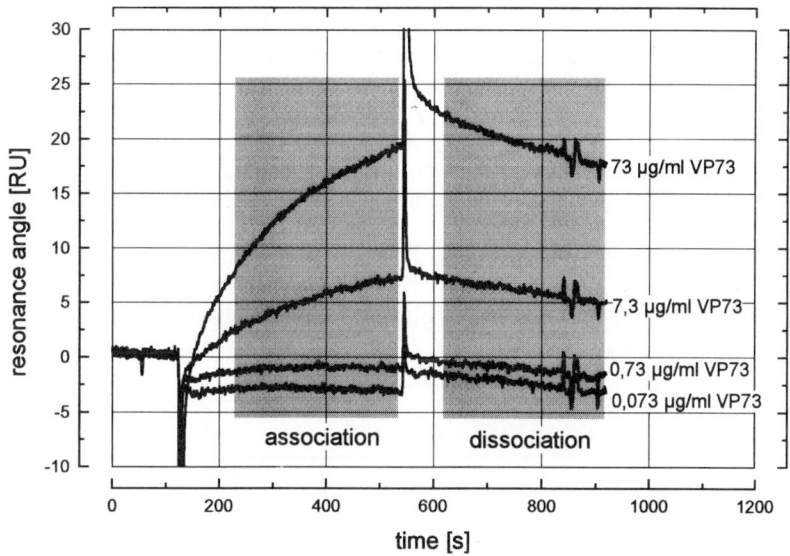

Figure 2: The change of the resonance angle of the SPR due to different concentrations of vp73.

3. Results

Figure 1 and figure 2 show the change of the resonance signals of the QCM and the SPR due to different concentrations of vp73.

The curve sections with grey background can be fitted with mathematical models which describe association (equation 1) and dissociation (equation 2). Association and dissociation rate constants (k_a and k_d) can be obtained with non-linear curve fitting of the time dependent signals of the adsorption and desorption processes.

The fitting result of the *QCM* data for k_a was $1.1 \cdot 10^4$ (Ms)$^{-1}$. The result for k_d was $3.1 \cdot 10^{-4}$ s^{-1}. From that, we obtained an equilibrium dissociation constant K of $2.8 \cdot 10^{-8}$ M.

The fitting result of the *SPR* data for k_a was $5.6 \cdot 10^4$ (Ms)$^{-1}$. The result for k_d was $7.9 \cdot 10^{-4}$ s^{-1}. From that, we obtained an equilibrium dissociation constant K of $1.4 \cdot 10^{-7}$ M.

3. Discussion

If we use the equations 1 and 2, we must be aware that these formulas are idealised models and are subjects to some restrictions.

1. mass transport effects may not occur,
2. the mathematical models do not take dispersion into account,
3. steric hindrance is not included in the model,
4. rebinding effects may not occur.

Nevertheless, all these effects occur during the measurement more or less. In case of the SPR these restrictions were fulfilled better. The reason was, that it was not possible to immobilise a high concentration of antibodies in the dextran layer. A consequence of this is that steric hindrance and rebinding effects are reduced. In contrast to that, the quartz crystal microbalance antibody coating was optimised to high density to achieve lower detection limits [4]. This results in a lower dissociation constant due to rebinding. The association constant is also lower because of rebinding.

The measurements show that fitting results must be discussed carefully. The obtained results are highly dependent from measurement conditions. Furthermore, according to our opinion the QCM has the potential to be an alternative to the BIAcore device.

Acknowledgments
We greatfully thank Frank Bier from Max Delbrück Zentrum in Berlin Buch for his kind support during the SPR measurements.

References

[1] Aberl F, Wolf H, Kößlinger C, Drost S., Woias P, Koch S, *Sensors and Actuators B*, 18 - 19 (1994) 271 - 275.

[2] Liedberg B, Nylander C, Lundström I, *Sensors and Actuators*, 4 (1983) 299 - 304

[3] C. Kößlinger, S. Drost, F. Aberl, H. Wolf, *Fresenius J. Anal. Chem.* (1994) 349:349 -354.

[4] E. Uttenthaler, C. Kößlinger, S. Drost, *submitted to Biosensors and Bioelectronics.*

Signals and Circuits

Fourier Transforming Sensors: a recondite source of major error

B S Chowdhry and J E Brignell

Department of Electronic Engineering, Mehran University of
Engineering and Technology Jamshoro, Pakistan

Department of Electronics and Computer Science
University of Southampton, Southampton SO17 1BJ, UK.

Abstract. It is now possible to devise miniaturised sensors with
substantial internal computing power. This makes commonplace the use of such
mathematical procedures as Fourier Transformation within the sensor. It does not
seem to be widely understood that, because of the properties of the well-known
Fourier delay function, very large errors can be introduced by a normally negligible
delay in the signal acquisition process. It is therefore essential to apply a special
calibration process to such sensors when they are used in the stimulus-response
mode.

1. Introduction

In the application of stimulus-response sensors a great deal of information can be
gleaned from the frequency domain. For example, utilisation of the imaginary part of the
response can produce very accurate linear displacement sensors based on magnetic loss,
while dielectric loss can be used to identify particular chemical species. Digital methods
in intelligent sensors also offer a whole new raft of valuable techniques by allowing
unorthodox stimuli to be generated (Brignell, and White, 1996). A decade ago we
described a technique that exploited one particular such signal, the perfectly filtered step
function (Chowdhry et al, 1988). This was realised in the form of a portable instrument
(Chowdhry and Brignell, 1993). Now that such a technique can be implemented in a
miniaturised sensor, it would seem timely to revisit the subject and add a footnote about
gross errors that can occur due to signal delays of a magnitude that are a small fraction of
a sample interval. It proves extremely difficult to predict the exact value of delay that can
occur in the signal acquisition system; so, as we demonstrate below, it is necessary to
resort to a calibration technique.

It is well known that, in conventional linear continuous systems, the perfectly
band pass filtered is unrealisable. It is a consequence of Fourier transform theory that any
function finitely limited in one domain must be infinitely extended in the other. As a
result any perfect filtering process must be 'non-causal', since its response is infinitely
extended in the negative as well as positive direction of time, and therefore unavailable in
the analogue world. However, in digital systems, by displacing the time-origin we can
quite simply generate such functions starting at some point where they show amplitude

variations less than one bit, and hence produce a waveform which is equivalent to the ideal stimulus. As discrete convolution is cyclic the origin can be restored by a process of rotating the time record, but this assumes that there are no extra delays in the signal acquisition process.

The impulse response of a perfect band pass filter of cut-off frequency ω_c can be written:

$$f(t) = \frac{1}{2\pi} \int_{-\omega_c}^{\omega_c} 1 \cdot \exp(j\omega t)\, d\omega = \frac{\omega_c}{\pi} \operatorname{sinc}(\omega_c t)$$

Since half the response occurs before the impulse, the system is non-causal and therefore unrealisable by traditional continuous methods. Any signal passing through such a filter will be convolved with this function, and will also exhibit non-causal response before the stimulus. Of particular interest in sensor applications is the step response, which is related to the impulse response by the linear process of integration, so the step response of the filter can be written:

$$F(t) = \int_{-\infty}^{t} f(t)\, dt = \frac{\omega_c}{\pi} \int_{-\infty}^{t} \operatorname{sinc}(\omega_c t)\, dt = \tfrac{1}{2} + \frac{1}{\pi} \operatorname{Si}(\omega_c t)$$

where Si is the well known sine integral, a transcendental function that is tabulated and available in such packages as Mathcad. It can be generated by means of series or more easily stored as a look up table. Because of its slow convergence, an incremental look up table of time intervals for a one bit change is much more storage-efficient (Chowdhry and Brignell,1993).

2. The Fourier Delay Function

The treatment of delay in Fourier Transform theory is deceptively simple. The transform of a Function delayed by a time τ is given by:

$$F_d(\omega) = \exp(-j\omega\tau)\, F(\omega)$$

Since this exponential multiplier is a function of unit magnitude it is frequently ignored, but in sensor applications we tend to require the extraction of the imaginary part and this is where gross errors can enter, even for a very small delay. In order to extract the real and imaginary parts we have:

$$\operatorname{Re}\{F(\omega)\} = \operatorname{Re}\{F_d(\omega)\}\cos(\omega\tau) - \operatorname{Im}\{F_d(\omega)\}\sin(\omega\tau)$$

and

$$\operatorname{Im}\{F(\omega)\} = \operatorname{Im}\{F_d(\omega)\}\cos(\omega\tau) + \operatorname{Re}\{F_d(\omega)\}\sin(\omega\tau)$$

Now, while it is theoretically easy to make such a calculation, τ arises in a real intelligent sensor from combined delays in the amplifier, anti-aliasing filter and sample-hold circuit. It therefore emerges that the only satisfactory procedure is a special calibration process as described below.

3. Magnitude of the error

In order to illustrate just how large an error can be generated by a small delay, consider the case where the primary sensor is replaced by a perfect capacitor subjected to a perfect unit step stimulus. First we note that the current at the origin of the step is infinite; which is why we have to resort to the Si function in the first place, as any practicable electronic amplifier would be saturated. The current is a unit impulse, which in the frequency domain is a real constant of unit value. However, when a delay, τ, is added this constant becomes the complex quantity:

$$\cos(\omega\tau) - j\sin(\omega\tau)$$

Now, if our sensor were based on a capacitor of area A and thickness d containing lossy dielectric material with complex relative permittivity ε_r the current resulting from a step of magnitude V would be:

$$I(s) = V\frac{\varepsilon_0 A}{d}\varepsilon_r(s)$$

and we could extract the loss information from the imaginary part of I, but this is where the damage occurs, since the sine term in the above quantity has exactly the same effect as the loss. It is easy to see that if τ is greater than half the sampling interval a spurious massive peak will occur within the measurable range (half the sampling frequency). When the Si function method is used, a false peak occurs at the cut-off frequency and in the presence of measurement noise this can look just like a genuine peak. The relative height of this peak is given by:

$$\sin(2\pi r\frac{\omega_c}{\omega_s})$$

where r is the delay as a fraction of the sample interval, ω_c is the cut-off frequency and ω_s is the sampling frequency. Taking typical values of the cut-off at a quarter of the sampling frequency and a delay of half a sampling interval this produces a false peak of about 0.7 of the magnitude of the ideally flat real part of the response of an ideal capacitor.

4. A calibration procedure

As stated above, it is virtually impossible to predict accurately the total delays in a practical intelligent sensor data acquisition system. It is therefore necessary to invoke a calibration procedure on a trial and error basis. The capacitive primary transducer is replaced by a high quality ceramic capacitor of similar value (or the inductive element by an air cored device). Then the following routine is implemented.

1. A delayed sinc voltage waveform is applied to the capacitor and the current waveform captured.
2. An advance (or delay) rotation is applied to the captured function with values at the sample points restored by a process of interpolation.
3. The waveform is rotated by an amount equal to half the total number of samples in order to restore the origin.
4. A Discrete Fourier Transform is implemented and the imaginary part extracted
5. A positive or negative peak is observed.
6. The advance (or delay) is changed in a direction likely to reduce the peak.
7. The whole process is repeated until there is no peak remaining.

At the end of this process the software advance (or delay) is equal to the hardware delay (or advance) and is correct for any real primary sensor that is substituted for the ideal capacitor. Fortunately, this whole process is a lot easier than might seem, as the optimum can be quite simply approached by a process of interpolation.

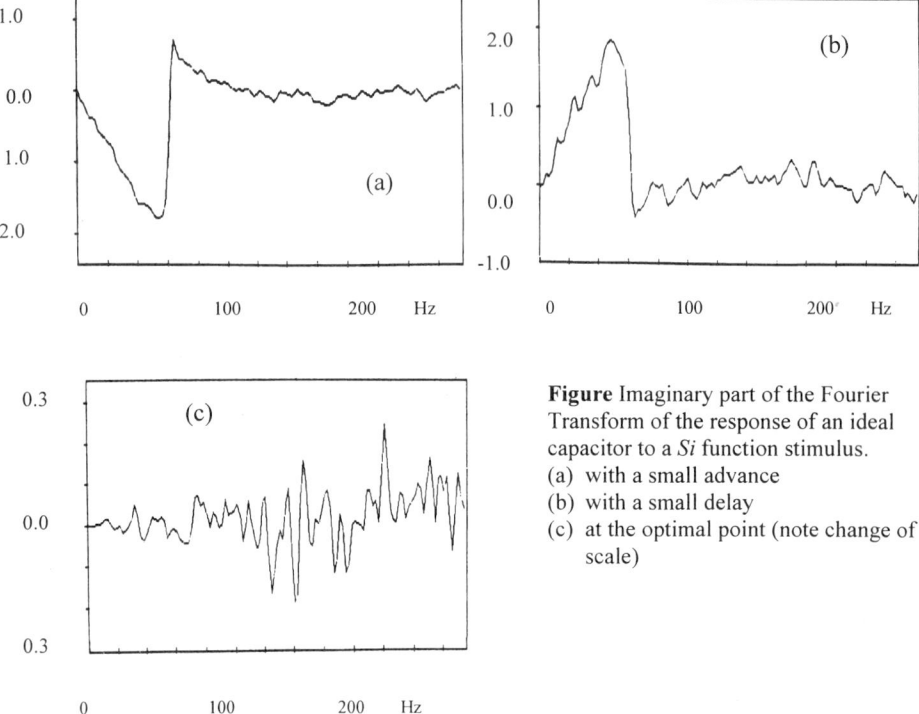

Figure Imaginary part of the Fourier Transform of the response of an ideal capacitor to a *Si* function stimulus.
(a) with a small advance
(b) with a small delay
(c) at the optimal point (note change of scale)

5. Conclusion

Fourier Transformation can be a powerful tool within intelligent sensors, but careful treatment of signal delays is imperative. A simple calibration procedure enables problems of spurious peaks, caused by the nature of the Fourier delay function, to be overcome.

References

Brignell J and White N, 1996, *Intelligent Sensors Systems*, 2nd Ed.(IOP Publishing Bristol) ISBN 0 7503 0389 1.

Chowdhry B S and Brignell J E 1993, *Design and Implementation of Programmable Si/sinc Function Generator*, Mehran University Research Journal of Engineering & Technology **12** No1.

Chowdhry BS, Shahi SS and Brignell JE, 1988 *Applications of Unorthodox Stimuli in Computer Measurement: Dielectric Relaxation as an Example*, J.Phys. E: Sci. Instrum. **21** 259-263.

Signals and Circuits
Paper presented at Eurosensors XII, 13–16 September 1998

Low impedance sensing technique for vibrating structures

A Weinert, M Berggren, G I Andersson

Chalmers University of Technology, Department of Microelectronics, Solid State
Electronics Laboratory, Göteborg, S-412 96, SWEDEN

Abstract. We report a refined version of the resonant gate transistor (RGT) [1]
which we have denoted the Resonant double gate transistor (RDGT). The
important difference between RGT's reported earlier and our new design is that in
addition to the oscillating gate we also have a fixed floating gate. According to
simulations this gives a spatial resolution in the range of 10^{-13} m/√Hz with an air
gap of 2 μm and a voltage of 10 V on the moving gate.

1. Introduction

The major advantage of the Resonant double gate transistor (RDGT) compared to a capacitive detection technique is the low impedance of the output achieved by the direct capacitance to current conversion of the transistor. Thus the RDGT is especially suitable for
sensor designs using diffused feed-throughs in order to make sealed encapsulation to the environment possible.

2. Detection principle

A schematic of the structure is shown in Figure 1. The detection principle for sensing the
motion of the oscillating gate (2) is capacitive. The RDGT (1) can be subdivided into a time
varying capacitance C_{AIR} and an FET-transistor (3).

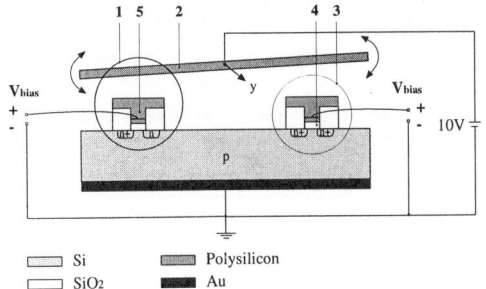

Figure 1.
*Sketch of two Resonant double gate
transistors:*

(1) RDGT
(2) Oscillating gate
(3) Intrinsic FET-transistor
(4) Thin oxide layer
(5) Floating gate.

As seen in Figure 1 a voltage change Δv_G can be generated at the floating gate (5) of the FET-
transistor by applying a DC-voltage to the oscillating gate:

$$v_G = V_{BIAS} + \Delta v_G = V_{BIAS} + V_{DC} \cdot \frac{\Delta C_{AIR}}{C_{IG} + C_{AIR}}. \tag{1}$$

Here C_{AIR} is the time varying capacitance between the oscillating gate and the floating gate. The capacitance C_{IG} represents the input to ground capacitance of the FET-transistor composed of the oxide capacitance, C_{OX}, the pad capacitance and the parasitic capacitance.

Experiments showed that the transistor in a "pure" floating gate configuration drifted. Thus a need for high impedance biasing of the floating gate arose. To enhance the overall signal-to-noise ratio and to enlarge the output signal of the RDGT we designed a differential amplifier stage based on two RDGT's.

2.1. Time varying capacitance

The spatial measuring function of the RDGT is provided by the double gate construction. The movable gate together with the lower floating gate constitutes a distance dependent planar capacitor. The capacitance C_{AIR} changes as the oscillating gate moves perpendicular to the plane of the substrate and the transistor. The lower floating gate determines the gate area of the FET-transistor. Thus, when the potential of the floating gate changes, the current in the channel of the MOSFET-transistor is modulated.

To improve the performance of the RDGT the area of the floating gate was extended over the transistor gate area. Also the oxide thickness under the extended part was increased compared to the transistor gate oxide thickness. This is to minimise the parasitic capacitance to the substrate that degrades the efficiency of the capacitive voltage divider.

3. Assumptions, made when determinating the resolution of the detectionsystem

System sensitivity was compared to the total noise generated by the biasing circuit and the amplifier. The overall signal to noise ratio was calculated for two different biasing methods: biasing via resistors and biasing using cascade amplifiers. The minimum detectable deflection of the oscillating mass was then calculated using the simulated sensitivity of the RDGT. Calculations show that very high sensitivity can be achieved if the biasing method is chosen carefully, see Table 1. The calculations were performed under the following assumptions:

- The operating frequency of the RDGT was 7 kHz and the output signal was evaluated over a bandwidth of 100 Hz.
- The expected sensitivity of the RDGT was 0.68 µF/m, the input to ground capacitance C_{IG} was 10.5 pF and the time varying capacitance C_{AIR} was 0.95 pF.
- The DC voltage applied to the moving gate was 10 V.
- The power supply voltage of the preamplifier is ±3 V and the input stage bias current was 100 µA.
- All noise sources except those in the preamplifier and the biasing circuits were ignored.

The transistors integrated in the bulk material were NMOS devices with a gate oxide thickness of 10 nm. These transistors were 50 µm wide and 10 µm long and had a transconductance, g_M, of 255 µA/V.

3.1. Equivalent input noise voltage

The input-referred RMS noise voltage v_{EQ} at the floating gate of the RDGT can be approximated [2, 3] by

$$v_{EQ-FET}^2 = 4k_B T \cdot \left(\frac{2}{3g_M} \right) \Delta f + \frac{K_F}{C_{OX} WLf} \Delta f + \frac{2qI_{DS}}{(g_M)^2} \Delta f. \tag{2}$$

The three terms in equation (2) represent thermal noise, 1/f noise and shot-noise in [3], respectively. The bandwidth over which the noise voltage is measured is denoted Δf. The value of the transconductance, g_M of the FET-transistor is well-known but the flicker noise factor K_F can only be estimated from measurements. Using the results of Vandamme et al. [4] and Chang et al. [5], we found the factor K_F to be between 4.8×10^{-24} Ws and 4.8×10^{-23} Ws. The contribution from the flicker noise was calculated to be between 4.0×10^{-14} V^2 and 4×10^{-13} V^2, while thermal noise and shot-noise contributed 4.3×10^{-15} V^2 and 4.9×10^{-14} V^2, respectively.

This means that the amplitude of the input-referred noise voltage was in the range of 0.5 μV to 1.0 μV.

3.2. Noise generated by the biasing stage

A noise amplitude of approximately 1.2 μV is generated at the input of the preamplifier if biasing was carried out with a 10 MΩ resistor connected to a DC voltage. The contributions to the noise spectrum from the flicker and shot-noise can be ignored since an extremely small current flows through the 10 MΩ resistors. Thus we consider only the thermal noise contribution of the resistors derived from [3, 6]

$$v_{TH}^2 = 4k_B T R \Delta f. \tag{3}$$

At frequencies where the capacitive input impedance is much smaller than R the RMS noise voltage can be described by

$$v_{TH}^2 = i_{TH}^2 \cdot \left| \frac{1}{j\omega C_{IG}} \right|^2 = 4k_B T \frac{R}{\left(\omega R C_{IG}\right)^2} \cdot \Delta f. \tag{4}$$

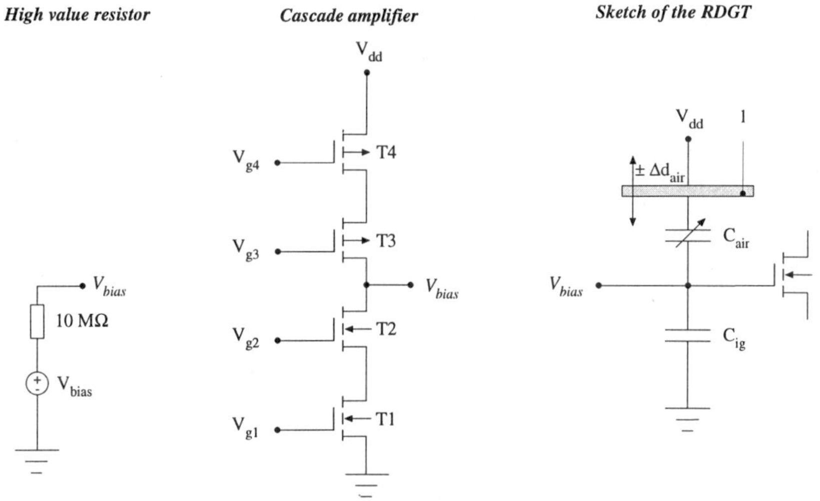

High value resistor *Cascade amplifier* *Sketch of the RDGT*

Figure 2. *Two Alternatives for the biasing stage.*

If instead cascade amplifier stages are used the theoretical value of the amplitude of the noise voltage at the inputs of the preamplifier would be 43 μV if the quiescent current is 10 μA and the length of the transistors in the biasing stage is 100 μm. The noise currents are dominated by flicker-noise and shot noise. Thermal noise has little effect due to the low transconductance g_M of the transistor T1 in the cascade amplifier stage. At frequencies where the input impedance of the preamplifier is governed by the input-to-ground capacitance C_{IG} the RMS noise voltage becomes

$$v_{N-T1}^2 \approx i_{N-T1}^2 \cdot \left| \frac{1}{j\omega C_{IG}} \right|^2 \approx \frac{2I_{DS1}}{(\omega C_{IG})^2} \left(\frac{k_{P1}K_{F1}}{C_{OX1}L_1^2 f} + q \right) \Delta f \ . \tag{5}$$

6. Results and Conclusions

In this paper we discuss a refined version of the resonant gate transistor (RGT) [1]. With the additional floating gate of the resonant double gate transistor (RDGT) problems caused by drift and differences in the transistor parameters in the differential input stage, can be minimized.

The calculated smallest detectable change in the air gap capacitance C_{AIR} and the corresponding deflection of the oscillating mass for the RDGT are listed in Table 1.

Biasing	None	Resistor 10 MΩ	Cascade amplifier
Noise voltage	0.5 µV to 1.0 µV	1.2 µV	25 µV to 45 µV
Detectable capacitive resolution	0.6 aF to 1.2 aF	2.0 aF to 2.6 aF	29 aF to 52 aF
Detectable mechanical resolution	1.1 10^{-12} m to 2.0 10^{-12} m	3.0 10^{-12} m to 4.0 10^{-12} m	4.0 10^{-11} m to 8.0 10^{-11} m

Table 1. *Resolution of the RDGT at 100 Hz bandwidth derived from the amplitude of the input-referred noise voltage, calculated for the FET-transistor without biasing and biased via a high value resistor or a cascade amplifier*

Circuit simulations of the overall system RDGT´s, biasing circuit and differential amplifier agree with the analytical calculations of an equivalent input noise voltage amplitude of 1,2 µV (see Table 1) indicating the possibility of achieving a spatial resolution in the range of 10^{-13} m/√Hz.

A discrete differential amplifier stage including the two RDGT´s has been designed and tested. Currently the differential amplifier stage and the biasing circuit are being integrated and manufactured in an ASIC.

In summary, the high spatial resolution, the low impedance of the output as well as the possibility of integrating the RDGT in a standard CMOS-process makes the RDGT highly interesting for micromechanical sensor applications.

References

[1] Nathanson H C, Newell W E ,Wickstrom R A and Davis J R 1967 *IEEE* Trans. on Electron Devices **3** 117-133

[2] Sze S M 1994 *Semiconductor sensors* (New York: Wiley)

[3] Berggren M 1997 *A preamplifier for a sensor using a novel detection method: "The mechanical transistor"* (Göteborg: Master Thesis at the Department of Solid State Electronics-Chalmers University of Technology)

[4] Vandamme L K J, Li X and Rigaud D 1994 *IEEE* Trans. on Electron Devices **41** 1936-1945.

[5] Chang J, Abidi A A and Vishwanathan C R 1994 *IEEE* Trans. on Electron Devices **41** 1965-1971

[6] Ott H W 1988 *Noise reduction techniques in electronic systems* (New York: Wiley)

Signals and Circuits
Paper presented at Eurosensors XII, 13–16 September 1998
© *1998 IOP Publishing Ltd*

Micromechanical cantilever as a sensor for vibrations measurement

V.Snitka, V.Mizariene
RC"Vibrotechnika",Kaunas University of Technology,Kaunas,LT-3031,Lithuania

Abstract. This study presents a novel method to measure an ultrasonic surface vibrations and acoustic fields in solids using scanning microsensor. The vibrations of sample surface in frequency range of 10-200 kHz and the influence of the ultrasonic surface vibrations on the cantilever tip - surface interaction were investigated.
Cantilever vibrations were investigated in the time and phase plane.Experimental results show that the cantilever in the case of normal surface vibrations has a complex vibration mode caused by nonlinear interaction at the contact point and strongly depends on the setpoint force and frequency. The adsorbates on the surface strongly influence the nonlinear dynamic of the vibrating cantilever.The second harmonic generation by V-shaped cantilever was observed.
The investigations have shown the possibility to use micromechanical cantilevers for vibrations and acoustic fields measurement with nanoresolution

1. Introduction

Atomic Force Microscopy, based on micromechanical cantilevers, which originally was intended to image the surface and measure its topography is also extensively used to probe different properties of the surface such as microadhesion, surface forces etc. Its principle of operation is based on sensing an interaction between the investigated surface and very sharp tip integrated on a flexible cantilever which is kept at a small distance from the surface.

Recently AFM started to extend its operation to the range of high frequencies and importance of precise measurement of dynamics of tip - surface interaction is constantly increasing. Rabe et al. investigated vibrations of free and coupled to surface rectangular AFM cantilevers in a spectral range of 100 kHz to 10 MHz [1]. The same group used AFM at high frequencies to image surface areas with different elastic properties [2]. Burnham et al. employed similar technique called Scanning Local Acceleration Microscopy [3]. Yamanaka et al. used ultrasonic surface vibrations for nanometer subsurface imaging [4].

In this paper we present the results of the study of dynamics of the tip and vibrating surface interaction . A new method for the determination of cantilever vibration was used.

2. Experiment

Schematics of measurement system is shown in Fig.1. Vibrations of microcantilever in frequency range 10 Hz - 2 MHz were excited using the piezoceramic transducer. Laser beam reflected from the cantilever

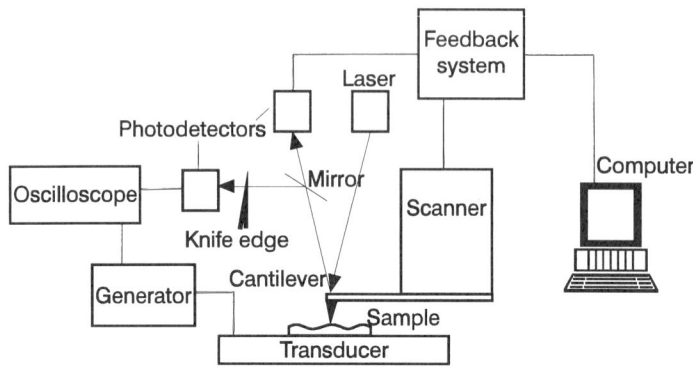

Fig.1. Schematics of measuring system

is incident on one of the photodetectors. Microcantilever vibrations in the range 2 Hz - 15 kHz were registered using the interface to the personal computer. High frequency vibrations were registered using the fast oscilloscope.

The vibrations of microcantilever were investigated when it is in contact and out of contact with the surface and also using the force-displacement mode, in which the cantilever is scanned in vertical direction and dependence of force acting on cantilever versus cantilever-sample displacement is recorded (Fig.1). Both triangular and rectangular cantilevers with 0.05 and 0.01 N/m spring constants respectively were used for the experiments. The lengths of rectangular and triangular cantilevers were 180 micron and 320 micron respectively were used for the investigation. Lisageau figure method was used to determine the frequency and phase differences between the cantilever normal and lateral vibrations and exciting signal.

Cantilever vibration was investigated using an oscilloscope mode of the AFM. It allowed us to plot the microcantilever signal in time domain or to plot a Lissajous figure which is derived summing the top - bottom and left - right signals of cantilever.

Surface vibrations were excited by applying high frequency voltage from a generator to a piezoplate. The investigated sample was glued on the plate. Normal, lateral and bimodal vibrations in frequency range of 10-200 kHz were investigated during the experiments. Surface vibrations were detected and tip - surface interaction dynamics was investigated using force-displacement curve scanning the sample in z-direction. All experiments were performed in the air, in normal laboratory conditions.

3.Results and discussion

Force-displacement curves for vibrations of different modes are presented in Fig. 2-3. Interaction between cantilever tip and vibrating surface was observed at separation distances up to few micrometers (Fig. 2). It is seen that vibrations of different modes (Fig. 3) can be clearly distinguished up to 70 kilohertz, and that amplitude of vibrations does not change appreciably with separation between cantilever and surface.

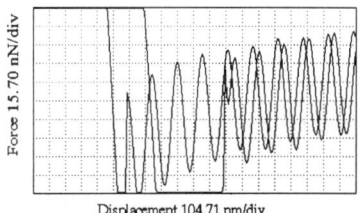

Fig.2.Force-displacement curve of normal
10 kHz frequency vibrations.

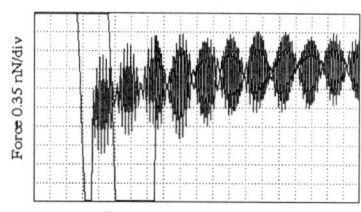

Fig.3.Force-displacement curve of bimodal
44 kHz frequency vibrations.

It shows that vibrations of cantilever are activated presumably by direct transmission of vibrations of the surface to the cantilever through an air gap. The same time non - linear interaction between tip and surface is seen as bending of cantilever to the surface. Besides that, it was noticed that amplitude of vibrations of cantilever is higher than that of the surface. This can be attributed to electrostatic interaction between electric field of vibrating piezoplate and charge on the tip. The experiments show that low frequency vibrations do not effect the tip - surface adhesion force.

Cantilever motion was investigated using an oscilloscope mode. The results are given in Fig.4 - 7.

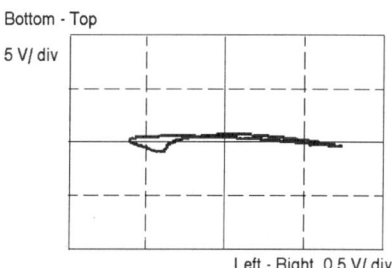

Fig.4. Triangular cantilever vibration at
5 kHz frequency.

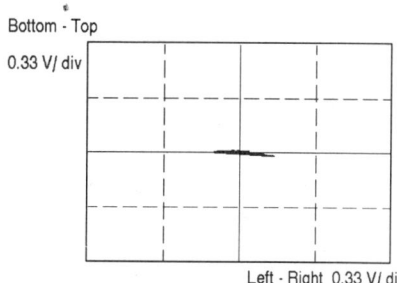

Fig.5.Rectangular cantilever vibration
at 8 kHz frequency.

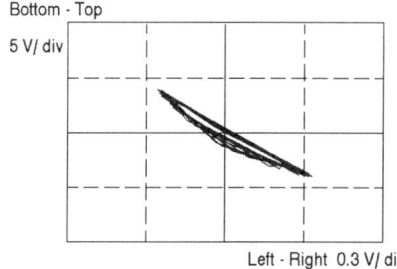

Fig.6.Rectangular cantilever vibration
at 2 kHz frequency.

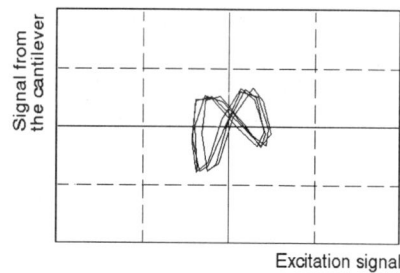

Fig.7. Lissajous figure of the triangular
cantilever at 10 kHz frequency.

These figures show that the motion of the cantilever consists of coupled normal and lateral vibration modes. Figure 4 shows the triangular cantilever motion at 5 kHz frequency. It is clear that even at low frequency lateral vibration mode takes place in cantilever's motion. According the Lissajous figure method one can notice that the normal vibration frequency is almost two times higher than the lateral vibration mode frequency. Changing exciting frequency in a wide range (20 Hz - 10 kHz) of frequencies does not significantly change the figure, only the phase shift between the normal and lateral vibrations is changed which is seen as the Lissajous figure's rotation. The amplitudes of normal and lateral modes depend significantly on setpoint force of cantilever. When cantilever is lowered down to the surface so that elastic force acting on cantilever increases, amplitude of normal vibrations decreases and the phase shift between normal and lateral modes changes. It is taking place because of changing in damping and spring constant of surface-cantilever mechanical system.

The rectangular cantilever exhibits slightly different kind of motion (Fig. 5-6). The lateral vibration has the same frequency as the normal but there is a 30^0 phase shift between them. The phase shift changes when the frequency is changing. Figure 6 shows the results of the rectangular cantilever vibration at 8 kHz frequency. It is clear that the lateral vibration mode has significantly bigger amplitude than the normal vibration mode.

Figure 7 shows the oscilloscope plot where the horizontal axis represents the exciting signal and the vertical axis signal from the cantilever. As it is seen from the figure the cantilever vibrates in double frequency in comparison with the exciting signal. This frequency doubling phenomena was observed for the triangular cantilever in a wide frequency range 10 Hz - 20 kHz. The rectangular cantilever does not exhibit such behavior. It was noticed that the frequency doubling significantly depends on the tip surface distance. The reason of the frequency doubling we found is the lateral mode of the cantilever vibrations.

4.Conclusions

We have performed experiments measuring interaction of the vibrating surface and micromechanical cantilever. The few methods to investigate the influence of the surface vibrations to the cantilever and image formation were used. The force - distance curves allow to evaluate the amplitude of the vibrations and to find the influence of non-linear interaction at the contact to the cantilever vibration.. Lissajous figures give us information about complex cantilever and cantilever - surface vibrations.

Our results show that the cantilever vibration exited by the vibrating surface very strong depends on the non-linear dynamics of the contact between cantilever tip and surface.

The frequency doubling for V-shaped cantilevers caused by non-linear interaction was observed. The amplitudes of the normal and lateral modes and phase shift between them strongly depends on interacting force between tip and surface and adsorbates on the surface strongly influence the contact mechanic.The possibility to use micromechanical cantilevers as vibration and acoustic fields microsensors have shown.

References
[1]   Rabe U, Janser K and W. Arnold, 1996 *Rev. Sci. Instrum.* 67 **9** .3281-6
[2]   Rabe U, Scherer V, Hirsekorn S and Arnold W, 1996 *Proc. Nano IV*, Beijing ,186-9.
[3]   Burnham N A, Kulik K , Gremaud G, Gallo P J and Oulevey F, 1996 *J. Vac. Sci. Technol.* **B 14**, . 794-8.
[4]   Yamanaka K, Ogiso H, and Kolosov O, 1994 *Appl. Phys. Lett* **64** January , 178.

Measurement of the response of resistive sensors using digital logic oscillators

R Stone and P A Payne

Department of Instrumentation & Analytical Science, UMIST, PO Box 88,
Manchester M60 1QD.

Abstract. This paper describes preliminary experiments in the use of a digital logic oscillator to measure resistance changes in a semiconducting polymer sensor. The basic stability of the system was evaluated using a fixed resistance and by exposing the sensor to dry air. Finally, the response of the sensor to the headspace above a 1% butanol/water mixture using a dry air carrier was measured to evaluate the possibility of using this circuit in a gas/odour sensing instrument.

1. Introduction

Conducting polymer sensors have found many uses including gas and odour analysis. In addition, by using an array of such sensors, each with slightly different specificity, the absorption of a volatile onto the array causes a unique pattern of responses which can be classified by an artificial neural network (ANN) [1-3]. A number of companies are marketing this use of sensors for applications such as quality monitoring in the food industry.

In many, if not all of these instruments, the resistance of the sensor is measured in a conventional manner. This can be done in two ways. Firstly, a fixed or programmable *voltage* can be applied across the sensor and the resultant current measured (usually by current to voltage conversion). Secondly, a fixed or programmable *current* can be applied to the sensor and the resultant voltage measured. These systems of measurement require analogue-to-digital converters (ADCs), digital-to-analogue converters (DACs) and support circuitry. Furthermore, when an array of sensors is used, analogue multiplexers are employed to increase the number of channels. This leads to sensor interfaces which are complicated to design, have high power dissipation and are large in size.

The motivation for this research was not to improve on the sensitivity of such systems but rather to simplify the electronics design and the physical size of the measurement circuit. This has been attempted previously by the use of application specific integrated circuits (ASICs) [4]. However, the main problem with ASICs is their high cost when produced in low volumes. The work detailed in this paper relies on low cost digital ICs with the possibility of integration for a commercial instrument.

2. The digital logic oscillator

The oscillator circuit chosen is shown in Figure 1. The main benefit of this circuit is that it requires only one component other than the logic gates and the sensor. The type of logic used was 5 V HC [5], which resulted in a duty cycle of around 50% and a frequency of approximately 1/1.8 RC [6] where R is the resistance of the sensor (ohm). The value of the capacitor was chosen so that the circuit operated in the low kilohertz range (between 1 and 10 kHz) with an appropriate sensor. The output of the oscillator was buffered by a further inverter and the time period, which is proportional to resistance, was measured using a Hewlett Packard HP53132A frequency counter connected to an IBM PC using a IEEE 488 interface. To minimise temperature effects on the period of oscillation the oscillator circuit and the sensor were placed in a temperature controlled environment held at 30°C ± 1°C.

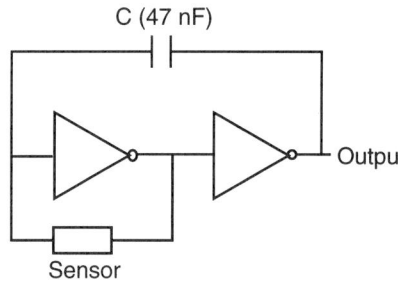

Figure 1. *Digital logic oscillator using 5 V HC technology*

3. Experimental

Three experiments were conducted to assess the system. The first experiment examined the basic stability of the oscillator by using a fixed 2 kΩ resistor in place of the sensor. Figure 2 shows the results of this stability test, and as can be seen the system is stable to a few thousandths of one per cent. Note that the percentage change in time period, (dT/T)%, is equivalent to the (dR/R)% of the sensor.

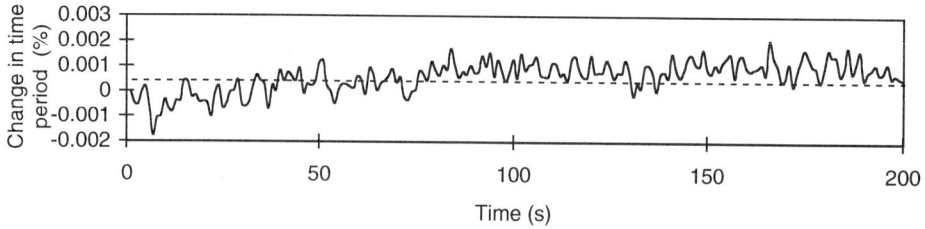

Figure 2. *Stability of the oscillator using a fixed resistor*

In the second experiment, the fixed resistor was replaced by a conducting polymer sensor. The sensor was connected to a gas delivery system that could be switched from dry air to dry air which had flowed over a 1% butanol in water mixture. The arrangement of the apparatus is shown in Figure 3. In a similar manner to the fixed resistor experiment, the stability of the sensor in the dry air stream was evaluated. The results can be seen in Figure 4.

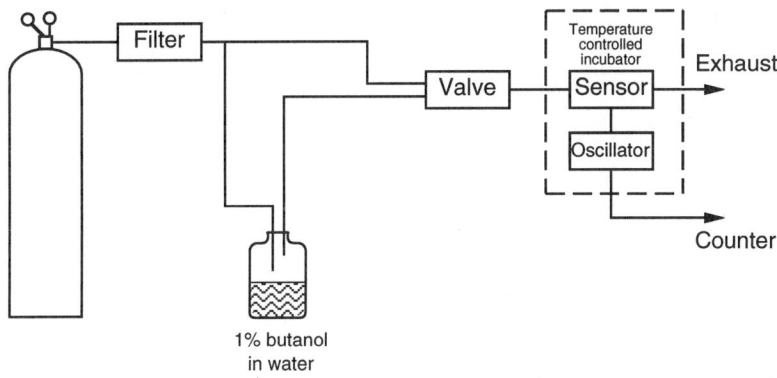

Figure 3. *Gas and vapour delivery system.*

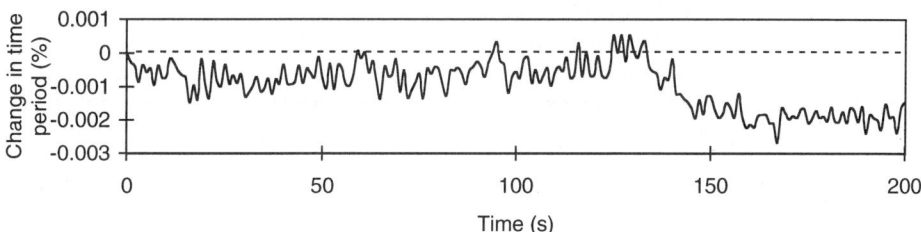

Time (s)

Figure 4. *Stability in response using a conducting polymer sensor exposed to dry air.*

As the graph shows, the output is a little less stable. However, this deviation is within acceptable limits for such a simple gas delivery system. The low frequency components of the noise in the system are mainly due to temperature fluctuations in the sensor, as in this experiment, the sensor substrate was not heated directly but from the convection of warm air within the incubator. This caused low frequency deviations in the time period as the temperature in the laboratory changed. The cooling effect of the gas stream also made the sensor more sensitive to flow rate. The main feature to note is that the high frequency components of the drift in Figures 2 and 4 are similar in amplitude, although the gas sensor noise is a little higher. This is possibly due to the longer lead lengths needed to connect the sensor to the oscillator.

The gas flowing across the sensor was switched from dry air to air which had been passed over the butanol solution and the response of the sensor measured. The data recorded is shown in Figure 5. The graph depicts the standard shape of response of a reversible gas sensor to a volatile. The baseline is recovered between the two pulses of solvent and the response is repeatable. In this particular case, the signal due to the adsorbed volatiles, is much greater than the noise in the system (around 2500 times greater) so the effect of the electrical noise is not seen in Figure 5.

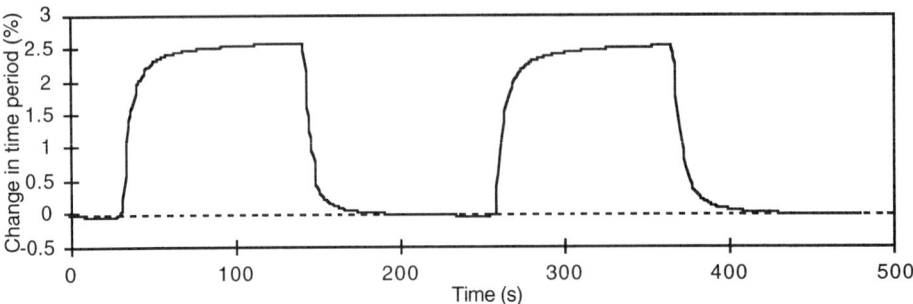

Figure 5. *Response of a conducting polymer sensor to pulses of a 1% butanol/water mixture*

4. Conclusions

The possibility of measuring the response of resistive sensors has been investigated as a possible route to simplify multi-channel instruments. The results obtained show that an un-optimised circuit, assembled on a small prototyping board, can be used to perform resistance measurement. The main advantage of this approach is that the number of analogue components can be reduced. In addition, the possibility of integrating the logic onto a programmable device, such as a field programmable gate array (FPGA), now exists. Further work needs to be carried out to assess the use of multiple channels, integrated frequency measurement and temperature compensation, for this type of device. It is expected that this idea would have particular application in the hand-held and remote odour sensing market.

Acknowledgements

Richard Stone is in receipt of an EPSRC Total Technology CASE Award which is supplemented by AromaScan plc.

References

[1] Amrani M E H, Dowdeswell R M, Payne P A and Persaud K C 1997 *Sensors and Actuators B* **44** 512-516

[2] Persaud K C and Travers P J 1991 *Intelligent Instruments and Computers* **9** 240-243

[3] Payne P A and Persaud K C 1996 *Sensor and Transducer Conference Proceedings, Gas and Chemical Sensors, MTEC,* Birmingham, 21-27

[4] Hatfield J V, Neaves P, Hicks P J, Persaud K C and Travers P J 1994 *Sensors and Actuators B* **18-19** 221-228

[5] Horowitz P and Hill W 1994 *The Art of Electronics* (Cambridge: Cambridge University Press) 569

[6] Graf, R F 1992 *Oscillator Circuits* (Boston: Newnes)

Signals and Circuits
Paper presented at Eurosensors XII, 13–16 September 1998
© 1998 IOP Publishing Ltd

Improvement of SAFT images by Real-Time Digital Processing of UT-signals

O Martínez, M Parrilla, M A G Izquierdo, L G Ullate
Inst. de Automatica Industrial (CSIC), La Poveda (Arganda del Rey), 28500 Madrid, SP

Abstract
The elaboration of UT-images using SAFT techniques is a usual practice in Non Destructive Testing systems. However the existence of side lobes and grating lobes introduce artifacts that distorts the images. Joining different techniques of digital signal processing is possible to reduce these effects and improve lateral and axial resolution.

1. Introduction

The use of ultrasonic arrays to obtain UT-images is a usual practice for medical diagnostic, where the medium conditions are delimited and well known. However, development of array systems in the area of Non Destructive Testing (NDT), where each application presents a great variety of designing factors, is rarely found. The high cost of parallelism required for real-time systems and the specific design required by each array application justify this fact. Synthetic Aperture Focusing Technique (SAFT) is an alternative that tries to reduce the high grade of parallelism of array systems.

Conventional SAFT using a linear array [1] provides images with good lateral resolution, but it introduces relatively high grating and side lobes compared with non-synthetic focusing. In this paper a method based on digitally processing of UT signals is proposed to improve SAFT. Each element transmits and receives in a sequential way. Once each UT signal has been analog conditioned, digital signal processing is carried out by a pipeline of specially designed processing modules. The architecture executes DSP algorithms in parallel operating at 10Msamples/s [2]. The proposed system operates in two stages:
(1) The first one executes over each received UT signal the following algorithms: apodization, deconvolution, and delay lines for quadratic dynamical focussing in a normal direction. Then signals are stored for the second step.
(2) Signals are combined in order to generate the UT image lines. In this stage the following algorithms are executed: delay lines for steering, digital envelope detection and image composition.

2. System operation

The block diagram of figure 1 shows the reception processing chain, which includes the following operations:

- STEP 1: Received traces from each transducer are processed and stored:
Apodization: A conventional technique for reducing side lobes is aperture apodization, where the signal corresponding to the ith element is attenuated by a factor $A(i)$.
Deconvolution: it is used to improve axial resolution and, consequently, a reduction of grating lobes is obtained. Actually any deconvolution technique could be employed to compute the inverse filter coefficients. A method to implement deconvolution in real time can be found in [3]. The output signal is:

$$y(t) = c(t)*x(t),$$

where (*) means convolution and c(t) is the inverse Fourier transform of :

$$C(f)=W(f)/Y_P(f).$$

$Y_P(f)$ being the ultrasonic signal spectrum and W(f) normalization filter to limit the bandwidth of the inverse filter $[1/Y_P(f)]$.

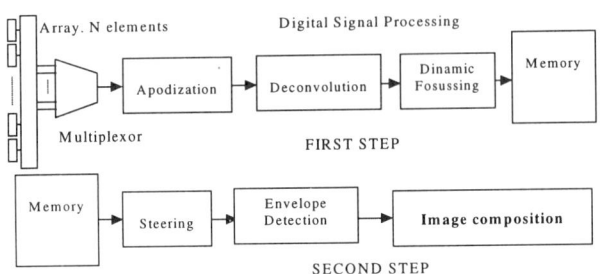

Figure 1. Diagram block of the receiving processing chain

Focussing: Cylindrical focussing is conventionally used to improve lateral resolution and for SNR enhancement. Cylindrical focussing applied to a spatial point R_f requires to delay the signal received at the ith transducer by an amount T_i:

$$T(i) = (R_f - (R_f^2 + Q(i)^2 - 2R_f Q(i)sin(\theta))^{1/2})/c$$

where $Q(i)$ is the distance of the i element to the center of the array, R_f is the depth of the focus lens, θ is the steering angle and c is the sound propagation velocity. Varying the focus R_f as the pulse propagates ($R_f = t/2c$) performs dynamic focussing. To facilitate the integration of the focussing technique into the architecture the Fresnel approximation of the previous equation [4] has been used:

$$T(i) \approx Q(i)sin(\theta)/c - Q(i)^2/(2R_f c)$$

Then:

$$T_f(i) = -Q(i)^2/(2R_f c); and \quad T_\theta(i) = Q(i)sin(\theta)/c$$

the first term representing the delays T_f, for cylindrical focussing in a normal direction and the second one indicating the delays T_θ for steering.

Using this approximation, the amount of memory and the complexity of hardware circuits involved are reduced. For example, in the case of dynamic focussing of signals from N array elements, with m samples per signal, and generating L lines in the image, the memory size required to store the control delays for a conventional system is $N*m*L$ elements. Applying the Fresnel approximation the memory requirements are lowered to $N*m$ elements for normal cylindrical focussing and $N*L$ elements for steering.

Normal dynamic focussing: Time delays T_i are expressed in number of samples of the unfocussed signal from the sampling rate. In order to increase the time resolution a linear interpolation processing stage computes:

$$X_i(k) = x_i(T_i(k)) + \frac{(x_i(T_i(k)+1) - x_i(T_i(k)))}{I_i(k)}$$

where $x_i(k)$ is the unfocussed signal of the ith element, $X_i(k)$ is the focussed signal, $T_i(k)$ and $I_i(k)$ are vectors with m elements each indicating the delay index and the interpolation factor for each sample.

Memory: In this module, the processed signals from each array element are stored, waiting for the second processing step.

- *STEP 2: traces from memory are processed to form the image.*

Steering: The signal from every element i must be delayed a time $T_i(\theta)$. The operation of this module is similar to dynamic focussing process:

$$X_i(k) = x_i(k + T_i(\theta)) + \frac{(x_i(k + T_i(\theta) + 1) - x_i(k + T_i(\theta)))}{I_i(\theta)}$$

Where $X_i(k)$ is the kth sample of the processed signal, $x_i(k)$ the incoming signal, $T_i(\theta)$ the index delay for the angle θ and $I_i(\theta)$ its interpolation factor. Signal of all array elements are delayed and added together to form the deflected line of the UT image.

Envelope Detection: The envelope of UT signals will contribute to a better quality of B-scan images. Hardware methods use a FIR or IIR filter, designed as a Hilbert transformer, to obtain an approximation to the analytical signal. For good results, high order FIR filters must be used, while lower order IIR Hilbert transformers may be unstable. The development of a new non-linear filter method [5] (Envelope Detection Filter) avoids these problems increasing at the same time the processing speed.

Image composition: In a final step, signals are visualized composing the images.

3. Experimental Results

Figures (a), (b) and (c) show an experiment of SAFT and digital processing over the signals acquired by 16 elementary transducers (0.5*8mm) forming an array aperture 8mm wide and 8 mm high. The image corresponds to a wire 0.5mm wide located at the spatial point (X = 0, Z = 35 mm). The central frequency of the UT pulses is 5Mhz. 81 steering lines, from -40° to +40° form the image. Images (a.1), (b.1), (c.1) have been formed by applying SAFT to the signals from the 16 array elements.

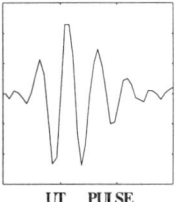

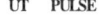
UT PULSE

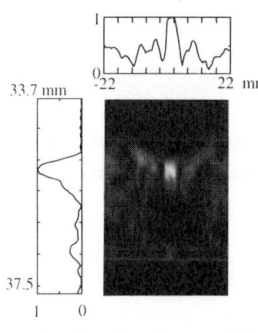

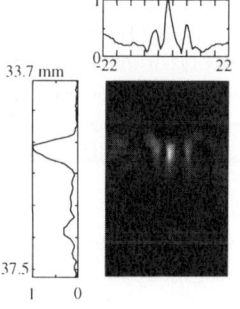

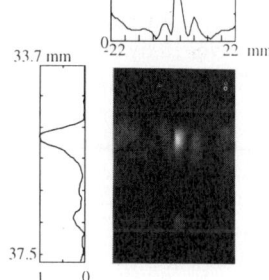

(a.1) Steering image [-40° +40°]. Side lobes and grating lobes are clearly visible.

(b.1) Steering image with dynamic focussing. Lateral resolution has increased.

(c.1) Steering image with dynamic focussing and apodization. Side lobes has been reduced but grating lobes still remain.

Images (a.2), (b.2), (c.2) have been obtained by applying an inverse filter to every one of the signals before SAFT. The figures include B-scan images, and lateral and axial profiles centered at the wire position.

870

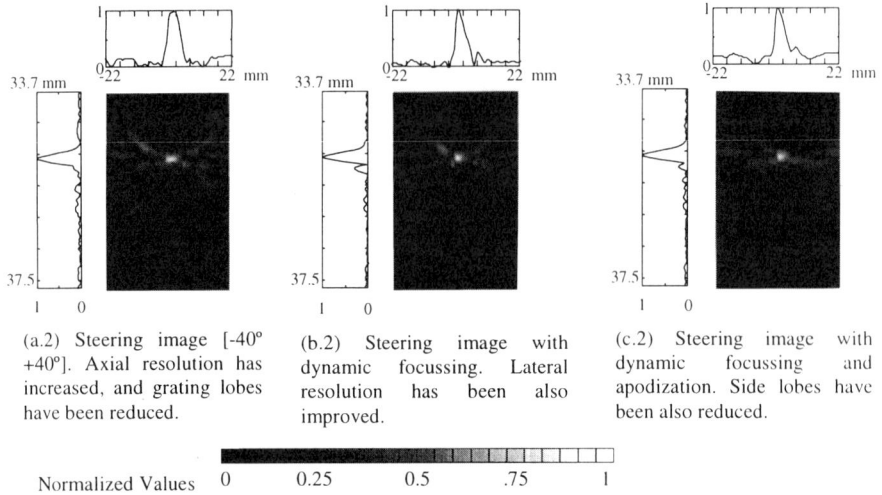

(a.2) Steering image [-40° +40°]. Axial resolution has increased, and grating lobes have been reduced.

(b.2) Steering image with dynamic focussing. Lateral resolution has been also improved.

(c.2) Steering image with dynamic focussing and apodization. Side lobes have been also reduced.

Normalized Values   0   0.25   0.5   .75   1

4. Final Discussion

Experimental images from figures (a), (b) and (c) show the benefits of digital processing in ultrasonic synthetic focusing operating onto the radio-frequency traces:
- Apodization reduces side lobes although reduces lateral resolution.
- Deconvolution improves axial resolution and reduces grating lobes
- Dynamic focusing improves lateral resolution and SNR
- Steering forms the image lines
- And finally, envelope detection smoothes ultrasonic images.

All these operations can be performed in parallel by using a pipeline hardware architecture. In particular, our group has developed an architecture called SENDAS [2] which incorporates modules to execute the digital processing chain. The global processing speed attained by SENDAS is 10Msamples/s. Using this architecture, the time T_p inverted by the SAFT procedures is for the two processing steps:

$$T_1 = N*m/10 \ (\mu s), \text{ and } T_2 = L*N*m/10 \ (\mu s); \ T_p = T_1 + T_2$$

For N=16 elements, m=1000 samples, and L=100 lines, T_1= 1.6ms, T_2=160ms, and T_p=161.6ms. Therefore, after processing, a global scanning rate of 6 images/s is obtained.

Acknowledgements: This work has been supported by CICYT TAP97-1128-C02-02 and TAP97-0662-C02-01 of the Spanish Ministry for Science and Education.

References:
[1] Peterson D.K., Kino G.S., 1984 IEEE Trans. **SU-31, (4)**, pp. 337-351.
[2] Fritsch C et al., 1995 Proc. IEEE Ultrasonics Symposium, 833-836.
[3] Anaya J.J. et al., 1992 IEEE Trans. **IM-41, (3)**, pp. 413-419
[4] Macovski, A., 1979, Proc.IEEE **67, (4)**, pp. 484-495
[5] Fritsch C. et al. *A digital envelope detection filter for Real-Time operation*, IEEE Trans. on Instrumentation and Measurement, (in revision)

Signals and Circuits
Paper presented at Eurosensors XII, 13–16 September 1998
© 1998 IOP Publishing Ltd

Fast discrete wavelet transform for B-H loop tracing

E.Hristoforou, H.Chiriac[1], V. Nagacevschi[1]

Laboratory of Metrology, TEI of Chalkis, Psahna Euboea 34400, Greece
[1]National Institute of R&D in Technical Physics, Iasi 6600, Romania

Abstract. An advanced and repeatable experimental arrangement, using the Fast Discrete Wavelet Transform and the Lifting Scheme for computational signal denoising, is presented in this paper concerning applications for magnetic material characterization, with examples related to computational techniques for B-H loop tracing,. The advantages of this method are also illustrated with respect to the classic methods of magnetic measurements and especially B-H loop tracing.

1. Introduction

B-H loop tracing is a key factor in magnetic measurements and characterization of magnetic materials. The commonly used experimental setup for such tracing, concerning soft magnetic materials, involves two pairs of coils, for excitation and detection [1]. The exciting and receiving coils in each pair must be identical to prevent external field influence as well as to successfully subtract the exciting field, such exact set-up being almost non-rebuildable for inter-comparison test purposes. Another problem is that the integrating fluxmeter could distort the results of measurements by introducing undesirable phase shift, which should be subject of compensation. However, tracing B-H loops at low magnetic field includes high inherent noise. Having the motivation of realizing such an arrangement, we developed a computational technique, according to which the signal received out of the simple receiving coil, is interpreted by filtering, subtracting, and integrating, by software techniques based on Fast Discrete Wavelet Transform and the Lifting Scheme.

2. Experimental configuration and procedure

The experimental set-up is illustrated in Figure 1. Two coils, one for producing magnetic field, called the exciting coil and a smaller measuring one, called the receiving coil, are used. Inside the receiving coil, the magnetic material to be measured is to be set. The coil system is electrically connected to a programmable waveform generator for producing the magnetic field, and a programmable digital oscilloscope for signals acquisition. All the instruments are driven by a PC through a controller and all are IEEE 488.2 (GPIB) and SCPI compliant.

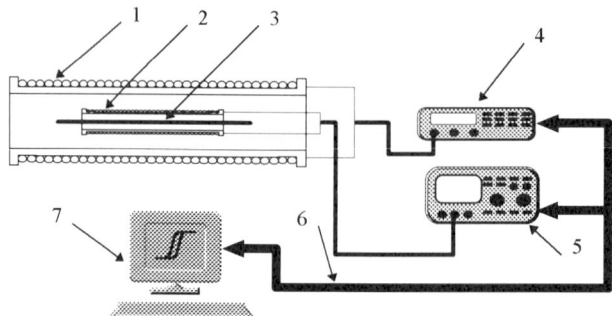

Figure 1. Experimental set-up: 1 - exciting coil; 2 - receiving coil; 3 - sample; 4 - function generator; 5 - digital oscilloscope; 6 - instrumentation buss; 7 - computer

Applying a sinusoidal waveform to the exciting coil, in the absence of the magnetic material, a phase shifted sinusoidal waveform of the input is to be detected at the receiving coil. In the presence of a magnetic material inside the receiving coil, a pulsed signal will be added to the output sinusoidal waveform, with the same frequency but shifted from the previous one. The output signal is to be acquired by means of an analog to digital converter (ADC), in our case the digital oscilloscope. The remaining B-H loop tracing procedure is to be performed by software. In order to obtain the B-H loop, the superimposed pulsed signal, is to be separated from the output waveform. Consecutively, the pulsed signal is to be integrated and composed with the input sinusoidal component. As the added pulsed signal has a high-frequency bandwidth, it could be eliminated from the output waveform by means of filtering, thus isolating the sinusoidal signal component. Then the pulsed signal component can be obtained by subtracting the sinusoidal one from the output waveform. Before the integration step the pulsed signal is to be filtered in order to eliminate all the noise that can affect this process.

It is expected that the pulsed signal to be integrated is a zero offset quasi-rectangular one. But in reality, because of the inherent noise and possible offset, it will result in a signal having ascending or descending tendency depending on the amount of the noise added in the integration process. So, before the integration of the pulsed signal, it is necessary to perform filtering process. The noise is a sum of high-frequency signals added to the measured signal, mainly caused by interference appearing during the acquisition process. Generally, to eliminate the noise from a signal, one can use a low-pass filter. Using digital signal processing (DSP) routines, software digital filters can be implement, so the filtering process could be performed by the computer software. But the process will distort the non-run integrated pulsed output by cutting all the components having the frequency higher than the filter limit, regardless of the amplitude. The main problem here is that the filter works in the frequency domain only, necessitating the use of an instrument for de-noising that could identify useful information and cut only the noise.

To solve this problem, we developed a new technique that involves the Wavelet Transform and it acts in the frequency as well as in the time domain, thus having the ability to cut the components with frequency that exceed a certain limit but with amplitudes below a threshold. The output sampled signal to be filtered, as above mentioned, is a sum of a sinusoidal and a pulsed signal. One can consider the pulsed signal as noise, so for both cases, we will use the same de-noising technique.

3. Discrete Wavelet Transform and the Lifting Scheme

Despite the fact that the basic principles of the wavelet theory were introduced by Gabor in 1945, the field of Discrete Wavelet Transform (DWT) is a quite recently developed one. DWT is an orthogonal function which can be applied to a finite group of data, very much like the Discrete Fourier Transform, both being convolutions. In a way, the DWT is another tool in the toolbox of digital signal processing. It is not our purpose to get deeply in this field, but to see how this tool will work in denoising the signal.

The used technique is called wavelet shrinkage and threshold method. When decomposing data using wavelets, one use filters that act as averaging and others that produce details. The small wavelet coefficients from the high-pass filter, corresponding to details, could be omitted without substantially affecting the data set. By the threshold method, one can set to zero all the coefficients that are less than a particular value. These coefficients are used in an inverse wavelet transform to reconstruct the signal. This kind of de-noising method is appropriate because it is carried out without smoothing the sharp structures of the signal.

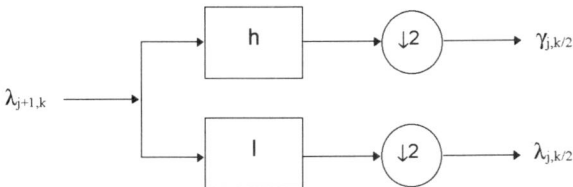

Figure 2. The block chart illustrating the "pyramid" algorithm for DWT.

The DWT "pyramid" algorithm is a computationally efficient method for implementing the wavelet transform using a two-band subband transform scheme (Figure 2). In each step j, the sampled signal α of length k is split in high-pass and low-pass band and then subsampled. The step is repeated recursively only for the low-pass band output. In our work, we need a direct and inverse DWT, first one to isolate and annul the high-frequency components below a threshold (the results of each high-pass band) and the second one to reconstruct the signal. Despite the easy implementation of the algorithm, the inverse DWT is not connected to the direct DWT. For this procedure we also need extra computational space for intermediate data. De-noising the signal we use another method based on wavelets, called Lifting Scheme. The lifting scheme is a new method for constructing bi-orthogonal wavelets, which is not relied on Fourier Transform. Its algorithm have some advantages as compared to the classical DWT one, including the in-place calculation with no extra space required, and faster implementation using the similarities between the high-pass and the low-pass filters for quick computation. In this way the number of floating operations per second is reduced by a factor of two. The inverse transform is immediately found by undoing the direct one, in other words, by changing each "+" with "-" and vice versa.

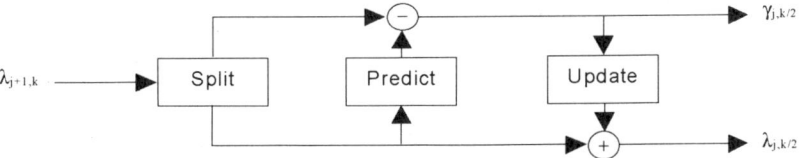

Figure 3. The block chart illustrating the lifting algorithm.

The lifting scheme algorithm consists in three simple phases as illustrated in Figure 3. Splitting the sampled data into the odd and the even subsets, Predicting, i.e. computing the wavelet coefficients as a failure to predict the odd set based on the even one, and Updating the even set using the previously computed coefficients to calculate the scaling function ones. If the sampled signal is l of length k then the code algorithm is:

$$\gamma_{\varphi,\kappa} := \lambda_{\varphi+1,2\kappa+1} - \Pi(\lambda_{\varphi+1,2\kappa})$$
$$\lambda_{\varphi,\kappa} := \lambda_{\varphi,\kappa} + Y(\gamma_{\varphi,\kappa}) \qquad (1)$$

where Π and Y are the Predict and Update operators respectively. The results in the Predict phase are computed using the polynomial interpolation, thus indicating the failure of the data to be smooth, in other words, the high-frequency components. So, for de-noising the sampled signal we must perform the lifting scheme algorithm, annul the data that are smaller than a threshold resulted in every Predict phase, and then reconstruct the signal by undoing the lifting scheme. In our particular case it is sufficient to use the Predict phase as a failure to be linear, by means of the linear interpolation, and the Update phase as a preservation of the mean. Substituting the Predict and the Update operators, the resulting code in this case is:

$$\gamma_{\varphi,\kappa} := \lambda_{\varphi+1,2\kappa+1} - (\lambda_{\varphi+1,\kappa} + \lambda_{\varphi+1,2\kappa+1}) / 2$$
$$\lambda_{\varphi,\kappa} := \lambda_{\varphi,\kappa} + (\gamma_{\varphi,\kappa-1} + \gamma_{\varphi,\kappa}) / 4 \qquad (2)$$

The policy to be used in the threshold method could be a static one as well as a time-adaptive one related on the level detail of the algorithm. Because the noisy signal has also some large spikes that cannot be attributed to noise, the high energy data propagate locally across the coefficients to the deep level of detail, thus dominating the disturbance caused by the noise. The thresholding method proposed here will not affect the detail coefficients in the deeper levels and will conserve the high-frequency information.

For comparison we used a classical system made by two pairs of exciting and receiving coils. The coils of each pair are electrically connected in serial opposition mode to prevent external field influence as well as to successfully subtract the exciting field. We use the same programmable instruments and an additionally integrating fluxmeter connected between the output of receiving coils and the digital oscilloscope. Both systems were calibrated using pure iron and nickel samples. Uncertainty of measurements indicate a noise level of at least 2% and 0.1% concerning the cpnventional and wavelet set-up.

It is worth extending these software capabilities, by adding new functions to perform different magnetic measurements. An important application is the introduction of such technique in signal conditioning of the output of sensors based on magnetic materials. More especially, considering the sensors based on the magnetostrictive delay line technique, one can see that despite their good repeatability and absence of hysteresis, a main problem is the poor sensitivity level due to the arbitrary magnetic domain noise. We have introduced DWT for such application. The performed signal conditioning indicated a reduction of the noise level to sub-μV, resulting in a significant and competitive improvement of the sensitivity.

References

[1]   Kulik T et al 1993 *J. Appl. Phys.* **73** 6855-6857

Real time monitoring multisensor system using shifting temporal window technique

Dominique REBIÈRE, Christian CAZAUBON*, Hervé LÉVI, Christophe BORDIEU, Jacques PISTRÉ and Roger PLANADE**

IXL - Université Bordeaux I (UMR CNRS 5818)
351, Cours de la Libération - F-33405 TALENCE Cedex, France
Phone : (33) 05 56 84 65 40 - Fax : (33) 05 56 37 15 45
email : rebiere@ixl.u-bordeaux.fr

* CRED - Université Bordeaux I
rue Naudet Gradignan - F-33405 TALENCE Cedex, France
Phone : (33) 05 56 84 57 58 - Fax : (33) 05 56 84 57 83
email : cazaubon@elec.iuta.u-bordeaux.fr

** CEB, Centre dEtudes du Bouchet, DGA-DCE
Le Bouchet BP n° 3, F-91710 Vert Le Petit, France

Abstract. An autonomous and portable multisensor system has been constructed, which comprises four SAW devices, four semiconductor metal oxide sensors, and a microcontroler in order to realize the sensor array data treatment using a back-propagation neural network. These processing modules were based on a shifting temporal window technique, and thus permit to implement a dynamic detection unit. Using the RS232 port, the system can be recalibrated and thus the network topology and the learning parameters can be adapted. To validate this approach, GB gas SAW responses have been applied and permit to demonstrate the performances of the system.

1. Introduction

To fulfil the increasing requirements of chemical analysis of gases, chemical microsensor systems are needed [1]. In some applications, such as military or safety, the gases involved are extremely dangerous. The detection speed becomes an essential parameter. So, to allow a real time and selective detection of gaseous analytes in air, we present a versatile microsensor system. This system is based on Surface Acoustic Wave (SAW) and semiconductor metal oxide sensors. A data acquisition module consisting of a microcontroler PCB with four SAW frequency counting channels and with four semiconductor voltage measurement channels performs the signal handling (figure 1). The signal patterns, obtained at different temperatures, are presented to a processing module which allows to detect the gas presence.These processing algorithms are implemented using neural network and a shifting temporal window technique. The network outputs are compared to a threshold to decide whether or not to turn an alarm on. This portable system is completely autonomous but a serial transmission line permits to send the digital sensor data to a personal computer. In this paper, the implementation of the neural network based detector and the shifting temporal window technique are described. A presentation of the portable microsensor system based on a microcontroler (Motorola MC68HC11F1) is given and discussed and some conclusions are drawn from them.

876

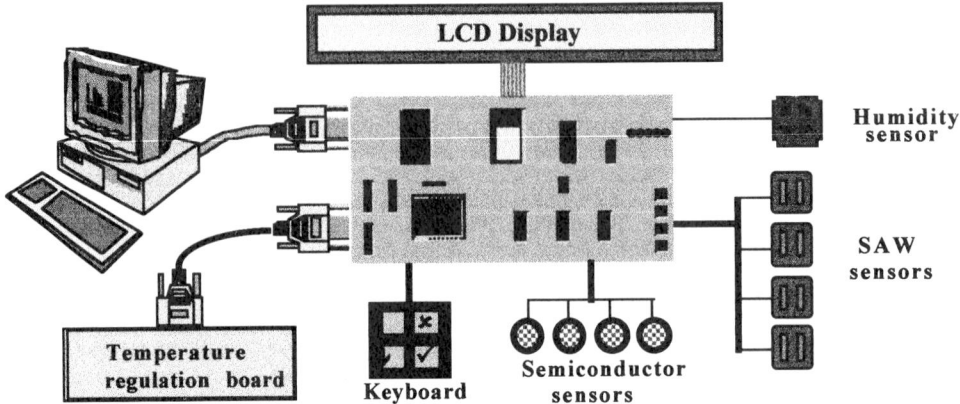

Fig. 1. Microsensor system based on an array of Surface
Acoustic Wave (SAW) and Semiconductor sensors

2. Implementation

After signal conditionning, a processing module allows to detect the gas presence. This unit
uses a shifting temporal window technique [2]. The temporal windows constitutes the inputs of
the gas detector which is implemented with a back-propagation neural-network. Its output is
compared to a threshold to turn on an alarm.

2.1 Shifting temporal window technique

This technique is well-suited to the real time processing of sensor responses. Let (s_i), with i in
$[1, s]$, a set of sampled data. A temporal window can be defined by a continuous sequence of n
samples. These windows are numbered (w_j) with j in $[1, w]$. At time $s+1$, one new sample is
acquired. The window w_{w+1} is then obtained by shifting forwards the window w_w. So, two
consecutive windows overlap each other having in common $n-1$ samples and differing only by
two samples : the first sample of the window w_w and the nth sample of the window w_{w+1}. The
figure 2 shows typical windows of a SAW sensor : one with no gas (window w_{off}) and another
with pollutant gas step (window w_{on}).

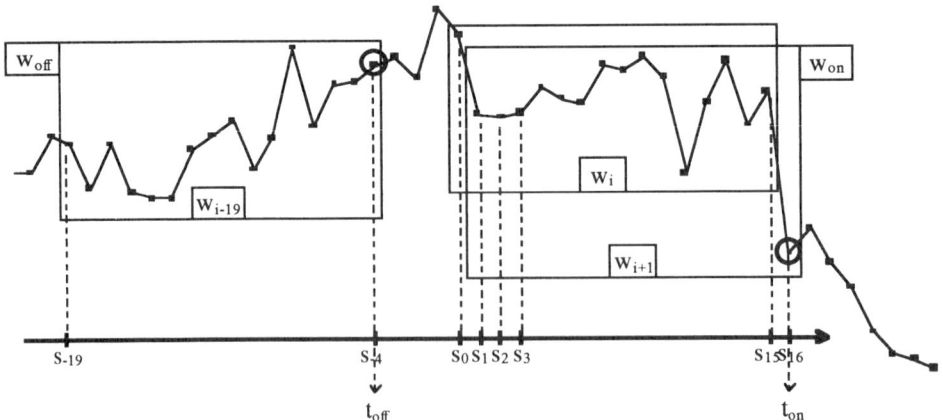

Fig. 2. The shifting temporal window system and the response reference samples.

2.2 Neural network based detector

Artificial neural networks (ANNs) are constituted of processing units linked together by a network of connections. A neuron consists of an input vector associated with a vector of adaptable weights and an activation function which calculates the neuron output [1,2]. The neurons can be interconnected in a multilayer architecture. In this case, an input layer, one hidden layer and an output neuron define the network.

Using a learning algorithm, ANNs adapt their weights to the data which are presented to their inputs so that the outputs are pertinent from the point of view of the developed application. The general form of the learning rule is :

$$W_{ij}(t+1) = f_{ij}[a_i(t+1), a_j(t+1), W_{ij}(t)]$$
$$\Delta W_{ij}(t+1) = W_{ij}(t+1) - W_{ij}(t)$$

with W_{ij} the weight between neurons i and j, a_i and a_j the activation states of respectively neuron i and j, and f_{ij}, the local procedure which updates the weight. t and $t+1$ indicate that all values are computed after the presentation of the tth and the $(t+1)^{th}$ learning examples respectively.

The temporal windows constitute the inputs of our neural network based detector. It was designed with a backpropagation neural network. This multilayer ANN is hetero-associative and employs an error backpropagation algorithm as supervised learning rule which is the Back-propMomentum algorithm implemented in the Stuttgart Neural Network Simulator (SNNS) [3]. The algorithm minimizes the square of the difference between the calculated and desired outputs by the gradient descent optimization method. Weights are updated from the output layer to the input layer, hidden layer by hidden layer, back propagating the output error.

Before feeding the network, the temporal windows are normalized between 0 and 1. Thus, the windows are expected to be independent of gas concentration. Since, whatever the concentration considered, the response shape is the same apart from a scaling factor, we have only selected SAW responses to GB gas steps (0.1, 0.2, 0.5, 1, 5 and 10 ppm). For each of the 12 recorded responses of our data set, two references samples are chosen. t_{off} allows us to create 10 windows with alarm off and t_{on} 10 windows with alarm on. So, 240 examples are available to construct the data sets. The learning data set is made of 112 examples and 128 examples constitute the test data set.

Once the network has converged, the learning phase is ended. Then it can be used in the real time mode.

Our aim is to realize a smart sensor. So, this is important to optimize the detector neural network. The influences of temporal window width, number of hidden neurons, weight initialisation, and learning parameters have been studied [2]. The detector performances in terms of convergence average speed (CS) and network generalisation properties (GP) were evaluated.

2.3 Experimental validation

To validate this new dynamic approach, SAW gas sensors have been chosen to detect GB gas using the previous results [4]. For our application, we use temporal windows with 75 samples (window size) for each surface acoustic wave sensor. We feed the detector simultaneously with 3 windows of 3 different SAW sensors. So, we use a ANN with 225 neurons for the input layer (3x75), 5 neurons for the hidden layer and one output neuron. This output is compared to a threshold to decide to turn on an alarm.

3. System

The network program, developed in C language, is implemented on the single PCB (figure 1) using the microcontroler M68HC11F1 working in extended mode. The neural-network topology is stored on a EPROM. RAM is automatically initialized with weights and threshold by the microcontroler reset, but the use of RAM allows in situ sensor recalibration. Thus, the system can be adapted and recalibrated to an other specific application via an external personal computer, using RS232 port.

3.1 Frequency measurement

An oscillator circuit is formed using a SAW sensor ; its typical resonance frequency is 100 MHz. A presence of gas induces a frequency shift of about 1 Hz up to 100 Hz and more. A variety of methods may be used to measure the oscillator frequency, in our case we build a digital frequency counter. Taking into account the response time of the sensor, we selected an integration time of 1 s, thus it is necessary to realize an effective 27-bit binary counter.

The frequency counter is built around a 16-bit binary counter (Lattice Pal ispLSI1016) and a microcontroler (Motorola MC68HC11F1). The outputs of the 16-bit binary counter are multiplexed and connected to the 8-bit data bus ; the carry is connected to an Input Capture (ICx) of the microcontroler. Two 8-bit registers are used to count the number of events on the Input Capture. At the end of the counting operation, the final result is available on a long word which is then processed in a C program. A signal of 1 s duration is available at the Output Compare (OCx) of the microcontroler, and drives the Enable (En) input of the 16-bit binary counter. This signal is obtained by counting and using an internal clock of 4 MHz. An external 16 MHz oven oscillator, with a short stability of $5x10^{-11}$/s, is used to insure the accuracy.

Four Input Capture and Output Compare are effectively available and offer the possibility to control simultaneously four SAW sensors.

3.2 Temperature control

A digital temperature control of the SAW sensor is provided by : a heating resistor, a T type thermocouple and a PID algorithm implemented in a microcontroler MC68HC11E9. The output signal of the thermocouple is amplified and it is connected to the 8-bit analog-to-digital microcontroler converter. An external 8-bit digital-to-analog converter is used to drive the heating resistor. Finally, the temperature accuracy is about 0.2°C in the range 20°C to 120°C.

4. Conclusion

We have presented a microsensor system for chemical analysis. This portable and versatile sytem is based on a new dynamic detection technique. Indeed, the shifting temporal window approach has been implemented on a microcontroler PCB and has showed its functionality. Applied to GB gas SAW sensor responses, the system has demonstrated these potentialities and the efficiency of the neural network based detector. One of most important advantages is the possibility to recalibrate the system, using the RS232 port, via a personal computer, and to adapt the network topology and the learning parameters to a specific application.

Acknowledgements

This work was supported by the French Ministry of Defense through a specific program from the Centre d'Etudes du Bouchet (DCE-CEB).

References

[1] W. Göpel , New materials and transducers for chemical sensors, Sensors and Actuators B, 1994, 18-19,1-21.

[2] Ch. Bordieu, D. Rebière, Jacques Pistré and R. Planade, Temporal window system : a new approach for dynamic detection, Sensors and Actuators B35-36 (1996) 52-59.

[3] SNNS, Stuttgart Neural Network Simulator, User Manual, Version 4.1, University of Stuttgart, Institute for parallel and Distributed High Performance System (IPVR), report No. 06/95.

[4] D. Rebière, C. Déjous, J. Pistré and R. Planade, SAW organophosphorus sensors coated with fluropolyol isomers : Role of humidity and temperature on the sensitivity, Transducers'97, june 16-19, 1997, Chicago USA.

Signals and Circuits
Paper presented at Eurosensors XII, 13–16 September 1998

HARMONIC MODULATION OF IMPURITY IN GAS FLOW AS A METHOD TO

STUDY THE SEMICONDUCTOR SENSORS

D.Yu.Godovsky, J.H.Kim*, S.S.Yakimov

RRC"Kurchatov Institute", 123182, Moscow, RUSSIA

*KIER, Taedok Science Town, Taejon, Republic of Korea

Abstract. In the given paper a method of harmonic modulation of impurity concentration in a gas flow as a method to characterize semiconductor sensors is presented. The method similar to any spectroscopic technique and consists in the analysis of the sensor conductivity response on the harmonic modulation of impurity, introduced into the gas flow.

1.Experimental

Propane in air with average value of 1000 ppm and various amplitudes and frequencies of harmonic wave is described here. Harmonic dependence of an active impurity in a gas flow was obtained by means of mixture of two flows, programmed by MFC RRG-9 with time of the positioning of the value less than 0.2 sec. The concentration of an impurity was set as $F1/ (F3 + F1)$, where the flow of an impurity $F1 = Co + Asin (\omega T + \phi)$, $F3 = N1 - Asin (\omega T + \phi)$ - massflow of air for dilution and $F1 + F3 = const$.

The tin-dioxide thick film semiconductor sensor was taken as an object of investigation, which was obtained as a result of joint project between KIER and

Kurchatov Institute and is described elsewhere [1].

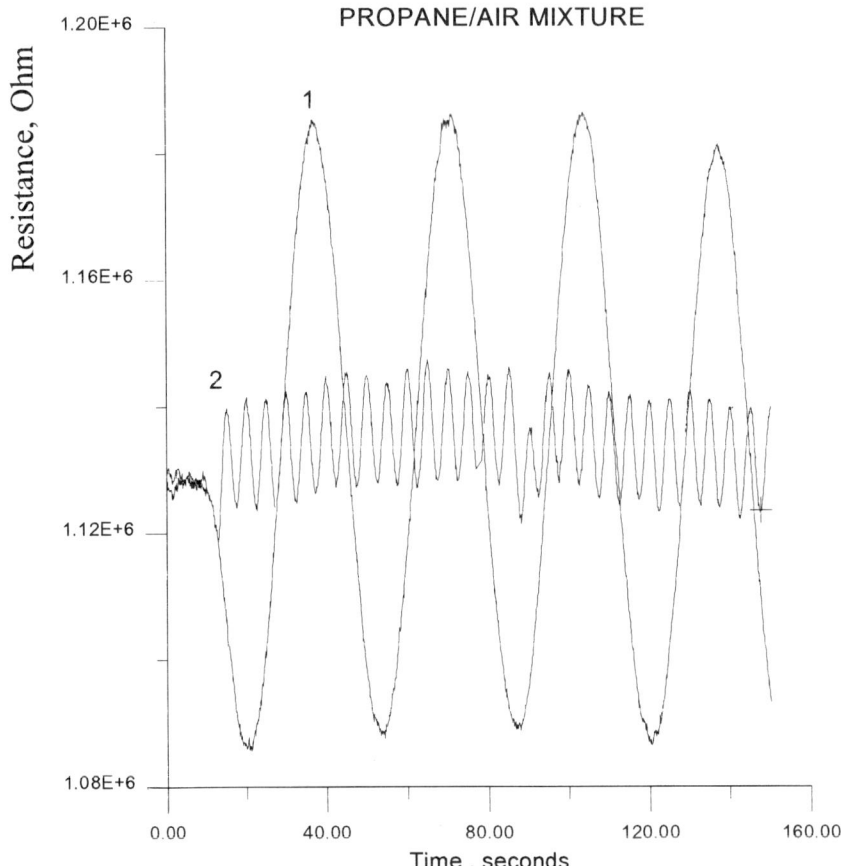

Fig. 1 Dependence of the sensor response to propane at average concentration 1000 ppm, amplitude - 100 ppm and frequency 1 - 0.03 Hz,,2 - 0.2 Hz

The Fig.1 demonstrates real experimental data of the sensor response for the harmonic modulation of propane concentration in a flow. The ranges of amplitude and frequency were investigated within which the response of the sensor exhibits liner character on the driving wave, and all further measurements were carried out

within the mentioned limits.

Both phaseshift and amplitude dependencies of the response conductivity wave on frequency of the driving harmonic modulation were investigated,
and the most interesting were the amplitude dependencies on frequency of modulating wave

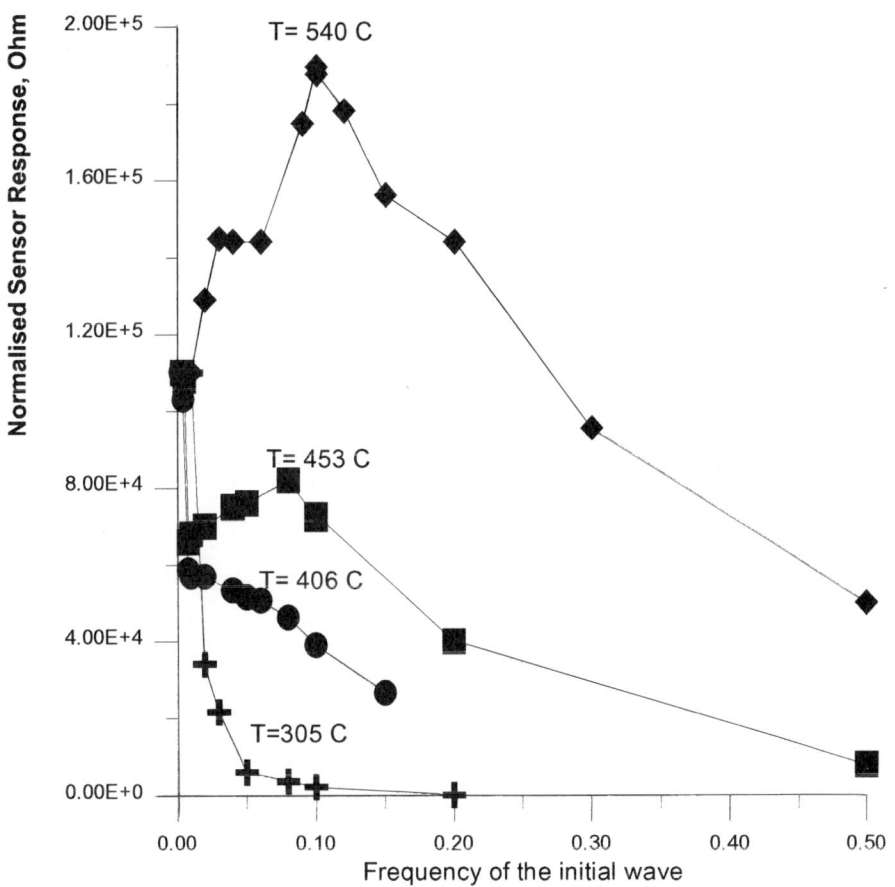

Fig.2 The dependence of the amplitude of sensor response on frequency of driving wave at bias propane concentration 1000 ppm and amplitude - 100 ppm.

at fixed amlitude and bias of the driving wave and for different sensor layer temperatures (Fig.2).

2.Results and discussion

It can be seen that the dependencies have the tendency to change from exponential ones to apex-like, the later fact deserved, from our point both attention and interpretation. The phase dependences in the apex area have also abnormal character .The possible explanation of the apex-like amplitude behavior for elevated temperatures region lays, from our point of view in the simultaneous processes of electron exchange between the surface acceptor states as O^- and vacancies in oxygen sublattice of tin dioxide $Vo^{\cdot}$, the concentration of which is becoming high at 450 - 550°C. The quantitative analysis of the obtained data and numeric modelling is now in a progress.

3.Conclusions

The results described above show, that the harmonic modulation technique certainly gives some new information, not accessible when one measures just the response to step-like concentration pulse, and also points out to the complex kinetics of tin dioxide sensor response formation. The method described could also be used to recognition of the unknown gas mixtures, since it seems that the amplitude and phase dependencies on frequency of driving wave are unique for each gas species.

References

1. J.H.Kim, D.Y.Godovsky, A.V.Pislyakov et al, 1996 "Propane/Butane semiconductor gas sensor with low power consumption", Sensors and Actuators B, Spec. issue devoted to Eurosensors X.

Signals and Circuits
Paper presented at Eurosensors XII, 13–16 September 1998
ⓒ *1998 IOP Publishing Ltd*

New Processing Methods for Microcontrollers Compatible Sensors with Frequency Output

Nikolay V. Kirianaki, Sergey Y. Yurish, Nestor O. Shpak

290013, Lviv, Ukraine, Institute of Computer Technologies, Bandera str.,12.
Tel./fax: (+38 0322) 97 16 41, e-mail: syurish@polynet.lviv.ua

Abstract

Two novels adaptive processing methods for microcontrollers compatible sensors with frequency output are proposed. Both of them guarantee the constant quantization error in all specified frequency range. The first method with non-redundant reference time interval provides the high frequency-to-code conversion speed, the second method with non-redundant reference frequency provides the low power consumption and dissipation in the sensor's elements. The methods are program-oriented and realised in the functional-logical architecture of the computing power (microprocessor or microcontroller) on the virtual level without additional hardware.

1. Introduction

Very fast advances in integrated circuit technologies brought new challenges in the micro-system design. The state-of-art of smart sensors has shown the more and more wide distribution of integrated sensors with the frequency output for various physical values. It is for example, a new monolithic temperature sensors [1, 2], intelligent opto sensors [3], integrated micro-sensor for position and speed [4], acoustic wave gas sensors [5], sensors based on monocrystal string tensoconvertors [6], sensor of rotation parameters [7] and many others. The generalised frequency range of such sensors is very wide (from 0.01 Hz up to some MHz). Some of them have high accuracy (up to 0.01 %). The physical value X in such sensors is transformed into the sequence of rectangular pulses, the frequency or duty-cycle of which is directly proportional to the X.

The essential advantages of frequency as informative output parameter of sensor are:
• the opportunity more precisely to be measured and integrated in comparison with simple means;
• the absence of distortions at its switching;
• the high noise immunity;
• the ease of scale change and reception of any code, required for input of result of measurement in computer and transmission it in the communication channels.

For the further evaluation of the information that is carried by the sensor's frequency output it is necessary to use the method of frequency-to-code conversion. In spite of the fact, that the frequency can be measured most precisely in comparison with the other informative parameters of signal, in practice, it is by no means the triviality task *"a simple time-window counting"*. The use of traditional methods of frequency-to-code conversion (the standard counting method, described, for example, in [8]) can result in essential quantization errors, comparable with the accuracy of the sensors itself because of increase of quantization error Δ_q by hyperbolic dependency at reduction of the frequency f_x. On the other hand, the modern methods of conversion on the basis of ratiometric counting processing technique [8] (sometimes called reciprocal counting) despite on the constancy of a quantization error in the whole frequency range, have the redundancy time interval. In turn, it reduces the processing speed. Beside that it is necessary to take into account the specific features of the sensors (for example, the strong jitters in the temperature sensors [1] due to the thermal noise). Hence, the development of novel uniform accurate methods for frequency-time conversion, taking into account the specific features of integrated frequency-output sensors is the urgent problem.

2. Method with Non-Redundant Time Interval

The novelty of the fist proposed method consists in the following:
1) It has constant relative quantization error in the all specified frequency range and, in the same time, non-redundant reference time interval, which is specified by the beforehand given error of conversion;
2) The possibility to transform the frequency f_x, superior the reference frequency f_0;
3) The method is program-oriented, e.g. realised in the functional-logical architecture of computing power (microprocessor or microcontroller) on the virtual level.
The timetable of the proposed method is shown in Figure 1. The method consists in the following. With arrival of the impulse of signals with the lower frequency f (this corresponds to the moment t_1 in Figure 1), the counters start to calculate the impulses of the both signals. The number of impulses N_i, stored in the counter, which calculates the impulse with the frequency F, is compared with the N_{put}. This number is set previously in the counter by microcontroller. At some instant of time (this corresponds to the moment t_2 in the Figure 1), when the number of impulses, calculated by this counter, will be N_{put}, e. g. $N_i = N_{put}$. With the arrival of next impulse (after the moment t_2) with the lower frequency f (moment t_3), the impulse count will be stopped. The number of impulses (of signal of lower frequency) counted by one of the counters is n, and the number of impulses (of signal of higher frequency), counted by the other one, is N $=N_{put} + \Delta N$. The conversion time t_x equals always to an integer number of periods of a signal with the lower frequency f:

$$t_x = \tau \cdot n = \frac{n}{f} \tag{1}$$

This interval can be given as:

$$t_x = T \cdot N = \frac{N}{F} = \frac{N_{put} + \Delta N}{F} \tag{2}$$

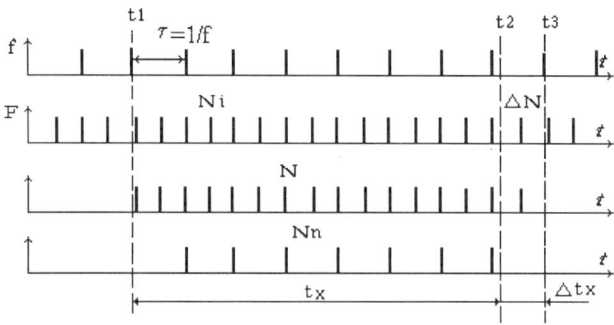

Figure 1. Timetable of the method with non-redundant time interval

From the equations (1) and (2) follow, that $f = F(n/N)$ or $F = f(N/n)$. If the measured frequency f_x is lower frequency, e.g. $f_x = f$, and reference frequency f_0 is high ($f_0 = F$),

$$f_x = f_0 \frac{n}{N} \quad \text{or} \quad f_x = f_0 \frac{N}{n}, \tag{3}$$

when $f_x = F$ and $f_0 = f$.

The microprocessor calculates the unknown frequency from the formulas (3) or (4). The program for calculation is determined by a command, which to be prepared on the basis of earlier entered the microcontroller information which frequency is lower. For the period τ or T the conversion is carried out similarly. The microcontroller calculates it values from:

$$\tau = \frac{N}{f_0 \cdot n} \quad \text{and} \quad T = \frac{n}{f_0 \cdot N} \tag{5}$$

We shall determine the quantization error for the considered conversion. This error is caused by that, the interval t_x (time of measurement) is not equal to an interval, which is determined by the integer number N of periods of a signal with high frequency, e.g.

$$t_x \neq N \cdot T \neq N/F. \tag{6}$$

At change of lower frequency in the known limits, the interval t_x will be changed (or, on the contrary, will be changed the N/F at change of higher frequency). It will result in change of the number of impulses N from $\Delta N = 0$ up to $\Delta N = \Delta N_{max}$ calculated by the counter. Here ΔN_{max} is the number of impulses in the interval $\Delta t_{xmax} = \tau$. Taking into account the fact, that the period of these pulses is equals T, we shall receive

$$\Delta N_{max} = \frac{\tau}{T} = \frac{F}{f}. \tag{7}$$

The maximum quantization error arises in that case, when the number of impulses N counted by the counter is minimum and equal to the N_{put} (one of frequencies is changed). Then

$$\delta_{max} = \frac{1}{N_{min}} = \frac{1}{N_{put}}. \tag{8}$$

Hence, the maximum error is determined by the value N_{put} only and practically does not depend on the measured frequency. The minimum value of an error will be at $N = N_{max}$. But as $N_{max} = N_{put} + \Delta N_{max}$, then

$$\delta_{min} = \frac{1}{N_{max}} = \frac{1}{N + \Delta N_{max}}. \tag{9}$$

3. Method with Non-Redundant Reference Frequency

The creation of modern integrated sensors brings to the problem of power consumption and dissipation in the sensor's elements becomes rather actual.

The dynamic average power consumption of a digital circuit can be given as

$$P_a = \sum \frac{a_i}{2} \cdot C_i \cdot V_{DD}^2 \cdot f_{clc}, \tag{10}$$

where V_{DD} is the supply voltage; f_{clc} is the clock frequency; C_i, a_i are the total output capacitance and transition activities associated with logical node i accordingly. The V_{DD} and C_i are constant for the specific integrated circuit and technology. The reduction of the f_{clc} is in contradiction with the necessary to increase the metrological performances in precise measurements. With the aim to eliminate this inconsistency, the second adaptive method for frequency-time conversion with the non-redundant reference frequency f_o is proposed. The essence of the offered method consists in the following. The embedded microcontroller or arithmetic unit calculates the value of necessary reference frequency according to the given quantization error. The reference frequency, which is received by division/multiplication of the clock frequency f_o and the measurand frequency of a quartz generator is calculated by the counters n and N accordingly. The time of a measurement is equal to the integer number of periods of the f_x. This frequency is calculated similarly, as for the first method mention above. The quantization error also does not exceed the given.

Similarly to the first method, the second method ensures a constancy of the quantization error in the whole specified frequency range from infralow up to high frequencies. Besides that, the reference frequency f_{oi} is non-redundant and determined by the given error of conversion. Thus, at the usage of the offered method of measurement in integrated sensors, it will be possible to have receive the next benefits: a high accuracy of conversion, constancy of the quantization error and reduction of the power dissipation during the conversion by use of lower (on the average by 1-2 order) reference frequency.

4. Conclusions

The main advantages and features of the proposed processing methods for integrated frequency-output sensors are the following:
- the methods have universality. They provides the precise frequency-to-code conversion for the all known sensors with frequency output in the wide specified frequency range with constant quantization error;
- the methods provides really digital sensor's output;
- the fabrication of such sensors with the built-in frequency-to-code converter does not require extra technological steps and can be fabricated by the usual CMOS digital processes;
- realization of the sensors together with the novel program-oriented methods implemented in the same chip can create the basis for new smart sensors for systems with distributed intelligence.

The first proposed method was used in the intelligent sensor for rotation speed [9]. The experimental results have shown the high accuracy of the method. The interval of estimation for the processing error is $\delta_{sum} = (-1.63 \div 2.3) \cdot 10^{-5}$ with the probability P = 0.98.

References

[1] V. Szekely, M. Rencz. A new monolithic temperature sensor: the thermal-feedback oscillator, In Proceedings of TRANSDUCERS'95 & EUROSENSORS IX, pp.124-127.
[2] W. Woiciak, A. Napieralski. An Analogue Temperature Sensor Integrated in the CMOS Technology, In Proceedings of THERMINIC'95, pp. 32-37.
[3] Intelligent Opto Sensor. Data Book. Texas Instruments, 1996.
[4] Inductive Micro-Sensor for Position and Speed MS1020/21. Preliminary Technical Data 0200.10.97. CSEM Center Suisse d'Electronique et de Microtechnique, Switzerland.
[5] Paul R. van der Meer, Gerard C.M. Meijer, Michiel J. Vellekoop, Harry M.M. Kerkvliet, Ton J.J. van den Boom. A Low-Cost Temperature-Control System for Surface Acoustic Wave Gas Sensors Using Smart Temperature Sensors. In Proceedings of THERMINIC'97, pp.229-234.
[6] R. Baitsar, S. Varshava, I. Ostrovskii, V. Rak. Application of Monocrystal String Tensoconvertors for Various Physical Parameters Control, In Proceedings of PDS'98, pp. 25-27.
[7] V.P. Deynega, N.V. Kirianaki, S.Y. Yurish. Microcontrollers Compatible Smart Sensor of Rotation Parameters with Frequency Output, In Proceedings of ESSCIRC'95, pp. 346-349.
[8] Burr-Brown IC Applications Handbook, 1994.
[9] V.P. Deynega, N.V. Kirianaki, S.Y. Yurish. Intelligent Sensor Microsystem with Microcontroller Core for Rotating Speed Measurements. In Proceedings of, EMAC'97, pp.112-115.

Signals and Circuits
Paper presented at Eurosensors XII, 13–16 September 1998
© 1998 IOP Publishing Ltd

A 1.4 V Oscillator For CMOS Capacitive Sensor Interfaces

Giuseppe Ferri, Pierpaolo De Laurentiis

Universita' di L'Aquila, Facolta' di Ingegneria, Dipartimento Ing. Elettrica
Localita' Monteluco Roio, 67040 Poggio di Roio, L'Aquila, Italia
Phone: +39 862 434446, Fax: + 39 862 434403, E-mail: ferri@ing.univaq.it

Salvatore Pennisi

Universita' di Catania, Dipartimento Elettrico, Elettronico e Sistemistico
Viale Andrea Doria 6, 95125 Catania, Italia
Phone : + 39 95 339535 , Fax + 39 95 330793, E-mail : spennisi@dees.unict.it

Abstract. A novel CMOS oscillator to be used in integrated capacitive sensor interfaces is
here presented. The proposed topology is based on a hysteresis comparator adapted for low-
voltage single supplies, the minimum being 1.2 V. A 1.4-V design with an overall power
consumption of only 23 μW is given. A capacitive range of 10 fF - 100 pF, to which
corresponds a frequency span of about 300 Hz - 3 MHz, has been considered. The circuit shows
low sensitivity to supply discharges and good output linearity for temperature variations. These
characteristics have been verified to be better than the corresponding performance of a typical
solution presented in literature.

1. Introduction

Research in analog integrated circuits has pushed towards the direction of low-voltage low-
power structures working in battery-operated portable systems and, in particular, in sensors
and microsystems, where the sensing elements and the processing circuitry have to be
integrated on the same chip to realize "smart sensors" [1-5]. In this context, CMOS has been
proved to be the main sensor technology [6-9] and for sensor interface electronics [10-13].

Moreover, capacitive sensors are suitable transducers exhibiting reduced current
consumption. In fact, they have a high impedance up to reasonably high frequencies and high
signal levels. They are often interfaced with a read-out circuitry realizing a capacitance-to-
frequency conversion, such as oscillators and phase shifters, which provide an output
proportional to the frequency [14-15]. The main characteristics have to be the following : high
dynamic range, good linearity and precision, low input noise and offset, long-term and
temperature stability, reduced area, low effect of parasitic capacitances, calibration and
compensation of the transducer characteristics. These constraints can be satisfied by interface
circuits which, if designed with low-voltage low-power techniques, can be utilized in portable,
remote and wireless systems for industrial, medical, automotive and consumer applications.

2. The novel low-voltage oscillator

Fig.1 shows the principle of the proposed oscillator. The sensor capacitance C is charged and
discharged with the constant currents I. The output voltage V_{out} drives the switch. The
hysteresis comparator has been realized with two traditional comparators having different

threshold voltages V_{tr+} and V_{tr-}. The aim of the comparator is to convert the triangular voltage V_C on the capacitor in a squared-wave, whose duty cycle is set by the choice of the current sources values. The output frequency is the following:

$$f = \frac{I}{2\ (Vtr + -Vtr-)} \frac{1}{C}$$

The outputs of each comparator are sent to a two-input multiplexer, with the output fed back to the strobe terminal. By doing so, it can be easily shown that for V_C decreasing, the output of the multiplexer is determined by comp1, whereas for V_C increasing, it is determined by comp2.

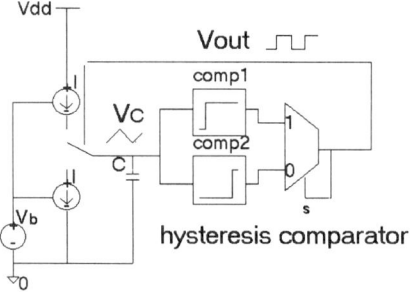

Fig.1- Oscillator block scheme

Fig.2 shows the basic topology of the multiplexer, formed by a combination of inverters and transmission gates. If s=1 (and -s=0), transistors of the transmission gate driven by comp1 are ON while those related to comp2 are in high impedance condition, so that $V_{out}=V_{comp1}$.

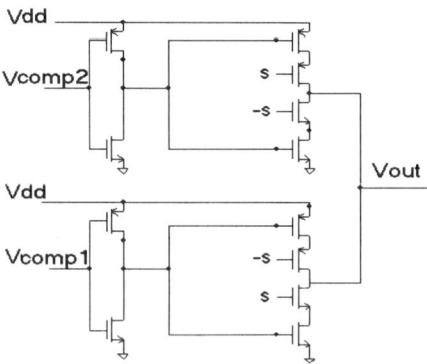

Fig.2 - Basic multiplexer topology

The simplified schematic of the oscillator is in fig.3. The comparators are formed by simple differential structures.

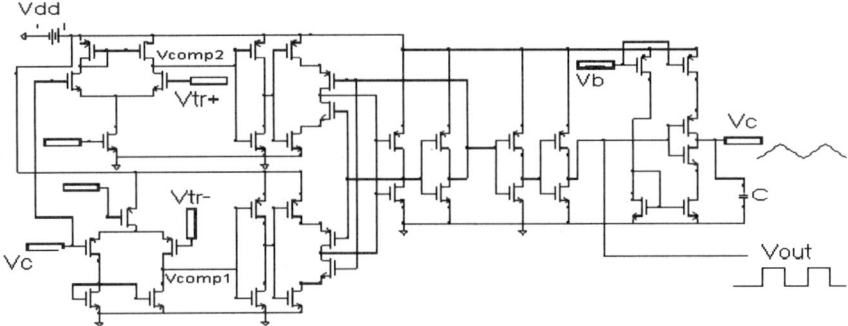

Fig.3 - Oscillator schematic

3. Simulation results and comparison with literature

Fig. 4 shows the output frequency vs. sensor capacitance for a temperature variation in the range (0-70) °C while in fig.5 the same curve is reported at 1.4 V and 1.2 V supply voltage, respectively. From these figures, it is evident a high insensibility to temperature variations and supply discharges. The circuit performance have been compared with a traditional solution, presented in literature [16] and shown in fig.6.

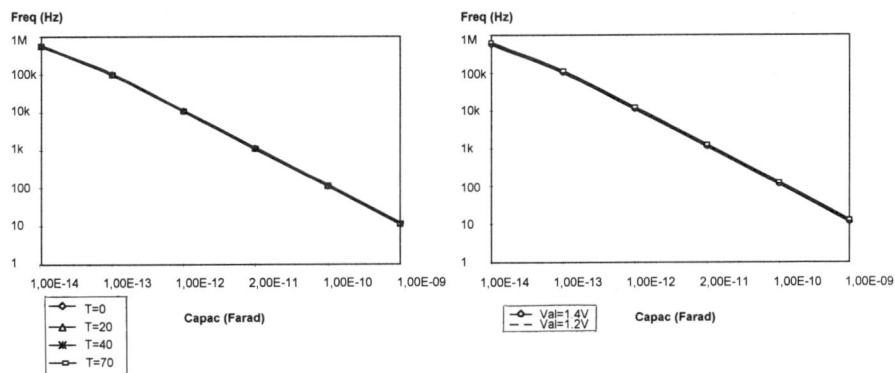

Fig.4 - Frequency vs.capacitance
(variation with temperature)

Fig.5- Frequency vs. capacitance
(variation with supply discharge)

The circuit resolution Δf/f is 0.05 % in our topology, while the same performance parameter for the circuit in fig.6 is 0.5 %. This last topology has also a higher power consumption (70 μW, if the same current is generated) and has a higher dependence on the supply voltage variations and especially on the temperature. In particular, the heavy dependence on temperature prevents the fig.6 topology to be used in sensor applications.

890

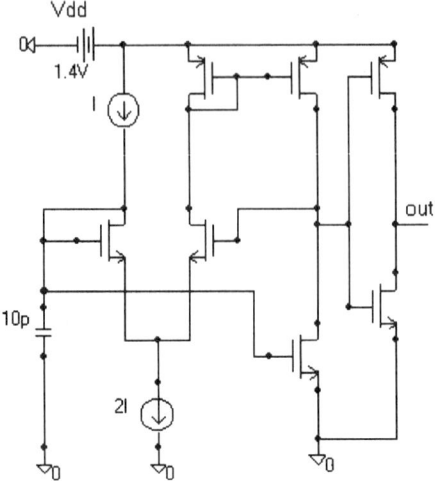

Fig.6 - A typical oscillator presented in literature

Acknowledgement

The authors want to thank Prof. Arnaldo D'Amico, Universita' Tor Vergata, Roma, for his invaluable advice.

References

[1] Smith T, Bardyn J, DeGeeter B and Nys O 1997 *Proc. Advances on Analog Circuit Design*
[2] Wouters P, De Cooman M and Puers R 1994 *IEEE Trans. on. Solid State Circuits* **29-8** 952-956
[3] Baltes H, Haberli, A Malcovati P and Maloberti F 1996 *ISCAS*
[4] Huijsing H 1992 *Sensors and Actuators* **A 30** 167-174
[5] Sansen W, De Wachter D, Callewart L, Lambrechts M and Claes A 1990 *Sensors and Actuators* **B1** 298-302
[6] Baltes H 1993 *Sensors and Actuators* **A 37-38** 51-56
[7] Hosticka B 1997 *Proc. Advances on Analog Circuit Design*
[8] Weber J et al. 1995 *Sensors and Actuators* **A46-47** 137-142
[9] Paul O and Baltes H 1995 *Sensors and Actuators* **A 46-47** 143-146
[10] Yamasaki H 1996 *Sensors and Actuators* **A 56** 129-133
[11] Baltes H 1996 *Sensors and Actuators* **A 56** 179-192
[12] Maloberti F, Malcovati P and Baltes H 1995 *Sensors for Domestic Applications* (Singapore : World Scient. Publ.) 201-216
[13] Malcovati P, Maloberti F and Baltes H 1996 *Sensors Update* VCH 143-171
[14] Middelhoek S, French PJ, Huijsing JH and Lian WJ 1988 *Sensors and Actuators* **15** 119-133
[15] Haberli A and Baltes H 1997 *Proc. Advances on Analog Circuit Design*
[16] Hwang C, Bibyk S, Ismail M and Lohiser B 1995 *IEEE Trans. on. Solid State Circuits* **42-11** 962-966

Signals and Circuits
Paper presented at Eurosensors XII, 13–16 September 1998
© *1998 IOP Publishing Ltd*

Development of a novel printed circuit board technology for inductive device applications

Olivier Dezuari, Scott E. Gilbert, Eric Belloy and Martin A. M. Gijs

Institute of Microsystems, Swiss Federal Institute of Technology, CH-1015 Lausanne, Switzerland

Abstract. This paper describes the fabrication and characterisation of 2-dimensional transformers integrated inside printed circuit boards. The transformers basically are composed of three layers of which the outer layers bear the printed coil patterns and the inner layer is a high permeability ferromagnetic sheet core. Both magnetic metal and copper layers are patterned using standard lithographic techniques. Electroplated interconnects between the outer layers complete the windings. These transformers are about 600 μm thick, with lateral dimensions of approximately 1 cm, and exhibit typically inductances of 1-10 μH at a frequency of 1 kHz.

1. Introduction

Inductive and transformer-like devices are at the heart of numerous sensing and actuating applications. With the increasing trend of miniaturisation of electronic devices, often the inductive components are the determining factor limiting further size reduction [1,2].

We present in this paper a new method for the fabrication of planar microtransformers (*ca.* 2 cm x 1 cm x 0.6 mm) using a hybrid printed circuit board (PCB) / flexible foil technology. In this process, a lithographically structured, amorphous magnetic foil core is laminated between thin planar coil windings on epoxy boards, bonded together using standard PCB techniques. The magnetic core material has an extremely high relative magnetic permeability (μ_r=100000) and a fairly large thickness (25 μm). These numbers are one or two orders of magnitude greater than the corresponding properties of films made using classical vacuum deposition methods. Consequently, for a given inductor size, the inductance value can be much higher than obtained before. Moreover, in the present technology, the core material can be selected from a variety of commercially available amorphous magnetic foils, amenable to lithographic patterning.

2. Experimental

2.1. *Choice of the magnetic core*

For purposes of obtaining a large magnetic flux in a small volume, a planar transformer requires a sheet core having a very high relative permeability μ_r. Among commercially available materials, the Vitrovac® magnetic metals [3] are amorphous, soft magnetic CoFeMo alloys, featuring very low coercivities and extremely high permeabilities at low frequencies. Moreover, they exhibit high mechanical strength and stress resistance, rendering them amenable to integration in multilayer PCBs using conventional fabrication techniques. For this work, Vitrovac® 6025 was used for the planar core. Its frequency-dependant permeability is reproduced in figure 1.

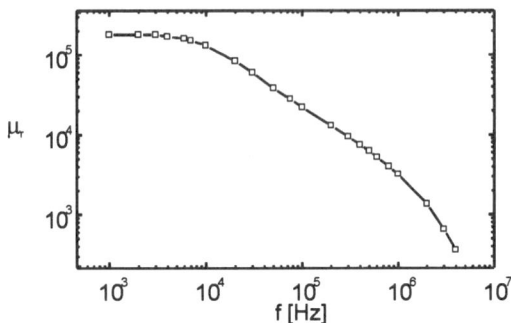

Fig.1: Permeability versus frequency of Vitrovac® 6025

2.2. Fabrication process

The proposed structures were fabricated based on conventional PCB processes. Two copper-clad epoxy boards were used as the outer layers. A liquid epoxy adhesive was used to glue Vitrovac® foil on the interior non copper-bearing epoxy board, forming the magnetic core of the transformer.

Vitrovac® was patterned photolithographically to form the transformer cores. Coil patterns were printed onto the copper laminates using standard photolithographic methods. The conductor width was 100 μm, the thickness 35 μm and the distance between adjacent turns approximately 250 μm.. Alternate half-windings were structured into each outer laminate layer, and corresponding top and bottom coil windings were connected by copper-filled vias. Patterned laminates were finally assembled and bonding by hot pressing with Prepreg® epoxy sheets to produce the multilayer PCB structure. The final PCB structure is schematized in figure 2, and photographs of structured magnetic cores and of the complete transformers are shown in figures 3 and 4.

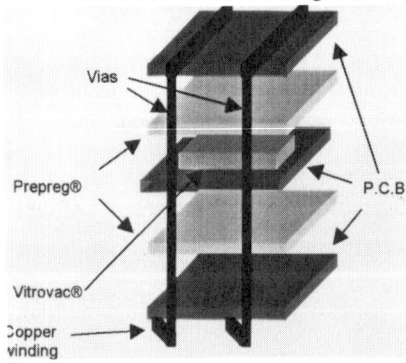

Fig. 2. Schematic diagram of transformer fabrication process.

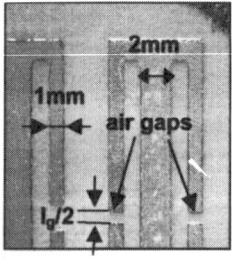

Fig. 3. Details of structured Vitrovac® E-cores, showing the two air gaps.

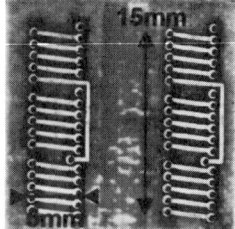

Fig. 4. Top view of finished transformers with E-cores

3. Results and Discussion

We show in figure 5 the frequency dependence of the primary self-inductance, L, and coil resistance, R, measured on a 10 turn primary rectangular transformer with an open secondary winding. L attains its highest values at low frequencies, and in this case a maximum of about 3.5 µH is measured at 1 kHz. The inductance characteristic follows the intrinsic permeability dependence (figure 1) to some extent, but is also degraded by leakage flux. The rise in coil resistance starting at about 50 kHz is primarily due to eddy current and magnetic core losses.

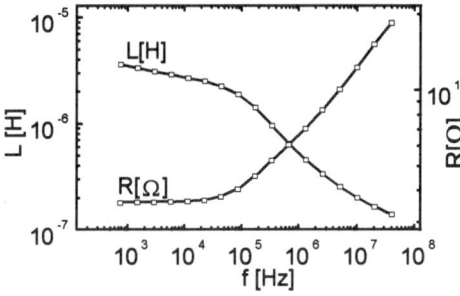

Fig.5: Inductance and resistance as a function of frequency for a rectangular transformer with no gap

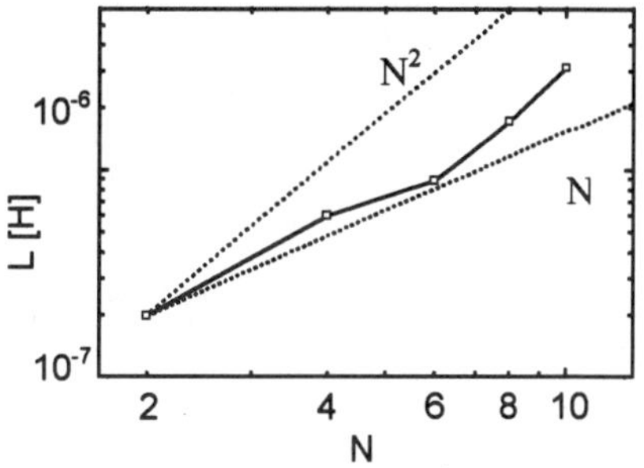

Fig.6: Inductance versus number of turns for rectangular transformer

Measurements of L as a function of the number of turns, N, is shown in the plot of figure 6. These measurements were made on a gapless rectangular transformer. The log-log plot reveals a power law for N of about 2 for a sufficient number of windings.

The effect of introducing air gaps in the outer legs of rectangular E-core type transformers (see figure 3) is demonstrated in figure 7 (curve a). The inductance L falls as the gap increases, thus following the classical behaviour of conventional transformers. However, L maintains a rather constant value after the gap reaches 400 µm. To a first approximation, L versus the gap width can be simulated by using equation 1:

$$L = \frac{\mu_o N^2 A}{l_g + \dfrac{l_m}{\mu_r}} \quad (1)$$

where L is the calculated inductance, μ_o is the permeability of free space, A is the cross sectional area of the core legs, N, the number of windings, l_m, the path length in the magnetic material, μ_r, the relative magnetic permeability, and l_g, the gap width. This equation is plotted as curve b in figure 7, based on an effective μ_r of 16000, and l_m of 33.4 mm. However, the experimental values are grossly underestimated by this theoretical curve, which predicts a more rapid fall-off of L than is observed. Flux fringing around the gaps, an effect which becomes more evident as the gap grows larger, can decrease the overall reluctance of the magnetic circuit, thereby offsetting the effect of the increasing gap. Its effect can be simulated by introducing a multiplicative correction factor, F, into equation 1, where $F = 1 + l_g/A^{1/2}\ln(2G/l_g)$ [4,5]. G is the vertical dimension of the core window ($G = 14$ mm in our case).

We have plotted equation (1) with the correction as curve c in figure 7, demonstrating that there is better agreement with the experimental data. Thus, fringing flux at least partially accounts for the observed behaviour, but a secondary effect, that being the closure of the flux lines across the window space as the gap width becomes very large (1 mm), may also play a significant role. This effect, which would be more important in microtransformers than in conventional ones, would reduce the mean length of the magnetic circuit l_m. This point remains to be clarified, however.

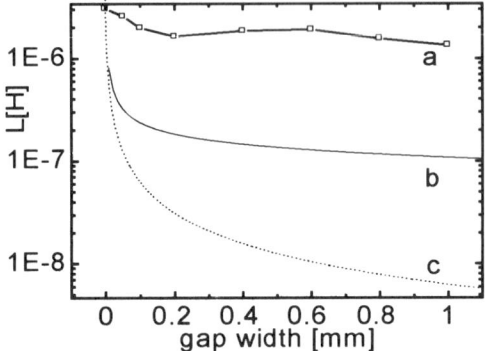

Fig.7: Inductance as a function of gap for a rectangular shaped core at 1kHz: Curve (a), measured values; Curve (b), theoretical curve, (eq. 1); Curve (c), equation (1) corrected by fringe factor F.

4. Conclusion

In this paper, we have presented a new planar inductor fabrication technology based on a low cost hybrid PCB-flexible foil process, along with preliminary results on prototype microtransformers. The use of an amorphous metal foil as a magnetic core permits the facile development of high inductance planar components. In comparison with existing planar inductor technologies, this new process virtually eliminates the need for time consuming and expensive thin film deposition techniques, while retaining compatibility with standard electronic packaging schemes.

5. References

[1] Yamasawa K, Maruyama K, Hirohama I and Biringer P 1990 *IEEE Trans. Mag.* **26** 1204

[2] Tsujimoto H and Koiso T 1996 *IEEE Trans. Mag.* **32** 4980

[3] Boll R 1993 *Soft Magnetic Materials* (Hanau: Vacuumschmeltze GmbH)

[4] McLyman W T 1988 *Transformer and Inductor Design Handbook, 2nd ed.* (New York: Marcel Dekker) p 150

[5] Lee R 1988 *Electronic Transformers and Circuits, 3rd ed.* (New York: John Wiley & Sons)

Sensor Applications

Sensor Applications
Paper presented at Eurosensors XII, 13–16 September 1998
© *1998 IOP Publishing Ltd*

Micromachined Resonator for Cavitation Sensing

E Peiner[a], R Mikuta[b], T Iwert[b], H Fritsch[b], P Hauptmann[b], K Fricke[a],
A Schlachetzki[a]

[a] Institut für Halbleitertechnik, Technische Universität Carolo-Wilhelmina zu
     Braunschweig, PO Box 3329, D-38023 Braunschweig, Germany
[b] Institut für Prozeßmeßtechnik und Elektronik, Otto-von-Guericke-Universität
     Magdeburg, PO Box 4120 D-39016 Magdeburg, Germany

Abstract. A micromachined resonator is investigated with respect to its response on short
uncorrelated impacts. The measured amplitude spectrum is dominated by resonances
caused by the fundamental and first order bending modes as well the torsional mode of
the resonator with its amplitude ratio depending on the direction of excitation. Using this
sensor solid borne vibrations generated by a centrifugal pump under different conditions
of operation were investigated. The obtained results show that cavitation can be
monitored at output data rates which can be easily transmitted by a field bus.

1. Introduction

State-dependent maintenance of machines is necessary to minimize down-times and to avoid
an accidental shutdown of highly productive industrial lines. This requires continuous
monitoring of moving parts of machinery with respect to solid-borne vibrations. One or several sensors mounted in a close distance to the vibration source are connected to a programmable logic controller by a field bus. The data rate which can be transferred is typically
around 500 kbit/s. However, machine vibrations comprising spectral components in the
frequency range of several 10 Hz up to above 10 kHz lead to data rates of Mbit/s for real-
time conditioning of the sensor signal. Especially for a control system comprising more than

one sensor the capacity of a field
bus may be exceeded. Thus, a selection of significant parts of the spectrum characterising the machine state is necessary before data readout.
During operation of a rolling bearing such characteristic frequencies
occur due to periodic impacts induced by the overrolling of flaws on
the surface of the rolling elements
or the raceways (Fritsch et al. 1997).
These frequencies can be selected
from a potentially complex spectrum
using a spring-mass resonator operated close to its resonance frequency.
Figure 1 shows the amplitude
spectrum of a micromachined reso-

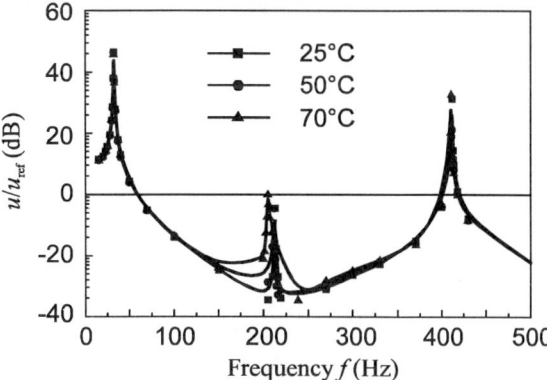

Fig. 1 Amplitude spectrum of a micromachined resonator.

nator periodically excited in its most sensitive direction. The resonances at 38, 210 and 410 Hz correspond to the fundamental bending vibration, the torsional mode and the first higher bending mode of the resonator, respectively. In this study the response of the resonator on uncorrelated impacts lacking a defined direction of excitation is described. In this context solid-borne vibrations generated by a centrifugal pump are investigated especially with respect to cavitation which may adversely affect the pump's performance.

2. Sensor principle

The spring-mass resonator used as the sensing element consists of a seismic mass supported by two thin suspensions as the spring (Fig. 2, upper part). It was fabricated using silicon bulk micromaching techniques (Peiner et al. 1997, 1998a, b). Excitation of the mass m by a short impact pulse $p = ma\Delta t$ of a duration Δt at the time $t = t_0$ leads to an exponentially decaying oscillation $\delta(t)$ (Chen et al. 1997):

$$\delta(t) = \frac{a\Delta t}{\omega_0}\exp\left[-\frac{\omega_0}{2Q}(t-t_0)\right]\sin\left[\omega_0(t-t_0)\right] \tag{1}$$

with the resonance frequency $\omega_0 = 2\pi f_0$, the quality factor $Q = \omega_0 m/b$ ($Q \gg 1$) and the damping coefficient b. The stress generated at the spring support is converted into an electrical signal $u(t)$ by a symmetric Wheatstone bridge using the piezoresistive effect of four diffused p-type resistors (Peiner et al. 1997, 1998b).

A short impact causes the spring-mass system to vibrate in resonance. In addition to the dominant fundamental mode torsional and first higher bending modes can be excited (Fig. 2). Their relative contributions to the vibration signal depend on the direction of the impact. By finite element model calculations (FEM) we found that in case (b) the torsional mode

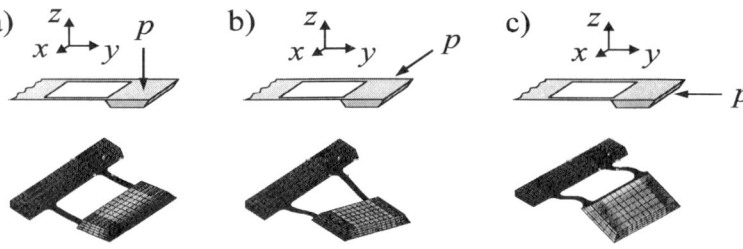

Fig. 2 Fundamental bending (a), torsional (b) and first higher bending modes of a micromachined spring-mass resonator.

and in case (c) the first higher bending mode are excited at relative probabilities of 30 and 20 %, respectively.

3. Vibration measurements

The sensor response on impact pulses of 2.4 ms duration was measured in dependence on the tilt of the resonator about its main axis (y-axis, Fig. 2). Bending and torsional modes were found as expected according to the FEM calculations. At a tilt of 90° with respect to the normal position, i. e. p is applied in x-direction (Fig. 2b) the fundamental mode was suppressed by a factor of 20. Simultaneously, the torsional mode was increased by factor of 8 leading to a relative probability of 35 % of this mode with respect to the fundamental mode as expected by the FEM calculations. We found that the described spring-mass resonator can detect impacts at high selectvity of the direction of excitation.

We used the described spring-mass resonator to investigate the complex spectra of the solid-borne sound emitted by a centrifugal pump (Etaline 100-125, KSB Frankenthal, Germany). The micromachined resonator and a broadband reference sensor were positioned at the exhaust tube of the pump. In the course of these measurements we increased the rotational speed from 1500 to 3000 min^{-1} leaving the mass flow cross-section constant. Figure 3 shows the output signal and its Fourier transform of the broadband sensor in dependence on the rotational speed of the pump. At 1500 Umin^{-1} short pulses were generated corresponding to

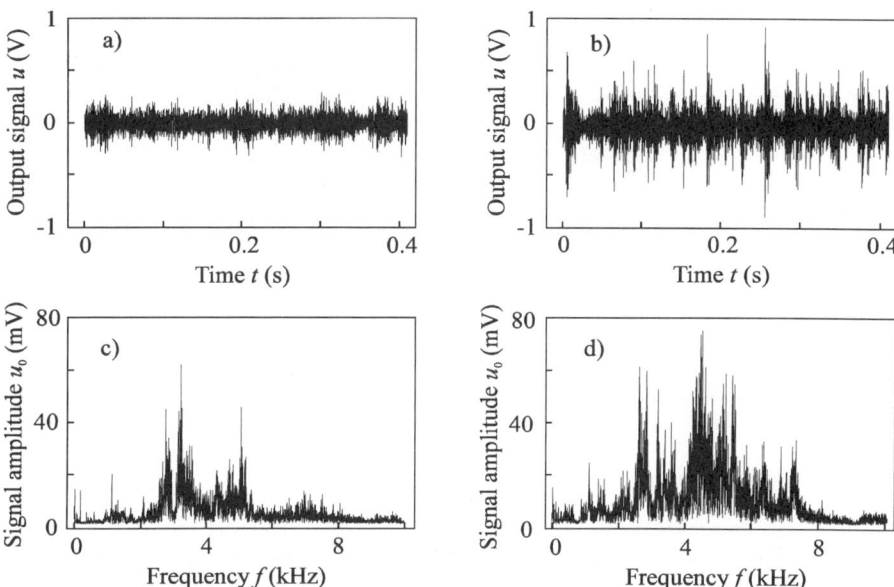

Fig. 3 Vibration signal (a, b) and amplitude (c, d) of a centrifugal pump at a rotational speed of 1500 min^{-1} (a, c) and of 3000 min^{-1} (b, d) measured using a broadband sensor.

broad bands between 2 and 5 kHz (a and c). Additional bands at 4 to 7.5 kHz which were related to cavitation appeared with increasing rotational speed (b and d).

Figure 4 shows the results corresponding to Fig. 3 measured with the micromechanical resonator. At 1500 Umin^{-1} both the output signal (a) and the amplitude spectrum (c) indicate by the dominant appearance of the fundamental mode that the excitation was essentially along the z-axis of the sensor (cf. Fig. 2(a)). With increasing rotational speed torsional and first higher bending mode increase at the expense of the fundamental mode (b and d). We conclude that the cavitation effect causes short uncorrelated impacts essentially directed perpendicular to the z-axis thus suppressing the fundamental vibration mode of the sensor.

This effect can be exploited for an immediate evaluation of the sensor signal and thus for a substantial reduction of the amount of data to be read out. For this purpose we used band passes to extract the contributions of fundamental and the torsion mode. After rectification and integration the ratio of both values was used as an indicator for the occurence of cavitation. Thus the bandwidth requirements for the data readout could be reduced to a range given by the reaction time which is necessary for an efficient control of the operation conditions of the pump.

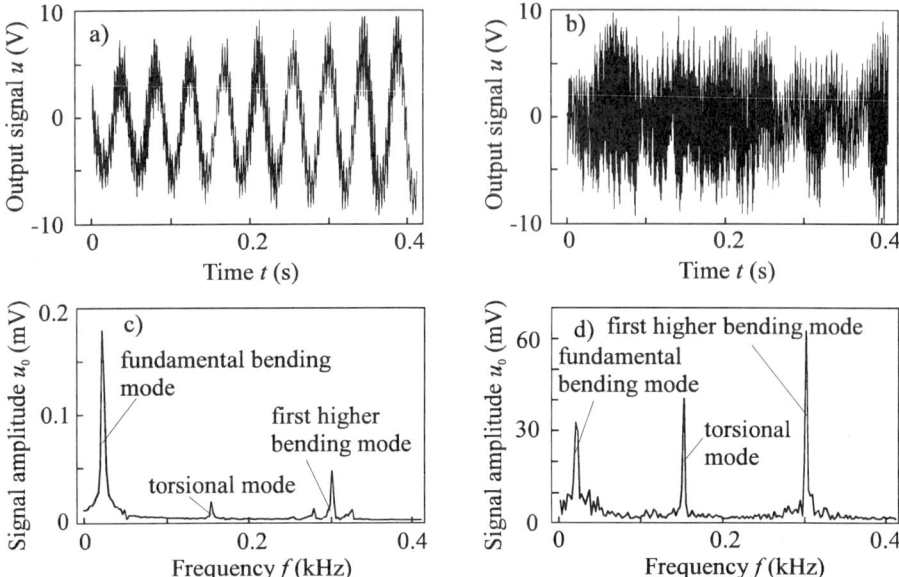

Fig. 4   Vibration signal (a, b) and amplitude (c, d) of a centrifugal pump at a rotational speed of 1500 min^{-1} (a, c) and of 3000 min^{-1} (b, d) measured using a spring-mass resonator.

4. Conclusions

A spring-mass resonator fabricated by Si micromachining and its response on short uncorrelated impacts was described. The measured amplitude spectrum is dominated by vibration modes corresponding to bending and torsion of the spring. The ratio of the respective amplitudes depends on the direction of the exciting impacts. This effect was used to monitor the operation of a centrifugal pump with respect to cavitation. The obtained results show that by a micromechanical resonator complex solid borne spectra observed by broadband analyses can be effectively compressed to the significant information. Thus the rate-capacity requirements for the subsequent data transfer can be considerably reduced to the range easily provided by a conventional field bus.

References

Chen H, Shen S and Bao M 1997, *Sensors and Actuators A* 58 197-201

Fritsch H, Lucklum R, Iwert T, Hauptmann P, Scholz D, Peiner E and Schlachetzki A 1997 *Sensors and Actuators A* 62 616-620

Peiner E, Fricke K, Schlachetzki A and Hauptmann P 1997, *Proc. Eurosensors XI, Warsaw* 293-296

Peiner E, Scholz D, Schlachetzki A and Hauptmann P 1998, *Sensors and Actuators A* 65 23-29

Peiner E, Scholz D, Fricke K, Schlachetzki A and Hauptmann P 1998, *J. Microelectromechanical Systems* 7, to be published

Porous Silicon Membrane for Humidity Sensing Applications

G.M. O'Halloran*, J. Groeneweg**, P.M. Sarro**, and P.J. French*

Electronic Instrumentation Laboratory*, DIMES**, Faculty of Information Technology & Systems, Delft University of Technology, Mekelweg 4, 2628 CD Delft, The Netherlands

Abstract. This paper reports on the development of a capacitive humidity sensor based on a porous silicon dielectric membrane. The membrane construction allows a heating resistor to be placed under the porous region which can be used for resetting the device. Initial results show an improved response time when compared to bulk type sensors.

1. Introduction

Porous silicon is a well known material for humidity sensing [1-3]. Despite the many advantages of using porous silicon for this application such sensors have had little commercial success to date. The aim of this work is to make a commercially viable integrated humidity sensor based on porous silicon.

Recently a capacitive type humidity sensor based on a porous silicon dielectric membrane was proposed [4]. The advantage of using a membrane is that it allows the placement of a heating resistor underneath the membrane for resetting the device. Also, the porous region forms the full dielectric layer in contrast to the bulk type devices, see Figure 1. This paper follows the progress of the development of these type of humidity sensors and presents new results on device performance.

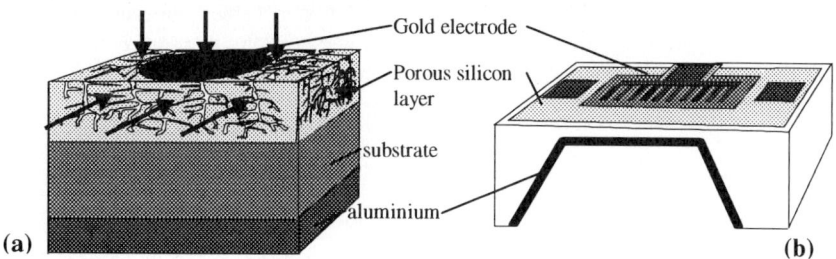

Figure 1   Capacitive type humidity sensors based on porous silicon (a) bulk type sensor (b) membrane sensor

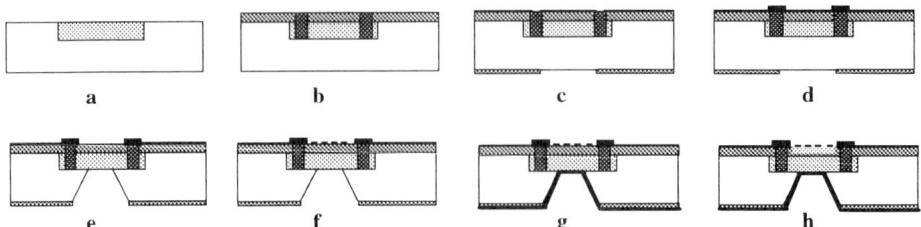

Figure 2   Membrane sensor fabrication sequence

2. Experimental

The substrate material is p<100> Si with a resistivity of 2-4 Ω.cm (International Wafer Services). The fabrication sequence (shown in Figure 2) is as follows:

a) A buried n-type layer is implanted and annealed.

b) A p-type epilayer is grown (1μm, $10^{16}/$ cm^3). Deep n-type contacts are diffused to the buried layer. A 20 nm oxide layer is grown .

c) LPCVD low stress SiN layer (30 nm) is grown, followed by a 300 nm low stress poly protection layer. The poly layer is then stripped from the frontside. Plasma etching of the SiN to form contact windows on the frontside and define membrane on the backside,

d) Dip etch to open contact windows. Chromium/Gold (60 nm Cr, 400 nm Au) deposition to form contacts to the buried resistor. Patterning and wet etching of Cr/Au layer.

e) Cleaning. Electrochemical etching of the membrane in 50% KOH solution at 85°C.

f) Cleaning. Chromium/Gold deposition (60 nm Cr, 400 nm Au) for the top electrode. Electrode patterning and wet etching of the excess Cr/Au.

g) Deposition of aluminium on the backside of the wafer (600 nm).

h) Electrochemical porous silicon formation.

The silicon is made porous using a standard anodisation technique. The formation parameters are chosen in accordance with previous results to give a layer with optimised humidity sensing characteristics [5]. The parameters chosen were 30% HF (1:1 with a Triton X100 solution), with an anodic bias of 30 mA/cm^2 applied for 30 seconds. This gives a porous layer of roughly 1 μm thick with a porosity of approximately 65%. Four types of top electrode geometry were chosen for the initial design - these are shown in Table 1. In each case the area is 2.25 mm^2. The Cr/Au was patterned using standard wet etchants - an iodine based etchant in the case of gold and a commercially available chromium etchant (Merck). A 10 second dip was sufficient to remove the layers.

3. Results

The devices were tested in a Heraeus HC7020 humidity chamber. A HP 4274A LCR meter operated at 100 kHz is used for capacitance measurements under computer control. The response characteristic of the devices were measured by allowing the device to stabilise at

10% rh and then ramping the humidity to 95%rh. The results for the four electrode types are shown in Table 1. Clearly the top electrode with the largest perimeter shows the largest response while that with the least perimeter shows the poorest response. It seems that the gold is too thick in this case to allow diffusion of moisture through the gold, so the moisture intake is from around the electrode perimeter. The time response shown is the time taken for the device to reach 63% of its final value going from 10%rh to 95%rh. In this case the fastest

Table 1: Summary of Test Results from Membrane Sensors

Top Electrode Type	Perimeter / mm	Response Characteristic	Response Time / min
Square	6		4.0
Grid	82		8.5
Comb	95		9.0
Serpentine	60		12.5

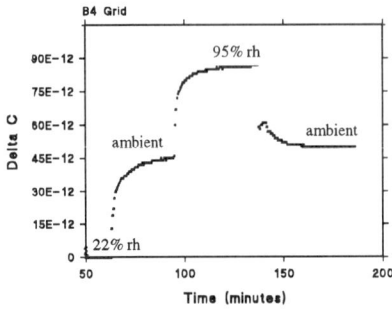

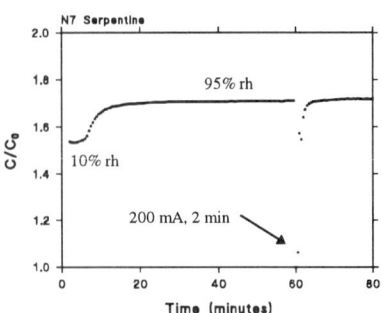

Figure 3 Time response   *Figure 4 Reset response*

response is from the device with the smallest change - the square electrode, with a total change in capacitance of 60 pF in 4 minutes - while the slowest device is the serpentine structure with a total change of 149 pF in 12.5 minutes.

Due to the large volume of the humidity chamber (1 m^3) it takes minutes for the change in humidity to be registered by the device. For this reason tests were carried out by letting the device stabilise inside the chamber at 22%rh at room temperature. Once stable the device was brought to ambient humidity (approximately 55%rh) - left to stabilise, and then replaced in the chamber where the humidity was now stable at 95%rh. The results are shown in Figure 3. In this case the time response of the device is more representative and is approximately 1 minute for a humidity change from ambient to 95%rh, and less than 0.5 minute for a humidity change from 95%rh to ambient.

The device was also tested for reset. Here a pulse of current was applied to the heating resistor with the capacitor placed in a 95%rh environment. The result is shown in Figure 4. It can be seen that while the current is applied the device effectively resets to a 0%rh level, and once the heat pulse is removed the output returns to the 95%rh level.

4. Conclusions

The results show the potential of a porous silicon membrane as a material for humidity sensing. At present devices are being designed to further improve the response by using a thinner porous metal top electrode, and also investigating the electrode shape.

Acknowledgment   This project is financially supported by the Dutch Technology Foundation in collaboration with Mierij Meteo B.V.

References
[1]. Burkhardt P.J. and Poponiak M.R. 1977 *U.S.Patent* No. 4,057,823
[2]. Anderson, R.C. et al, 1990 *Sensors and Actuators* **A21-23** 835-839
[3]. Richter, A., 1993 *7th Int.Conf.Solid.State.Sens.Act.* June 7-10, Yokohama Japan
[4]. O'Halloran G.M. et al, 1997 *Proc. Transducers '97*, Chicago June 16-19 563-566
[5]. O'Halloran G.M. et al, 1997 *Sensors and Actuators* **A61** 415-420

Sensor Applications
Paper presented at Eurosensors XII, 13–16 September 1998
© *1998 IOP Publishing Ltd*

Fast, efficient bulk micromachined heater for integrated sensor applications

I. Bársony[1], M. Ádám[1], S. D. Kolev[2*], Cs. Dücső[1], É. Vázsonyi[1], I. Szabó[1] and A. van den Berg[3]

[1] Research Institute for Techn. Phys. and Mat. Sci., P.O. Box 49, H-1525 Budapest, Hungary
[2] Faculty of Chemistry, University of Sofia, 1 James Bourchier Ave., BG-1126 Sofia, Bulgaria
[3] MESA Research Institute, P.O.Box 217, NL-7500 AE Enschede, The Netherlands

Abstract. The major obstacle a/o in the fabrication of combustive gas sensors (pellistors) is the realization of a fast and efficient heating by moderate power consumption. The scheme introduced in this work overcomes this problem by a suspended, thermally isolated, single crystal silicon heater formed by porous silicon bulk micromachining. The catalyst can than directly be embedded in the rough conditioned surface of the heater, providing a most efficient temperature transfer for combustion.

Keywords: porous Si, micromachining, microheater, pellistor

1.Introduction

Dynamic performance of sensor structures is often limited by the thermal response of the high temperature heaters used. Low mass, low loss heaters of minimum power consumption are therefore a prerequisite in integrated applications. Combustive type gas-sensors e.g. operate by the catalytic action of metal particles (Pt, Pd) at temperatures well beyond 400°C. Therefore, discrete pellistors usually fail to fulfill the standard power consumption limits in transducers. The readout signal is provided by the drop in the input power required for maintaining the catalysis i.e. the operational temperature set. This principle of operation lends itself to miniaturization, however, the thermal capacitance and the thermal losses of the device have to be minimized. This is contradicting with the requirement of large surface area needed for the catalytic process. In the optimum design both criteria have to be addressed: the device must be reduced to a simple resistor having minimased thermal conductivity in the suspension membrane or bridges, and a large heated surface, which is acting as host matrix for the catalysts.

2.Experimental

A single side processed bulk micromachining technique with sacrificial porous silicon was developed for optimum heat isolation[1] according to the process flow in Fig. 1. In the low doped p-type (100) Si starting wafer the heater resistor is selectively formed by P ion-implantation. The implant is driven-in to a depth of 2.0 μm, the final sheet resistivity is set to ca. 400Ω/□ to obtain the heater resistor of ca. 16 kΩ after completion of the process *(a)*.

Non-stoichiometric CVD nitride is used as protective masking layer, and to form the two suspension beams. The heater is contacted by Pt metallization through vias at both ends of the meander-resistor, followed by a second stress-free nitride layer to provide isolation from the Pt thermometer formed in a subsequent metallization step. The thermometer can be skipped later for optimum thermal performance *(b)*. The structure is encapsulated in a third nitride

---

* Present address: School of Chemistry, La Trobe University, Bundoora, Victoria 3083, Australia

deposition process. The etching windows for the masked selective anodization step are opened in the nitride. The p-type silicon is etched electrochemically in aqueous HF to a relatively thick (>20μm) sponge-type porous layer protruding under the nitride mask and undercutting the n-type regions *(c)*. The sacrificial layer thickness determines the final isolation gap between the heater and the bulk wafer *(d)*. Integrity of this thick porous layer is maintained by a tight control of the porosity profile during anodization by the current density [1]. After sacrificial layer removal in diluted KOH the suspended structure stays in place *(e)*. The bottom side of the active area will be the surface of the heater resistor attached to the nitride membrane. This has a ragged texture after sacrificial etching, which might even be further enhanced by additional conditioning. Following an adequate passivation of the silicon surface the catalyst can be deposited e.g. by a self-lithgraphic CVD process *(f)*.

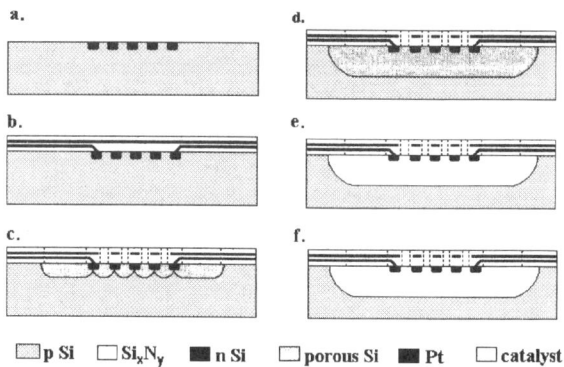

Fig.1. *Process sequence for the micropellistor fabrication.*

In the experimental design a series of resistors were realised with different geometries in order to analyse their thermal performance (Fig. 2.) All the suspension bridges contain two Pt wires embedded contacting the heater and thermometer respectively.

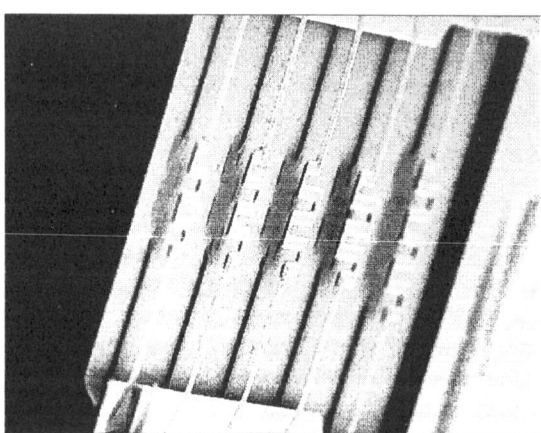

Fig.2. *SEM cross section of the pellistor-heaters suspended across the cavity. Cavity depth is 20um*

The bottom side of the bridge structure is shown in Fig.3. The shape of the released single crystalline Si resistor reflects the diffusion profile used for selective porous Si etching.

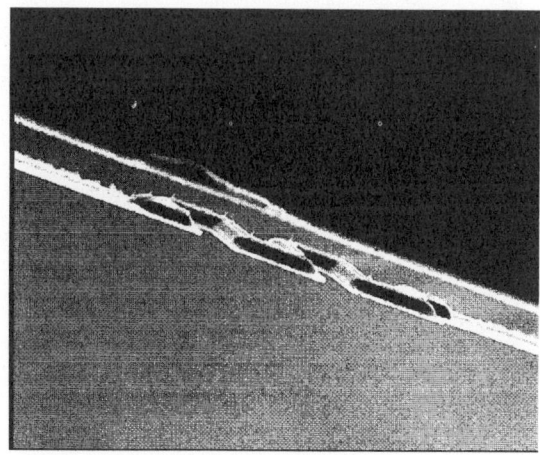

Fig.3. SEM view of the bottom side of the suspended heater. Resistor thickness is 2μm.

3. Thermal performance

Ab-initio calculations of the thermal behavior were performed for the suspended bridge structures, treating losses by radiation and conduction in the structure and in the air [2]. The heater meander was approximated by a square silicon slab of the thickness in the range of 0.1 to 1.5 μm and the linear size in the range of 16 to 64 μm. The area of the slab ranging from 16x16 to 64x64 $μm^2$ roughly corresponds to the surface of the meander

Medium	Thermal diffusivity [m^2/s]	Thermal conductivity [W/m/K]	Emissivity	Heat capacitance [Ws/kg/K]
Si	$2.204 \cdot 10^{-5}$	45	0.5	880
Si_3N_4	$6.937 \cdot 10^{-6}$	21	0.5	880
Pt	$2.395 \cdot 10^{-5}$	75	0.1	146
Air	$5.564 \cdot 10^{-5}$	0.0404		0.00103

Table 1: Physical constants valid at 773 K used in the model.

From the simulations we can conclude, that increasing the depth of the cavity below the meander seems to have no significant effect on the thermal performance. Generally, steady-state temperature was reached for all technologically feasible dimensions within 0.4 ms. The influence of the input power, the area and the thickness of the silicon slab and that of the suspension width on the maximum achievable temperature at the silicon/air interface (where the catalysis is going to take place) is illustrated in Fig. 4. If not stated otherwise the size and thickness of the silicon slab as well as the suspension width are 16 μm, 1.5 μm and 5.7 μm, respectively, and the input power is set to 20 mW. Reduction of the suspension width corresponds to the case after possible omission of two of the present four suspension beams.

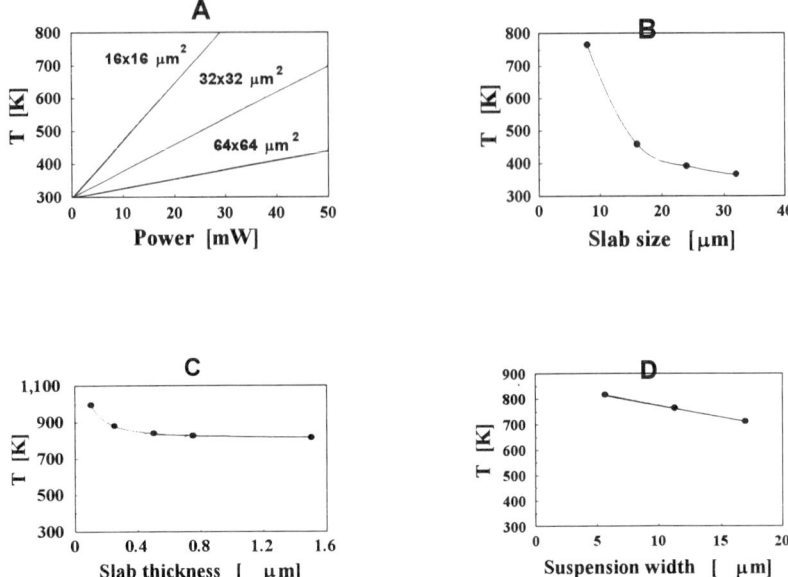

Fig.4. Maximum steady-state temperature (T) vs.: (A) input power for different sizes of the square-type silicon slab; (B) slab size; (C) slab thickness; and (D) suspension width.

4.Conclusion

The suggested micromachined, suspended heater structure is capable of producing a steady-state temperature in excess of 500°C at a consumption level of approximately 20 mW with a settling time of <0.4ms. This attractive thermal performance makes it suitable to be used e.g. in micro-pellistors for standard transducers. Moreover, the process allows the manufacture of integrated sensor arrays, even in combination with other, thermally less demanding kinds of sensor elements.

5. Acknowledgement

This work was supported by the Copernicus "PORSIS" project CP 940963 of the EU. The help of Ms. M. Erős and Ms. M. Payer with processing and Dr. J. Mizsei with SEM analysis is greatly appreciated.

6. References

[1] Cs. Dücső, E. Vázsonyi, M. Ádám, I. Bársony, J.G.E. Gardeniers, A.v.d. Berg, Sens. &Act. A 60 (1997) pp. 235-239
[2] S.D. Kolev, M. Ádám, I. Bársony, A.v.d. Berg, C. Cobianu, S. Kulinyi, Microelectr. J. (in press)

Sensor Applications
Paper presented at Eurosensors XII, 13–16 September 1998
© *1998 IOP Publishing Ltd*

Sensor applications of amorphous glass-covered wires

H Chiriac, C S Marinescu, and T-A Óvári

Nat'l. Inst. of R&D for Technical Physics, 47 Mangeron Blvd., 6600 Iaşi 3, Romania

Abstract. In this paper, we discuss the main operating principles for the applications of amorphous glass-covered wires based on their magnetic behavior. We also present two examples of new magnetic sensors based on such wires – a magnetic field sensor and a torsion sensor, which reveal the advantages of using these wires in sensor applications.

1. Introduction

The magnetic properties of amorphous wires with diameters of 80 to 300 μm, the so-called conventional wires (CW), have been widely studied lately, and they represent the basis for their applications in sensors and transducers [1]. These applications are mainly based on the large Barkhausen effect (LBE) that appears at low axially applied fields in positive magnetostrictive CW (e. g. $Fe_{77.5}Si_{7.5}B_{15}$ with the magnetostriction constant $\lambda = 35 \times 10^{-6}$) and in negative magnetostrictive CW (e. g. $Co_{72.5}Si_{12.5}B_{15}$ with $\lambda = -3 \times 10^{-6}$), but also on other effects like the Matteucci and inverse Wiedemann ones. Applications based on the giant magneto-impedance (GMI) effect in nearly zero magnetostrictive CW (e. g. $Co_{68.15}Fe_{4.35}Si_{12.5}B_{15}$ with $\lambda = -1 \times 10^{-7}$), have been recently developed [2].

A new class of amorphous wires, the thinner glass-covered ones, offers new perspectives on the applications of such magnetic materials. Amorphous glass-covered wires (abbreviated AGCW) consist of a cylindrical metallic core with diameters of 2 to 40 μm, covered by a glass insulation with a thickness of 1 to 20 μm. Their magnetic properties have been extensively studied recently, and they were found to be suitable for applications, either by replacing the CW, or in other new ones [3]. These properties depend on composition and on the wire dimensions – metallic core diameter, glass cover thickness, and their ratio [4].

The aim of this paper is to present some of our results concerning the main operating principles for the applications of AGCW, the tailoring possibilities for their magnetic properties in order to improve the sensitivity of the devices based on such wires, and two examples of new magnetic sensors based on AGCW – a dc field sensor and a torsion sensor – which reveal the advantages of using these wires in sensor applications.

2. Operating principles for magnetic sensors based on AGCW

The basic operating principles that can be used to develop different applications using AGCW are based on the outstanding magnetic behavior of these wires.

Positive magnetostrictive AGCW display a *large Barkhausen effect* (LBE), that is an abrupt magnetization reversal in the axially magnetized inner core of these wires, which occurs at low values of an axially applied field, called the *switching field*. Thus, the main characteristic of such wires is that they display a sharp rectangular hysteresis loop. The appearance of the LBE is favored by the specific magnetoelastic anisotropy distribution, which is a consequence of the rapid in-glass solidification process. The remanence to saturation ratio – called *reduced remanence* – of $Fe_{77.5}Si_{7.5}B_{15}$ AGCW is close to 1, as compared to 0.5 for CW with the same composition. The LBE in AGCW appears even in 1-2 mm long samples, while in CW it does not appear for sample lengths below 5 cm. The presence of the LBE is maintained after the glass cover removal, but the reduced remanence decreases by 50%. The switching field decreases as well, by one to two orders of magnitude, depending on the wire dimensions [4]. The main application of the LBE in positive magnetostrictive AGCW and wires after glass removal is in *pulse generator elements*. The magnitude and width of the induced voltage peaks is easily controlled through an accurate tailoring of the main characteristics of the LBE – reduced remanence and switching field. These two quantities can be modified directly from the preparation process, through a rigorous control over the wire dimensions, or indirectly through different types of post-production treatments – gradual or complete glass removal, annealing, stress-annealing, etc.

AGCW with positive magnetostriction also display another remarkable effect – that consists in the generation of an induced voltage across the ends of a twisted wire in the presence of an alternating magnetic field – known as the *Matteucci effect*. The appearance of this effect is favored by the helical magnetoelastic anisotropy that exists in these wires [5], which determines a change in the circumferential magnetization of a sample subjected to an alternating axial field. The Matteucci voltage can be detected even in the absence of an applied torque, but it significantly increases when the wire is subjected to torque. The Matteucci voltage induced at the ends of an AGCW is one order of magnitude larger than in CW (100 mV as compared to 20 – 30 mV). The magnitude of the Matteucci voltage increases by 100% after glass removal, mainly due to stress relief.

The recently discovered *giant magneto-impedance* (GMI) effect [2] – an abrupt change of the high frequency impedance of a magnetic conductor subjected to a small dc magnetic field – can be used to develop *field and current sensors* based on nearly zero magnetostrictive AGCW and wires after glass removal (e.g. $Co_{68.15}Fe_{4.35}Si_{12.5}B_{15}$). The GMI response of such wires is comparable to that of cold-drawn tension annealed CW with the same composition, having diameters reduced down to 30 μm, in which the most sensitive GMI effect was reported. The appearance of this effect in CoFeSiB AGCW and wires after glass removal is favored by the very good soft magnetic properties of these wires, and by their circumferentially magnetized outer shell, a peculiarity of their domain structure. A large and sensitive GMI effect is also obtained in Fe-based nanocrystalline glass-covered wires and wires after glass removal, due to their outstanding soft magnetic properties that originate in their vanishing magnetostriction. The sensitivity of the GMI effect in amorphous and nanocrystalline glass-covered wires and wires after glass removal can be tailored through their dimensional characteristics, as well as through different types of annealing.

3. Torsion and field sensors based on AGCW

The proposed *torsion sensor* allows a direct and simultaneous determination of the torsional angle and its direction by means of the synchronic-detection technique, using a 10 mm long FeSiB amorphous wire obtained after glass removal, with 16 μm in diameter.

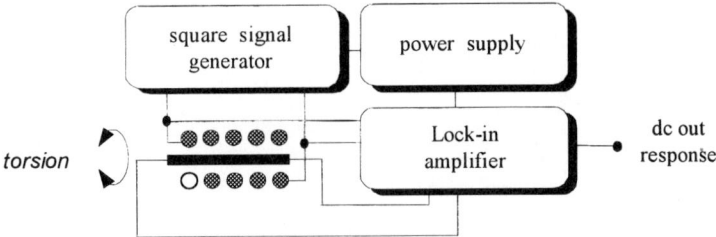

Figure 1. Block-diagram of the torsion sensor based on Matteucci effect in FeSiB glass-covered amorphous wires obtained by the glass-coated melt spinning method after glass removal.

The torsion sensor presented in the block-diagram from Figure 1 consists of a sensitive element, a square signal generator (10 kHz), and a synchronic-detection amplifier. The sensor's characteristic presented in Figure 2, shows a good linearity of the response in the range ±45°. In the torque linearity range of the Matteucci voltage, V_M, (± 50 rad/m), the dc signal obtained after peak-detection is proportional to the torsional angle and does not require any correction. The value of the torque sensitivity of V_M, about 2.5 mV/deg, ensures an angular precision better than 20".

Due to the relatively high value of the torque sensitivity and the good linearity of the Matteucci voltage induced in $Fe_{77.5}Si_{7.5}B_{15}$ wires after glass removal, a torsion sensor based on this effect requires very simple electronics. This fact allows the miniaturization and cost reduction of the device. A possible application of this torsion sensor is a sensitive control of mechanical arms movement. Such a device was designed and tested, and was found to work well on a viscosity measurement apparatus based on the torque pendulum method.

Based on the Matteucci effect, we proposed a new *field sensor* using a 50 rad/m twisted 10 mm long $Fe_{77.5}Si_{7.5}B_{15}$ AGCW with 16 μm in diameter, which allows a direct and simultaneous determination of a dc field – its value and direction – by means of the differential-detection technique. The field sensor is presented in the block-diagram from Figure 3. It consists of a sensitive element, a square signal generator (10 kHz) and a differential-detection amplifier. The sensor's characteristic (Figure 4) shows a good linearity of the response in the range of ±250 A/m.

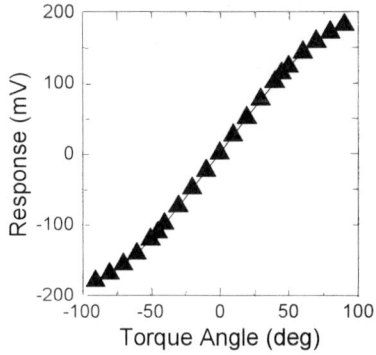

Figure 2. The response of torsion sensor versus the torsional angle.

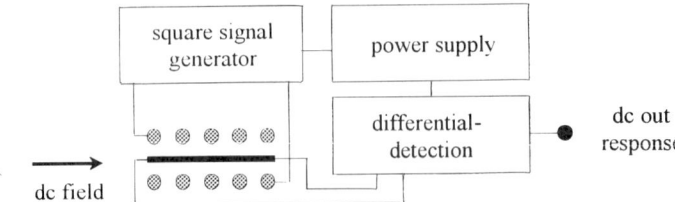

Fig. 3. Block-diagram of the dc field sensor based on the Matteucci effect in twisted FeSiB AGCW.

The advantage of these wires is that they do not require stress-relief pre-annealing like the CW. It is worthwhile to mention that sensors based on field sensitivity of Matteucci voltage do not require pick-up coils, and the minimum length of the wires can be around 2 mm, that is the minimum length at which the Matteucci effect still appears in such wires.

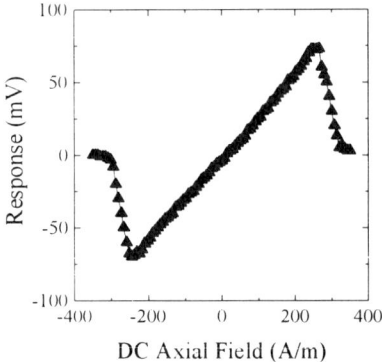

Fig. 4. The response of the field sensor versus the applied axial dc field.

The utilization of AGCW and wires after glass removal in applications offers several advantages:

- They are obtained in a one-step process (two-step for wires after glass removal) at smaller diameters in comparison with the thin amorphous wires obtained by cold drawing in several steps starting from "conventional" in-rotating-water quenched amorphous wires;
- They offer miniaturization opportunities for sensing devices working with "conventional" amorphous wires - e.g. the critical length at which the LBE still appears in such wires is about 1-2 mm;
- They have more degrees of freedom for tailoring their magnetic properties due to the existence of the glass cover, and thus, one can adjust their basic magnetic properties - coercive force, switching field, permeability, remanence - in a wide range of values.

References

[1] Squire P T, Atkinson D, Gibbs M R J and Atalay S, *J. Magn. Magn. Mater.* **132** 10-21.
[2] Panina L V, Mohri K, Bushida K and Noda M, *J. Appl. Phys.* **76** 6198-203.
[3] Chiriac H, Óvári T-A, Pop Gh, Barariu F, *Sensors and Actuators A: Physical* **59** 243-51.
[4] Chiriac H, Óvári T-A, Pop Gh and Barariu F, *IEEE Trans. Magn.* **33** 782-7.
[5] Chiriac H, Óvári T-A, Marinescu C S and Nagacevschi V, *IEEE Trans. Magn.* **32** 4755-7.

Sensor Applications
Paper presented at Eurosensors XII, 13–16 September 1998
© *1998 IOP Publishing Ltd*

Ammonia detector for emissions of agricultural origin

P Boeker†, G Horner‡, T Rechenbach†,U Schramm§, P Schulze Lammers† and J Bargon§

†Institute for Agricultural Engineering, Nussallee 5 &

§Institute for Physical Chemistry, Wegelerstr 12
Rheinische Friedrich-Wilhelms-Universität Bonn, D-53115 Bonn (Germany)

‡HKR Sensorsysteme, Gotzinger Strasse 56, D-81371 Munich (Germany)

Abstract. Emissions and immisions of malodours and climatically active pollution gases from intensive livestock farming have become a matter of public awareness. Those biogene emissions not only pollute the environment, but can also effect the health and well-being of the livestock and therefore cause ecological as well as economical problems. A sensor system has been developed in order to evaluate the emission rates of some of the most important pollution gases, such as ammonia [1]. The low-cost sensor system is suitable for multi-point detection, e.g. to control the ventilation systems in farm buildings. Results from field tests are presented, monitoring the ammonia emission from pig slurry.

1. Measurement Principle

The sensor principle is based upon quartz crystal microbalance devices. Quartz crystal microbalance (QCM) sensors are highly mass-sensitive devices. Combined with chemically active layers they can be used as chemical sensors. The interaction of analyte molecules with the gas sensitive layer material leads to a mass increase of the piezoelectric material. This induces a shift of the resonance frequency of the quartz resonator. Thus, quartz crystal resonators are ideal transducer devices for chemical sensing as a vast range of materials can be used as gas sensitive layers, e.g. all stationary phases from gas chromatography [2][3].

Most gas-sensitive layers are non specific, i.e. they react to a wide range of gases and vapours. One disadvantage of using these layers on a QCM sensor is the poor sensitivity to gases with low molecular weight. A promising approach to extend the sensitivity of QCM sensors to latter analytes is the use of host-guest complexes as sensitive layers [5].

These organic crystallites show a definite interaction with the analyte and are therefore more specific to certain gases or vapours.

The quartz crystal arrays are coated with highly selective host molecules for the selective detection of specific guest analytes. Remaining cross sensitivities of the sensor array to other gases will be compensated by methods of pattern-recognition and multivariate regression [4]. The integration of all sensor elements onto one single quartz crystal wafer helps to decrease temperature effects. The desorption process is increased by heating pulses from a heater element integrated onto the quartz substrate. By using cryptophanes (CPH) with different functional groups as sensor-active layers, the system can selectively detect ammonia in the presence of other gases.

The molecular cages can be synthesised with different functional groups. By this means the remaining cross sensitivities of the material can be modified in an appropriate manner [6]. Several sensors coated with such differently modified layers are combined to a sensor array. The signals of the different sensors can be processed by methods of pattern-recognition in order to compensate the cross sensitivities.

2. Experimental Set-up

A calibration unit for generating a multi-component gas mixture was used to evaluate the sensor characteristics. The unit can adjust the flow rates of the different gas components via eight mass flow controllers. The simulated biogene gas composition is obtained by calibrated pre-mixed gases and a humidifier. A thermostatted measurement chamber can take up to 12 QCM sensors. In this experiment a six element QCM sensor array was used. The highly integrated electronics include resonator circuits with ASIC-oscillators, six 24-bit counters implemented on a FPGA device and an RS232 serial interface. An automated data acquisition program stores the sensor data on PC via the serial interface. The data can be displayed on the PC-Monitor on-line during the measurement.

3. Results

Functionalised cryptophane compounds (CPH) were used as sensor-active layers for detecting ammonia emissions from pig slurry. The CPH acts as the host component for host-guest complexes. The reversible adsorption and inclusion of ammonia in the sensor-active layer material and the dependence of the ammonia concentration on the sensor characteristics was investigated. The sensor signal of a QCM sensor is the shift of the resonance frequency between clean air (or inert gas) and the test gas. Therefore, the gas flow was switched between two lines. While connected with the first line the sensor cell was purged with clean air, whereas the second line was carrying the test gas to the measurement chamber.

The sensor signals were calculated from the frequency shifts of the resonators resulting from a switched mode operation (see Figure 1). The behaviour of the sensors to humidity variations is very important in this application. It appears that ammonia and water have different interactions with the various sensor materials. Therefore, the impact of the two gases on the sensor signals can be separated. The absorption of water vapour in the CPH does not change the characteristics of the ammonia inclusion. The

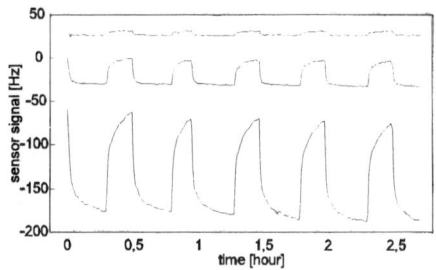

Figure 1. Switched mode signals of different QCM sensors

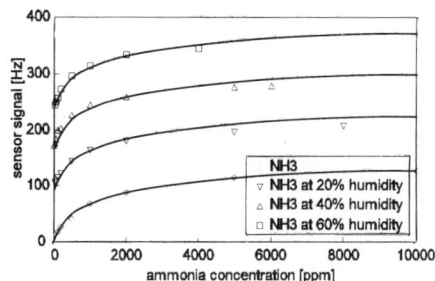

Figure 2. Sensor characteristics of CPH-1 at various levels of humidity

sensor signals can therefore be described as the superposition of the ammonia sensitivity and the humidity sensitivity.

The cross sensitivity of the host material CPH to other biogene gases is negligible. Figure 2 shows the characteristics of a sensor coated with CPH-1 to ammonia. The ammonia concentration was varied over a wide range while the water concentration was kept constant at various levels. Ammonia concentrations in most agricultural applications vary between 1 and 200 ppm. The absorption sites in the sensor material for ammonia are not saturated at these concentrations. Therefore, the sensor responds linearly within this whole concentration range.

By modifying the functional groups of the CPH the sensor characteristics to ammonia and humidity are changed. The ammonia concentration can be accurately calculated from the signals of the differently coated sensors in the presence of varying water concentration. In case the sensor signal is dominated by only two gas components (e.g. ammonia and water vapour) the ammonia concentration can be accurately predicted using only two sensor elements.

Figure 3 shows first results obtained under typical agricultural conditions. The polluted air to be measured was taken from a manure heap emitting ammonia in the range between 25 ppm and 60 ppm. The diagram shows the present ammonia concentration measured by an FT-IR spectrometer and the NH3-prediction from the CPH-sensor. The variation of the water concentration during the measurement period (see Fig. 3, dashed line) reflects in the uncorrected sensor signal (Fig. 4, left). This variation can be corrected, applying a simple linear model to the sensor signals (Fig. 4, right).

916

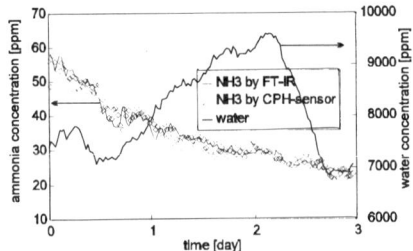

Figure 3. CPH-sensor measurement of biogenous ammonia emissions

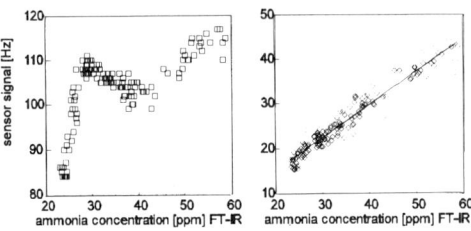

Figure 4.　　left :　Uncorrected signal from a single sensor element;
right : Corrected signal from sensor array

4. Conclusion

QCM sensor arrays and gas sensitive layers from host-guest chemistry are a promising tool for the detection of biogene emissions. Based on this technology compact and low-cost monitoring systems for agricultural emissions can be realised. By modifying and optimising the inclusion compounds most relevant gas components of biogene emissions can be monitored.

5. Acknowledgement

This work was supported by a research grant from the German Ministry of Science and Technology BMBF.

References

[1] Boeker P 1997 *KTBL-Arbeitspapier 244* (Darmstadt : KTBL) 134–140

[2] Schmautz A 1992 *Sensors & Actuators B6* 38–44

[3] Schierbaum K D Göpel W 1993 *Synthetic Metals 61* 37–45

[4] Horner G, Hierold C 1990 *Sensors & Actuators B2* 173–184

[5] Ehlen A Wimmer C Weber E Bargon J 1993 *Angew. Chem 105 Nr.1* 116–117

[6] Weber E 1995 *Inclusion Compounds Kirk-Othmer Encyclopedia of Chemical Technology 4.Ed* 122 154

Sensor Applications
Paper presented at Eurosensors XII, 13–16 September 1998
© *1998 IOP Publishing Ltd*

Absolute-Orientation Sensor for Mobile Robotics

Eduardo Iriarte - J M Martín Abreu - Leopoldo Calderón Estévez

Instituto de Automática Industrial, CNIII Km 22.8, 28500 Madrid, Spain

Abstract. Vehicle heading acquires special interest when the challenge is an accurate control over the trajectory. In this line, we present an orientation system applied on mobile robot control. The system consists of an ultrasonic transmitter at well-known position and a sensor head, composed by an array of eight receivers, on board the vehicle. The measurement of the phase between the received signals combined with the position of the robot allows us to calculate the orientation of the sensory head and therefore the orientation of the vehicle relative to the static transmitters.

After calibration, accuracy better than 1 degree is obtained in the whole range of measurement up to 15 meters.

Keywords: Robotics –Heading sensors- Ultrasonics - absolute measure

1. Introduction:

Heading sensors are considered important for compensate the foremost weakness of odometry and other positioning methods. In addition, they are basic devices on positioning techniques based on triangulation. However, when the task is to control the motion for accomplishing well-defined trajectories, is necessary to know position and velocity vector not only in module but also in phase, consequently orientation acquires a new dimension.

Many methods and techniques are applied to know the orientation of mobile robot. The most widely employed sensors for this task are compasses and gyroscopes. Ample revision of the current techniques applied on gyroscopes and compasses is made in [1].

Gyroscopes in all the variants give relative measures and suffer drift during navigation [1]. Until recently, the cost grew drastically on low drift systems, making them impractical in many cases. The most recent fiber-optic gyroscopes offer low drift (18°/hr rms) at medium-cost, and become suitable for low-autonomy applications.

Compasses give absolute measures but they are susceptible of interference of power lines and field distortions due to steel structures. This inhibits its use on indoor applications. Moreover, less expensive models have poor dynamic performance, with settling times around one second. [3][4]

Other systems employ laser scanning beacon [5] or ultrasonic rotary receivers [6] on board the vehicle, and passive or active (transponder) targets for triangulation positioning. In these, if the basic triangulation problem is solved, heading is inherent and static accuracy of orientation measurement is very high, but they have moving parts as the mechanical variants of compasses and gyroscopes.

The system here described provides an absolute measure of orientation with no moving parts, and works embedded in an ultrasonic positioning system taking the signals of the same sensor.

We will describe here only the orientation subsystem, the fundamentals of method and basic considerations (Section 2), its implementation with standard electronic circuits (Section 3), practical remarks for enhance the accuracy (Section 4) and experimental results using a *Robuter* platform.

918

2. Fundamentals of Method

The delay between the signals in two points p1 and p2 affected by a same ultrasonic wave front that moves on the direction $\vec{v}$ .is:

$$(1) \qquad T = d \cdot \cos(\alpha) \cdot v^{-1}$$

where d is the distance between the points p1 and p2, v is the velocity of sound in the medium, and α is the angle that forms the direction of propagation $\vec{v}$ with the vector $\vec{d}$, which is defined as directed from p1 to p2, as shows the Figure 1.

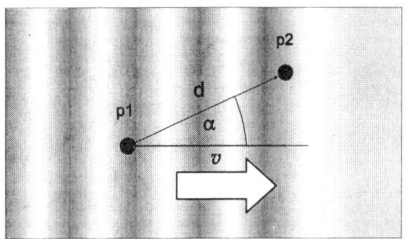

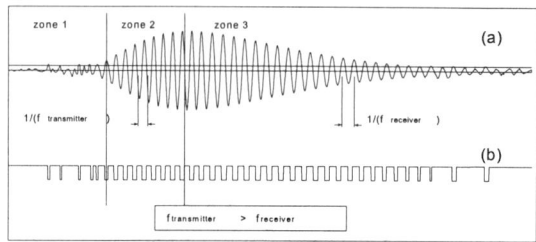

Figure 1: Plane wave front.　　　　　　　**Figure 2**: Typical echo signal,amplified(a) and binarized(b)

If the ultrasonic wave is periodic and its wavelength is longer than $d \cdot \cos(\alpha)$ the orientation α referred to the transmitter could be calculated with (1) directly by measuring the phase [7].

The form of an ultrasonic wave generated by means of the pulse-echo technique with resonant transducers is complex [8], its frequency is initially performed by the transmitter and next shifts to the frequency of resonance of the receiver. Due to directivity and inequalities on the sensibility of the transducers, in the practice is not possible to detect the first cycle, of very small amplitude. Since we could not know the start time of both signals, it is not possible to measure differences longer than one wavelength. Techniques based on frequency-division for phase measurement beyond one wavelength [9] are not applicable because is very difficult to find homologue signal edges.

In the sensor of the system here described (see Figure 3) eight receivers are placed forming an octagon of such side that the wavelength is never exceeded.

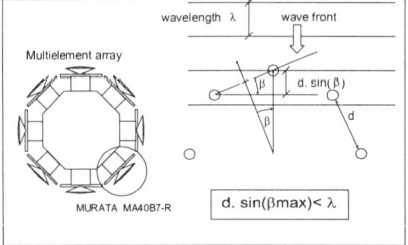

Figure 3: Place of receivers on the system sensor　　　**Figure 4**: Hardware of the orientation system

3. Orientation System

The measurement of the phase between two signals could be more easily carried out if such signals are digital. For distant emitters the received signals are weak and noisy. Several amplifier-filter stages are necessary, but the phase between channels should be preserved. In addition, binarisation should be done taking a threshold of 0 volts, in order to minimise phase distortion by unequal levels of signal. We describe the characteristics of each module implemented (see Figure 4)

3.1. Signal conditioning

For simplifying the circuits, neither time-gain-control (TGC) nor integrated filters are used. Filter-amplifier stages are designed with Op-Amps, low drift passive components, low sensitivity configuration and low Q to minimise group-delay dispersion between the channels. For expand the range, instead of TGC, clamping and over-amplification is used. Thus, ranges up to 15 meters are obtained. Digitalisation is done with hysteresis thresholds of 0.7 and 0 volts. Therefore, the cycles of the digital signals have a valid edge for phase measurement, obtained with the threshold of 0 volts.

3.2. Events handling and phase measurement.

Events handling is done by a µC PIC16C61 operating at 20 MHz. It determines the two-of-eight adjacent channels with the firsts signals. It discards spurious pulses by checking the period of the first signal, cycle by cycle, until the second signal appears. The phase between the selected signals is achieved by one PAL22V10 operating at 10 MHz together with a 12 bit HCMOS counter.

3.4. Angle calculation

The angle between the sensor and the emitter in the range of 360 degrees is obtained from

$$(2) \qquad \alpha = n \cdot 45 + s \cdot asin(k \cdot c)$$

where c is the phase count, n (0 to 7) is the code of the pair of receivers, and s (-1,1) depends of the arrival order. The value of k depends of sound velocity, clock and transmitter frequencies, number of measured periods, and distance between receivers.

Absolute orientation ω is calculated by combining the angle α with the positions of the transmitter $(x_t, y_t,$ known a-priori) and the sensor $(x_s, y_s$ obtained by the ultrasonic ranging system), therefore:

$$(3) \qquad \omega = \alpha + atan2\left(\frac{y_t - y_s}{x_t - x_s}\right) \qquad \text{where } -\pi < atan2\left(\tfrac{a}{b}\right) \le \pi$$

4. Practical Remarks

4.1. Valid zones for phase measurement.

How we already say, the analysed signals have a frequency that is initially the characteristic of the received wave, but that shifts to the frequency of resonance of each receiver.

Phase measurements on the beginning of the wave are uncertain because the signal is yet weak and therefore unstable to the comparator output. On the other hand, due to difference, although small, between the frequencies of resonance of the receivers, is unreliable measure many cycles later where receiver frequency prevails. Some cycles before the maximum amplitude, where the signal is clear and the transmitter frequency prevails, we obtain reliable phase measurements.

4.2. Other error sources.

Current sensor consists of eight separate channels, and small differences in frequency of resonance, distance between sensors and sensibility exist. Besides the disparity between channels, other physical phenomena may affect the measurement [10].

The use of inappropriate cover grid for sensor protection may cause scattering and diffraction effects. With a homogeneous thin grid, no influential effect was produced at 40 kHz.

Near to the transmitter, the wave front is not plane. The resulting error is smaller than 0.5 degrees at distances longer than 50 cm ($0.14°$ at 200 cm), and it could be compensated by combining the distance measured with a spherical wave front approximation.

Temperature affects the velocity of sound, and causes errors of -0.04 degrees/°K. Since the only local temperature affects, it could be compensated with a thermal sensor in the receivers housing.

Doppler shift occurs when the vehicle moves toward the transmitter. The relative error in phase measurement due to Doppler effect is proportional to the vehicle speed toward the transmitter. For velocities up 4 m/s the angle error is lower than 0,3 degrees and could also be compensated.

4.3. Static Calibration

The whole range of 360 degrees is achieved by chaining one pair with another, but discontinuities between pairs and errors of 5 degrees are obtained without calibration. To reduce the error we modified the equation (2) with three calibration parameters. Only 16 measurements are necessary to adjust these parameters in order to reduce the error at less than one degree. This calibration process could easily be automated.

5. Conclusions

After calibration and temperature compensation, static errors smaller than 0.7 degrees were achieved over distances up to 15 meters. Successive measures gave dispersion smaller than 0.2 degrees, except in the limits between pairs where the dispersion grows up to one degree. Averaging over 4 to 6 cycles was tried in order to diminish the dispersion, but due to the differences in the frequency of resonance how we mentioned, the results were worse.

The measurement time is typically shorter than 500 [μs] computed since the first signal arrives. The attainable sample rates are limited by the positioning system and secondary echoes. In tests in a zone with many obstacles, due to secondary echoes and over-amplification schema, we should lower the sampling rate down to 40 measures per second. Increases of this sampling rate could be obtained with TGC in all the channels.

Finally, dynamic performance was checked by means of navigation traces using a *Robuter* platform, and the heading error stayed below observable.

References:

[1] Borenstein J Everett H.R. and Feng L 1996, *Sensors and Methods for Mobile Robot Positioning.* Univ. of Michigan.19-29 32-64 221-233
[2] Barshan B and Durrant-Whyte, H F 1994 *IEEE Int. Conf. on Intelligent Robots and Systems* 1867-1874.
[3] Maenaka K. et al 1990 *Sensors and Actuators* 747-750.
[4] KVH Industries *C100 Compass Engine Product Literature*, 110
[5] NAMCO Namco Controls, 7567 Tyler Blvd. Mentor, OH 44060.
[6] Arai Tatsuo, Nakano Eiji1983 *Journal of Dynamic Systems, Measurements, and Control.* (ASME Hq)
[7] Martín Abreu J M Calderón L Ceres and R Pérez L A 1997 *Meas. Scientific Technology* 1279-1284
[8] Martín Abreu J M 1990 *Ph.D. Thesis*, Univ. Complutense de Madrid. 53-58
[9] Figueroa J F and Barbieri E. 1991 *Acustica* Vol. 73 47-49.
[10] Elsminger D1988, *Ultrasonics: Fund Tech App.* 2nd. Ed., 19-20, 42-47, 50-63, 142-150 (Dekker, Inc.)

Sensor Applications
Paper presented at Eurosensors XII, 13–16 September 1998
© 1998 IOP Publishing Ltd

Application of gas sensors to detection of underground cable faults

D. E. Miller, M. P. Bertinat* and P. A. Payne

Department of Instrumentation and Analytical Science, University of Manchester Institute of Science and Technology, PO Box 88, Manchester M60 1QD, UK.

*EA Technology Limited, Capenhurst, Chester. CH1 6ES

Abstract. This paper considers the suitability of electronic gas sensors for detection of underground cable faults. The proposed detector would work by detecting the gases produced during a cable fault. The methods used to determine the products of electrical insulation degradation are outlined and the results are presented. The target molecules for the sensor system are identified and metal oxide sensors are proposed for incorporation into the detector.

1 Introduction

This work considers a new application for electronic gas sensors as detectors for locating faults on Low Voltage[1] (LV) underground electrical cables. Such faults are notoriously difficult to locate, cause considerable disruption to supply, and are very unpopular with both electricity companies and customers. The proposed detector would facilitate location of these troublesome faults by using electronic gas sensors to sniff out the characteristic smell of burnt insulation that often accompanies electrical arcing [1].

The main types of insulation used on LV cables are cross-linked polyethylene (XLPE), oil-impregnated paper, and poly vinyl chloride (PVC) [2]. The first step for development of a detector for underground cable faults was to determine the gases produced by degradation of these insulation types. Three methods were used for identification of the degradation products, samples of insulation were pyrolysed in a furnace [3], cable samples were subjected to electrical degradation and soil samples from around cable faults were collected and analysed to determine which gases persist in the soil. Only the results from the pyrolysis experiments are presented here.

---

[1]Low Voltage power cables are designated in British Standards as 600/1000 Volts, where 600V is the voltage between the conductor and earth and 1000V is the voltage between conductors for which the cable is designed.

2 Experimental

The three experimental methods used for identification of the insulation degradation products are outlined below.

2.1 Pyrolysis

Pyrolysis experiments were carried out to allow the thermal degradation products of electrical insulation to be determined. A sample of insulation was placed inside a quartz glass tube which fits into an electrical furnace. A thermocouple was placed in the tube beside the insulation sample to enable the temperature of the sample to be monitored. An inlet was present to allow air or inert gas to be supplied to the sample and the gases produced were removed through a sampling port at the opposite end of the tube.

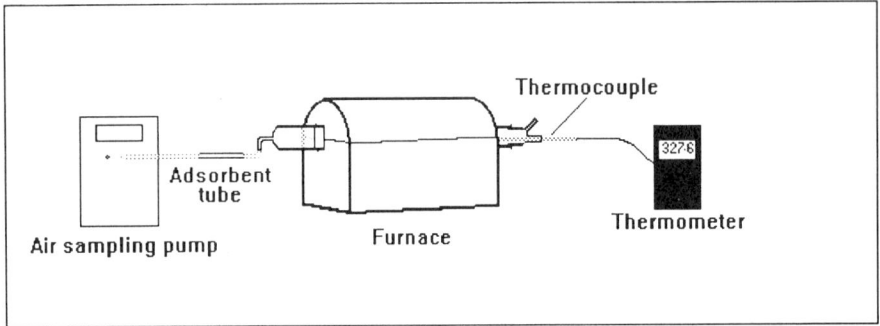

Figure Experimental set-up for pyrolysis experiments

2.2 Cable Degradation Experiments

A fault was induced in a sample of cable by drilling a small hole part-way through the cable insulation. The cable was then energised allowing the fault to break down causing electrical degradation of the insulation. The fault was enclosed in a glass vessel and the gases produced removed through a sampling port and concentrated in a Tenax-TA adsorbent tube which was then desorbed and analysed using Gas Chromatography and Mass Spectrometry (GC/MS).

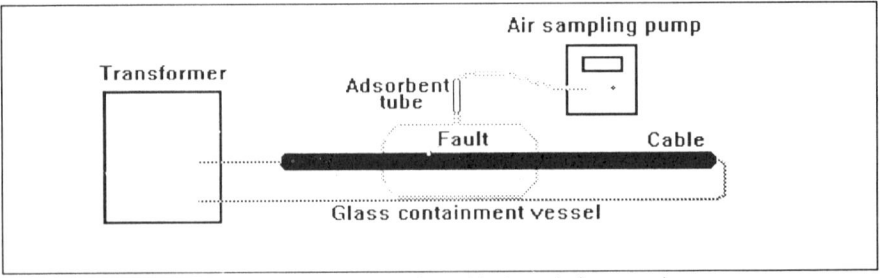

Figure 2: Experimental set-up for cable degradation experiments

2.3 Soil Analysis

Soil samples were collected from around known cable faults directly before they were excavated for repair. The soil samples were heated to 98°C in a water bath and air passed over the sample and drawn out through a Tenax-TA adsorbent tube to concentrate the volatiles. The adsorbent tubes were then desorbed and analysed using GC/MS.

3 Pyrolysis Experiments

The samples were allowed to pyrolyse in air at temperatures up to 800°C and the gases produced were drawn out of the sampling port through Tenax-TA and Carboxen adsorbent tubes at a flow rate of one litre per minute.

The gases trapped by the adsorbent were then desorbed using 1ml of acetone and the extract analysed using GC/MS.

3.1 Apparatus

The GC/MS instrument used was a Hewlett Packard 5890 Gas Chromatograph with the output being fed directly into a Hewlett Packard HP5970 Mass Selective Detector. The column used was a HP1 cross-linked methyl silicone gum capillary column with dimensions of 12m x 0.2m x 0.3μm film thickness. The injector was used in splitless mode at a temperature of 150°C and an injection volume of 1μl. The column temperature was programmed from 40°C (1 min isothermal) to 150°C at 10°C/min.
The MS conditions were: Solvent Delay - 2 mins, Voltage - 2400V, Mass range 50-300 amu, Scans per second - 1.71.

3.2 Results

From the analyses carried out, the main compounds formed during pyrolysis of electrical insulation in air have been determined. The most significant compounds for each insulation type have been identified for use as target molecules for detection of cable faults and are summarised below.

Table 1: Pyrolysis products of cable insulation

Insulation Type	Target Compounds
XLPE	1-Decene 1-Methyl-2-pentyl cyclopropane
Oil-impregnated paper	2-Furancarboxaldehyde 5-Methyl-2-furancarboxaldehyde
PVC	4-Octene, (E)- 2-Ethyl-1-hexanol

4 Implications for the sensor system

From the target molecules identified during insulation degradation and soil analysis, the appropriate sensor type for incorporation into an electronic sensing device for detection of cable faults was identified.

It can be seen from the types of target molecule identified that conducting polymers are not likely to be a viable option as little or no response will be obtained from conducting polymer compounds when exposed to these molecules. Mass sensitive devices such as Quartz Crystal Microbalances (QCM) or Surface Acoustic Wave devices (SAW) may be suitable for this application, but the most obvious option, considering the target molecules, is to use metal oxide semiconductor sensors (MOS). MOS sensors have previously been used for detection of paper degradation products in transformer oil [4].

An array of three or four metal oxide semiconductor sensors with differing sensitivities, combined with the appropriate sensor interrogation software, should be capable of detecting each of the six target gases identified from the cable degradation experiments. Pattern recognition software will be required to extract the pattern of response from the sensors and compare with the known response patterns of the target molecules.

5 Conclusions

Metal oxide sensors have been found to be the most suitable for incorporation into a sensor system for location of underground cable faults by detecting the gases produced during electrical arcing.

Acknowledgements
This paper is published with the permission of the EA Technology Director of Research and Development. The project is funded by the Electricity Companies participating in Optional Module 3, Cable Networks, of the Strategic Technology Programme (STP). The project is also part of the EA Technology-UMIST Postgraduate Training Partnership Scheme (PTP) and is supported by the DTI and EPSRC.

References

[1]   Clegg, B.: Underground cable fault location, McGraw-Hill Book Company, (1993).

[2]   Electric cables handbook (Second Edition), Ed. Bungay, E.W.G. and McAllister, D., BSP Professional Books, Oxford 1990.

[3]   Miller, D.E., "An analysis of breakdown products from pyrolysis of low voltage cable insulating materials", EA Technology Report Number 4386, January 1998.

[4]   Mitchell,A.: "The application of multi-element gas sensor arrays to condition monitoring," PhD Thesis, UMIST, 1996.

Sensor Applications
Paper presented at Eurosensors XII, 13–16 September 1998
© *1998 IOP Publishing Ltd*

The integration of pressure and position sensors into an intelligent control valve diagnostic system

Mohamed A Sharif, and Roger I Grosvenor

Systems Engineering Division, Cardiff School of Engineering, University of Wales
Cardiff, PO Box 688, Newport Road, Cardiff CF2 3TE, Wales, UK

Abstract. A microprocessor based control valve diagnostic system was used to study the effects of a number of faults, that were deliberately introduced into a globe valve under controlled conditions, on the overall valve performance. Faults considered and reported in this paper are the gradual blockage of the actuator's vent hole and the increase in frictional forces due to a damaged valve packing. These faults were selected since they are common in petrochemical plants where control valves are exposed to harsh industrial environments. Similar tests were also carried out on control valves in an industrial plant, where real faults developed over a period of time. The results obtained from the experimental work and that from the industrial site were correlated and provided consistent results.

1. Introduction

Conventional sensors such as strain gauge pressure sensors are widely used in monitoring the condition of critical process plant components. These sensors are usually integrated into data acquisition systems in order for the monitored signals to be conditioned, processed and analysed accurately. A comprehensive study was carried out [1] to establish the capabilities and the limitations of the latest available sensors that can be used as part of process plant condition monitoring and fault diagnosis systems.

A control valve is the final control element in any process loop where fluid is regulated through pipelines. The most common type of fault that develops in such a vital element is internal leakage, which is not easily detectable at its early stages. Control valve internal leakage has been investigated using vibration analysis techniques [2]. The reported results indicated that a leaking valve produces vibration in the adjacent pipeline as well as in the valve, and the vibration signal in the range of 0 - 20 kHz provides vital information about the leakage rate. Valve internal leakage was also detected by conducting an industrially based research programme using acoustic emission techniques [3].

A real time rule-based expert system was developed for diagnosing faults in a steel plant after evaluating the plant's parameters and involving plant maintenance specialists [4]. Condition monitoring of large plant components such as pumps and valves was considered to be an important part of an energy efficiency management system [5]. Critical studies were carried out to compare the merits of the first and the second generation expert systems for fault diagnosis [6]. The reported analysis described the type of instrumentation required for developing a reliable and comprehensive diagnostic system.

Simulated faults in control valves were successfully diagnosed using multivariate statistical techniques together with a knowledge-based system (KBS) [7].

The incentives for conducting a systematic programme of diagnostic tests on control valves, and trying to develop a novel expert diagnostic system which incorporates the results of individual tests, are the significant savings that can be made by early detection of faults.

In addition, vital information for devising maintenance strategies to optimise the plant availability and product quality are then available. Some results of the initial research work on a globe type control valve were reported by the authors [8]. This paper concentrates on new results that were obtained from different types of tests conducted to relate certain waveform characteristics to specific faults.

2. Experimental Methodology

The schematic diagram of the test rig used for the research work is illustrated in figure 1. It consists of a 25.4 mm (1 in) globe valve with a diaphragm type actuator. The digital valve controller (DVC) is an intelligent device consisting of a current to pressure converter, pressure, temperature and position sensors which are connected to a microprocessor. The microprocessor is configured by downloading the parameters from a personal computer, which holds the diagnostic software as well as the valve's database. The pressure sensor was used to monitor the variation in the applied pressure to the actuator while the position sensor was used to monitor the valve movement, and the temperature sensor to monitor the ambient temperature.

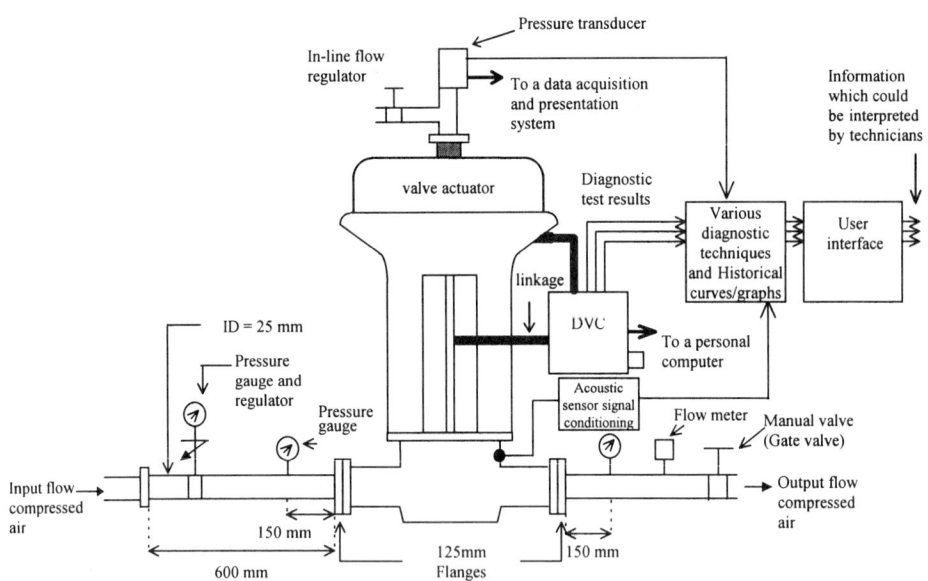

Figure 1 Schematic diagram of the Control Valve test rig

3. Results

Dynamic error band: This test was used to establish the dynamic behaviour of the control valve under normal as well as faulty conditions. The diagnostic test involved applying an input signal, in the range of 4 - 20 mA dc, to the current to pressure converter of the DVC and a relationship between the percentage valve travel and the percentage input signal was established. Results of some of the tests that were carried out on the test rig are illustrated in figures 2 and 3 as well as the summary table (Table 1).

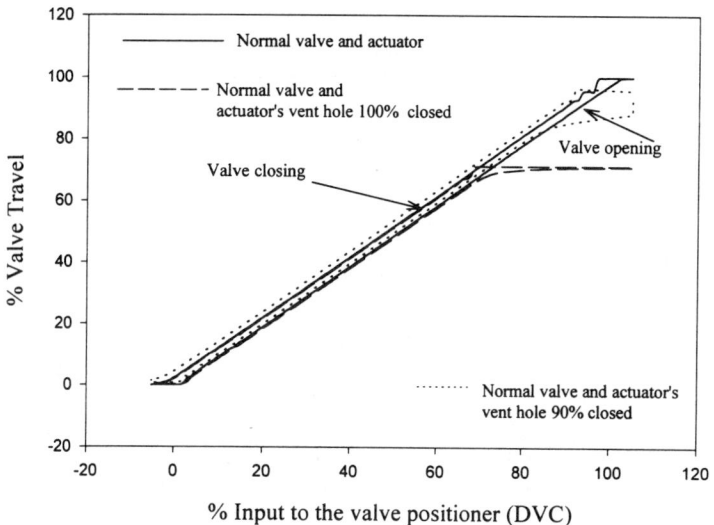

Figure 2 Graph of the variation in the Valve Travel versus the input signal to the positioner

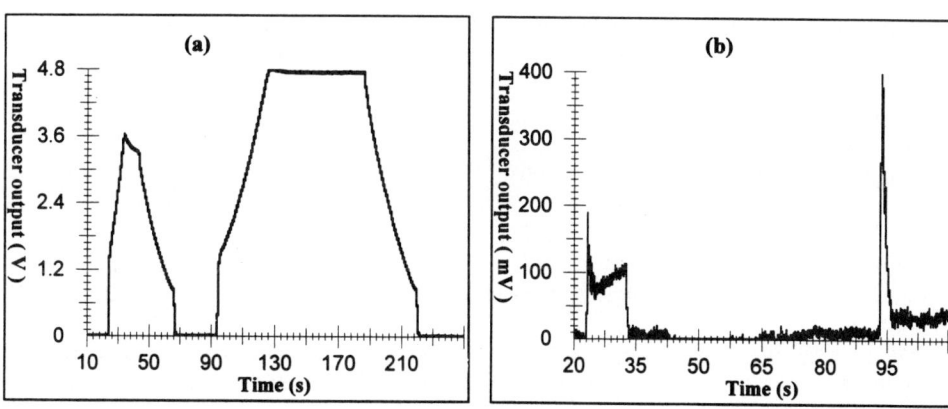

Figure 3 Graph of pressure transducer output versus time for: (a) normal valve and actuator's vent hole 100% closed and (b) for the valve with a section of its stem coated with a thin film of cyanoacrylate adhesive and its actuator's vent hole 100% open.

Given that the Pressure transducer's input and output ranges were 0 - 60 kPa and 0 - 10 V dc respectively, the maximum pressure that was developed in the upper diaphragm casing was 28.8 kPa.

Table 1 Summary of the results of the dynamic error band test

Parameter	Normal Valve				Valve stem with cyanoacrylate adhesive film			
	Actuator's vent hole 100% open	Actuator's vent hole 50% open	Actuator's vent hole 10 % open	Actuator's vent hole 100% closed	Actuator's vent hole 100% open	Actuator's vent hole 50% open	Actuator's vent hole 10% open	Actuator's vent hole 100% closed
Minimum dynamic error %	2.58	2.77	3.88	3.40	2.77	2.91	3.70	4.09
Maximum dynamic error %	3.00	3.12	4.47	3.74	4.08	3.47	4.42	4.39
Average dynamic error %	2.78	2.93	4.03	3.56	3.57	3.20	3.92	4.25
Dynamic linearity (indep.) %	0.13	0.17	0.72	0.11	0.13	0.10	1.02	0.09

4. Conclusion

The experimental results indicated that, the gradual degradation in the performance of control valves and actuators can be monitored on a continuous basis using pressure and position sensors, which are integrated into a data acquisition and processing system. Furthermore, it was deduced that in order to diagnose faults reliably, it is not sufficient to consider the results of one type of diagnostic test. This was the main reason for installing a pressure transducer in the actuator's vent hole, which provided vital information about the variation in the magnitude of the pressure that was developed in the upper diaphragm casing, as the orifice of the vent hole was reduced under controlled condition.

If the vent hole of a valve actuator in an industrial plant is blocked and is not detected, then it could cause disruption to the production line. In addition, if a valve does not travel to its full range after applying the maximum control signal to the valve positioner, plant technicians usually assume that the supply air pressure to the positioner is not sufficient and tend to increase it until the valve travels to its full range. This process temporarily masks the cause of the fault but may have undesirable effects in the long term.

Work has already started in combining the results of various diagnostic tests to develop an expert valve diagnostic system which could be easily utilised by plant technicians.

References

[1] Sharif M A and Grosvenor R I April 1998 *J.IMechE. E: Process Mechanical Engineering*
[2] Thompson G and Zolkiewski G April 1997 *J.IMechE. E: Process Mechanical Engineering, 195-207*
[3] Germain J L, Granal L, Provost D and Touillez M 1996 Proc. International Conference on Condition Monitoring and Diagnostic Engineering Management (COMADEM), Sheffield, UK, 411-420
[4] Mile R 1991 Proc. International Federation of Automatic Control (IFAC) symposium, Fault Detection, Supervision and Safety for Technical Processes (SAFEPROCESS), Germany, 7-9
[5] Johnston J 1996 *J. InstMC 143-146*
[6] Tzafestas S 1991 Proc. International Federation of Automatic Control (IFAC) symposium, Fault Detection, Supervision and Safety for Technical Processes (SAFEPROCESS), Germany, 1-6
[7] Zhang J, Martin E B and Morris A J 1996 Trans. IChemE Part: A 89-96
[8] Sharif M A and Grosvenor R I May 1998 Proc. IEEE International Conference on Instrumentation and Measurement Technology, St. Paul, Minnesota, USA

Sensor Applications
Paper presented at Eurosensors XII, 13–16 September 1998
© *1998 IOP Publishing Ltd*

Non-amplified pyroelectric PVDF sensors on ceramic and PCB substrates

K Benjamin, A F Armitage and R B South

The Department of Electrical and Electronic Engineering, Napier University, 219
Colinton Road, Edinburgh EH14 1DJ
Tel: 0131 455 4638   email: k.benjamin@napier.ac.uk

Abstract: This paper describes infrared sensing measurements using thin films of the
pyroelectric polymer PVDF on inexpensive substrates, without extra amplification of the signal
beyond that provided by the usual lock-in amplifier. PVDF is usually used on silicon substrates,
which enables integrated circuit amplification close to the signal source. That is not possible with
the present substrates, ceramic (96% Al_2O_3 alumina) and PCB (FR-4), but they are cheaper and
significantly simpler technologically. Further, their considerably lower thermal conductivities
compared with that of silicon may lead to better crosstalk performance.

1. Introduction

Pyroelectric materials possess an electric polarisation in the absence of an external electric field.
When a pyroelectric material is heated, an electric charge proportional to both its pyroelectric
coefficient and to the rate of temperature change with time develops on its surface, and may be
detected in an external circuit. When using a steady thermal (infra-red) source, a mechanical
chopper - usually a rotating slotted disc - must be used, in order to simulate continuous change. The
common PIR security detectors also change the incoming radiation, but in a slightly different way.
As a person moves across the field of view of the detector, fresnel lenses or a faceted mirror
modulate the incoming radiation so that a temperature change is induced, typically in differential
elements.

Infra-red sensors may be classified according to whether they are based on a single element
detector, such as photodiodes sensitive to the IR range of wavelengths, or on multiple pixels such as
are used in infra-red astronomy and night vision systems. Pyroelectric materials can also be
classified into single crystals, such as $LiTaO_3$, TGS and $NaNO_2$ or ceramic ($PbTiO_3$, modified lead
zirconate) or polymers (PVDF and its copolymer P(VDF/TrFE)).

This study concentrates on the use of PVDF films on inexpensive ceramic and PCB
substrates, with the long-term goal of producing cheap medium-resolution imaging arrays.

2. The PVDF samples

Each sample formed a simple 2x2 array of pixels. Each pixel was square, of side 4mm. The
spacings between the pixels varied from sample to sample, and were nominally 0.5mm, 1mm, 2mm
and 5mm. Samples were made on both ceramic and PCB substrates, and using both 9μm and 28μm
thick PVDF. The PVDF, which was metallised on the top side to provide an earth contact, was
attached to the substrate using a measured quantity of Loctite 358 UV curing insulating epoxy. The
lower metal contact was silver and copper for the ceramic and PCB substrates respectively. A
PVDF square of side approximately 18mm was enough to cover all four pixels on any one sample.

3. Experimental arrangement

A schematic diagram of the experimental arrangement used is displayed below in Figure 1. Infra-red
radiation from a Nernst filament is applied to the pyroelectric sample. The mechanical chopper
modulates the radiation to provide an ac signal for the sensor and reference signal for the lock-in
amplifier. The signal processed by the lock-in amplifier is measured by the digital voltmeter. The

computer has two-way communication with the digital voltmeter and the lock-in amplifier, through the graphical programming environment HP VEE.

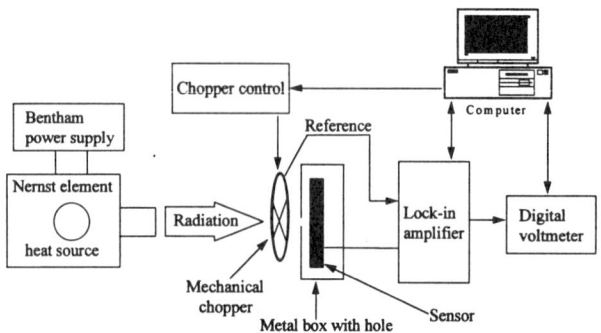

Figure 1: Experimental arrangement

4. Crosstalk measurements, pixel separation 2mm

Initial problems with some inconsistency in the measurements were resolved by placing the sample in a metal box which provided electromagnetic screening. Crosstalk measurements of the sensor voltage against chopping frequency were then made. The metal box had a small circular (4mm diameter) hole cut in it so that radiation could reach one of the pixels. The signal from this pixel was measured, and as a control the signal was also measured with the radiation blocked. The radiation was then allowed to reach the sensor once more, and the signals from the other three pixels were also taken. The other three pixels were not directly lit, and so these measurements would show whether any heat was spreading from the irradiated pixel to the other pixels. These measurements were taken for all substrate/material combinations, and at two power levels of infra-red radiation - the Nernst filament current was either 6.0A or 8.5A. In general, comparing the 9µm with the 28µm PVDF for the same infra-red power, the thinner material gave a higher voltage and also a greater frequency at which the response peaked. Both of these findings accord with theory.

Also, in order to determine if the sensors reacted predictably to varying infra-red input power, the irradiated pixel 3 alone was measured as the current through the Nernst filament was increased, at 0.5A intervals, from 2.0A to 8.5A. It was found that for all thickness/substrate combinations, the sensor voltage when plotted on a logarithmic scale increased uniformly with current once the current exceeded approximately 3.5A. Thus the sensors may be used to assess the strength of an unknown infra-red source.

For these samples with a pixel separation of 2mm, a typical set of crosstalk results is shown in Figure 2, for 28µm PVDF on a ceramic substrate.

The reading from the irradiated pixel 3 is the dominant trace on the graph. The readings from the three darkened pixels are indistinguishable from the blocked reading of the pixel that is normally irradiated. Hence the inference is that no heat is spread from the irradiated pixel to the other pixels, at least for this pixel separation of 2mm, the pixels being 4mm square. This was true of both 9µm and 28µm thick PVDF, of both ceramic and PCB substrates and of both 6.0A and 8.5A Nernst filament current levels.

5. Crosstalk measurements, pixel separation 0.5mm

The next samples investigated were those with a pixel separation of 0.5mm. The samples examined were 28µm PVDF on a ceramic substrate, 28µm PVDF on a PCB substrate and 9µm PVDF on a PCB substrate.

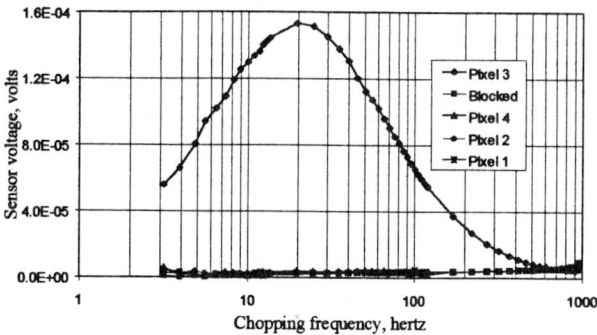

Figure 2: Crosstalk measurements at 8.5 amps Nernst filament current for 28micron PVDF on ceramic, pixel separation 2mm.

It was found that, like the previous result, there was no crosstalk at both Nernst current levels for the 28μm PVDF material. However, this time there *was* some crosstalk for the 9μm material, at both 6A and 8.5A Nernst filament current. The 6A results are shown in Figure 3. The trace for the irradiated Pixel 3 is again dominant, but this time the traces of the other three pixels are significantly above the trace of Pixel 3 when the radiation is blocked. At frequencies above about 260Hz, the traces for pixels 4, 2 and 1 fall below the blocked trace. This is thought to be a feature of the lock-in amplifier: it is capable of retrieving very small signals in the presence of noise, down to -60dB. The traces from these three pixels are genuine signals, and hence give coherent information to the lock-in amplifier, even at very low levels. However, the blocked trace contains no regular information, and is simply random noise, of an arbitrary level. Thus it is possible for the crosstalk traces to be below the blocked trace. Figure 2 confirms this indirectly: its traces for pixels 4, 2 and 1, because there is no crosstalk, are random, thus at the same general level as the blocked trace. In Figure 3, the traces for pixels 4, 2 and 1 *are* coherent, and thus may go below the blocked trace.

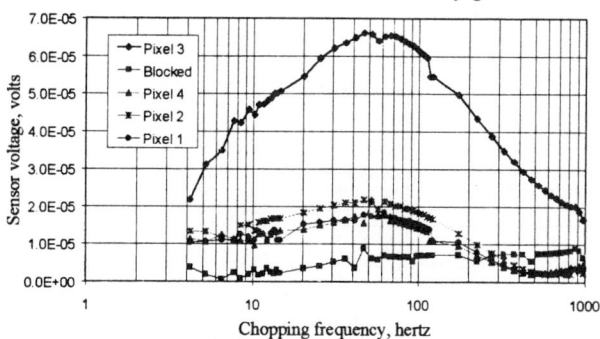

Figure 3: Crosstalk measurements at 6.0 amps Nernst filament current for 9 micron PVDF on PCB, pixel separation 0.5mm.

6. Discussion

Work proceeds on clarifying why the 9μm PVDF should show crosstalk at a pixel separation of 0.5mm, but not the 28μm material. Since the 28μm material on a PCB substrate showed no crosstalk, whereas the 9μm material also on a PCB substrate did, it can be deduced that the crosstalk is due to the 9μm pyroelectric material itself, rather than to the substrate. It is also intended to calculate the *amount* of crosstalk. This has been done before, for instance by Schopf *et al*[1], but their calculations were for an instance of a pyroelectric material of considerably greater thermal conductivity than its substrate, which is not the case here. The crosstalk calculations will be further complicated by a degree of uncertainty about the thickness of the bonding layer and the thickness of the metallisation.

It is additionally observed that the form of some of the measurements taken here departs from previous work. If η is the emissivity and p the pyroelectric coefficient, the sensor output varies with angular chopping frequency according to the **responsivity equation** (for example, see Hammes and Regtien[2])

$$R_V = \eta p \omega \frac{R_T}{\sqrt{1 + \omega^2 \tau_T^2}} \frac{R_E}{\sqrt{1 + \omega^2 \tau_E^2}} \tag{1}$$

where R and τ are resistance and time constant respectively, both thermal and electrical. This gives a smooth curve that peaks at a frequency

$$\omega_p^2 = 1/\tau_T \tau_E \tag{2}$$

with 20dB/decade slopes on the rising and falling edges. However, the present work has shown several instances where the rising and falling edges have different slopes. One such trace is shown below.

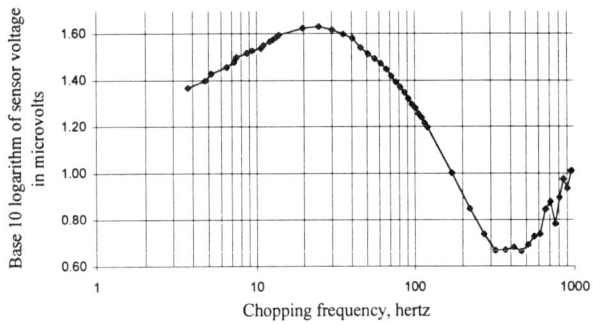

Figure 4: 28 micron PVDF on PCB, 6.0 amps current

The slope on the rising edge towards the peaking frequency of 25Hz is approximately half of the slope on the falling edge. One explanation may be that above the peaking frequency, a third time constant comes into play. It appears unlikely to be a shunt capacitance of the test equipment, because that would show on all the measurements. Not every sample exhibited different rising and falling slopes, but those that did it showed it repeatedly. This effect was most marked on the 28μm material, but earlier work in this Department[3] has shown a similar effect for 9μm PVDF. Indeed, that earlier work displayed not only different rising and falling slopes, but a falling slope that continued to steepen beyond the peaking frequency. This effect is also being investigated with a view to explaining, eliminating or exploiting it.

References

[1]   H. Schopf, W. Ruppel and P Würfel: A 16-element linear pyroelectric array with NaNO₂ thin films; *Infrared Phys.* Vol. **29** No. 1 pp 101-106, 1989

[2]   P.C.A. Hammes and P.P.L. Retgien: An integrated infrared sensor using the pyroelectric polymer PVDF; *Sensors and Actuators A,* **32** (1992) pages 396-402

[3]   S. Webster: Application Specific Infrared Sensor, Napier University Report, 27.9.1993

Silicon diode temperature sensor with nearly linear response curve down to helium temperatures

Yu.M. Shwarts, V.L. Borblik, N.R. Kulish, E.F. Venger

Institute of Semiconductor Physics, pr. Nauki, 45
252650 Kiev, Ukraine

Abstract

By means of increase in doping level of thermodiode base the silicon temperature $p^{++}n^+$-type sensors have been got which response curves do not exhibit anomalously steep sections in cryogenic temperature region. High doping did not worsen their sensitivity. Temperature response curves for thermodiodes with high doping levels point to effect of band gap narrowing in diode material.

Introduction

In temperature response curves U (T) of the majority of diode temperature sensors (U-voltage drop across diode when constant direct current I flows through it) anomalously steep sections are observed in cryogenic temperature region. Such sections in temperature response curves (TRCs) occur for diode made in Ge, Si and GaAs; the latter must to have p-type base; in diodes with n- GaAs base such sections are absent [1-3].

Above mentioned TRC sections are connected with freezing out of carriers at impurities and as a consequence increase in series resistance of diode base across which a large voltage drop is created by exciting current I.

Presence of abrupt kink in TRC is undesirable due to many reasons. Above all the abrupt kink complicates utterly a procedure of interpolation table derivation for diode temperature sensor because polynomial representation of such response curve will suffer from large error. Then it is this range of temperatures in which the most scatter in data from device to device occurs, i.e. individual properties of thermodiodes are exhibited. At last, such abrupt kink of TRC is undesirable from point of view of the secondary registering apparatus which transfer coefficient has to be equal in whole temperature range.

It is important also that in the kink range new mechanisms of current transport are involved which have been investigated slightly up to date; they are impurity band conductivity and hopping conductivity.

Basic idea

With increase in the base doping level a thermal ionization energy of impurities decreases and freezing out of majority carriers (and anomalous section of TRC) shift to more and more low temperatures (fig. 1, curves 1 and 2). One can expect that if the base doping level N is brought up the critical one N_c corresponding to Mott's dielectric-metal transition [4]

$$N_c a_B{}^3 \approx 0.02, \tag{1}$$

where $a_B = \hbar^2 \epsilon / m^* q^2$ - Bohr radius of shallow impurity center, $\hbar$ - Planck's constant divided by 2π, m^* - majority carrier effective mass, q - electron charge, ϵ - dielectric constant, the anomalous section of TRC will not appear at all. For silicon Eq. (1) gives $N_c \approx 1.2 \cdot 10^{18}$ cm^{-3} in n-type base and $N_c \approx 2 \cdot 10^{19}$ cm^{-3} in p-type base. [1]

Experimental

In order to verificate this suggestion we have investigated emitter-base $p^{++}n^+$-junction of the silicon transistor with base doping level of $2.5 \cdot 10^{18}$ cm^{-3}. The typical TRC of such diodes is represented in fig. 1 by curve 3 which is nearly linear down to T = 4.2 K.

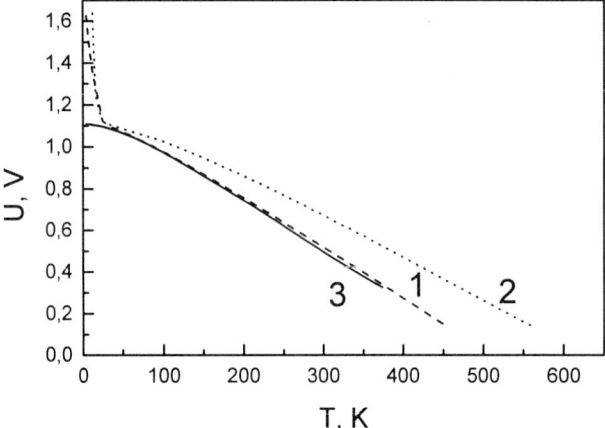

Fig. 1. Temperature response curves of silicon diodes produced by Lake Shore Cryotronics, Inc. (1), and our data for n$^+$p-diodes with acceptor concentration of $2 \cdot 10^{17}$ cm^{-3} (2) and $p^{++}n^+$-junctions of transistors with base donor concentration of $2.5 \cdot 10^{18}$ cm^{-3} (3) at excitation current of 10 μA.

---

[1] Difference of critical concentrations for n- and p-type materials is due to difference in majority carrier effective masses. Perhaps it is why anomalous sections of TRCs take place in thermodiodes with p-GaAs base, but are not available in diodes with base made in n-GaAs.

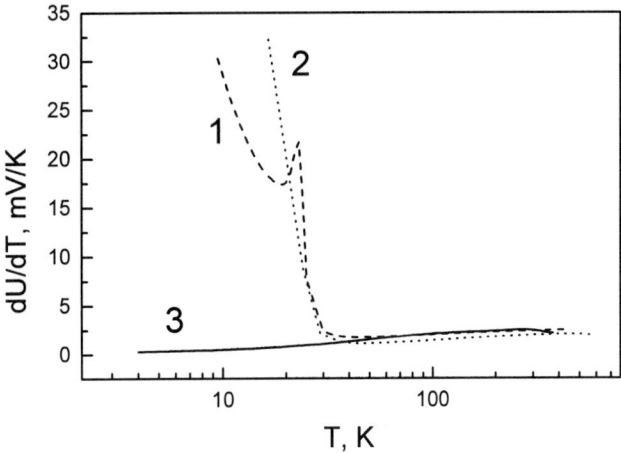

Fig. 2. Sensitivities dU/dT of thermodiodes which response curves are represented in fig. 1.

The sensitivity curves dU/dT for all of sensors which response curves are given in fig. 1 are represented in fig. 2. It is seen that sensitivity of the sensor with base which is doped up to metallic conductivity is the same in the whole temperature region and coincides practically with sensitivity of lighter doped Lake Shore Cryotronics diode at T > 30 K. [2]

Discussion

In connection with absence of free carriers freezing out current - voltage characteristics of the highly doped diodes have exponential character with a quality factor n (T); therefore their TRC may be represented as

$$U(T) = \frac{E_g(T)}{q} - \frac{nkT}{q}\ln(\frac{A(T)}{I}),$$
(2)

where $E_g(T)$ - temperature - dependent energy gap in diode base and saturation current $I_s(T)$ is taken in form of product of $\exp\left(-\dfrac{E_g}{kT}\right)$ and slowly varying function A (T), which includes the all other parameters

$$I_s(T) = A(T)\exp\left(-\frac{E_g(T)}{kT}\right).$$
(3)

---

[2] We had at our disposal sensitivity date for thermodiodes produced by this firm only.

As follows from Eq. (2) at T→ 0 diode voltage goes to $E_g(0)$ - energy gap (in Volts) at T = 0. In this respection fig. 1 demonstrates plainly the effect band gap narrowing with increase in base doping level, because TRCs corresponding to more and more high doping level lead to more and more low voltage value at T = 0 .

The simplest formulas which describe effect of band gap narrowing $\Delta E_g(N)$ in silicon (as is also in a number of other semiconductors) and are suitable for device use, are presented in paper [5] (for concentrations starting from N = 10^{18} cm^{-3}).

In particular, for n-Si

$$\Delta E_g(N) = 10.23\left(\frac{N}{10^{18}}\right)^{1/3} + 13.12\left(\frac{N}{10^{18}}\right)^{1/4} + 2.93\left(\frac{N}{10^{18}}\right)^{1/2} meV . \tag{4}$$

At N=$2.5\cdot10^{18}$ cm^{-3} ΔE_g = 35 meV follows from Eq. (4), that agrees with $E_g(0)/q$ = 1.12 V to which U(T) "looks" in the case of metalized base diode. The rest curves in fig. 1 corresponding to smaller doping levels lead to U (0) values which are more and more close to $E_g(0) /q$ = 1.166 V.

Conclusion

Thus we managed to make the silicon temperature sensor with nearly linear response curve in the whole temperature range 4.2 ÷ 400 K. It allows to reduce temperature measuring error in low temperature range due to simplification of U(T) approximation procedure as compared with the case of kink-containing response curves. Furthermore measurement of temperature response curves of diodes with high base doping levels gives an information about narrowing of diode material band gap.

References

1. N. Sclar, D.B. Pollock. Sol. St. Electr., 15, 473, 1972.
2. S.P. Logvinenko, T.D. Aluf, T.M. Zarochintseva. Criogennaja i vakuumnaja teknika, N 2, 63, 1972 (in Russian).
3. Temperature measurment and control, 1995, part 1 of 2. - Lake Shore Cryotronics, Inc., USA.
4. N. Mott, E. Davis. Electron processes in non-crystalline materials. 2 edit., Clarendon Press, Oxford, 1979.
5. S.C. Jain, D.J. Roulston. Sol. St. Electr., 34, 453, 1991.

Sensor Applications
Paper presented at Eurosensors XII, 13–16 September 1998
© *1998 IOP Publishing Ltd*

A method for sub-pixel measurement of 3D surfaces based on structured light

Jorge Gimenez, Alberto Ibañez, Alberto Alvarez, Teresa Sanchez, Luis G. Ullate.

Instituto de Automatica Industrial (CSIC), La Poveda, Arganda, 28500 Madrid, SP

Abstract. A method for sub-pixel measurement of 3D surfaces based on vision cameras and structured light is presented. The method includes algorithms for sub-pixel calibration, laser stripe detection and 3D reconstruction. The accuracy obtained is around 0.2 pixels for the 3D coordinates.

1. Introduction

Many industrial processes need an accurate 3D measurement of their products, especially in the final step of quality control. This is particularly important for parts of complex geometry, which are present at several industries such as ceramics, plastics, casting, shoemaking, aeronautics, etc. Systems based on vision cameras and structured light are very suitable, as they allow automated and non-contact 3D measurements. This technique currently uses a laser source provided with cylindrical optics, which projects a sheet of light onto the measured surface. A video camera captures the image of the laser stripe formed over the surface from a known angular direction. Then, the 3D position of the stripe points can be determined. By this way, the sensor movement is partially replaced by the electronic scanning of the video camera, and the system is qualified for high-speed operation.

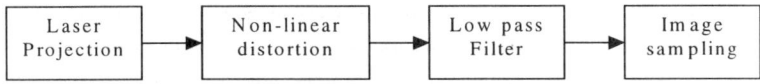

Figure 1. - Simplified model of a laser-based measurement system

The system accuracy is primarily determined by the quality of components, but also by the transforming algorithms used for the measurement process. In fact, there are several aberration effects, which prevent the system from behaving ideally. Figure 1 shows a model with four stages representing the image distortions. They include: (1) the widening of the laser stripe, (2) non-linear distortion due to optics and to a wrong estimation of the camera parameters (principal point position, etc), (3) a low-pass effect, representing the smoothing on the image edges even when the subject is in focus, and (4) digitisation and synchronism errors produced during the image sampling.

The image resolution R is the distance represented by two consequent pixels:
R= 2*Z*tan $(\Delta\alpha)$/N

being $\Delta\alpha$ half of the camera aperture angle, Z the subject depth, and N the number of pixels per camera line. Let us consider the following example: $\Delta\alpha=45°$, Z=1000 mm and N= 500 pixels, gives R= 4mm, which can be too poor for certain industrial applications. Increasing the number of pixels per line N will improve image resolution, but at a highest cost. Moreover, technology will limit the maximum value of N available. An alternative solution is to use methods and algorithms, which allow implementing measurements with sub-pixel precision. These algorithms should be considered for every one of the steps involved in the measurement process: (a) camera calibration, (b) location of the laser stripe in the image, (c) 3D reconstruction.

2. Sub-pixel camera calibration

Camera calibration in the context of 3D machine vision is the process of determining the internal camera geometric and optical characteristics (intrinsic parameters) and/or 3D position and orientation of the camera frame relative to a certain world coordinate system (extrinsic parameters) [1]. To this purpose, several methods can be found in literature. The simplest one considers an ideal lens and calculates the camera parameters performing a simple perspective projection onto the image plane ("pin-hole"). By this method, the standard deviation of errors between the model and the calibration pattern is around a pixel, although deviations of several pixels can also be found. Other procedures [1,2] consider that lens distortion causes shifting of image points mainly along the radial direction from the optical centre (radial distortion). The algorithm is simple and involves a relatively low computation cost. However, it may produce errors as large as the pin-hole method when the image centre is not correctly estimated. Recently, more complex algorithms have been presented [3], which consider several parameters to compensate both radial and tangential distortion. Applying such algorithms the precision attained is around one tenth of pixel. This is the method followed in the present work. As it is said before, a set of parameters are obtained during calibration from which the camera model is determined:

$(x_c, y_c, z_c, x_u, y_u)=g(f, k_1 ,k_2, p_1 ,p_2, c_x, c_y, \theta, \varphi, \psi, t_x, t_y, t_z)$

where the transforming matrix "g" is determined by an optimisation method indicated in [3] from the 3D coordinates (x_c, y_c, z_c) of the circle centres of the calibration pattern and the coordinates (x_u, y_u) of the centres in the image given in pixels. This is a non-linear function relating the spatial coordinates as a function of the camera intrinsic parameters (focus (f), principal point coordinates (c_x, c_y), radial and tangential distortion correction constants (k_1, k_2, p_1, p_2)), and extrinsic parameters indicating the camera position with respect to a known coordinate system $(\theta,\varphi,\Psi, tx, t_y, t_z)$.

The centres of the circles (x_u, y_u) are calculated by the following procedures: first the image is converted to binary by threshold, then circles are dilated, and finally, the centre of each grey scale enlarged circle is calculated by the next expression.

$$\overline{u} = \frac{\sum_i u_i * c_i}{\sum_i c_i}; \overline{v} = \frac{\sum_i v_i * c_i}{\sum_i c_i}$$

Once the circle centres have been obtained, a fitting procedure by non-linear optimisation techniques is carried out to determine the camera model.

Figure 3 shows the little difference between points from the true pattern image (+) and those calculated by the camera model (*). Parts (b) and (c) show the histogram of the standard deviations from errors in horizontal and vertical directions, whose mean value is 0.099 pixels. The maximum error is not greater than 0.2 pixels.

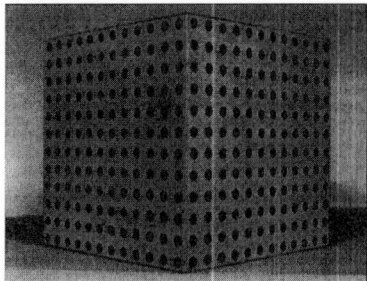

Figure 2. Calibration pattern

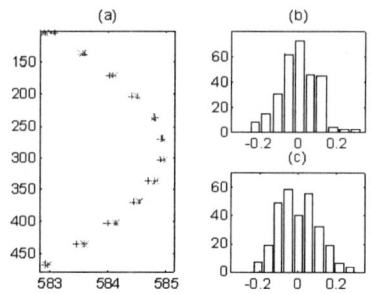

Figure 3. (a) Circle centre (+) and model estimation (*). (b) and (c) xu and xv differences between data and model.

3. Three-dimensional measurement.

3D measurement is the final object of this work. Once obtained the camera model, the next step is to detect the measurement point, with sub-pixel accuracy.

3.1. *Laser stripe sub-pixel detection.*

Due to physical reasons, the source emits a widened laser beam, with an almost Gaussian distribution in both longitudinal and transversal directions. This fact facilitates to compute the centre by centroid estimation of each transversal section from the laser stripe [4]. Once centroids have been calculated, a polynomial interpolation is made to obtain the stripe centres as a function with continuos derivatives.

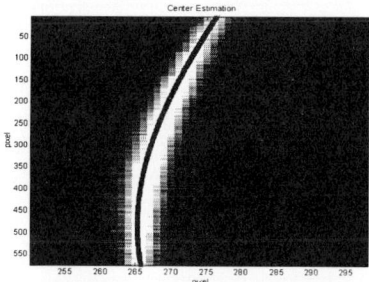

Figure 4. Laser sub-pixel detection.

Figure 4 shows a stripe projected over a plane surface, and its corresponding sub-pixel positioning curve.

The standard deviation of the differences between the centroid and the curve point in each line is 0.37 pixels.

3.2. *3D Reconstruction.*

Reconstruction process has two steps: (a) undistorted coordinates determination, and (b) 3D coordinates computing. The first step corrects the lens distortion from the 2D image-space, and the second one recovers the 3D coordinates be means of the laser stripe restriction.

(a) Undistorted coordinates (x_u, y_u) are obtained from the laser points (x_d, y_d) by:

$$x_d = x_u + x_u * (k_1 * r^2 + k_2 * r^4) + 2 * p_1 * x_u * y_u + p_2 * (r^2 + 2 * x_u^2)$$

$$y_d = y_u + y_u * (k_1 * r^2 + k_2 * r^4) + p_1 * (r^2 + 2 * y_u^2) + 2 * p_2 * x_u * y_u$$

$$r = \sqrt{x_u^2 + y_u^2}$$

To obtain these unknowns a non-linear search from Levenberg-Marquardt [5] can be used that solves the problem by minimising the error.

(b) 3D computing. Once (x_u, y_u) are known, (x_w, y_w, z_w) related to the exterior coordinate system can be found from:

$$x_u = f * \frac{r_{11} * x_w + r_{12} * y_w + r_{13} * z_w + x_0}{r_{31} * x_w + r_{32} * y_w + r_{33} * z_w + z_0}$$

$$y_u = f * \frac{r_{21} * x_w + r_{22} * y_w + r_{23} * z_w + y_0}{r_{31} * x_w + r_{32} * y_w + r_{33} * z_w + z_0}$$

r_{ij} and (x_0, y_0, z_0) being the rotation and translation parameters respectively. All these parameters are determined during the camera calibration process. To solve the undetermined system is necessary to include the equation of the laser plane:

$$Ax_w + By_w + Cz_w = D$$

As any illuminated point belongs to this plane. A, B, C, D represent the geometric laser position in function of the exterior origin.

A test of the 3D-reconstruction method was made for the straight line shown in figure 4 and its 3D reconstruction. The standard deviation of the errors computed was 0.194 pixels.

4. Conclusions.

The accuracy of 3D measurement systems based on camera video and structured light can be improved by using proper algorithms in the three procedures involved: camera calibration, laser stripe detection, and 3D reconstruction. By this way, resolution can be increased by one order of magnitude. Experiments give errors around 0.1-0.2 pixels for the reconstruction of a straight-line.

References:

[1] Tsai R Y 1987 IEEE Journal of Robotics and Automation **4** 323-344
[2] Pedersini F 1997 IEEE Trans. PAMI **11** 1278-1284.
[3] Heikkilä J and Silvén O 1996 Int. J of Patt Recogn & Artificial Intelligence 10
[4] Welch S S 1993 NASA Technical Papers **3331**
[5] Press W H 1992 2nd Edn Cambridge University Press

Sensor Applications
Paper presented at Eurosensors XII, 13–16 September 1998

Six-band infrared pyrometer

Z Bielecki, K Chrzanowski

Military University of Technology, Institute of Optoelectronics, 01-489 Warsaw, Poland,
e-mail:zbieleck@wat.waw.pl

Abstract. Multiband pyrometer MBP 98A developed for non-contact temperature measurement of objects with unknown wavelength depended emissivity is presented in this paper. The pyrometer was designed using single PbS detector of spectral band 1-2.5μm and 6 narrow-band optical filters. It enables temperature measurement of objects of temperature within range 500°C-1200°C with speed of 75 Hz. High measurement frequency allows the users to test fast thermal phenomena. Small, comparing to typical pyrometers, field of view gives opportunity to measure temperature of small details. Generally the obtained parameters enable precise temperature measurement of the objects with small angular dimensions in a real time. It can be used for control of various industrial-technological processes, in research works, as well as for control of classic singleband pyrometers.

1. Introduction

Division of systems for non-contact temperature measurement using radiation emitted by tested object is based on number of spectral bands used by measuring system; and there are generally three group of such systems: single-, dual- and multiband systems. Singleband systems determine object temperature on basis of signal measured in one spectral band; dual systems - in two spectral band, multiband systems - in at least three bands.

Nowadays, at least 90% of systems present on market are singleband system; dualband systems are rather rarely met; multiband systems are still in laboratory stage of development. However, it can be noted an increasing interest in multiband systems [1,2,3,4] as they can potentially bring significant improvement of non-contact temperature measurement accuracy, particularly of so called difficult objects. As such objects are considered mostly objects whose emissivity depends on wavelength and time measured in hot background conditions. These cases are met in many industrial applications; particularly often in semiconductor industry.

This paper presents results on the development of a multiband pyrometer for non-contact temperature measurement of objects with unknown and wavelength-dependent emissivity. This type of objects presents a particular challenge as errors of temperature measurement with classical sigleband systems are often quite high, and what is even more important - it is difficult to estimate these errors.

2. Design of the pyrometer MBP 98A

The general diagram of the MBP 98A pyrometer type is presented in Fig.1. Principle of operation of the pyrometer is as follows.

Infrared radiation from the tested object is focused, using glass BK 7 – silica achromat type optical system on photo-conductive type PbS infrared detector. The optical system was optimised to have the aberration blur smaller than the diameter of the detector. Moreover the optical system is characterised by small F-number that enables to obtain high signal-to-noise ratio. Signal from the object is modulated by rotary plate on which six optical filters are fitted. Spectral bands of most of the filters were chosen with aim to minimise influence of atmospheric absorption on signal from the object coming to the detector.

The detector absorbs infrared radiation with wavelengths shorter than its cut-off wavelength. Absorption of such radiation causes an increase in the materials electrical conductivity and a corresponding decrease in the detector resistance. This effect enables us measurement of the radiation coming to the detector. However, the resistance is changed only by a small fraction, typically less than 0.1 %.

Photo-conductive type PbS infrared detector of spectral band 1- 2.5 µm. was chosen for application in the pyrometer because of a few reasons. First, from simulations it was deducted that 1-2.5 µm. is an optimum spectral band for the required temperature measurement range. Second, low cost materials can be used to design the optical system for 1- 2.5 µm. spectral band.. Third, this type of IR detectors is characterised by relative low price in comparison to HgCdTe detectors.

Two-stage thermoelectric cooler that ensures detector temperature about – 25°C, when ambient temperature is about +25°C is used in order to increase detector sensitivity. The thermoelectric cooler is biasing by subminiature proportional temperature controller of the HY-5600 type, of HYTEC firm.

According to manufacturer recommendations, this detector should be connected to constant voltage power supply. The bias voltage required for optimum performance depends on the size, shape and composition of the detector used. We use low noise reference source of REF-01 type, for supply of the PbS photoresistor. When load resistance R_L and dark resistance R_d of PbS detector are the same the signal of the highest value can be obtained.

A preamplifier is used to amplify a very small signal at the output of the PbS detector. The preamplifier is characterised by low noise, ultra low input current. Total gain of this preamplifier is set 1000 V/V. The preamplifier has typical gain–bandwidth products from DC to 20 kHz. The signal from the output of the preamplifier is sent both to the amplifier in the main measurement channel and to an additional analogue output.

Signal from the object is modulated by rotary plate on which six optical filters are fitted. Spectral bands of most of the filters were chosen with aim to minimise influence of atmospheric absorption on signal from the object coming to the detector. The detector absorbs infrared radiation with wavelengths shorter than its cut-off wavelength. Absorption of such radiation causes an increase in the materials electrical conductivity and a corresponding decrease in the detector resistance. This effect enables us measurement of the radiation coming to the detector. However, the resistance is changed only by a small fraction, typically less than 0.1 %.

Photo-conductive type PbS infrared detector of spectral band 1- 2.5 µm. was chosen for application in the pyrometer because of a few reasons. First, from simulations it was deducted that 1-2.5 µm. is an optimum spectral band for the required temperature measurement range. Second, low cost materials can be used to design the optical system for 1- 2.5 µm. spectral band.. Third, this type of IR detectors is characterised by relative low price in comparison to HgCdTe detectors.

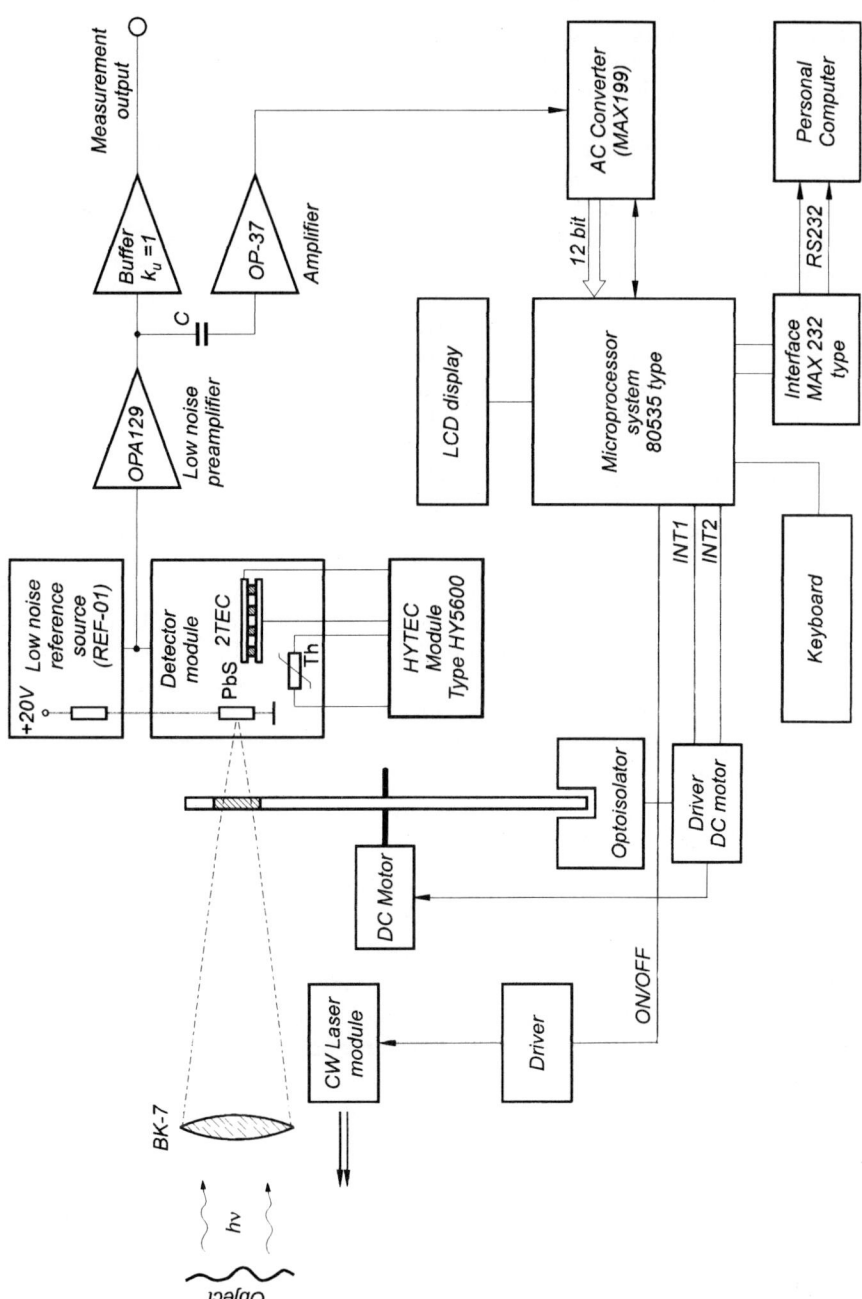

Fig.1. Diagram of the developed pyrometer

Two-stage thermoelectric cooler that ensures detector temperature about – 25°C, when ambient temperature is about +25°C is used in order to increase detector sensitivity. The thermoelectric cooler is biasing by subminiature proportional temperature controller of the HY-5600 type, of HYTEC firm. This device is intended for "cool only" fixed temperature applications where front panel controls and digital readouts are not required. The HY-5600 operates in conjunction with a thermistor bridge to precisely measure and regulate the temperature of a device affixed to the TEC with resolution below 0.1 K. The Th thermistor is located as close to the TEC as possible in order to avoid thermal lag.

According to manufacturer recommendations, this detector should be connected to constant voltage power supply. The bias voltage required for optimum performance depends on the size, shape and composition of the detector used. We use low noise reference source of REF-01 type, for supply of the PbS photoresistor. When load resistance R_L and dark resistance R_d of PbS detector are the same the signal of the highest value can be obtained

A preamplifier is used to amplify a very small signal at the output of the PbS detector. The preamplifier is characterised by low noise, ultra low input current. Total gain of this preamplifier is set 1000 V/V. The preamplifier has typical gain–bandwidth products from DC to 20 kHz. The signal from the output of the preamplifier is sent both to the amplifier in the main measurement channel and to an additional analogue output.

The analogue signal from the amplifier is next converted to digital 12-bit word by IC MAX 199 type converter. The signal after digitisation is registered in computer memory.

All functions of the pyrometer are controlled by microprocessor system, single chip computer 80535 of the Siemens company. This microprocessor system groups and processes data from each filter. Next, the information is sent to microcomputer system. Interface between microprocessor system and the slot RS 232 of personal computer is realised by MAX 232. Additionally, the µP system controls the driver of DC motor. LCD display allows for presentation of measuring data. Rotation of the plate is assured by DC motor, which is controlled by driver and microprocessor system 80 535 type. Speed of rotation was optimised to have situation when signal from the object is modulated with frequency 600 Hz.

CW laser module is used as an indicator, which allows operator to specify place of measurement temperature. The module is controlled by a microprocessor system.

3. Conclusions

An experimental 6-band pyrometer for non-contact temperature measurement of objects with unknown and wavelength-dependent emissivity was developed. It enables temperature measurement of objects of temperature within range 500°C-1200°C with speed of 75 Hz. High measurement frequency allows the users to test fast thermal phenomena. Small, comparing to typical pyrometers, field of view gives opportunity to measure temperature of small details. Generally the obtained parameters enable precise temperature measurement of the objects with small angular dimensions in a real time. It can be used for control of various industrial-technological processes, in research works, for control of thermal imagers indications as well as for control of classic singleband pyrometers.

4. References

[1] Tank V., Infrared temperature measurement with automatic correction of the influence of emissivity, *Infrared Phys.* **29**, 211-212 (1989).
[2] Tank V, Dietl H. Multispectral Infrared Pyrometer For Temperature Measurement With Automatic Correction Of The Influence Of Emissivity, *Infrared Phys.* **30**, 331-342 (1990).
[3] Hunter G. B. et al., Multiwavelength pyrometry : an improved method,*Opt. Eng.*, **24**, 1081-1085 (1985).
[4] Kosonocky W. F., Kaplinsky M. B., McCaffrey N. J. , Multi-Wavelength imaging pyrometer,*SPIE* **Vol. 2225**, 26-43 (1994).

Sensor Applications
Paper presented at Eurosensors XII, 13–16 September 1998
© *1998 IOP Publishing Ltd*

Feasibility study into the use of a micro-machined grating spectrometer to measure fat and protein content in liquid milk products

Desmond Brennan, John Alderman, William Lane
*Barry O' Connor, *Michael McGrath, *Avril O' Sullivan.
NMRC, Lee Maltings, University College, Cork, Ireland.
* Food Engineering Department, University College, Cork, Ireland

Abstract

A study was undertaken to determine the feasibility of building a system capable of measuring the protein and fat content in static dairy based samples. The system would be based upon a micro-spectrometer. Using Beers-Lambert law, spectral changes in absorption and offset due to protein and fat variation in dairy samples are obtained, over the spectral region 800nm to 1100nm. Samples representing a range of liquid milk products, manufactured in the dairy industry were fabricated. These samples spanned typical fat and protein ranges. Each sample after preparation, was chemically analysed using standard Mojennier and Kjeldahl, fat and protein tests. The optical set-up required to optimise the collection from the sample is discussed.

1. Introduction

Near infrared Spectroscopy is well established as a tool for laboratory analysis and is also being used more regularly for in line measurement of water content, fat content , protein content and solid content of products in the food industry. The non-intrusive and non contact nature of NIR spectroscopy makes this a promising technique for use in qualitative and quantitative analysis.

Present commercially available instruments used to carry out such analysis are expensive (from 50 k ECU to 500 k ECU). This spectrometer diffraction grating has been imaged onto a silicon base using the LIGA, X-ray photolithography technique. Standard structure micro-machining techniques are then used to write the grating, this makes such instruments relatively inexpensive. Such a system would be portable and rugged, compared to available FTIR systems.

2. The principle of optical absorption measurements

The nature of optical absorption measurements require that all measurements be made with respect to a zero reference. Absorption in beam transmission instruments follows the Beers-Lambert law.

$$I = I(0) . \exp(-c\,x)$$

$I(0)$ = Radiation intensity incident on the attenuating medium.
c = extinction coefficient of the medium.
x = optical path-length of the medium.
I = radiation intensity after travelling distance x in attenuating medium.

There will be a characteristic extinction coefficient for all material present in the sample. In the present case, the constituents of interest are fat and protein. The extinction coefficient is dependent on the wavelength and on the scattering and absorption properties of the sample constituents. Milk samples are typically composed of 80%-90% water, 5-15% solids, 0-5% fat, 0-5% protein.

3. The system set-up

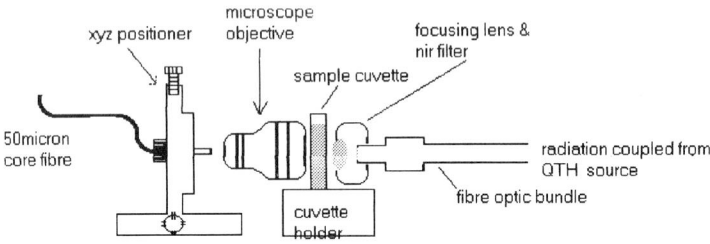

Figure 1 A schematic illustration of the experimental optical setup.

The micro-spectrometer uses a 50 micron core multi-mode fibre to couple light to the spectrometer, this fibre also acts as the system slit. This slit has a numerical aperture of 0.2. A microscope objective is used to couple light from the sample cuvette to the collection fibre. A optical path-length of 2mm was used in this set-up. Radiation was carried via a fibre optic bundle to the sample cell from a quartz tungsten halogen source, a combination of convex lenses were used to focus the incident radiation through the sample. The radiation from the reference and sample measurements are dispersed into spectral component wavelengths. The dispersed spectra is detected by a 256 pixel photo-diode array, suitable electronics sample the current from each pixel and absorption spectra are obtained. The micro-spectrometer is controlled by an rs232 port to a computer.

4. Results

Two sets of samples were prepared , (a) whole milk samples spiked with cream to modify fat levels from 2% to 7% , (b) whole milk samples spiked with whey protein concentrate to modify protein levels from 2% to 6%. The most significant change in each of the sample sets is offset due to scatter in the milk sample. As fat and protein content increases there is also an increase in solid content in the sample, as illustrated in figure 1. The spectral effect of increased solid content is to increase light due to scatter. This scattered light has been shown in closely related fields to be spectrally featureless, thus the decrease in light sampled by the system, due to scatter, results in a spectral offset.

Less subtle are the spectral absorption changes that occur due to fat and protein variation. The spectral region of interest runs from 800nm to 1100nm. Fat typically has an absorption peak at 925nm, with a broad shoulder that coincides with a water absorption peak at 975nm. The whey powder, used to vary protein content in samples, is composed of a broad absorption feature, commencing at 1micron. This broad absorption feature is composed of individual protein absorption bands.

5.1. Fat Variation

Fat was added to full milk samples, these samples were homogenised and chemically analysed. The optical pathlength of the sample was 2mm. Figure 1 illustrates the spectral change in offset due to increased scatter from fat globules.

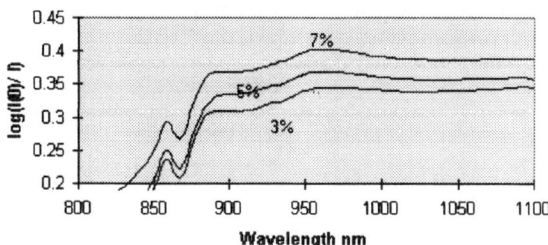

Spectral attenuation for 3%,5%& 7% fat in full milk.

Figure 1 Illustrated are spectra obtained from 3%, 5% , 7% full milk fat solutions. Offset due to scatter is the dominant spectral feature.

There is also an absorption change in the 900nm to 1050nm spectral region caused by fat. This absorption change also takes place in the spectral location of a water harmonic located at 975nm.

The above samples protein content varied little, thus spectral offset and absorption changes are due to fat variation in samples.

5.2 Protein variation

Protein content in full milk samples was varied using whey protein powder. Each sample was calibrated using Kjeldahl analysis, verifying protein content.

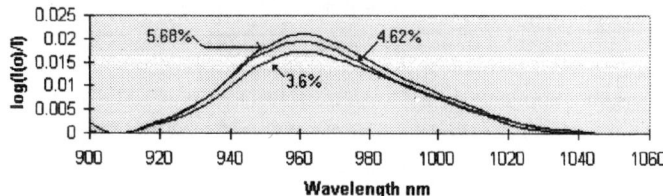

linear baseline removed from protein samples

Figure 2 The assumption of non spectral scatter contribution has been made, thus a linear offset has been removed from each of the sample spectra.

The observed spectral effect of protein variation in full milk samples has been observed for protein variation 3.6%, 4.62% and 5.68% protein. It has been noted that spectral offset due to fat variation in these samples dominate offset, thus the offset contribution due to protein appears to be masked. Spectral changes that occur due to protein, are observed in the spectral region 900nm to 1100nm. The effect is the emergence of a broad protein absorption shoulder around 1020nm. When a linear baseline is removed between 900nm to 1100nm the effect on the water peak is a reduced peak. This spectral effect due to protein is evident in figure 3. The samples in this data set were obtained from a local dairy. They represent protein concentration in a given dairy product. In this sample set fat content was less than 0.5% , thus spectral offset is entirely due to increasing protein content. The emergence of protein spectral features above 1000nm is evident in the higher protein samples.

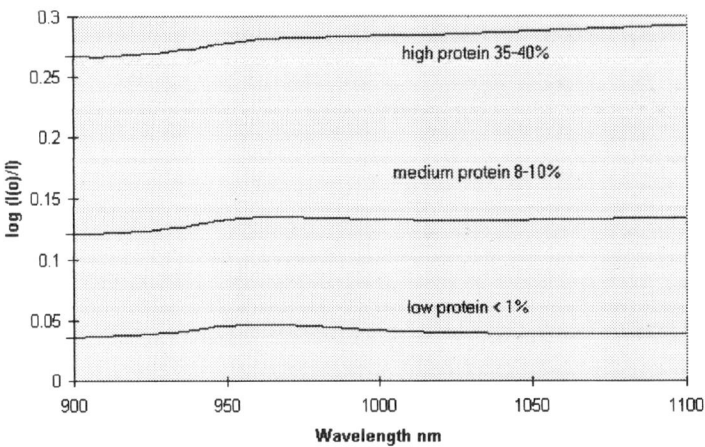

Figure 3 Spectra obtained from a dairy process from a low protein step < 1% , medium protein < 10% and from high protein concentration > 30% .

6. Discussion

The limitation on performance of this instrument is the 50 micron core fibre. It collects little of the forward scattered light. The signal to noise ratio is such that milli- absorbance units are the detection limit in the spectral range 0-1.5 absorbance units.

To detect the 0.1% protein and 0.1% fat changes typically demanded by the food industry, such a spectrometer would require a larger collection fibre typically 400micron to 600 micron. Future work will investigate other more transparent products in the food industry. Other applications for this micro spectrometer will also be investigated.

7. References

1.J.R.V Zanevelt & Bartz 1984 Beam attenuation and absorption meters, ocean optics vii, proc. Soc. Photo opt. Inst Eng 489: 318-324

2. NIRsys Process application notes.

3. Commercial Testing and product control in the dairy industry, J. Foley, M.F Murphy, J Buckley.

Sensor Applications
Paper presented at Eurosensors XII, 13–16 September 1998
© *1998 IOP Publishing Ltd*

Networking intelligent sensing for transport monitoring

D.Guinea, G.S.Mezquita

Instituto de Automática Industrial C.S.I.C. Ctra. Nal.III Km.22 La Poveda
Arganda del Rey 28500 Madrid Spain http//www.gpa.iai.csic.es

Abstract. Present paper offers a concise view of a transport monitoring system from sensor data to remote user applications. Sensing particular variables in mobile transport units or fleets through data communication networks offers a new and promising domain for multiple transducer types. Small, low power onboard autonomous systems must be able to capture, interpret and handle the information flow from sensor to concise meaningful messages through narrow band data links. The European dimension of aim covers multimodal transport with a tracking system, which is open to new sensing requirements, user needs or communication networks. Main design criteria have been: flexibility for remote programming of mobile data collection units including new types of sensors, processing capabilities or data carriers, safety for secure and selective management of the information flow and reliability to warrant data consistence through complex and diverse transmission paths.

1. Introduction and objectives

Mobile communication systems allow today wireless information transference between a vehicle and regular wire networks. Beyond classical radio links for ships or aeroplanes mobile phones, digital phone networks, satellite data links or short distance microwave electronic labelling offer expanding possibilities to information transference. Finding stolen cars, automatic control of trains, wagons or loads in railways, tracking or remote vigilance for fleets of vehicles are new and relevant application areas for these technologies. Tracking systems for transport fleets must be able to remote answer of three critical questions What is the mobile unit? Where is it? and How is the vehicle and its load? Thus, identification, location and monitoring are relevant information issues from a general scope of user domains.

Diverse automatic labelling systems have been developed for many specific purposes such as product identification by bar code readers, car plates recognition by artificial vision systems, electronic payment for highway access control, etc. Automatic location of the vehicle is often related to the particular reading point like motorway or railway tag readers. The entry point is known for the identification code o a particular vehicle offer its current position. However, remote monitoring of transport conditions goes further remote measurement from a remote sensor through a certain data link. Communication networks usually offer narrow band communication channels mostly due to price and technical limitations such as available reading locations, maximum length of the message or price per transmitted character.

Thus, no raw data directly from sensors reaches the network, only is allowed the transference of highly condensed information. Meaningful messages must be rooted on the real environment defining the state of the vehicle and the load by the available set of sensors. But, this flow of information must be filtered, condensed and interpreted by the monitoring goals defined for the system customer. Alarm calls, timing data or requesting services could be available by a remote potential user. Multimodal transport implies that a particular load must be followed from a railway wagon to a cargo ship and then to a certain truck. Load and vehicle are separated mobile units dynamically related in the database of the system, which tracking and monitoring considerations are only partially overlapped in time and shared resources.

Different loads, vehicles and transport conditions would require sensors of diverse nature and performances and also an onboard data processing system with multisensor integration capabilities able to autonomous local decision making. Remote firing of particular processing algorithms for each sensor and download of interpretation rules make a local system with flexible and dynamic data processing. Ideally, the system might be focused to the user needs or requirements using multiple data carrier by means of the existing communication networks with low maintenance requirements and high reliability.

2. From sensors to users: MULTITRACS, an open system

Existing communication networks are able to link mobile units to users in messages that include the identification code, the location and the perceived state. From user to units the message select the unit request for certain information or set the operating conditions for the onboard processor.

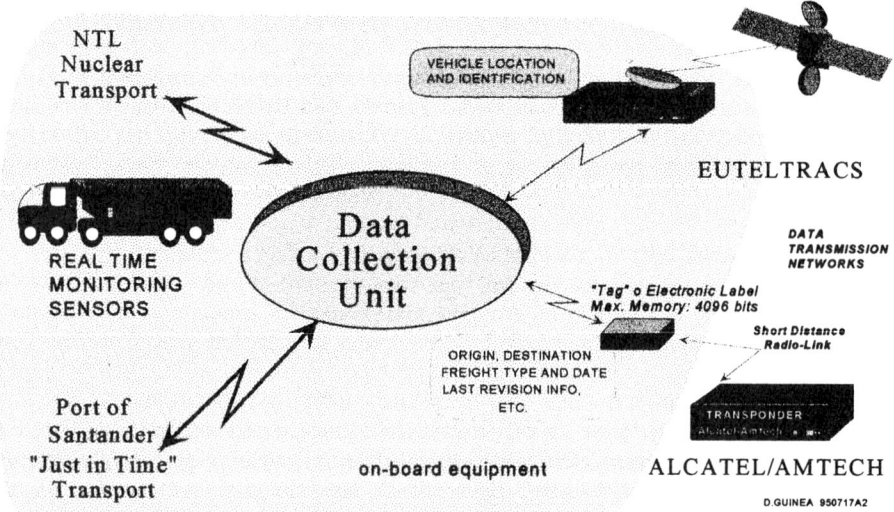

2.1. Mobile units: identification, location, sensing, the DCU. Tracked objects must be endowed able for data acquisition from required sensors, processing algorithms, capabilities for interpretation the perceived state and decision making and management of the communication modules. In this aim, low power, small size, efficient computer boards have been developed, running reliable approximate reasoning modules.

2.2. Data communication networks: Sensing conditions or perception results are already supported by different digital data carriers. These data links are "transparent" to the user application, and also, to the required set of sensors. Information bandwidth, response time, reliability, safety, etc. are global parameters to be optimised selecting dynamically the available resources of the whole system.

2.2.1. Euteltracs satellite communications network. Digital communication service using narrow band channel from mobile units through the Eutelsat satellite constellation. Oriented to human-human communication supports its own global positioning estimation and is able to receive and transmit autonomously from an onboard computer.

2.2.2. Alcatel Amtech railway and motorway electronic labelling network. Electric power is not available on standard containers or freight train wagons. Thus, autonomous electronic labels allow identification of transporting units passing near electronic readers located between the rails or close to the road gates. A bi-directional short distance microwave link allows the tag reader to read or write a limited amount of information to or from the tag memory buffer. Thus, identification of the mobile unit and sensed state are complemented with the location and time of a particular message.

2.2.3. GSM- Tetra. Digital mobile phone networks offer an excellent price/performance ratio for data communications restricted by the are in which the service is available. The fast and wide expansion of these areas over Europe makes GSM a relevant carrier for Multitracs transport monitoring system.

2.3. The "Tracking Centre" (TC): Regional communication centres condense information flowing through diverse networks to an *Information Centre (IC)* where databases are located. Correspondence between vehicles and loads, selection of network resources, data accessibility to users, etc. are decided.

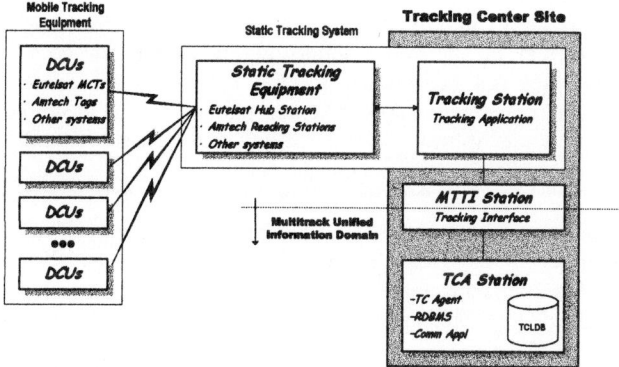

2.4. Internet gateway: Networking digital information to any potential user is at present time very close to INTERNET. Computer devices, software and servers reach almost everybody with easy access and excellent service/cost ratio.

2.6. User applications Able to deal with a particular vehicle and also with a certain group or with the complete fleet of vehicles. A standard local database structure drives inquires or commands from the user to the remote units and receive the answers, the alarm calls or the programmed reports as e-mail messages. This simple software organisation allows easily to

952

present mobile units in the map with visual representation of their sensing state, emergency situations, working parameters, etc.

3. Smart monitoring: the Data Collection Units

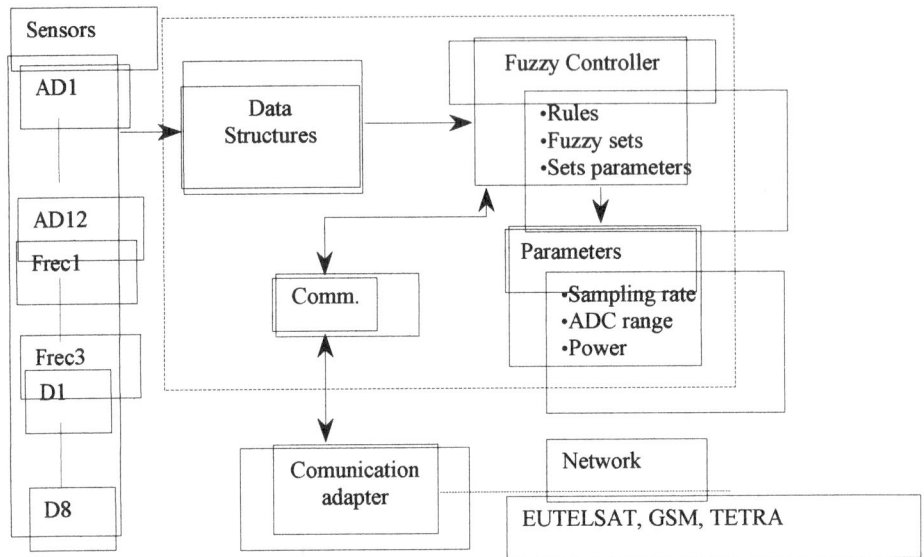

3.1. *The sensor connection*: An embedded computer acquires digital, analogue and frequency coded sensor signal from the vehicle (r.p.m., fuel flow and speed) or its load (temperature, pressure and vibration). Data capture can be driven by external events, programmed timing or particular request from the network. Acquisition parameter for each sensor (range, resolution, sampling rate) can be focused from a remote user to particular transport conditions. Data structures allow a flexible representation of the incoming information. Sensor signal descriptors are firmware modules capable to generate virtual sensors on the DCU memory.

3.2. *Fuzzy Logic inference machine* accomplishes the interpretation, decision making and communication manager in the DCU, scheduling reports or firing alarms messages if necessary. Membership Functions for sensing variables and Fuzzy Logic Rules can be compressed and download through the MULTITRACS network from the user application.

Acknowledgements

Telematics programme of the E.U. has been the main support of present work under MULTITTRACS project . A partial funding of CAM help to complete some aspect of the DCU development in project SEDAM. Thanks to our partners GIGA-BIS, MARBEN-ATOS, AMTECH, RONDA, TÜV, NTL, SEMAT and in particular to Puerto de Santander for their helpful co-operation.

Sensor Applications
Paper presented at Eurosensors XII, 13–16 September 1998
© *1998 IOP Publishing Ltd*

High Temperature Measuring System for MIS Sensor Investigation in Aggressive Gas Atmosphere

A.A.Vasiliev**, W.Moritz*, V.V.Filippov**, A.A.Terentjev**, L.Bartholomäus*, U.Roth*

* Humboldt University, Bunsenstr.,1, 10117, Berlin, Germany
** RRC Kurchatov Institute, 123182, Moscow, Russia

Abstract. A measuring system including high temperature gas cell and gas equipment for the measurement of gas sensitivity of SiC based MIS gas sensors at t temperatures up to 400^0C is developed. The performance of this system in the measurement of sensor properties in fluorine, hydrogen fluoride and fluorohydrocarbons is discussed.

1. Introduction

The measurement of fluorine and hydrogen fluoride concentration in air seems to be an important problem because these gases are dangerous at low concentration and can arise in different technologies such as aluminum and polymers production. But the most important group of gases which requires high temperature measurements is the group of fluorohydrocarbons. These gases are known to be responsible for the destruction of the ozone layer of the Earth atmosphere. HF is a product of their pyrolysis and pyrohydrolysis which could take place in a high temperature measuring cell.

Recently [1] it was suggested to use for these measurements the SiC based MIS structure gas sensor with LaF_3 F^- - conducting solid electrolyte layer between the gate and semi-conductor.

The aim of this abstract is to present our experience in conducting the measurements of MIS structure high temperature gas sensors in aggressive gas, first of all, fluorine containing.

One of the most important questions for such kind of measurement is the proper choice of construction and materials used for measuring cell. This cell should assure the high stability of the capacity measurement of the MIS structure, stable temperature of the sample, should permit to measure the response times of the order of several seconds and not to disturb strongly the gas concentrations at operation temperature up to 400^0C.

2. Experiment

2.1. MIS Structure Gas Sensor

The construction of the MIS gas sensor used in our experiments was described in details in the work [1]. It consists of silicone carbide substrate with further subsequent layers: epi-SiC layer (n_d-n_a = $5\cdot10^{15}$ ... $5\cdot10^{16}$ cm^{-3}, thickness about 5 µm), SiO_2 isolating layer (30 - 40 nm), LaF_3 solid electrolyte layer (200 nm), platinum gate (usually 30 nm thick).

954

2.1 Cell Construction

The design of the cell is presented in Fig.1. All the cell was placed in the external oven used for heating and stabilization of the cell temperature.

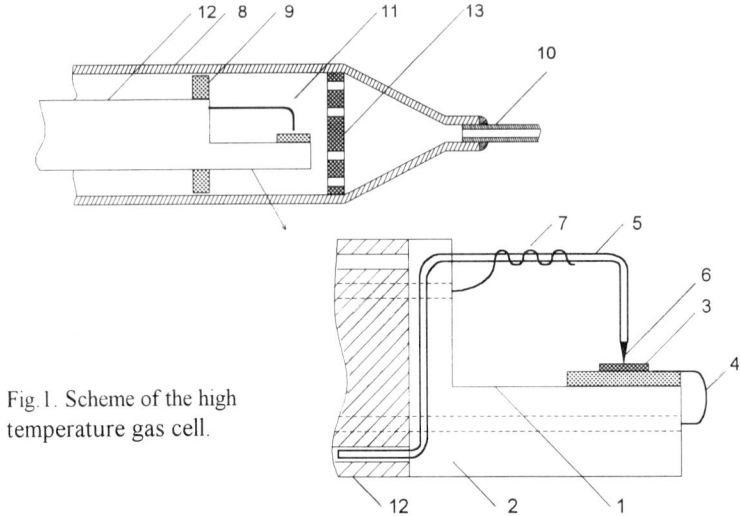

Fig.1. Scheme of the high temperature gas cell.

We were restricted in the use of materials because we should work in fluorine containing gas at temperatures up to 400^0C. Therefore the materials used for the cell were pure nickel (external tube and ring), alumina (isolating stick), platinum and gold (contacting elements) and graphite (top contact needle). Graphite was used for the connection to the gate electrode of the sensor because all other materials damage the 30 nm thick gate at high temperature.

The measuring part of the cell was made in a further way. For the base of it we used the 8 bore alumina stick with 8 mm external diameter. The end of it was treated by diamond saw to form the table (1) and the slot (2). The back side contact (3) was made of platinum and pressed into the slot (1). The external platinum wire (4) was connected to the contact (3) by soldering with gold at the temperature 1150^0C. The soldering was carried out in the oven in room air atmosphere.

The top side contact (5) was made of 1 mm platinum wire. It was fixed in the bore of the alumina stick and prevented from rotation by the slot (2). The graphite needle (6) (0.5 mm in diameter) was pressed to the bore made in the end of contact (5). The external wire (7) was connected to the contact (5) by soldering with gold.

The connection to the top and back side contacts of the sensor were assured by stringiness of the 1 mm platinum wire. It was sufficient to assure a good quality low noise contact. The total noise of the contact did not exceed 0.01 pF when measured 100 pF capacity of the sensor element.

The measuring part was placed to the nickel tube (8). The nickel ring (9), fixed on the alumina stick (12) and place to the nickel tube with a narrow gap, separated the volume of gas measuring cell from the other parts of the tube. The free volume of the cell (11) was about 0.5 cm^3 that at the typical gas flows used in the work of about 4 cm^3/sec assured the gas exchange time in the volume of the cell less than 1 sec. The perforated nickel disk (13) was pressed into the tube (8) and assured the homogeneity of the gas flow.

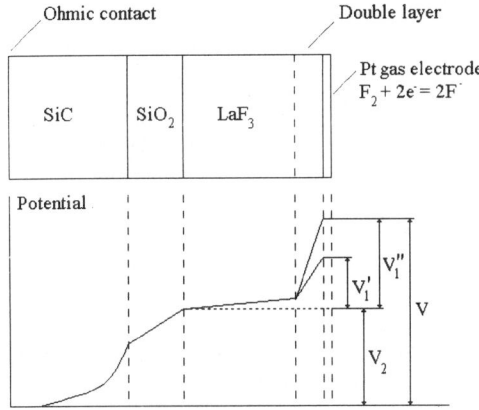

Fig.2. Potential distribution in the MIS stricture with LaF₃ solid electrolyte layer. Constant capacity mode of measurement. V'_1 or V''_1 - potential drop in the electrochemical part, V_2 - potential drop in the „semiconductor part of the sensor.

The gas was supplied to the cell with a nickel capillary tube (10). The internal diameter of the capillary tube is 0.8 mm, the length is 25 cm. These dimensions are enough to heat the gas up to the oven temperature, i.e. the characteristic time of heat transfer process $\tau_{heat} = r^2/a$ is much smaller than the gas residence time in the capillary tube $t_{res} = l/v$ (r is the radius of the capillary tube, l - the length, v - linear velocity of the gas, a - coefficient of temperature conductivity, close to the diffusion coefficient in gas). Indeed, in the conditions of our experiment even at room temperature ($a \approx 0.3$ cm²/s) the difference between these two times is about one order of magnitude and the geometry of the cell assure the gas heating before the cell. At higher temperatures this difference increases.

2.3. Measurement Procedure

The potential distribution in the MIS structure device is presented in Fig.2. The MIS structure used in our experiments consists of two connected parts. The first (electrochemical) part contains the layers of platinum gas electrode, lanthanum threefluoride solid electrolyte and LaF₃/SiO₂ interface. The „semiconductor" part consists of SiO₂ layer, epi-SiC, bulk SiC and back side ohmic contact layer (Fig.2).

The increase of fluorine concentration in gas leads to the increase of the potential drop between top Pt layer and LaF₃ bulk accordingly to the Nernst law (V_1 in the Fig. 2 is the sum of this potential drop and small potential drop in the LaF₃ bulk). The other electrochemical interface (LaF₃/SiO₂) is considered in this scheme as blocking.

Two different schemes of measurement could by realized.

More simple and more direct is the constant capacity scheme. It means that the certain capacity C_0 of the MIS structure is fixed by measuring device (Fig.3). Increase of the electrochemical potential on the Pt/LaF₃ interface leads to the shift of the C-V curve of the structure to the positive direction and this shift is a measure of sensor sensitivity.

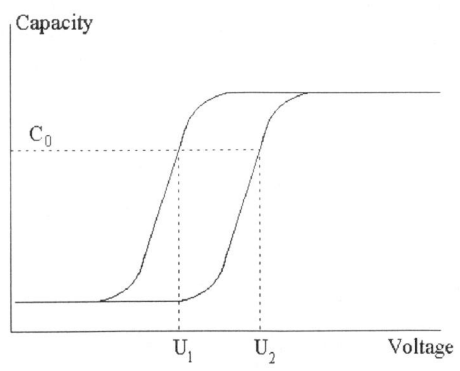

Fig. 3. Shift of the C-V curve under the action of fluorine. C_0 is the fixed value of capacity in the constant capacity mode of measurement. $U_2 - U_1 = V'_1 - V''_1$

From the practical point of view sometimes it is more convenient to use the constant voltage scheme of measurement. In this scheme first the C-V curve of the structure is

956

measured. Then at the constant voltage supplied to the structure $(V = V_1 + V_2 = \text{const})$ its capacity is determined. The increase of the electrochemical potential drop V_1 leads evidently to the decrease of the potential V_2 supplied to the „semiconductor" part of the device and therefore to decrease of its capacity. After this, using the C-V curve measured before, the equivalent potential shift, which would be measured in the constant capacity experiment, is calculated. The constant voltage measurement permits to measure the sensor characteristics faster and with lower noise than it is possible in a constant capacity experiment.

3. Results and Discussion

The high temperature cell and the measuring system was used for the study of the gas sensitive properties of SiC based sensors to F_2, HF and different fluorohydrocarbons.

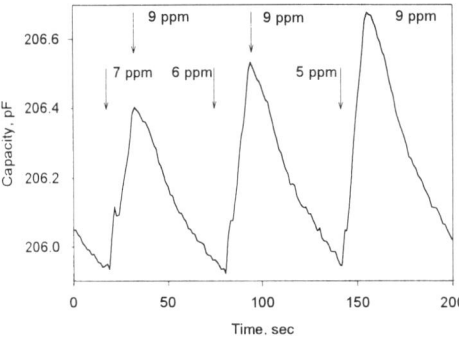
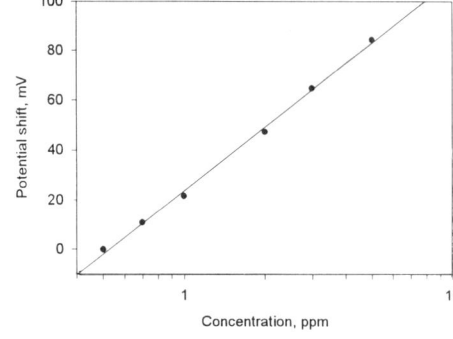

Fig. 4. Response time characterization of the high temperature measuring cell. Jumps of fluorine concentration at 340^0C.

Fig. 5. SiC based sensor sensitivity to HF at 370^0C. The equilibrium sensor signal. Sensitivity equals to 85 mV/dec.

Data presented in Fig. 4 characterize the response kinetics of the sensor and the influence of the cell to jumps in fluorine concentration. The duration of the low concentration jump (7, 6, and 5 ppm, respectively) was in this experiment 13 sec. The time constant corresponding to the gas cell is less than 1 sec and does not disturb the shape of the sensor signal.

Of course the duration of concentration jumps used here is not enough to reach the equilibrium state of the sensor, but it is sufficient for determination of the signal derivative and therefore for determination of the change of the concentration in comparison with a certain reference level, because the derivative is proportional to the concentration change.

The other illustration of the possibilities of our measuring system is the plot presented in Fig 5. The linearity of the plot confirms that the measuring system does not disturb the hydrogen fluoride concentrations at the temperature up to about 400^0C.

Acknowledgments. We acknowledge Volkswagen foundation, INTAS and DFG for the financial support of this work.

Reference

[1] W.Moritz, V.Filippov, L. Bartholomäus, et al. Gas sensors for fluorine using different semiconductor substrates. Proceedings of the 11-th European Conference on Solid State Transducers, „Eurosensors XI", Warsaw, Poland, Sept. 21-24, 1997, vol.1, p. 111.

Magnetic Sensors

Magnetic Sensors
Paper presented at Eurosensors XII, 13–16 September 1998
© 1998 IOP Publishing Ltd

Thin film thickness sensor based on a novel magnetostrictive delay line arrangement

E.Hristoforou, H.Chiriac[1]**, J. N. Avaritsiotis**[2]

Laboratory of Metrology, TEI of Chalkis, Psahna Euboea 34400, Greece
[1]National Institute of R&D in Technical Physics, Iasi 6600, Romania
[2]National Technical University of Athens, Athens 15773, Greece

Abstract. In this paper we demonstrate the application of a novel magnetostrictive delay line (MDL) arrangement in thin film thickness determination, during film production. According to this new set-up, the MDL arrangement can be miniaturized in the micrometer scale, without the use of coils and air gaps, allowing thus a simple and cost effective manufacturing process. Experimental results indicate absence of hysteresis in this sensor, thus allowing a good level of uncertainty.

1. Introduction

Measurements of thin film thickness during manufacturing are important for many reasons and applications. The state of the art concerning in situ thin film thickness sensors is the oscillating quartz technology. The frequency of an oscillating quartz is decreased as the thin film thickness grows up. Having the motivation to develop such a thin film thickness sensor, which could have a more sensitive dependence in the time domain, a smaller dependence on temperature, we have employed the magnetostrictive delay line technique (MDL), which has been used in the past for many sensing applications [1]. Following the basic principles of this technique, the target was the development of a new sensor, according to which pressure or force applied on an MDL, can result in distortion of the propagating elastic signal, so that the detected signal output decreases with respect to the applied pressure [2].

The basic idea was the realization of a delay line set-up, able to be set in the thin film preparation chamber, where thin film deposited on the magnetostrictive element ought to result in attenuation of the propagating elastic signal. Generating dynamic microstrain at the region of the magnetostrictive element covered by the excitation coil, results in an surface acoustic wave propagating along the wafer and detected at the receiving coil region due to the inverse magnetostriction effect. Depositing atoms on the magnetostrictive surface, results in applying static force and consequently static distortion on the propagating elastic wave. Such distortion results in a decrease of the MDL voltage output. The amount of atoms is proportional to the thickness layer deposited on the magnetostrictive element. Thus, such arrangement can result in a thin film thickness sensor used in situ. In order to make such sensing principle industrially applicable, we used a recently developed delay line arrangement, already pending for patentship, using no coils and air gaps to generate and detect elastic strain.

2. Sensor description and operation

The novel delay line set-up able to be used as thin film thickness sensor is illustrated in Figure 1. In this set-up two rectangular magnetostrictive elements (1) are set at the two ends of a glass substrate (2), which has two long parallel cuts, thus being able to act as long acoustic waveguide with rectangular profile. Two pairs of copper ribbons (3) are connected by silver paint (4) aside the magnetostrictive elements. The whole set-up can be controlled by a PC by generating and detecting the pulsed voltage input and output respectively.

Supplying pulsed voltage at the ends of the one pair of copper ribbons, results in pulsed current transmission at the surface of the material within a small depth, due to the skin effect, using the in-between-set magnetostrictive material as part of the current circuit. The transmitted pulsed current causes in turn magnetic field, the profile of which is controlled by the thickness of the magnetostrictive element. As a figure of merit, considering that the pulsed current peak is of the order of 0.1 A and the magnetostrictive element thickness of the order of 20 microns, simulating the case of magnetic ribbons, the resulting pulsed field is of the order of 50 Oe, which is reaching almost the saturation levels in most soft magnetostrictive elements. It is noted that the levels of peak current excitation in case of conductors orthogonal and non-conducting to the MDL is of the order of 10-20 A. The dispersion of the acoustic wave is also controlled by the pulsed field duration and conducting length of the magnetostrictive element, which is now the effective length of the MDL.

The pulsed magnetic field generates an elastic microstrain due to the magnetostriction effect, which is coupled with the glass substrate and then it propagates along the length of it. Consequently, when the propagating acoustic wave approaches the region of the second magnetostrictive element, it is coupled with it, causing elastic strain in the magnetostrictive element, and consequently a magnetic flux change due to the inverse magnetostriction effect. Such flux change induces a pulsed voltage output across the second pair of copper conductors, connected with the second magnetostrictive element, due to the Matteucci effect.

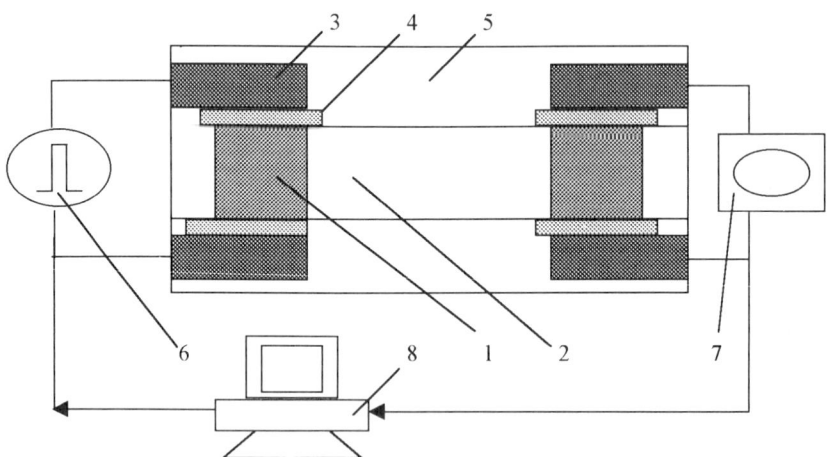

Figure 1. Thickness sensor arrangement. (1) Rectangular magnetostrictive element, (2) Glass substrate, (3) Copper ribbon, (4) Silver paint connection, (5) Sensor support, (6) Pulsed voltage generator, (7) Pulsed voltage detector, (8) Controlling PC.

Increasing and decreasing the pressure around the sensor arrangement, the sensor voltage decreases and increases respectively. Such effect is used by means of introducing pressure in the form of atoms deposited on the magnetostrictive element used as acoustic waveguide. After introducing the sensor in the thin film manufacturing chamber and starting making vacuum, the pulsed voltage output response of it start increasing until a maximum amplitude Vom. This maximized response is due to the presence of a minimized number of gas atoms within the vacuum chamber, beating the acoustic waveguide surface. Then, introducing the plasma gas, the pressure starts decreasing up to a nominal amplitude Von, which is larger than the unloaded sensor response in free atmosphere, Vo.

When the procedure of the thin film preparation starts, atoms of the target beat the deposition substrate as well as the glass substrate of the sensor. Beating the magnetostrictive elements is prohibited by using a metal mask. Thus, atoms are connected on the glass substrate, affecting the characteristics of the waveguide surface, resulting in the distortion of the propagating signal and the detected output. Beating atoms cover sequentially more and more this surface, until an atomic layer is created. Until the moment all atoms cover a single, atomic layer of deposition, the pulsed voltage output response is expected to decay approximately with the same rhythm, down to an amplitude of Vo_1, corresponding to a full single atomic layer. Then, deposited atoms start generate the second atomic layer, thus decreasing the voltage output down to Vo_2, before starting the third atomic layer. Concerning the second layer, the rhythm of the signal decaying is expected to be smaller, while in the third it is even smaller, this procedure being continuous until the deposition stops. Thus, it is expected that the sensor peak output response decreases exponentially with respect to the thickness of the deposited thin film on it.

3. Experiment and discussion

A Leypold sputtering thin film manufacturing facility was used as the means of characterizing the response of the new sensor. The calibrating means was the system quartz oscillator sensor, thus allowing a secondary standard characterization procedure of ± 1 nm uncertainty level. Our sensor was made by using hybrid thick film technology: The glass substrate was of 0,3 mm thickness, 2 mm width and 7 mm length was glued under heat reflow on an alumina support. Two rectangular $Fe_{78}Si_7B_{15}$ magnetostrictive ribbons of 25 microns thickness and 2 mm width and length, were connected at the two ends of the glass substrate, also using heat reflow process. The four Cu ribbons of 0,5 mm thickness 2 mm width and 4 mm length were similarly connected at the edges of the glass substrate and magnetostrictive elements. Silver paint was used to secure connections. The heat was 350°C, in order to obtain stress relief in the magnetostrictive elements, resulting in a higher degree of magneto-mechanical coupling factor. The whole sensor arrangement was set next to the quartz oscillator, to avoid any secondary effects due to deposition non-uniformities. The deposited material was a Si target. The input of the MDL sensor was controlled by a computer and a HP arbitrary waveform generator, driving an amplifier able to transmit pulsed current of up to 1 A p-p amplitude, 1 µs duration and 1 ms period. The pulsed output of our sensor was driven to an HP digital oscilloscope and then to the computer. Controlling the waveform generator and the oscilloscope output was realized by the HP VEE software package. The characterization of the sensor was obtained by plotting the dependence of the amplitude of the MDL sensor response on the quartz oscillator response, by taking independently the two responses with respect to time.

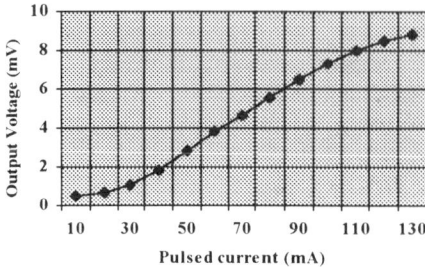

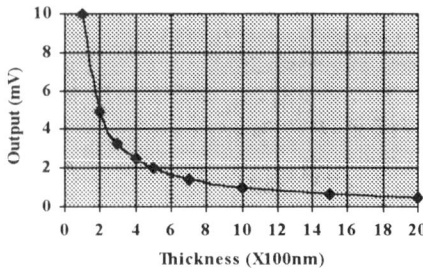

Figure 2. The dependence of the sensor voltage output on the exciting pulsed current input.

Figure 3. The response of the sensor concerning Si target deposition.

The dependence of the amplitude of the sensor voltage output on the exciting pulsed current input is given in Figure 2, while the response of the sensor is given in Figure 3, indicating a good agreement with the prediction of exponential response. This new delay line arrangement improves the state of the art in MDL design for sensing applications. It requires small power consumption to operate. Having a peak current excitation of 0.1 A, the power consumption is less than 1 mW during excitation. Provided that the duty cycle of excitation is about 1:1000, the total power consumption is of the order of 10 μWatts. Using such set-up, the MDL structure becomes simpler and repeatable in manufacturing without use of coils and air gap separation. This set-up is also competent in terms of response repeatability. The involving parameters, like pulsed field and sensor geometry, which determine the operation of such delay line are controllable by the thickness of the magnetostrictive element. Preliminary experiments on force, stress, field and magnet displacement show a monotonic decrease of the output signal from 100% down to zero. Controlling such parameters arrangement by magnetic annealing and geometrical arrangement, one could approach a linear or quasi-linear response.

Multipurpose sensors can also be realized, due to this technique. Preliminary experiments indicate a different bandwidth response concerning field and stress effect on the whole arrangement. Hence, a FFT design could be incorporated to show separate the two different sensing inputs, like field and stress. This observation could also be used to avoid the undesirable effect of ambient fields at the vicinity of the arrangement. Considering the ambient field shielding process another technique could be also used. A permanent magnet ribbon could be set below the magnetostrictive element in order to strongly bias the MDL set-up. Another possible arrangement could be the use of only one magnetostrictive element for generating and detecting the magnetoelastic wave, by using the reflected wave in the glass acoustic waveguide.

Future work is underway in order to deeply study the physics behind such arrangement, face up the problems mentioned in the previous sections and realize complete devices based on thick and thin film technology. The scaling down process is based on the fact that the ratio of the rectangular x-y and z dimensions are to remain the same.

References

[1]   Vazquez M and Hernando A 1996 *J Phys D Appl. Phys* **27** 1596
[2]   Hristoforou E 1996 *Sensors & Actuators* **69** 183-191

Magnetic Sensors
Paper presented at Eurosensors XII, 13–16 September 1998
© *1998 IOP Publishing Ltd*

A Low Cost Analogue Electrical Current Transducer

Alexander Pross, Christopher Lewis, Thomas Hesketh

Nonlinear Systems Design Group
Coventry University, School of Engineering, Q-Block
Priory St., Coventry, CV1 5FB, UK
Tel.:+44-1203-838842, Fax.:+44-1203-838949, email:pross@coventry.ac.uk

Abstract: In this paper a new concept for an electrical current transducer is described, which does not employ Hall effect sensing. In the new approach a limit cycle is used to enable the transmission of direct current, which is insensitive to temperature variation.

1 Introduction

Current transducers based upon Galvanic isolation are used in many industrial applications, e.g. industrial drive systems. Most electrical current sensors employ magnetic flux sensing using a Hall sensor in a ferromagnetic core. The sensor performance is improved by the use of a closed loop structure. The major disadvantage of Hall sensing is it uses a semiconductive device and consequently is subject to temperature variation. In order to overcome this effect a transducer which does not use a Hall sensing element has been developed. A carrier signal is provided by an internally generated oscillation (limit cycle) which tresults in transmission of direct current. The system uses a closed loop, nonlinear structure to meet this objective, the design leads to a reduction in the number of components. Other advantages of the new transducer are that the core is not cut and windings are continuous, resulting in improved magnetic efficiency, lower manufacturing costs, and insensitivity to temperature variations.

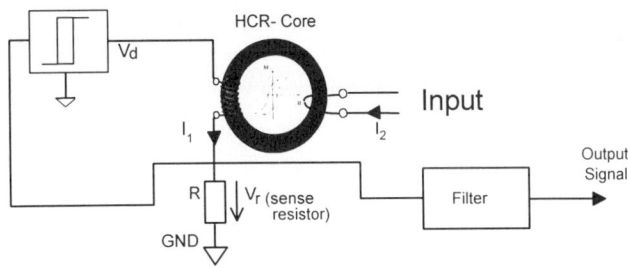

Fig.1 A simple model of the current transducer

2 Structure

The basic concept of the transducer is shown in fig.1. The system contains an amplifier with a rectangular hysteresis transfer characteristic which is used to excite the toroidal core which has a rectangular B/H magnetic characteristic. The sensed current is fed back to the amplifier to provide a current control system which results in a limit cycle. The input signal winding usually consists of a single turn. A low pass filter is used to recover the signal from the modulated signal at the ouput of the transducer.

3   System Operation

The device relies on the generation of a limit cycle and it was found necessary to develop a closed loop model of the transducer to assist in the desing process. The operation of the device may be explained by regarding the limit cycle as a high frequency dither. In the core the flux is triangular and it is known that under such conditions a linearised characteristic may be produced [1]. The dual input describing function concept is stricly not valid due to the low filtering within the loop, nevertheless, a 'feel' for the system behaviour may be obtained, since the measured current has a much lower bandwidth than the limit cycle frequency.

4   Simulation

The model of the current transducer was designed using MircoSim Design Centre V 8.0 (PSPICE for Windows). The nonlinear equation for the core model developed by Jiles- Atherton is provided by the PSPICE library. The core data was estimated by the method described in [2] [3] and [4], which gave an adequate model for the magnetic cores used. Fig. 2 shows the amplifier voltage Vd and the sense resistor voltage Vr in the unforced condtion, i.e. with no input current.; Fig.3 shows the same voltages with an input current signal.

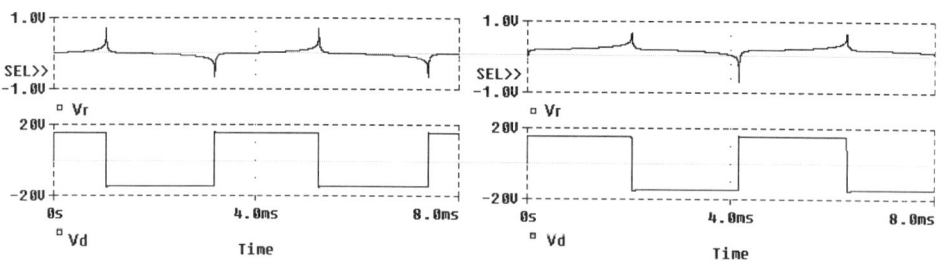

Fig.2 Vd and Vr without input current   Fig.3 Vd and Vr with input current

5   Measurement

The simulated voltage waveforms in the upper recordings were confirmed by observations on a practical device. In fig.4 and 5 the voltage waveforms Vd and Vr are shown in the unforced and forced conditions. The prototype sensor had a limit cycle of 200Hz and which was removed from the sense resistor voltage Vr by a low pass filter which had a corner frequency of 100Hz. A frequency response of the system is shown in fig. 6, which shows that the low pass filter dominated the overall system performance. The dc characteristic of the transducer was tested with two different cores as shown in fig.7.

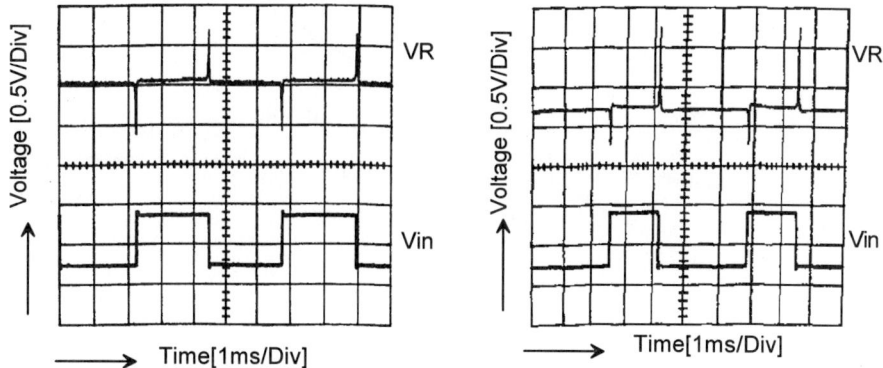

Fig.4 Voltage without input current　　　Fig.5 Voltage with input current

The gain of the dc characteristic was set by altering the winding ratio N2/N1 and the sense resistor R. The system bandwidth could be improved by increasing the limit cycling frequency. This can either be done by increasing the value of R or by decreasing the cross sectional area of the core.

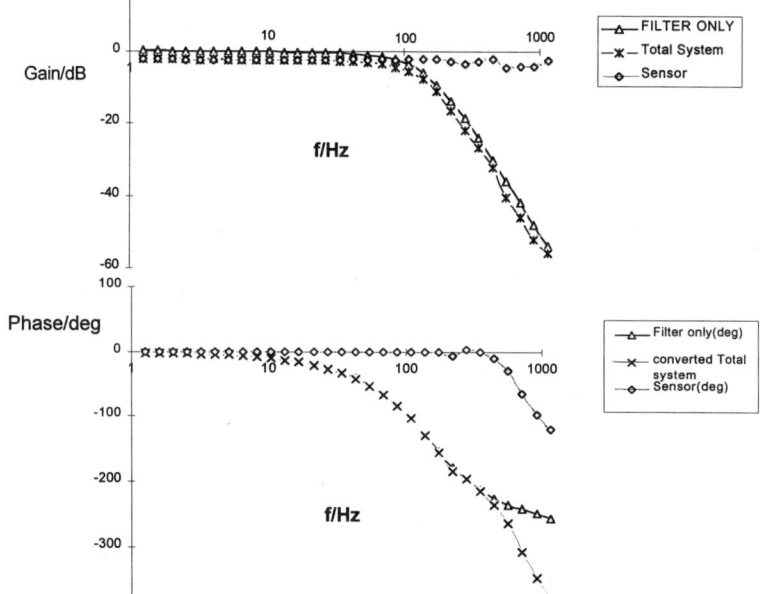

Fig.6 Frequency response

966

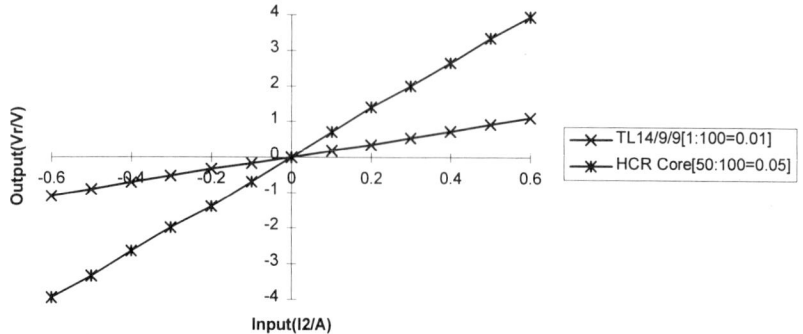

Fig.7 I/O characteristic for a variety of cores with the winding ratio 1:100 and 50:100

6 Conclusion

A transducer has been developed which can sense low frequency current down to dc and has a linear transfer characteristic. Because there is no requirement for a semiconductive sensor the device is insensitive to temperature variation and may be constructed simply since the winding is continuous, there being no need to cut the toroidal magnetic core. The bandwidth and the accuracy of the transducer is limited by the frequency of the limit cycle.

7 References

[1] Gelb A. and Vander Velde W., Multiple Input describing function,McGraw-Hill,1968

[2] Jiles and D.L Atherton, Ferromagnetic Hysteresis, Sep. 1983,IEEE Transactions on Magnetics ,Vol. , No.5,pp. 2183-2186

[3] Jonathan H.B. Deane, Modeling the Dynamics of Nonlinear Inductor Circuits, Sep. 1994 IEEE Transactions on Magnetics ,Vol. 30, No.5,pp. 2795-2797

[4] MicroSim Corporation, Circuit Analysis Reference, 1997 Version 8.0

Magnetic Sensors
Paper presented at Eurosensors XII, 13–16 September 1998
© *1998 IOP Publishing Ltd*

AC - driven AMR and GMR magnetoresistors

P. Ripka [1]**, M. Tondra, J. Stokes and R. Beech.** [2]

[1] Czech Technical University, Faculty of Electrical Engineering, Dept. of Measurement,
166 27 Praha 6, Czech Republic. ripka@feld.cvut.cz
[2] Nonvolatile Electronics, 11409 Valley Road, Eden Prairie, MN 55344-3617, USA.
markt@nve.com, http://www.nve.com

Abstract. Anisotropic (AMR) and Giant (GMR) Magnetoresistive sensors are attractive for industrial applications, as they are more sensitive and stable than Hall sensors. Their performance may be improved by AC excitation: flipping for AMR and AC biasing for GMR. AC excitation lowers the hysteresis and reduces the offset and in some cases also decreases sensor noise. The sensitivity to perpendicular fields is reduced in case of AMR sensors. AC-driven magnetoresistors are competitive to miniature fluxgate sensors and they are suitable for precise applications such as compass.

1. Introduction

Ferromagnetic (AMR and GMR) magnetoresistors are superior to any semiconductor magnetic field sensors for low-field applications [1]. Recent developments in this technology allow ferromagnetic magnetoresistors to compete with miniature fluxgate sensors [2]. Magnetoresistors has been used for proximity switches, speed sensors, angular and displacement measurement, compasses for automotive navigation systems, current meters, magnetic ink reading and security applications.

Semiconductor magnetoresistors are less sensitive than AMR and GMR, but newly developed multiple-element InSb magnetoresistors in combination with rare-earth permanent magnets are attractive for position-sensing in automotive applications as they are stable in required -60 to +200 °C temperature range [3].

2. Anisotropic Magnetoresistors (AMR)

Flipping field perpendicular to the sensing direction were first generated by an external coil [4]. It is necessary to fully saturate the sensor by flipping pulses having opposite polarity to avoid distortion of the sensor characteristics caused by perpendicular fields. The Philips KMZ10A1 sensor bridge output was sensed by a gated integrator for intervals between flipping pulses of 10 μs duration and 100mA amplitude which caused a field of 4 kA/m. This technique reduced the hysteresis below 100 ppm FS and resulted in improved offset stability on the order of 10 nT. Significant noise reduction was not observed for single-axis flipping [5]; however, suppression of Barkhausen noise was reached by using a rotating bias [6].

Newly developed AMR sensors such as Honneywell HMC 1002 and Philips KMZ 51 have integrated flipping and feedback coils . They are more compact and even if they require a large amplitude of flipping current such as 4 A, they still may work from a 5 V source when switched-capacitor techniques are employed.

3. Giant Magnetoresistance Sensor Construction

The GMR resistors are fabricated from a "sandwich" type material: two ferromagnetic layers separated by a nonmagnetic conducting layer. The structure is of the form NiFeCo / CoFe / Cu / CoFe / NiFeCo. The strength of magnetic coupling (both parallel and antiparallel) between the two ferromagnetic layers is strongly dependent on the thickness of the intermediate Cu layer. By making the Cu layer 35Å or more, coupling between the two magnetic layers is minimised, which results in significantly lower saturation fields. This lower saturation field comes at the expense of the total GMR signal, which is a relatively modest 5% in these films. Operation of the basic GMR sandwich stripe has been described in detail elsewhere [7].

Using standard semiconductor processing techniques, the material is patterned into resistors that are 1.8 μm wide and have a total resistance of about 1kΩ. The resistors are connected into a bridge configuration using a patterned 0.5 μm thick Al film. Over the top of the bridge are the integrated biasing straps. These are formed using a patterned 1μm thick Al film. The field from this biasing strap is approximately 0.1 mT / mA.

Without a biasing field the four magnetoresistors are symmetrical, so the bridge has, in ideal case, zero output, independent on the field. Biasing field causes the shift of resistor characteristics so that the bridge output is proportional to the field: large field characteristics is shown in Fig. 1 for several values of bias current: sensitivity is increasing with the current value, but at currents over 12 mA the sensitivity changes are small. However, bias current is limited to approx. 15 mA due to the heating of the chip. Fig. 2 shows that the characteristics

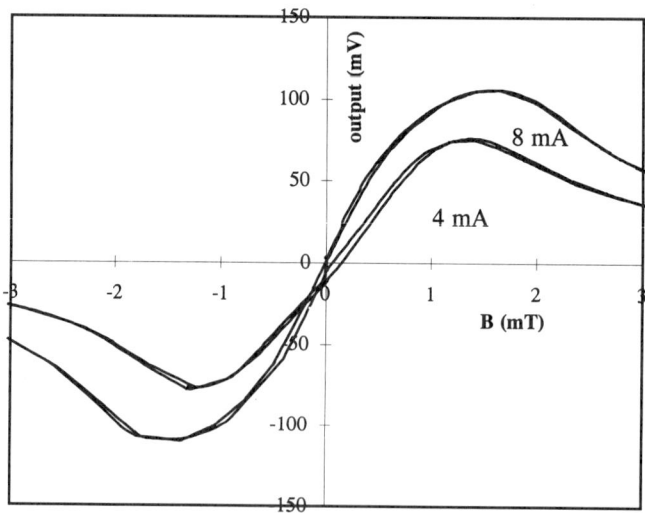

Fig. 1 GMR characteristics for bias current of 4 and 8 mA

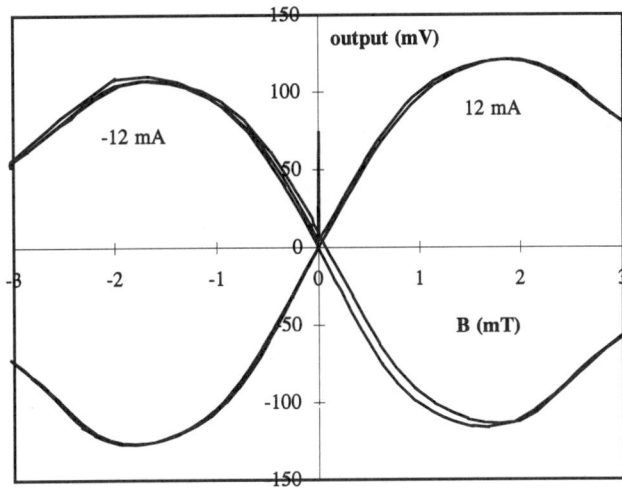

Fig. 2 GMR characteristics for +/- 12 mA bias current

is reversed for opposite bias polarity. Unlike the flipping in case of AMR, bias field should be present during the measurement period. The DC biased GMR sensor has 5% hysteresis in the 300µT range as seen from the Fig. 3.

MR sensors may be also biased by an AC current. For the first experiments the sensor was supplied from built-in source of SR 830 Lock-in amplifier and the bridge output was measured using the same instrument. The resulting sensitivity was independent on the bias frequency up to 100 kHz. An example of the measured characteristics is in Fig. 4 for 10kHz/ 5mA rms bias. The hysteresis was reduced to 1%, offset to 1 µT and we also observed reduction of the sensor noise by the factor of 3.

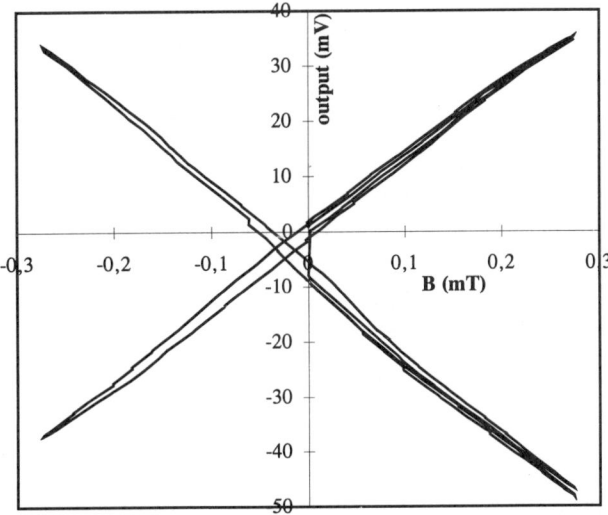

Fig. 3 Low-field characteristics for +/- 8mA bias

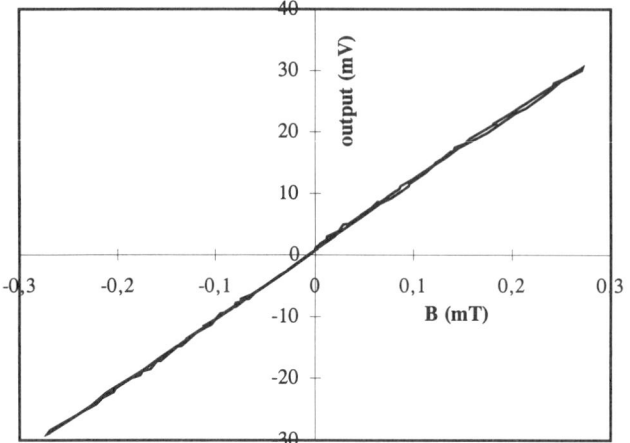

Fig. 4 Low-field characteristics for AC bias

4.Conclusion

AC excitation of magnetoresistors improve their characteristics. In the case of AMR the short flipping pulses of large amplitude cause changes of the sensor remanent magnetisation and the sensor output is measured between these pulses. GMR sensors can be biased by much lower current but during the whole measurement cycle. The performance of the bias waveforms other than sinewave which was used in the presented study might be of advantage: squarewave bias current with short duration may allow to use higher field amplitudes and keep the chip temperature low at the same time.

Acknowledgement:

This work was supported by the Grant Agency of the Czech Republic under No. 102/96/1251.

References

[1]   Ripka P. 1994 *Sensors and Actuators A* **42** 394-397.

[2]   S. Kawahito, H. Satoh, M. Sutoh, Y. Tadokoro 1996
       Sensors and Actuators A **54**, , 612-617.

[3]   Heremans J. 1997, *Proc. 16th Conference on Properties and Applications of Magnetic
       Materials, Chicago, IIT,* unpaged

[4]   Ripka P. J. Appl. Phys. **79** (8) , 1996, 5211-5213

[5]   Ripka P. 1996 *Journ. Magn. Magn.* **157/158** 424-427

[6]   E. Paperno E , Kaplan B.Z. 1995 *IEEE Trans. Magn.* **31** 3161-3163

[7]   J.M. Daughton, 1994 *IEEE Trans. Magn.* **30** 364-368

Magnetic Sensors
Paper presented at Eurosensors XII, 13–16 September 1998
© 1998 IOP Publishing Ltd

A novel non–plate like Hall Sensor

Randjelovic Z[1], Haddab Y, Pauchard A and Popovic R S

EPFL, Swiss Federal Institute of Technology, Department of Microengineering, Institute of Microsystems, 1015 Lausanne, Switzerland

Abstract. Conventional Hall devices are plate like, i.e. made in a thin semiconductor layer. We present here a Hall device made in a semi-infinite piece of semiconductor. Its active zone is limited by only one non-conductive surface. Although the base of the device is open, the part of the limited active zone filled by the current remains thin. The novel device exhibits a sensitivity of 300 V/AT. We have shown that the characteristics of the non–plate like Hall device when its size is small asymptotically approach the characteristics of the conventional Hall plate. It opens the door to the realisation of 3D Hall device in the technology of the vertical Hall devices. In order to use the novel sensor as a stand–alone device we improved it by an additional doping at the surface.

1. Introduction

Usually Hall plates have a very thin active zone. These Hall device structures are however not compatible with the vertical Hall devices [1], which need a very deep sensitive zone. In order to make possible the realisation of a 3D Hall device [2], we developed a new non–plate like Hall sensor. Unlike conventional Hall devices (Fig. 1.a)) the new Hall sensor has an "open bottom", i.e. it is not isolated at all from the bulk (Fig. 1.b)). Our sensor structure is completely compatible with the well established technology for vertical Hall devices. The advantages of the devices realised in this technology are their long–term stability, (<100 ppm over 20 years) as well as their good sensitivity due to the low doping.

2. Sensor structure and operation

Fig. 1 shows cross sections of a conventional and the novel non–plate like Hall device. The main difference between a classical Hall plate and the new sensor considered here is the form of the active zone. The standard Hall plate usually has a thin plate–like active

---

[1] E-mail: Zoran.Randjelovic@epfl.ch

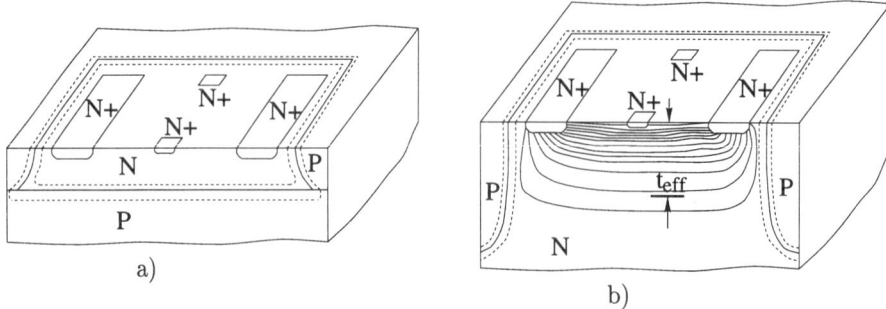

Figure 1. a) A conventional Hall plate. b) The novel non–plate like Hall device.

zone. In our sensor the active zone (N substrate) is still surrounded by the P-well ring, nevertheless the base of the sensor is open from the bottom side (Fig. 1.b)). The bias current flows only within the active region.

The non–plate like Hall sensor we designed is invariant for a rotation of $\pi/2$ [3]. All four contacts are equivalent and are situated within an octogon. This can be advantageously used to cancel the sensor offset, as explained in [4].

The supply–current-related sensitivity S_I of a conventional plate like Hall devices is given as:

$$S_I = G(\lambda)\frac{r_H}{q\,n\,t_{eff}} \tag{1}$$

where G is geometrical factor, r_H is the Hall factor, n is the majority carriers concentration and t_{eff} is the effective thickness of the active zone. The factor $\lambda = c/l$ is the ratio between the sum of the contact lengths c and the length of the circumference l.

For the octogonal geometry and given magnetic field, G has been calculated by Versnel [5]. A very important parameter which directly affects S_I is the thickness of the active zone. Any conventional plate like Hall device has a fixed thickness. However, our novel non–plate like Hall device has a variable effective thickness of the active zone referred to as t_{eff} (Fig. 1.b)). By measuring S_I one can calculate t_{eff}. In this way it is possible to find the optimal geometry of the non–plate like Hall devices, i.e. to design the sensor with t_{eff} comparable to the thickness of the active zone of conventional Hall plates.

3. Experimental results

In order to examine the behaviour of the devices, we made several variations of the design. First, we studied the current distribution within several different volumes of the shielded active zone. Then, we additionally doped the surface of the devices in order to make a more conductive shallow region and to prevent the penetration of the current into the bulk. Finally, using the devices with additional surface doping we examined the variations of G (Eq. 1).

3.1. Non–plate like Hall devices with the different active volumes

The non–plate Hall devices are fabricated by the well established technology for the vertical Hall sensors. The devices we have studied differ by the volume of the active zone and they have the same λ. We fixed λ because it affects $G(\lambda)$ in Eq. 1 [5].

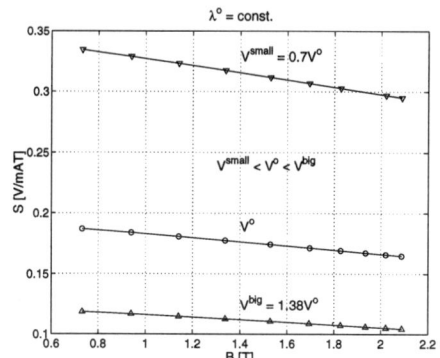

Figure 2. Measured sensitivity of the novel sensor with λ constant but different volumes of the active zone.

Fig. 2 shows the sensors sensitivity for different volumes of the active zone. Corresponding computed effective thicknesses of the active zone are listed in Table 1. By shrinking the active zone, t_{eff} becomes smaller, i.e. the current runs closer to the chip surface. For the smallest device only one third of the shielded active volume has been used effectively by the current (Table 1, second row). This shows that t_{eff} can be reduced by a factor 3 simply by reducing the size of the device.

Table 1. The effective thickness of the active zone

Device		$\| \lambda^o V^{big}$	$\lambda^o V^o$	$\lambda^o V^{small}$	$\| \lambda^{small} V^o$	$\lambda^o V^o$	$\lambda^{big} V^o$
Non-doped	t_{eff} [μm]	$\|$ 29.80	19.65	10.96	$\|$		
Doped	t_{eff} [μm]	$\|$ 9.83	8.52	7.08	$\|$ 10.49	8.52	6.64

3.2. Devices with an additional n–type shallow doping and different active volumes

In a variation of the sensor design, some devices have received an additional technological step during the fabrication process. Namely, the surface of the devices was additionally n–doped. This shallow n–layer increases the conductivity close to the chip surface. Hence, the current is forced to run through that layer. All devices have the same factor λ. The sensitivities of the devices are shown in Figure 3.a). The effective thicknesses of the active zones (Table 1, third row) almost do not change, which proves the existence of the well defined conductive region in spite of the open device base.

3.3. Devices with the additional surface doping, different factor λ and fixed volume of the active zone

In order to examine the geometry effects we designed several devices with the same active zone but different lengths of the contacts. The basic idea was to test the influence of the factor λ on G in Eq. 1, with respect to [5]. Fig. 3.b) shows the sensitivity of the devices. The sensitivities change in accordance with the variations of the geometrical factor G with an accuracy lower than 1%. However, due to the non–limited base there is still a small variation of t_{eff} (Table 1, third row).

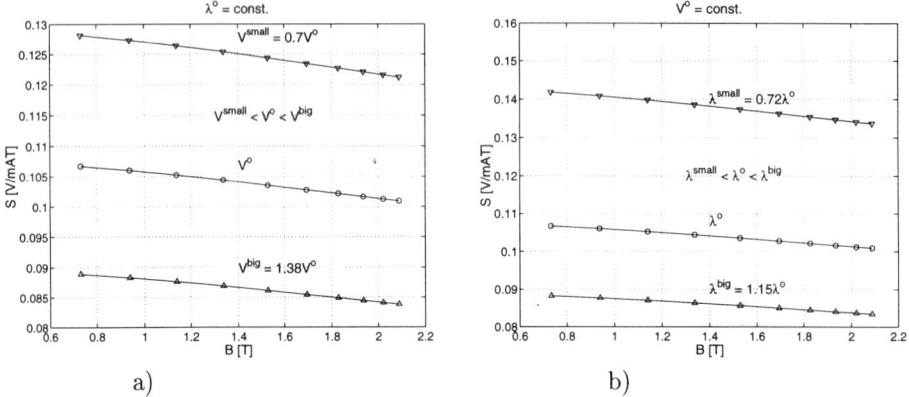

a)   b)

Figure 3. Measured sensitivity of the sensors with an additional doping at the surface. a) Sensors with the same λ and different volume. b) Sensors with the same volume and different λ.

4. Conclusion

A novel non–plate like Hall device has been presented. The device behaviour has been studied through several variations of the design. The characteristics of the device do not scale down linearly with its dimensions. In small devices the part of the active zone filled by the current is thinner. Therefore, the small sized non–plate like Hall devices asymptotically approach the characteristics of conventional Hall plates. We succeeded in reducing the effective thickness of the active zone by a factor three as compared to the thickness of the shielded active zone.

The non–plate like Hall device has been enhanced by the introduction of an additional surface doping. We have shown that the current flows close to the chip surface in spite of the open base, making it similar to a conventional Hall plate. The supply–current–related sensitivity as well as the input resistance are decreased due to the additional doping, whereas the supply–voltage–related sensitivity has been improved by a factor five.

References

[1] Popovic R S 1984 *IEEE Electron Device Letters* **EDL-5** 357–8

[2] Burger F *et al* 1997 *Proc. EUROSENSORS XI* **2P2–3** 631–4

[3] Falk U 1989 *Transducers* (Amsterdam: Elsevier) **2** 751–3

[4] Popovic R S 1991 *Hall Effect Devices* (Bristol: Adam Hilger)

[5] Versnel W 1981 *J. Appl. Phys.* **52** 4659–66

Magnetic Sensors
Paper presented at Eurosensors XII, 13–16 September 1998
© *1998 IOP Publishing Ltd*

Magneto-optic sensors for signalogram analysis in criminalistics

V GVishnevski, S V Dubinko, S V Levy*, A S Nedviga, A R Prokopov

"Domain" design bureau at Simferopol State University, Studencheskaya str.12, Simferopol, 333610, Ukraine; *National Technical University "Kiev Politechnical Institute", Pobedy ave.37, Kiev, 252056, Ukraine

Abstract. Development and utilization of magneto-optic converters based on epitaxial Bi-substituted garnet films for analysis of recorded signalograms are described. Variants of garnet layers for special tasks from technological points of view are discussed. The soft-hardware system, magneto-optic image analysis tools and their applications in criminalistic phonoscopy examinations and lost-data restoration usages are reported.

1. Introduction

Recorded pattern topology measurements is an important problem in science and technics (to analyze the recording process, for example). Numerous magneto-optic (MO) methods give us possibilities "to see" the stray field distribution and to make quantitative evaluations. Kerr effect microscopy is characterized by submicron spatial resolution, but requires reflecting properties of the media surface and it is impossible to obtain high-contrast images without the computer image processing technique [1]. Another MO visualization method is based on film converters whose functional advantages are conditioned by Faraday effect in Bi-substituted epitaxial garnet films (EGF) [2,3]. In order to obtain a contrasted image the magnetic pattern at the media has to be converted into an optical one: it is necessary to press the tape or disc surface against a MO transducer the magnetization structure of which is coupled to the recorded pattern and reproduces it in a given range of spatial frequencies. As a rule uniaxial anisotropy EGF with binary stripe domains are used, but it has been demonstrated experimentally that non-removable domain noises and limited range of forced domain restructuring prevent from reaching a dynamic range above 25-30 dB. Thus we have succeeded in the development of high-sensitive EGF with "facet" structure and "easy plane" anisotropy films for expansion of MO method capabilities. It now appears that the films utilization makes possible not only the composite signalogram analysis for criminalistics (phonoscopy examinations) but lost-data restoration.

2. The sensors and their features

In practice it is intricate to visualize recorded-pattern stray field with amplitude lower than 50 Oe with ~500 l/mm resolution when only bubble garnet films are used. For this reason "facet" garnet films were proposed to obtain maximum spatial resolution and sensitivity. Contrary to operative visualisation with the "standard uniaxial" films, these EGF must be heated up to the Curie temperature by "contact printing"; only in such a way it is possible to receive the dynamic range ~55 dB. The recorded patterns of the tape or disc are "printed" in the garnet film by the thermal transfer and then visualized after cooling down to room temperature. The magnetic image of signalogram has been investigated is obtained on the "facet" structure background. These patterns could be erased by the thermal treatment; both the saturation magnetization $4\pi M_s$ and coercivity H_c of magnetic media do not change drastically in the temperature range $T = 0\text{-}100$ °C. It is noteworthy that "contact printing" method was provided earlier with GdFe and TbFe alloy films [4] and observed by Kerr effect. To obtain the analog type MO conversion for reconstruction of the recorded signal form the "easy plane" anisotropy films were synthesized [5]. It was shown that the local remagnetization of monodomain "planar" anisotropy layer along normal-to-plane direction is of a linear character. Initially, the films were developed to input analog data from the media into the optic processor channel.

The sensors were multilayer plates composed of a glass substrate, a garnet film grown on the monocrystalline gallium-gadolinium garnet (GGG) substrate and a thin antiabrasive mirror. All film specimens were prepared by means of liquid-phase epitaxial growth. Base compositions were $(Bi,Lu,Ca)_3(Fe,Ge)_5O_{12}$ and $(Bi,Tm)_3(Fe,Ga)_5O_{12}$ for "easy plane" films. Main properties are: $4\pi M_s = 70\text{-}300$ Gs, film thickness 1-5 μm, GGG-substrate thickness 0.3-0.5 mm, Faraday rotation $\theta = 0.9\text{-}1.5$ °/μm, coercivity level $H_c = 0.1\text{-}0.3$ Oe. In-plane disorientation of "easy plane" films is lower than 5 %. Base composition of "contact printing" films was $(Bi,Lu,Sm)_3(Fe,Ga,Al)_5O_{12}$ where Sm was used to increase the coercivity and Al was used to reduce Curie temperature T_c. Al ions substitute not only for tetrahedral but, more extensively than Ga ions, for octahedral positions in the garnet. On the other hand, $4\pi M_S$ regulation was provided with Ga ions. The EGF lattice constant was changed directly in a range of 12.370-12.470 A to receive a film-substrate lattice mismatch. Since faceted film surface diffuses and depolarizes the light the specimens were polished to 0.5 μm depth. The measured H_c and T_c are 50-400 Oe and 40-60 °C correspondingly. Using the films we obtained spatial resolution better than 950 l/mm and sensitivity ~0.1 Oe. All the types of films were coated with thin (0.18 μm) reflective and antiabrasive layer TiN. In case of the "facet" films this layer functions additionally as a resistor, heating the films when current pulses are applied. TiN layer Ohmic resistance is about 20 Ohm/cm.

3. The soft-hardware system. Experiments

The soft-hardware system for signalogram analysis in criminalistics was developed [6,7] to provide for fully automatic searching and measuring of recorded signal parameters. The hardware consists of: an incident ligth polarizing microscope, a special tape/disc stretcher with a precision-mechanics control adapter, an MO sensor set, a CCD-camera, audio and video input adapters, a video control unit and an IBM PC/AT with peripheral equipment. The system is controlled via the computer's keyboard.

The visualized structure of induced magnetization is projected on CCD-matrix camera, and is stored in the IBM PC through an interface adapter with a frame-grabber. Image

These pictures show video record assemble traces remained on VHS cassettes while making insert editing using "record" mode of the video cassette recorder

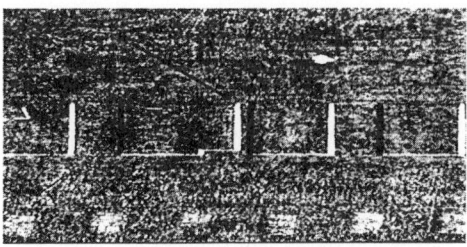

Pic.7. Non-recurring synchro-pulses of the control track

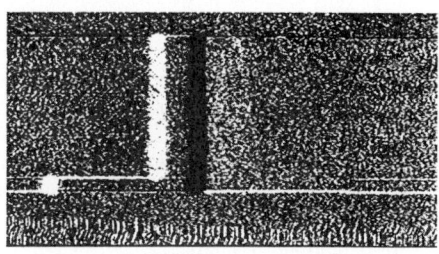

Pic.8. The same enlarged

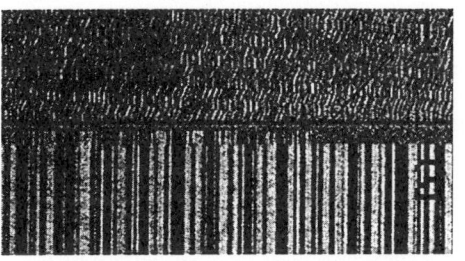

Pic.9. Overlap of sound tracks:
1 - old record; 2 - new record.

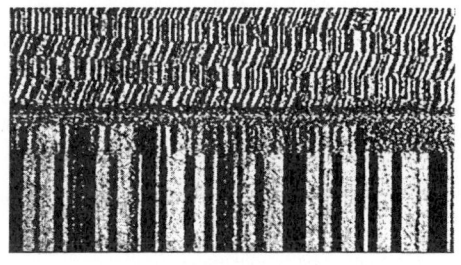

Pic.10. The same enlarged

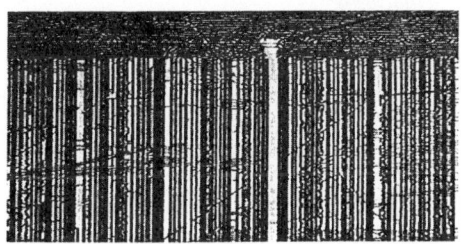

Pic.11. Magnetic pulse remained from turning the erasing head on.

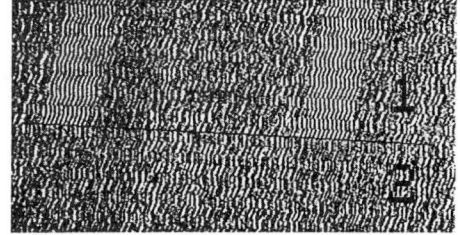

Pic.12. Video track synchronization breakdown: 1- old record; 2- new record.

processing software was developed to reconstruct two-dimensional stray field spatial distribution. The software includes video and audio signal processing programs and a special program part for criminal law researchers. The measuring accuracy for signalogram geometry characteristics is ~0.7 μm with the comparative error of magnetic field amplitude measuring within 5 % .

The application of MO method in criminalistics is motivated by needs in phonoscopic examinations (classification, diagnostics and identification of tape recorders, revealing of audio/video signalogram montage traces, lost-data restoration). Special methodics for such purposes have extensively been developed using our film sensors [8]. For identification of recording devices, 36 signs geometric parameters, topograms, pulses parameters, spectral components of distortions and noises - were proposed. For revealing of signalogram montage traces, 15 signs - transfer process pulses combinations, frequency of periodic components, spectral components of distortions - were proposed. And 31 signs, format geometry parameters, among them, were proposed to make recorder diagnostics investigations. To illustrate the capabilities of the sensors and the system selected magnetic images are shown on pic. 1-6 (taken from real expert examinations).

In addition our experiments have demonstrated excellent possibilities of MO converters in the restoration of lost (erased) data. We obtained an aircraft "black box" signalograms from the tape which was heated up to 500 °C. The "contact printing" images provided for a clear recognition of the flight information tracks against the magnetic noises background. To visualize such signalograms a "facet" film converter was used ($T_c = 60$ °C, T_c of the metal tape ~460 °C). Moreover, in some cases we managed to obtain up to 3 previous records "through" the following ones. Otherwise, recorded patterns are not erased completely but reduce down to 50-60 dB and can be restored by means of the MO method.

References

[1] Fukuzawa K, Ohkubo T, Kishigami J and Koshimoto Y 1992 *IEEE Trans. on Magn.* **28** 3420-2
[2] Neubert S, Lehureau J, Colineau J and Castera J 1988 *SPIE Proc.* **1018** 102-8
[3] Kubrakov N 1989 *SPIE Proc* **1126** 85-92
[4] Imamura N, Mimura Y and Kobayashi T 1977 J. Appl. Phys. **48** 2634-7
[5] Vishnevski V, Dubinko S, Nedviga A, Prokopov A and Groshenko N 1996 *Int. Conf. on Ferrites ICF7, Bordeaux France* p 439
[6] Levy S, Ostrovski A and Agalidi Yu 1993 *SPIE Proc.* **2108** 142-6
[7] Vishnevski V, Levy S, Dubinko S and Groshenko N 1994 *Int.Colloquium ICMFS'94, Dusseldorf Germany* 751-2.
[8] Vishnevski V, Dubinko S, Nedviga A, Prokopov A and Levy S 1997 *IX Int.Conf. on Sensors and Actuators Transducers97, Chicago USA* 2B1.10P.

Magnetic Sensors
Paper presented at Eurosensors XII, 13–16 September 1998
© *1998 IOP Publishing Ltd*

Fe-based amorphous thin films as sensing element for

miniaturized magnetostrictive delay line arrangement

H. Chiriac, Mirela Pletea and E. Hristoforou[*]

The National Institute of Research and Development for Technical Physics, Blvd. Mangeron 47, 6600 Iasi 3, Romania, [*]TEI of Chalkis, Faculty of Engineering, Greece

Abstract. A miniaturized arrangement based on magnetostrictive delay line (MDL) mechanism, able to be used as sensor, is realized using thin films technology. In order to obtain high-performance miniaturized magnetostrictive delay line arrangement we used $Fe_{70}B_{20}Si_6C_4$ amorphous thin films as magnetostrictive sensing element because of their better magnetic and magnetoelastic characteristics than conventional magnetostrictive thin films. A monolithic design has been achieved with the electric equipment and sensing element contained onto the same silicon substrate.

1. Introduction

Among various types of magnetoelastic sensors, those based on magnetostrictive delay line (MDL) technique attract an important attention because their numerous applications for sensing displacement, force, position, pressure and proximity [1, 2].

Due to the increased interest in magnetoelastic sensing applications, research has been focused on materials exhibiting two contradictory properties, namely, high saturation magnetostriction and low magnetic anisotropy. The experimental investigations have shown that Fe-based amorphous alloys posses these two characteristics and also better magnetic and magnetoelastic properties in comparison with other magnetoelastic sensing element [3].

The transition metal-metalloid amorphous alloys in the shape of ribbons and wires are suitable for magnetostrictive delay line, but can not be adapted for miniaturization. Magnetostrictive thin films are promising sensing materials for various magnetomechanical microsensors because they offer the following advantages: (1) magnetostrictive thin films require no power supply cable being able to perform measurements in a non contact manner, (2) magnetostrictive thin films can be adapted for miniaturization. In spite of this, only few works describing miniaturized MDL arrangement using magnetostrictive amorphous thin films as sensing element have appeared.

2. Fabrication and design of miniaturized MDL arrangement

The miniaturized arrangement size is 33×1.2 mm^2 and uses 33 mm long Fe-based amorphous thin films as magnetostrictive delay line (MDL) medium, with an exciting straight conductor (EC) deposited at one end of the delay line and a receiving coil (RC) at the other. The relative position of EC and RC is important in the design of the miniaturized magnetostrictive delay line arrangement, determining the delay time between exciting and receiving signals and consequently, the detection of a discrete MDL output signal. It has been found that the minimum distance necessary to obtain a discrete MDL output signal is of the order of 30 mm. The simplified operation principle of this miniaturized arrangement is based on the conventional MDL technique [4].

The schematic arrangement of the miniaturized magnetostrictive delay line fabricated on a silicon wafer using a multilayer-like structure X/Y/X, where X is a multilayer structure of SiO$_2$/Cu/SiO$_2$ type and Y is the magnetostrictive material used as delay medium, is presented in Figure 1.

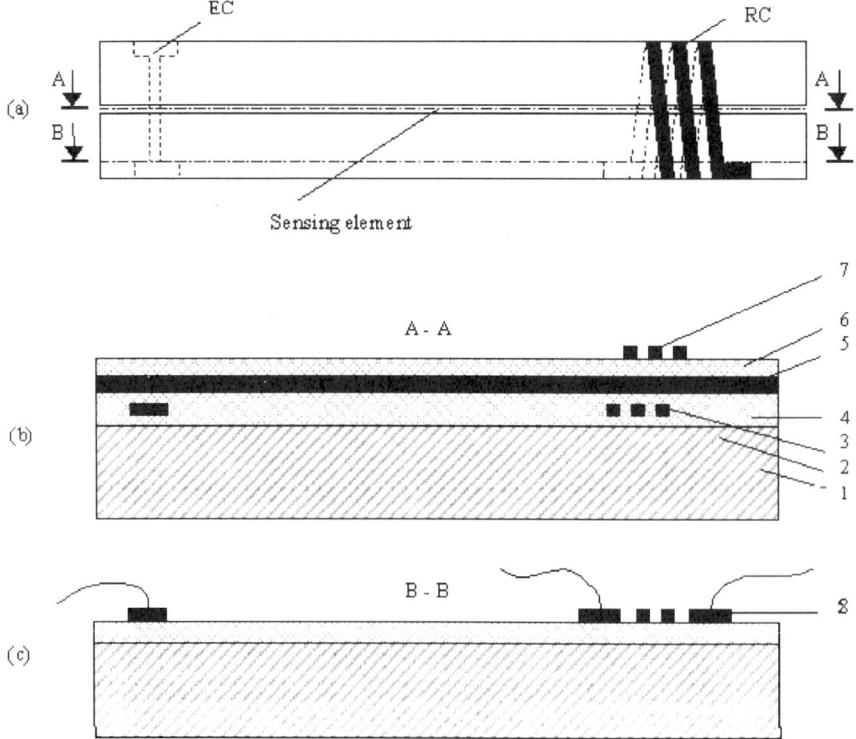

Figure 1. The miniaturized arrangement of MDL fabricated by thin film technology: *(a)* top view of the complete packaged thin film magnetostrictive delay line; *(b)* cross sectional view A - A: 1, thin film wafer; 2, insulating layer; 3, first layer of Cu; 4, insulating layer; 5, Fe$_{70}$B$_{20}$Si$_6$C$_4$ amorphous thin films operating as MDL medium; 6, insulating layer; 7, last layer of Cu; *(c)* cross sectional view B - B: 8, pads of exciting conductor and receiving coil.

The miniaturized MDL arrangement using $Fe_{70}B_{20}Si_6C_4$ amorphous thin films 9500 Å in thickness as sensing element is produced on silicon substrates using standard technological processes described in [5]. The deposited amorphous alloy, after mask formation in the shape of a narrow element, 0. 5 × 33 mm, is the delay line medium. The excitation element is a copper or aluminum straight line orthogonal to the amorphous alloy, set at one end of it, separated by a silicon dioxide layer placed under the whole deposited amorphous alloy. Additionally, the receiving element can be realized arranging two arrays of conductors situated above and below $Fe_{70}B_{20}Si_6C_4$ alloy, in order to form a receiving coil surround it. The insulating layers and amorphous thin films were deposited by r.f. sputtering method. The lithography process used to form the exciting and receiving coils was based on the ordinary positive photo resist application, exposure by ultraviolet contact photolithography and development. The pattern is defined by wet chemical etching.

3. Magnetic and magnetoelastic characteristics of sensing element used for miniaturized MDL arrangement

3.1. *Experimental procedure*

In order to use $Fe_{70}B_{20}Si_6C_4$ amorphous thin films as sensing element for miniaturized MDL arrangement presented in this paper, the magnetic and magnetoelastic properties of technical significance for this application have been investigated.

The magnetic measurements of the Curie temperature, saturation magnetization and magnetic anisotropy field were made by a torque magnetometer. The saturation magnetostriction was measured using a capacitive cantilever technique. Based on obtained values for saturation magnetostriction λ_s and anisotropy constant k_u, we estimated the magnitude of the ΔE-effect for investigated samples determining the values of the relative change of elasticity modulus $\Delta E/E$.

3.2. *Results and discussion*

The Curie temperature T_C, saturation magnetization M_s and magnetic anisotropy field H_k are forming a triad of magnetic characteristics which in combination with magnetoelastic characteristics (saturation magnetostriction λ_s, magnetoelastic coupling coefficient $b^{\gamma,2}$, magnetoelastic coupling factor k and ΔE-effect) determine the performance of the material used as sensing element for miniaturized MDL arrangement.

Among the crystalline materials, nickel thin film is a good one as sensing element for miniaturized MDL arrangement [5]. Various experiments show that the magnetomechanical coupling factor k and ΔE-effect for $Fe_{81}B_{13.5}Si_{3.5}C_2$ amorphous alloy have larger values in comparison with nickel or any other magnetostrictive material [6]. It has been established that the magnitude of the ΔE-effect depends on the ratio between the saturation magnetostriction constant λ_S and the magnetic anisotropy constant k_u. If E_s is the Young's modulus in magnetically saturated state and E is the Young's modulus in magnetically unsaturated state, the following equation gives information on the possibility of maximum changes of the sound velocities or the time of the signal delay for different materials used as sensing elements [7]:

$$\frac{E}{E_{s\ min}} = \left(1 + \left(\frac{9\lambda_s^2 E_s}{2k_u}\right)\right)^{-1} \tag{1}$$

The study of the magnetic and magnetoelastic properties of Fe-based amorphous thin films shows that they exhibit ideal characteristics for magnetostrictive delay line medium. The values of the relevant parameters of the Fe-based amorphous thin films as sensing element for miniaturized MDL arrangement are presented in Table 1.

Thickness (Å)	T_C (°C)	σ_s (emu/g)	$\lambda_s \times 10^{-6}$	k_u (erg/g)	$b^{\gamma,2} \times 10^6$ dynes/cm^2	$D \times$ $10^{-10}G^{-1}$	$\Delta E/E$
9500	338	132	32	2×10^3	-30	425	0. 58

Table 1. The values of the relevant parameters for sensing element performance of $Fe_{70}B_{20}Si_6C_4$ amorphous thin films: Curie temperature T_C, saturation magnetization σ_s, saturation magnetostriction λ_s, anisotropy constant k_u, magnetoelastic coefficient coupling $b^{\gamma,2}$, static stress-sensitivity D and ΔE-effect.

Other advantage of the use of amorphous thin films as magnetoelastic sensing element, applicable for the wide-range position sensors [8], is the large distance through which magnetoelastic wave propagates. This advantage arises as the effect of the high electrical resistivity of amorphous thin films, resulting in low eddy current losses and low value of the attenuation constant of the propagated magnetoelastic waves.

4. Conclusions

The conventional arrangements of miniaturized MDL based on nickel thin films as material for sensing element has been improved by using amorphous thin films. In terms of sensitivity, dynamics and versatility, it is possible to obtain high-performance microsensors if Fe-based amorphous thin films are used as material for magnetostrictive sensing element because of their better magnetic and magnetoelastic characteristics than conventional magnetostrictive thin films.

Further improvements in performance of miniaturized MDL using Fe-based amorphous thin films as sensing element can be achieved if two new miniaturized configurations of this device are used, targeting on its miniaturization process, which will be considered in a work in progress.

References

[1] Hristoforou E and Reilly R E 1994 *IEEE Trans. Mag.* **30** 2728-2733
[2] Meydan T and Elshebani M S M 1992 *J. Magn. Magn. Mat.* **112** 344-346
[3] Gibbs M 1994 *Physics World* January 40-45
[4] Hristoforou E and Niarchos D 1992 *J. Magn. Magn. Mat.* **116** 177-188
[5] Chiriac H Hristoforou E Grigorica M and Moga A 1997 *Sensors and Actuators* **59** 280-284
[6] Livingston J D 1982 *Phys. Stat. Sol. (a)* **70** 591-596
[7] Gibbs M R J 1995 *Sensors and their applications VII* (Bristol IOP Publishing)
[8] Takemura Y Masuda S Yamada T and Kakuno K 1995 *IEEE Trans. Mag.* **31** 3115-3118

Magnetic Sensors
Paper presented at Eurosensors XII, 13–16 September 1998
© 1998 IOP Publishing Ltd

Giant magnetoresistance Co/Cu multilayer sensors for use in magnetic field mapping

C Christides[1], S Stavroyiannis[1], G Kallias[1], A G Nassiopoulou[2], and D Niarchos[1].

(1) Institute of Materials Science, (2) Institute of Microelectronics NCSR "DEMOKRITOS", 153 10 Agia Paraskevi, Attiki, GREECE.

Abstract. Low hysteresis giant magnetoresistance multilayers made from sputtered Co/Cu were used to form a 2x4 array of sensing elements. Patterning was performed by optical lithography, using the lift-off technique. To assess the applicability of this sensor magnetic field mapping of a ring-magnet was performed with a computer controlled scanning system and the results were compared with those obtained from a Hall-sensor.

1. Introduction

The giant magnetoresistance (GMR) effect is of interest for applications in magnetoresistive recording read-heads or sensors and has stimulated much of the activity in this field. GMR arises in a variety of magnetic systems including multilayers, spin valves and granular materials [1]. Recent practical developments have favoured the spin-valve or the GMR-multilayer design for building ultrahigh density ($\sim$10 Gbit/in^2) magnetic heads [2] and magnetic field bridge-sensors [3]. To assess the applicability of GMR sensor elements the GMR-ratio $\Delta R_{max}/R_o$, (ΔR_{max} and R_o the maximum resistance change and the minimum resistance respectively), the switching field range and the magnetic coercivity are the only relevant parameters for dc-field measurements. As compared to sensors based on the anisotropic magnetoresistance (AMR) effect, GMR devices offer superior signal amplitude and can achieve very high linear resolution without shields [2,3]. However, fabrication of the spin-valve device requires a novel approach to setting the directions of the antiferromagnetic exchange layers that bias the sensor. Thus, mass production of GMR devices at acceptable yields in wafer fabrication requires very good process controls [4] that impose a yield limit of 99% for each processing operation in order to achieve an overall yield of $\sim$80%.

Today, microfabricated AMR or GMR sensing elements are based on magnetically soft permalloy ($Ni_{81}Fe_{19}$) films which are able to measure the field components lying in-plane of the film, in contrast to Hall-effect semiconducting sensors that are sensitive only in fields perpendicular to the film plane. Further effort is required in the fabrication of semiconducting microsystems to achieve accurate 2-dimensional (2D) or 3D magnetic field measurement [5].

Figure 1. An overview of the sensing device is shown on the left side while on the right is displayed a detail of the pattern definition after the lift-off process.

Very recently [6] thin, horizontal-plane Hall sensors were constructed from narrow-gap semiconductors for read-heads in magnetic recording and are shown to be competitive with spin-valve GMR heads. The difficulty of fabricating a commerially practical 2D or 3D magnetic field sensor, which requires numerous independent process steps with high (>95%) yield per step, is the only real barrier to use the above devices.

In our approach a GMR Co/Cu multilayer sensor is proposed that require less demand on the growth process. The sensing device is connected with a computer controlled system that performs magnetic field mapping in real time using an X-Y scanning system.

2. GMR device structure and fabrication.

Lately, the ongoing research in magnetron sputtered $[Co(1 \ nm)/Cu(2 \ nm)]_{30}$ multilayers brought forward a specific microstructure, in terms of grain size and degree of layer texturing, leading to GMR curves with small hysteresis [7]. To evaluate the appllicability of these multilayers as GMR sensor elements the design of a 2D array of 2x4 units was adopted. Conventional optical lithography and the lift-off technique were applied for the film patterning. Interconnections were assumed by aluminium stripes (fig.1). Aluminium pads were used as voltage contacts in a four-contact configuration, where a common current flow pass through all sensing elements and the induced voltage drop across each element is measured.

In this design, the magnetic field resolution of the sensor depends on the spacing between the voltage conductor leads overlapping the GMR elements. The two dimensional array of sensors is designed to detect magnetic field variations near the surface of magnetic materials with uncommon shapes. Since the shape of the GMR curves changes drastically upon changing the direction of the applied field relative to the film-plane, it is optimal to sense the gradient of the transverse signal field that gives a better contrast for the field changes in a horizontal plane. So, it is decided to measure the relative difference between the GMR output signal at zero field and its response over the magnetized surface and construct a map of the transverse field changes as a function of the sensor position at a constant distance above a magnet.

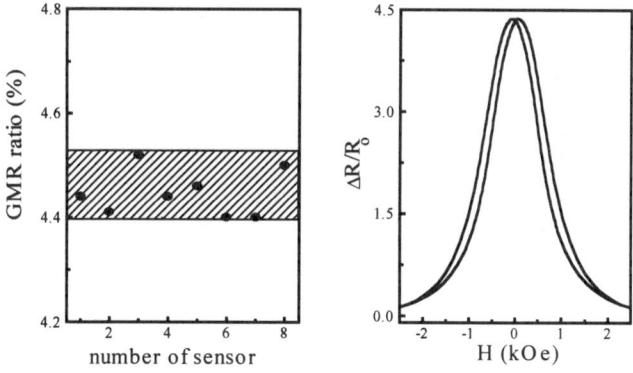

Figure 2. The variation of the $\Delta R_{max}/R_o$ output signal from each sensing element is shown on the left, while a representative GMR curve obtained with a homogeneous field applied perpendicular to film plane is shown on the right. The measurements were performed with the four-point-probe method, using a dc current of 10 mA.

3. Magnetic field mapping of a ring-magnet.

The *transfer function*, that determines the output-stimulus signal relationship for the GMR sensor, is given by the output voltage change with the applied field (stimulus). Thus, the sensor characteristics were defined by the shape of the GMR curves. The observed repeatability of GMR curves per sensing element (fig.2) indicates that a high yield factor is achieved from the patterning process. Fig.2-right shows that the highest accuracy and linearity occurs between 40 and 900 Oe. Accordingly it was selected as the full span range of the sensor. In this range the hysteresis error is taken into account by fitting a least-square line that defines the calibration function of the sensor. However, the GMR sensor can not distinguish changes of the field sign because the GMR response signal is an even function of the applied field. To evaluate the error that is introduced from this inefficiency we have performed simultaneous scans of the field of a ring-magnet, having a Hall-sensor attached next to the GMR sensor. The scanning system is fully computer controlled. An IEEE-488 interface unit, a computer controlled Parker scanning system (*X-Y* stage) with two independent stepping motors, a Keithley-220 programable current source and a Hewlett Packard 3457A multimeter were used. The test object was a sintered NdFeB ring-magnet that was placed beneath the sensors at a distance of 6 cm. Both sensors were mounted on a horizontal bar that was fixed on the moving part of the *X-Y* stage, and an area of 10x10 cm^2 was scanned over the magnet with a constant step of 0.25 mm along the *X-* and *Y-* directions. To detect the transverse field component a mu-metal shielding plate (~1 mm thick) was placed in front of the GMR sensor. Three separate scans were performed with the Hall-sensor arranged in such a way as to sense each time the x, y and z component of the magnetic flux density. The three magnetic field components were combined with a computer algorithm to give the total field mapping shown in fig.3. Two major differences can be seen between the two sensors: (i) The change of field sign (negative field) that occurs near the center and at the edges of the ring-magnet is sensed from the Hall-sensor but not from the GMR sensor. Instead, it causes a positive signal in the place of negative field thereby making the image of

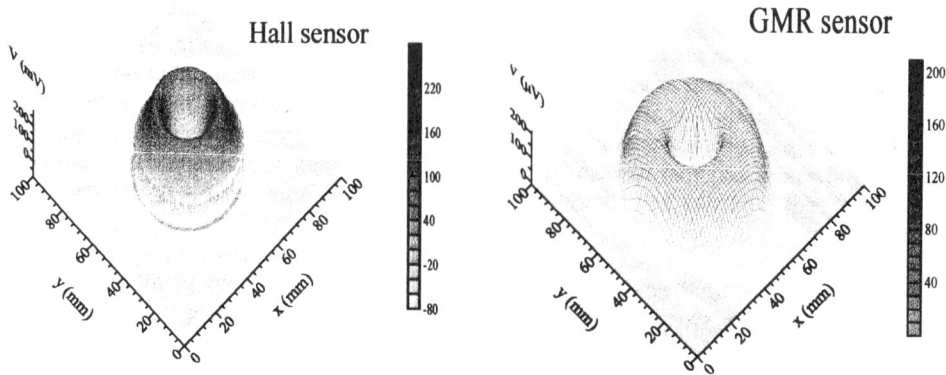

Figure 3. Magnetic field mapping of a ring-magnet with 4 cm outer, 3 cm inner diameter, and thickness of 1 cm.

the magnet wider. (ii) The output voltage is in mV for the Hall-sensor and in μV for the GMR-sensor.

In summary, using a simple magnetic sensor design it is demonstrated that the developed microstructure [7] in Co/Cu GMR multilayers complies with the qualifying factors needed for field-difference mapping in real time. The obtained results indicate that an advanced electronic design of the device might enable the application of the specific Co/Cu multilayer [7] in commercial sensors.

Acknowledgment

This work has been supported by the EKBAN-280 project of the General Secretariat for Research and Technology of the Development Ministry in Greece. Mrs P. Loupaki has performed the lithography processing.

References

[1] Parkin S S P 1994 *Ultrathin Magnetic Structures II*, edited by B. Heinrich and J.A.C. Bland, (Berlin: Springer-Verlag), Chap. 2.

[2] Smith N, Zeltser A M and Parker M R 1996 *IEEE Trans. on Magn.* **32** 135-141.

[3] Spong J K, Speriosu V S, Fontana R E, Dovek M M and Hylton T L 1996 *IEEE Trans. on Magn.* **32** 366-371.

[4] Baubock G, Dang H Q, Hinson D C, Rea L L and Kim Y K 1996 *IEEE Trans. on Magn.* **32** 25-30.

[5] Schott Ch, Manic D and Popovic R S 1997 *Proceedings of the 11th European Conference on Solid State Transducers.* (Warsaw: Eurosensors XI), 635-638.

[6] Solin S A, Stradling R A, Thio T and Bennett J W 1997 *Meas. Sci. Technol.* **8** 1174-81.

[7] Christides C, Stavroyiannis S, Boukos N, Travlos A and Niarchos D 1998 *J. Appl. Phys.* **83**.

On the saturation magnetostriction in low magnetostrictive

Co-rich amorphous wires

H. Chiriac, Maria Neagu and E. Hristoforou*

National Institute of Research & Development for Technical Physics, 47 Mangeron Blvd., 6600 Iasi 3, Romania, *Technological & Educational Institution of Chalkis, Psahna, Euboea, 34400 Greece

Abstract. The saturation magnetostriction (λ_s) for low magnetostrictive amorphous wires with nominal compositions $Co_{65.25}Fe_{4.5}Si_{12.25}B_{15}Cr_2Ni_1$ and $Co_{68.15}Fe_{2.35}Si_{12.5}B_{15}Cr_1Ni_1$ has been determined by means of the small angle magnetisation rotation method. λ_s has been evaluated as $+0.39 \times 10^{-6}$ for $Co_{65.25}Fe_{4.5}Si_{12.25}B_{15}Cr_2Ni_1$ and -1.15×10^{-6} for $Co_{68.15}Fe_{2.35}Si_{12.5}B_{15}Cr_1Ni_1$ amorphous wires in the as-cast state. Changes in saturation magnetostriction after stress current annealing were also analysed.

1. Introduction

Magnetic amorphous wires obtained by in-rotating-water spinning method presents a special interest for basic research as well as for their potential applications. These materials are prospective for numerous types of sensors due to their wide range of possibilities to control the magnetic and magnetostrictive properties [1,2].

The saturation magnetostriction constant λ_s is the basic magnetoelastic quantity that determines the magnetic and magnetoelastic behaviour of the magnetostrictive materials. Control of the sign and value of λ_s by means of alloy composition and by annealing the samples is very important with the view of applications.

This paper presents results concerning λ_s in the as-cast state as well as its dependence on the stress current annealing conditions in low magnetostrictive Co-Fe-Si-B-(Cr,Ni) amorphous wires. We obtained amorphous wires having low positive or negative saturation magnetostriction by replacement of Co or Fe with small amounts of Cr and Ni in nearly zero magnetostrictive Co-rich alloys. The saturation magnetostriction has been determined by means of the small angle magnetisation rotation method (SAMR) [3].

2. Experimental details

Amorphous wires with nominal compositions $Co_{65.25}Fe_{4.5}Si_{12.25}B_{15}Cr_2Ni_1$ and $Co_{68.15}Fe_{2.35}Si_{12.5}B_{15}Cr_1Ni_1$, 125 µm in diameter, were obtained by in-rotating-water spinning method.

The small angle magnetisation rotation is determined by applying simultaneously a small amplitude ac magnetic field (H_{ac}), high dc magnetic field (H_{dc}) perpendicular and parallel to the axis of the wire and a tensile stress σ. H_{dc} magnetic field must be high enough to produce the magnetisation saturation of the amorphous wire. The SAMR method is based on detection of the induced voltage V_{2f} (the second harmonic of the applied ac magnetic field) in the receiving coil set around the amorphous wire. When a tensile stress σ is applied on the wire (at constant H_{ac} and H_{ac} magnetic fields), V_{2f} increases or decreases as a consequence of the increment of magnetoelastic anisotropy having a transverse or an axial easy axis respectively. The change in V_{2f} can be compensated by an adequate modification of H_{dc}, so that V_{2f} takes the same value as before the stress was applied. The magnetostriction constant is obtained as:

$$\lambda_{s} = - (\mu_0 M_s/3)(\Delta H_{dc}/\Delta\sigma), \text{ for } V_{2f} \text{ and } H_{ac} = \text{const} \tag{1}$$

where $\mu_0 M_s$ is saturation magnetization.

The value of H_{dc} magnetic field for which SAMR method is suitable should be high enough so that $V_{2f}H_{dc}^2$ to be constant [4].

The transverse magnetic field H_{ac} was obtained by passing through the wire an electric current having 4 kHz frequency and the maximum value less than 50 A/m at the wire surface. The value of ac current was small enough to neglect the increase in temperature of the amorphous wire. The second harmonic V_{2f} was detected using a lock-in amplifier.

3. Results and discussion

In order to establish the measurement conditions for λ_s determination, we analysed the dependence of the second harmonic V_{2f} on the tensile stress σ, H_{ac} and H_{dc} magnetic fields.

Figure 1 (a) and (b) shows the dependence of V_{2f} on the applied stress σ, with H_{dc} field as parameter, for $Co_{65.25}Fe_{4.5}Si_{12.25}B_{15}Cr_2Ni_1$ and $Co_{68.15}Fe_{2.35}Si_{12.5}B_{15}Cr_1Ni_1$ amorphous wires, tested in the as-cast state. The V_{2f} value decreases when H_{dc} increases. When σ

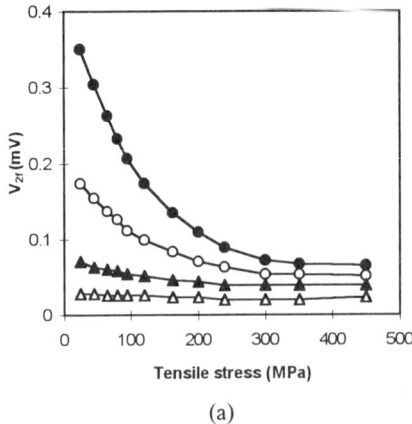 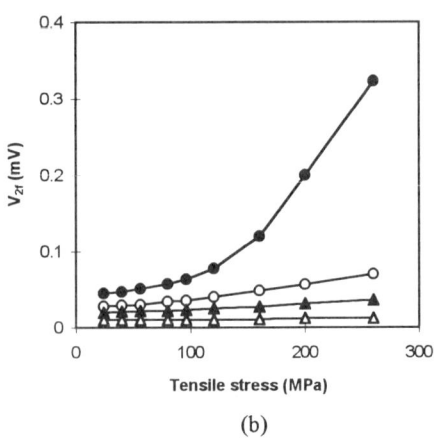

(a)   (b)

Figure 1. The V_{2f} dependence on the applied tensile stress σ, with H_{dc} field as parameter, for amorphous wires tested in the as-cast state: (a) $Co_{65.25}Fe_{4.5}Si_{12.25}B_{15}Cr_2Ni_1$ and (b) $Co_{68.15}Fe_{2.35}Si_{12.5}B_{15}Cr_1Ni_1$: • 700 A/m; o 1600 A/m; ▲ 2400 A/m; Δ 3200 A/m.

increases, V_{2f} decreases for low positive $Co_{65.25}Fe_{4.5}Si_{12.25}B_{15}Cr_2Ni_1$ and increases for low negative $Co_{68.15}Fe_{2.35}Si_{12.5}B_{15}Cr_1Ni_1$ amorphous wires.

The saturation magnetostriction λ_s has been evaluated as $+0.39 \times 10^{-6}$ for $Co_{65.25}Fe_{4.5}Si_{12.25}B_{15}Cr_2Ni_1$ and -1.15×10^{-6} for $Fe_{2.35}Si_{12.5}B_{15}Cr_1Ni_1$ amorphous wires tested in the as-cast state. The measurements were made at 7 kA/m of magnetic bias field for which $V_{2f}H_{dc}^2$ is constant.

Having in mind to control the saturation magnetostriction value for these materials we studied the influence of stress current annealing on λ_s. The elevated temperatures were obtained by Joule heating produced by an electric current passing through the amorphous wire. The current density and annealing time were increased up to 40 A/mm^2 and 35 minutes respectively. To evaluate the average temperature inside the samples during the annealing current flowing, the saturation magnetisation was measured as a function of the temperature during conventional furnace treatment and compared to that obtained by Joule heating [5].

Figure 2 (a) and (b) shows the changes in saturation magnetostriction for $Co_{65.25}Fe_{4.5}Si_{12.25}B_{15}Cr_2Ni_1$ and $Fe_{2.35}Si_{12.5}B_{15}Cr_1Ni_1$ amorphous wires respectively tested after stress current annealing for 30 minutes, under 100 MPa tensile stress.

For low positive magnetostrictive $Co_{65.25}Fe_{4.5}Si_{12.25}B_{15}Cr_2Ni_1$ amorphous wires λ_s value remains unchanged up to about 20 A/mm^2 and than strongly increases. A peak about five times higher than in the as-cast state was obtained for about 35 A/mm^2.

For low negative magnetostrictive $Co_{68.15}Fe_{2.35}Si_{12.5}B_{15}Cr_1Ni_1$ amorphous wires λ_s decreases about two times after stress current annealing.

The obtained results can be analysed taking into account the structural relaxation and induced transverse magnetic anisotropy after stress current annealing.

For Co-rich amorphous wires annealed up to about 20 A/mm^2 current density, the stabilised average temperature is up to about 450 K. For $Co_{65.25}Fe_{4.5}Si_{12.25}B_{15}Cr_2Ni_1$ and $Co_{68.15}Fe_{2.35}Si_{12.5}B_{15}Cr_1Ni_1$ amorphous wires, the Curie and crystallisation temperatures (T_C and T_x) are: 513K, 833 K and 533 K, 823 K respectively.

The effect of the structural relaxation is important at low temperature. The maximum value of λ_s is obtained for annealing temperatures $T_{ann} > T_c$, which denotes that these changes

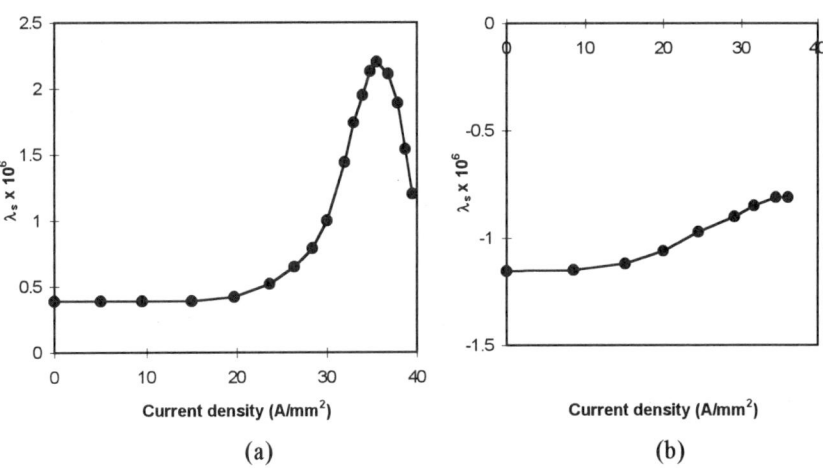

(a)   (b)

Figure 2. The dependence of saturation magnetostriction λ_s on the current density for 30 minutes of stress current annealing under 100 MPa (a) $Co_{65.25}Fe_{4.5}Si_{12.25}B_{15}Cr_2Ni_1$ and (b) $Co_{68.15}Fe_{2.35}Si_{12.5}B_{15}Cr_1Ni_1$

are related to the creep induced magnetic anisotropy that was observed in Co-rich amorphous wires [6].

4. Conclusions

Low positive or negative magnetostrictive amorphous wires were obtained by replacement of Co or Fe with small amounts of Cr and Ni in nearly zero magnetostrictive Co-rich alloys.

For $Co_{65.25}Fe_{4.5}Si_{12.25}B_{15}Cr_2Ni_1$ and $Fe_{2.35}Si_{12.5}B_{15}Cr_1Ni_1$ amorphous wires in the as-cast state the saturation magnetostriction λ_s has been evaluated as $+0.39 \times 10^{-6}$ and -1.15×10^{-6} respectively.

After stress current tretments at annealing temperatures $T_{ann} > T_c$, the saturation magnetostriction λ_s exhibit the following behaviour:
- increases about five times for $Co_{65.25}Fe_{4.5}Si_{12..25}B_{15}Cr_2Ni_1$ amorphous wires
- decreases about two times for $Co_{68.15}Fe_{2.35}Si_{12.5}B_{15}Cr_1Ni_1$ amorphous wires .

These materials are very attractive for a wide range of sensing applications especially for those based on the GMI effect and for security system sensors. The tailoring of magnetic and magnetostrictive properties for special applications can be also realised.

References

[1]   Squire P T, Atkinson D and Atalay S 1995 *IEEE Trans. Magn.*. **31** 1239-1248.
[2]   Vazquez M and Hernando A 1996 *J. Phys.D: Appl. Phys.* **29** 939-949
[3]   Narita K, Yamasaki J and Fukunaga H 1980 *IEEE Trans. Magn.* **16** 435- 439
[4]   Mitra A and Vazquez M 1990 *J. Appl. Phys.* **67** 4986-4988
[5]   Knobel M, Allia P, Gomez-Pola C, Chiriac H and Vazquez M 1995 *J. Phys.D: Appl. Phys.* **26** 2398-2403
[6]   Kraus L, Vasquez M and Hernando A 1994 *J. Appl. Phys.* **76** 5343-5348

An Optimised Integrated Bipolar Magnetotransistor

Marioara Avram, Otilia Neagoe, Monica Simion

National Institute of Research and Development for Microtechnology, Erou Iancu
Nicolae 32B, Bucharest, 72996, Romania

Abstract. An integrated bipolar magnetotransistor based on enhanced modulation of emitter
injection and carrier deflection is presented. The absolute magnetosensitivity of the device is
about 5 mV/mT. The relative magnetosensitivity is in the range 3-5 T^{-1} at collector currents of
the order of 0.5 mA. Between the magnetic field and the output signal exists a linear
dependence for certain biasing conditions.

1. Introduction

The bipolar magnetotransistors are active bipolar devices with current output and their
essential component is a current source in the form of a p-n junction, injected minority
carriers in the base region and one or more reverse-biased p-n junctions as collectors
picking up the useful signal. As the bipolar magnetotransistor is built on a Si substrate
that is a non-magnetic substrate, the external magnetic field B influences only the kinetic
processes in different regions of transistor structure. The sensitivity of the bipolar
magnetotransistor is based on two effects caused by the Lorentz force on the current
carriers: generation of a Hall voltage in the active region of the device, which causes a
modulation of the emitter injection current and deflection of the emitter injected minority
carrier current [1]. In the device designed and realised the both effects play a significant
role.

2. Device structure

The magnetic field is applied parallel to the z-axis, perpendicular to the device
cross section.

The application of voltage V_B between the two bases causes a majority carrier
current flow through the device's active region, which leads to a certain potential
distribution in the silicon substrate. With the increase of V_B, the voltage applied to the
emitter junction in the forward direction also increases and the emitter begins to inject.
The reverse-biased collector collects some of the injected carriers (electrons). The
application of the magnetic field B leads to a redistribution of the electrical potential and
therefore to potential changes in the volume and at the boundaries of the transistors. The
changes due to the majority carrier current have opposite sign for the area between the
two bases, and in the silicon substrate near the injecting side of the emitter injected
current and as a result the collector current changes too [2].

The direction of the applied electric field E makes the injected carriers drift towards the two collectors. Without the magnetic field, for the same emitter-base voltage, the collector currents have the same value. If a magnetic field B is applied in a perpendicular direction to the plane of the chip, in base region, the field E is turned at a Hall angle $\varphi \approx \mu_{maj}.B$ with respect to the direction of current j. The generated Hall field E_H is oriented across the structure. As a result, the potential in the base area it is no longer a constant. Its variation along the emitter-base junction causes the emitter injection modulation. The mechanism of deflection together with the emitter injection modulation operates in the base region. Both mechanisms are dominant due to the device geometry and the asymmetry of the collector position (fig.1).

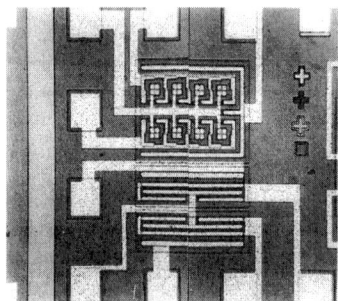

Figure 1. Bipolar magnetotransistor structure photo.

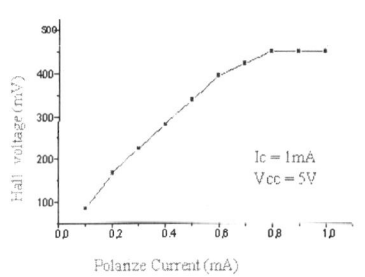

Figure 2. Hall Voltage as function of supply base current.

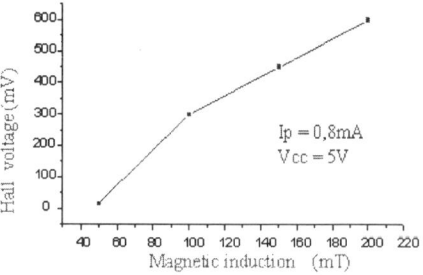

Figure 3. Hall Voltage as function of magnetic induction.

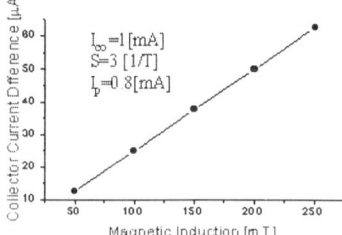

Figure 4. Collector current difference as function of magnetic induction.

One important characteristic of the presented devices is an exponential dependence between the output signal and the magnetic field exists at certain biasing conditions (fig.2, 3).

3. Experimental results and discussions

Our experiments were made with bipolar magnetotransistors structures fabricated by the standard planar technology on p-type silicon wafers: resistance $\rho \approx 7.5$ Ω. Cm ($n_0 \approx 10^{15}$ cm^{-3}), crystallographic (111) and (100) orientation and thickness ≈ 350 μm.

The bipolar magnetotransistor were realised in an n-type epilayer surrounded by deep p$^+$ diffusion on a p-type substrate. The doping level of epilayer is approximately $6 \cdot 10^{15}$cm^{-3}, and the epilayer thickness is approximately 12μm.

The depths of the heavily doped diffusion regions of the emitter, collector and bases are approximately 1.5 μm.

The differential collector current component due to carrier deflection is a linear function of the magnetic field B, it depend on the difference of the capture area [3, 4], and the differential collector current component due to the emitter injection modulation is an exponential function of the voltage Hall (fig.4), V_H:

$$\Delta I_C = I_{C1} - I_{C2} = \alpha(I_{C0}, B) \cdot V_T / V_H \cdot (e^{VH / VT} -1) \cdot (e^{EH \cdot (L-l) / VT} +1)$$

All the measurements are made in common base configuration (fig.5). A positive voltage is applied to base from a D.C. voltage source V_{CC}. The application of voltage V_B between the two bases of the device causes a majority carrier current flow through the device's active region, which leads to a certain potential distribution in the silicon substrate.

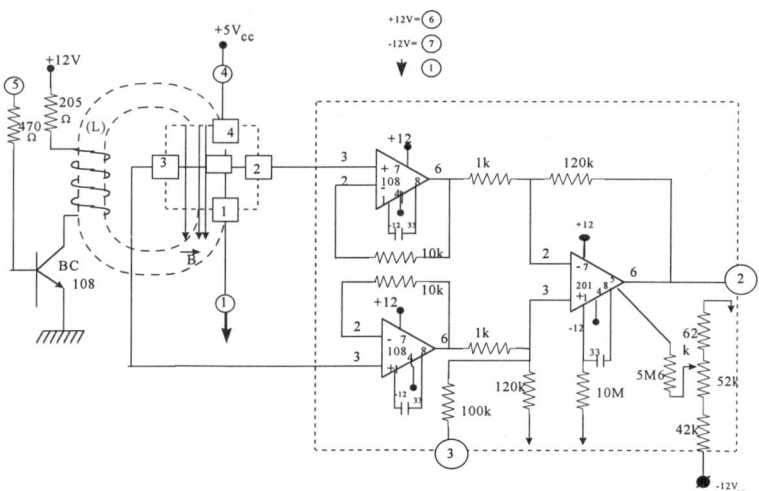

Figure 5. Schematic of the biasing and signal-conditioning circuit for bipolar magnetotransistor.

4. Conclusions

An integrated bipolar magnetotransistor for transversal magnetic field detection was presented. Their special geometrical shapes employ the combined phenomena of charge carrier deflection and emitter injection modulation in presence of the magnetic field.

The experimental results confirm the theoretical operating principle. An increasing of the relative magnetosensitivity is possible by increasing the emitter length. A layout with longer emitter was designed and will be experimented.

5. References

[1]   Ch. S. Roumein, "Magnetic Sensors Continue to Advance Towards Perfection", Sensors and Actuators A 46-47 (1995), p. 273;

[2]   R. Castagneti at all, "Dual Magnetotransistors with on -chip Bias and Signal Conductivity", Sensors and Actuators A 37-38, (1993), p. 698;

[3] Otilia Neagoe, Marioara Avram, "A New Bipolar Magnetotransistor with Combined Phenomena of Carrier Deflection and Emitter Injection Modulation", International Conference of Semiconductors, (1996), p. 456;

[4]   Marioara Avram, Otilia Neagoe, et. al. "Lateral Bipolar Magnetotransistors with Enhanced Emitter Injection Modulation and Carrier Deflection", International Conference of Semiconductors, (1997), p.519.

Magnetic Sensors
Paper presented at Eurosensors XII, 13–16 September 1998
© 1998 IOP Publishing Ltd

Tensile stress measurements based on ΔE effect in positive magnetostrictive amorphous wires

H. Chiriac, E. Hristoforou* and Maria Neagu

National Institute of Research & Development for Technical Physics, 47 Mangeron Blvd., 6600 Iasi 3, Romania, *Technological & Educational Institution of Chalkis, Psahna, Euboea, 34400 Greece

Abstract. In this paper experimental results concerning the dependence of longitudinal sound velocity on tensile stress and bias magnetic field in $Fe_{77.5}Si_{7.5}B_{15}$ amorphous wires used as magnetostrictive delay line are presented. Using magnetostrictive delay line technique, tensile stress applied along the length of the amorphous wire can be detected.

1. Introduction

Fe-rich amorphous wires prepared by the in-rotating-water spinning method are very attractive for basic research as well as for technological applications. The magnetic, magnetoelastic and mechanical properties of these materials make possible the production of sensors with high sensitivity, fast response, reliable operation, small dimensions and mechanical robustness [1,2].

The magnetoelastic effects arise from the interaction between magnetic moments and mechanical structure of the material. Knowledge of the dependence of magnetoelastic properties on different influence factors (external magnetic field, force, torque) offers the possibility to use such materials as new sensing elements [2-4]. The change of the elastic modulus (ΔE effect) in ferromagnetic materials originates in rotation of the magnetization under the influence of applied magnetic field and stress. According to the ΔE effect, the longitudinal sound velocity depends on the magnitude of the bias magnetic field and tensile stress applied parallel to the anisotropy axis.

In this paper results concerning the dependence of the longitudinal sound velocity in $Fe_{77.5}Si_{7.5}B_{15}$ amorphous wires on the bias magnetic field and tensile stress are presented. The amorphous wires were tested in the as-cast state and after heat treatments.

2. Experimental details

The magnetostrictive delay line (MDL) principle was used to determine the longitudinal sound velocity [3,4]. Magnetoelastic waves are generated at one end of the amorphous wire (1) by a pulsed current in an exciting coil (2). The waves are propagating through the

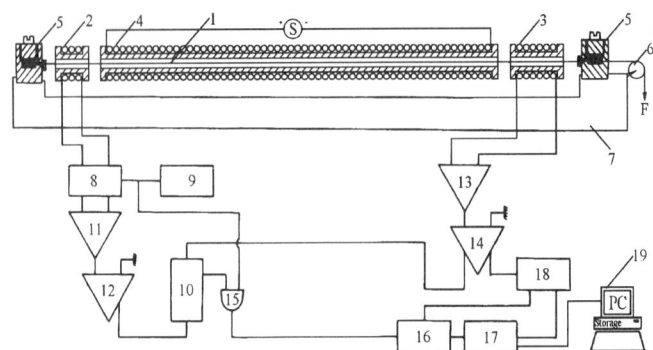

Figure 1. Basic diagram for the measurement system: (1) amorphous wire, (2) exciting coil, (3) receiving coil, (4) bias coil, (5) fixing device, (6) tensile stress device, (7) support, (8) pulse generator, (9) oscillator, (10) flip-flop, (11) amplifier, (12) comparator, (13) amplifier, (14) comparator, (15) gate, (16) counters, (17) bistables latches, (18) drive circuit, (19) computer.

amorphous wire and they are detected at the opposite end by a receiving coil (3). The amorphous wire was set in a solenoid (4) that generates the bias magnetic field. The tensile stress was applied at one end of the wire. The distance L between the centres of the exciting and receiving coils was kept constant (500 mm) during the measurements. A time counter (16) was used to measure the time interval T between the exciting and receiving MDL pulses.

Fe$_{77.5}$Si$_{7.5}$B$_{15}$ amorphous wires, 125 μm in diameter, were tested in the as-cast state and after heat treatments performed between 250-370^{0}C for 15 minutes to 3 hour, in a hydrogen atmosphere, in the absence of a magnetic field, in a non-inductive furnace which ensure an uniform heating of the wires all along their length.

The longitudinal sound velocity v is given by:

$$v = L/T \tag{1}$$

The experimental set-up was calibrated using a nickel wire as delay line. The uncertainty in measurement of the dependence of longitudinal sound velocity on tensile stress and bias magnetic field results only from the error of the time-counting system. This error is 0.5 x 10^{-7}s, resulting in a longitudinal sound velocity uncertainty of the order of 0.2 %.

3. Results and discussion

The obtained results show that changes in longitudinal sound velocity with bias magnetic field (H) are strongly dependent on thermal treatment temperature of the amorphous wires. In the as-cast sate, the change in longitudinal sound velocity is negligible.

The curves v = f(H) present a minimum that becomes more pronounced with increasing the annealing temperature. The value of the bias magnetic field corresponding to the minimum value of longitudinal sound velocity decreases with increasing the annealing temperature.

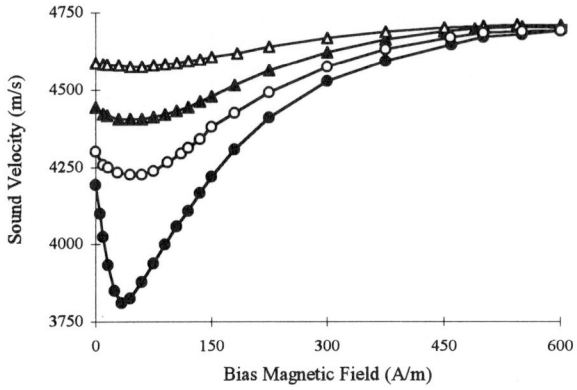

Figure 2. The dependence of the longitudinal sound velocity in $Fe_{77.5}Si_{7.5}B_{15}$ amorphous wire thermally treated at 370°C for 15 minutes, on the bias magnetic field under:• 0 MPa; o 1.6 MPa; ▲ 5.7 MPa; Δ 12 MPa applied tensile stress.

Figure 2 shows the dependence of the longitudinal sound velocity in $Fe_{77.5}Si_{7.5}B_{15}$ amorphous wire thermally treated at 370°C for 15 minutes, on the bias magnetic field, under 0, 1.6, 5.7 and 12 MPa applied tensile stress. For bias magnetic field up to 300 A/m, the longitudinal sound velocity strongly increases with increasing the value of the tensile stress. For about 16 MPa tensile stress and 600 A/m bias magnetic field the saturation value of the longitudinal sound velocity is reached.

The dependence of the longitudinal sound velocity on the applied tensile stress for $Fe_{77.5}Si_{7.5}B_{15}$ amorphous wire tested after thermally treatment at 370°C for 15 minutes, under 0 A/m and 90 A/m bias magnetic field are presented in Figure 3. A monotonic dependence is observed up to about 12 MPa. The measurements made by increasing and decreasing the value of the tensile stress show the absence of hysteresis.

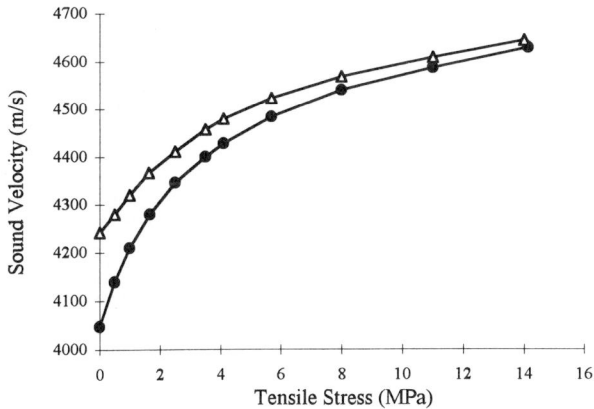

Figure 3: The longitudinal sound velocity dependence on the applied tensile stress for $Fe_{77.5}Si_{7.5}B_{15}$ amorphous wire tested after stress-relief process at 370°C for 15 minutes under: • 90 A/m ; Δ 0 A/m bias magnetic field.

The obtained results concerning the dependence of the longitudinal sound velocity on the tensile stress and bias magnetic filed applied along the amorphous wire used asmagnetostrictive delay line can be explained taking into account the ideal model of the domain structure for positive magnetostrictive amorphous wires [1].

4. Conclusions

A strong dependence of the longitudinal sound velocity on the applied tensile stress and bias magnetic field is observed, in good agreement with the results for the ΔE effect in these materials.

For about 16 MPa tensile stress and 600 A/m bias magnetic field, the saturation value of the longitudinal sound velocity is reached.

In the case of delay lines made of Fe-rich amorphous wires a monotonic dependence of the longitudinal sound velocity on the applied tensile stress up to 12 MPa was observed.

The monotonic and fast response as well as the absence of hysteresis suggest to use this technique in tensile stress measurements.

References

[1]   Squire P T, Atkinson Dm Gibbs M R J and Atalay S 1994 *J. Magn. Magn. Mater.* **132** 10-21.
[2]   Vazquez M and Hernando A 1996 *J. Phys.D: Appl. Phys.* **29** 939-949
[3]   Hristoforou E 1997 *Sensors and Actuators A* **59** 183-191
[4]   Hristoforou E, Chiriac H, Neagu M and Darie I 1994 *J. Phys.D: Appl. Phys.* **27** 1595-1600

Magnetic Sensors
Paper presented at Eurosensors XII, 13–16 September 1998
© *1998 IOP Publishing Ltd*

CMOS versus bipolar Hall plates regarding offset correction

Sandra Bellekom

Delft University of Technology, Information Technology and Systems, Electronic Instrumentation Laboratory, P.O. Box 5031, 2600 GA Delft, The Netherlands.

Abstract. Integrated silicon Hall devices are used to measure magnetic fields. Unfortunately, they can not be used in low field applications because they suffer from a large, unpredictable and drifting offset. Offset can be reduced to a few microtesla using the spinning-current method. This paper reports on the influence of the fabrication process, CMOS or bipolar, on the offset of spinning-current Hall plates. The depletion layer width of the junction, which is different for the two fabrication processes, can be influenced by the bias current and the substrate voltage. The spinning-current principle drastically reduces the offset, independent of the bias current and substrate voltage, and therefore independent of the depletion layer width and fabrication method.

1. Introduction

Magnetic sensors are very useful devices, not only for the direct measurement of magnetic fields, like geomagnetic fields or data on a magnetic tape, but also for indirect measurements. Indirect measurement of mechanical quantities, using a magnetic sensor combined with a small permanent magnet, can for example be applied in cars to determine the throttle angle or the rotational speed of a toothed wheel.

The advantage of Hall devices compared to, for example magnetoresistive sensors is that they can be realized in standard IC processes such as the bipolar or CMOS process. These integrated Hall plates show the advantages of IC fabrication: low cost, small size, batch fabrication etc. Also the realization of smart sensors is possible [1]. The disadvantage of Hall plates is that offset can cause large measurement errors.

2. Fabrication

The sensitivity of Hall plates depends on the mobility and the concentration of the charge carriers in the device. Because the mobility of electrons is larger than the mobility of holes, Hall plates are realized in n-type material. A lower impurity concentration gives a larger sensitivity. In a CMOS process, the n-well satisfies both properties. In a bipolar process, the n-type epitaxial layer can be used as the active material for the Hall plate. Figure 1 shows the two fabrication methods.

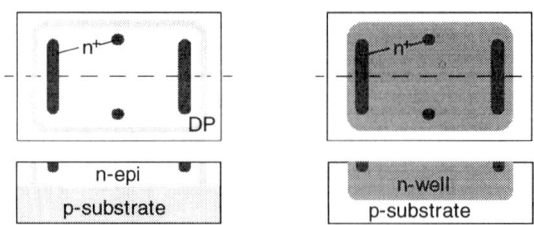

Figure 1. Integrated Hall plates fabricated in a bipolar (left) and CMOS (right) IC process. DP is the deep p-type diffusion used for isolation.

3. Offset correction

Unfortunately Hall devices show a large, unpredictable and drifting offset (i.e. they generate an output voltage even when no magnetic field is applied). This offset causes an error in the measured magnetic field strength. To reduce the offset, the spinning-current Hall plate [2] was developed and recently a continuous version [3] was studied.

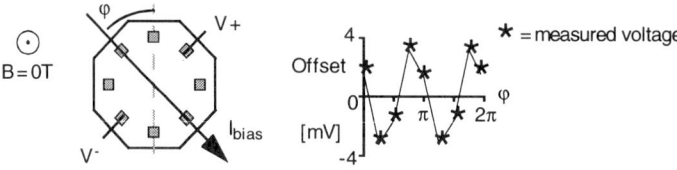

Figure 2. A spinning-current Hall plate (left) and the measured offset voltages versus bias current direction angle (right).

A spinning-current Hall plate is a symmetrical Hall plate in which the direction of the bias current I_{bias} is spun through 360° and the corresponding output voltage is measured. Most offset generating effects, like geometry errors, processing errors or the piezoresistive effect, generate a periodic offset voltage. The Hall effect itself does not depend on the direction of I_{bias} and can be separated from the periodic offset components by averaging the measured voltages. Figure 2 explains the spinning-current principle.

4. JFET effect and sensitivity

The most important difference between the two fabrication processes is the doping profile across the junction. The influence of the reverse biased pn-junction on the Hall plate is called the JFET effect; a useful application and explanation of this effect is given in [4]. In integrated Hall plates the thickness t equals the junction depth x_j minus the depletion layer thickness in the Hall plate's material:

$$t = x_j - \sqrt{\frac{2\epsilon}{q} \cdot \frac{N_{sub}}{N_h \left(N_{sub} + N_h\right)}} \cdot \left[V_{bi} + V_{sub} + \frac{1}{2}\left(I_{bias}R_{hp}\right)\right] \tag{1}$$

with ϵ the dielectrical constant of silicon, N_{sub} and N_h the doping concentrations of the substrate and the Hall plate's material, V_{bi} the built-in voltage between those materials, V_{sub} the applied substrate voltage and R_{hp} the Hall plate's resistance. From equation 1 it is clear that t depends on the processing, I_{bias} and V_{sub}.

Using $I_{bias} = 0.5\ mA$, the sensitivity $S = \frac{I_{bias}}{N_h qt}$ of CMOS and bipolar Hall plates equals $126\ mV/T$ and $102\ mV/T$ respectively. The inaccuracy of the process parameters probably causes a difference with the measured values, $140\ mV/T$ and $77\ mV/T$.

The influence of the thickness, and therefore of the processing, on the offset of spinning-current Hall plates cannot be predicted and will therefore be examined experimentally. The thickness can be influenced with the bias current and the substrate voltage, so these variables are included in the measurements.

5. Offset measurements

Because of the different sensitivities for CMOS and bipolar devices it is practical to use offset fields instead of offset voltages ($B_{offset} = V_{offset}/S$). For all devices the residual offset, the offset after applying the spinning current method, is only a few microtesla. Figure 3 shows two examples of these residual offset fields for a CMOS and a bipolar Hall plate. In the figure the average and the standard deviation are drawn. The figure

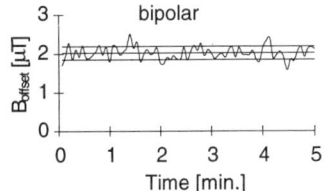 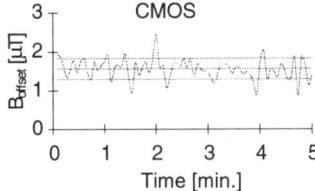

Figure 3. Measured residual offset field of spinning-current Hall plates realized in de bipolar and CMOS process.

shows that the residual offset field of the bipolar devices varies less than that of CMOS devices. The standard deviations are $0.2\ \mu T$ and $0.3\ \mu T$ respectively.

5.1. Bias current

The measured offset *voltage* depends linearly on the bias current. The offset *field* does hardly depend on the bias current (see figure 4) because the sensitivity is inversely proportional to the bias current. The residual offset field after spinning is only a few microtesla, over the entire bias current range, for both the CMOS and bipolar devices.

5.2. Substrate voltage

The influence of the substrate voltage on the sensitivity does not compensate that of the offset *voltage* and therefore the offset *field* still depends on V_{sub} (see figure 4). The measurements show that the offset caused by a variation in the substrate voltage is cancelled by the spinning-current method [5]. Measurement results on other devices also show a smaller residual offset for CMOS devices than for bipolar devices.

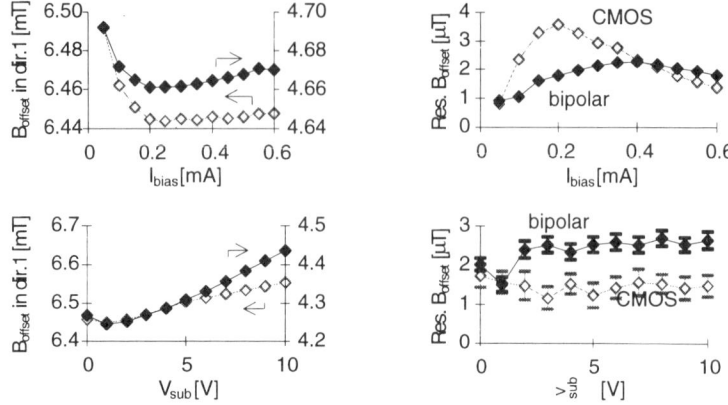

Figure 4. Offset field versus bias current (top) and substrate voltage (bottom), before the spinning-current method is applied (left) and after (right). Notice the difference in scale in the figures.

6. Conclusions

The offset voltage of traditional Hall plates depends linearly on the bias current, and therefore the offset field is not very sensitive to the bias current. The spinning-current method reduces the offset field to only a few microtesla, for the entire range of experimental bias currents, for both processing technologies.

The offset of traditional Hall plates also depends on the substrate voltage while the offset after spinning is again, independently of this parameter, only a few microtesla. The influence of the substrate voltage on the offset is significantly reduced by the spinning-current method.

For bipolar and CMOS Hall plates the spinning-current principle reduces the offset to only a few microtesla, even when the bias current or the substrate voltage is varied. Both processes are therefore equally suited to fabricate spinning-current Hall plates.

References

[1] Sandra Bellekom et al. Spinning-current Hall plate with integrated switches. In *Proceedings of ProRISC*, pages 43–48, 1997.

[2] P. J. A. Munter. A low-offset spinning-current Hall plate. *Sensors and Actuators*, A21–A23:743–746, 1990.

[3] R. Steiner et al. Offset reduction in hall devices by continuous spinning current method. In *Proceedings of Transducers '97*, pages 381–384, 1997.

[4] Ch. Schott and R. S. Popović. Linearizing integrated Hall devices. In *Proceedings of Transducers '97*, pages 393–396, 1997.

[5] Sandra Bellekom. Influence of substrate voltage on sensitivity and offset of (spinning-current) Hall plates. In *Proceedings of the Dutch conference on sensor technology 1998*, 1998. to be published.

Stability of magnetic microsensors under effect of ionizing irradiation

I Bolshakova, A Matkovskii, T Moskovets
State University "Lviv Polytechnic", 12 Bandera Str, Lviv, 290646, Ukraine

N Zamyatin, V Makoveev, M Bolshakov
Joint Institute for Nuclear Research, Dubna, Russia.

Abstract. Magnetic microsensors with the active area volume $10^{-5}\,\text{mm}^3$ are fabricated on the basis of III-V semiconductor microcrystals. Complex doping of semiconductor microcrystals used in fabrication of magnetic microsensors allows the improvement of radiation hardness of magnetic microsensors. In addition to a basic donor dopant with a concentration corresponding to its minimum change under effect of radiation, rare earth elements and special extras able to bind oxygen and act as drains for radiation defects are introduced in a certain proportion. Effect of irradiation by fast neutrons with energies from 100 keV up to 13 MeV, γ-quanta with energy 1.25 MeV and protons with energy 24 GeV on a carrier concentration and mobility in InSb microcrystals was considered. It has been shown that a carrier concentration and sensitivity change of InSb-based magnetic microsensors irradiated with neutrons does not exceed the measurement error 0.3%.

1. Introduction

Activity aimed at the development of magnetic microsensors reported at EUROSENSORS XI [1] is focused at improving their radiation hardness.

The insufficient radiation hardness of existing magnetic measuring instrumentation prevents one from measuring in radiation conditions in charged particle accelerators, orbital spacecraft and other sites with increased radiation.

There are facts [2, 3, 4, 5] evidencing that despite certain changes that appear in semiconductor sensitive elements under irradiation, III-V semiconductor magnetic sensors made with special technologies can be successfully used in radiation conditions.

2. Technology

The radiation hardness of Hall magnetic sensors is determined mainly by the change of its basic parameter - the sensitivity which is inverse proportional to the carrier concentration. Therefore the radiation hardness of these sensors can be evaluated by changes of carrier concentration in the sensitive element material. There are several approaches to solving the problem of minimizing the carrier concentration change under irradiation. One of these approaches is the creation of materials for sensitive elements of magnetic sensor with optimum carrier concentration by means of complex doping of semiconductor crystals during the growth (metallurgical doping). In comparison with ion and nuclear doping this method is lower-cost, and it is featured by better reproducibility of technology.

If during the growth of these III-V semiconductor compounds a complex doping is used when rear earth elements and extras taken in a certain proportion are introduced in addition to the basic donor doping impurity, the radiation hardness of those materials can be increased significantly. Such impurities can block oxygen which usually exists in III-V crystals and prevent it from reacting with primary radiation defects followed by formation of stable radiation defects of the acceptor type [6]. It is not unlikely that these impurities are good drains for both primary radiation defects, and various impurities [7].

The radiation hardness of InSb microcrystals used for fabrication of microsensors was improved with complex doping including:

- basic dopant (Te, Sn) with the optimum concentration $(2-3) \cdot 10^{18}$ cm^{-3} corresponding to its minimum change under irradiation;
- rare earth elements Yb, Gd which can bind oxygen;
- special doping extras Cr, Mn, Ge which are good drains for radiation defects.

Just such complex approach enables one to obtain with special technology radiation resistant materials for magnetic microsensors which can ensure magnetic field measurements under radiation and may become the basis for the development of radiation resistant magnetic measuring devices and systems.

3. Studies

Numerous studies [2-7] concerning the ionizing irradiation effect on III-V semiconductors and the investigation of their radiation defects performed in several research centers during last 10 years permit one to expect the successful development of radiation resistant magnetic sensors based on InAs, InSb and GaAs.

Results of the investigation of ionizing irradiation effect on characteristics of InSb microcrystals used as basis for fabricating microsensors are given in the Table.

Under irradiation by neutrons with energies from 100 keV up to 13 MeV, fluence 10^{14} n·cm^{-2} and intensity 10^{10} n·cm^{-2}·s^{-1} the concentration change in IS3 microsensor crystals with the initial concentration $n_0 = 3 \cdot 10^{18}$ cm^{-3} does not exceed the measurement error which is equal to 0.3%. For IS2 microsensors with the initial concentration $n_0 = 2 \cdot 10^{17}$ cm^{-3} the concentration change caused by neutron irradiation is larger in 4-5 times.

γ-quanta irradiation was performed in ^{60}Co radiation emitter. For γ-quanta energy 1.25 MeV and dose 0.8 Mrad the concentration change is considerably smaller in comparison with that caused by neutron effect, even in the case of IS2 low doped samples.

Under proton irradiation with energy 24 GeV and intensity $6 \cdot 10^9$ p·cm^{-2}·s^{-1} the considerable concentration change reaching several percents was observed.

Table - Effect of ionizing irradiation on characteristics of materials for sensitive elements

Irradiation type and dose	Change of material characteristics			
	Concentration change, $\Delta n/n$, %		Mobility change $\Delta\mu/\mu$, %	
	Sensor type		Sensor type	
	IS2	IS3	IS2	IS3
γ-quanta, dose 0,8 Mrad	0,5	0,2	0,6	0,2
neutrons, fluence 10^{14} n·cm^{-2}	1,2	0,3	1,5	0,3
protons, fluence 10^{14} p·cm^{-2}	2,0	1,0	6,7	4,4

In all investigated samples a carrier concentration is changed in a significant less degree than mobility. Charge carrier concentration is the semiconductor material parameter determining the main characteristic of magnetic sensors, i.e. their sensitivity. Therefore the most important problem related to the construction of radiation resistant magnetic sensitive devices is the development of semiconductor magnetic sensors with minimum concentration changes under irradiation. Though the semiconductor material disordering under fast neutron effect can really decrease the carrier mobility, the influence of carrier mobility on characteristics of current biased magnetic sensors is the second order effect [4]. The mobility change influences on the zero field voltage and maximum permitted sensor current. But, in general, the decrease of mobility because of formation of additional scattering centers (point defects and disordering areas) has no significant effect on the main characteristic of a magnetic sensor, i.e. its sensitivity.

4. Discussion. Conclusions

By means of complex doping III-V semiconductor materials can be made more radiation resistant. On the example on InSb microcrystals it has been shown that the carrier concentration change appearing in these crystals under fast neutron irradiation can be reduced to values corresponding the measurement error 0.3%. Results of performed studies have confirmed that works focused at further improving the radiation hardness of magnetic microsensors can be successfully continued for both InSb, and other III-V compounds.

Acknowledgements

This research was supported by the Science and Technology Center in Ukraine (Project 320).

References

[1] Bolshakova I, Gurjeva T, Izhnin I, Klimenko A, Matkovskii A and Moskovets T 1997 *Proceedings of EUROSENSORS XI* 2 651-654

[2] Kolin N, Osvensky V, Rytova N and Yurova E 1986 *Physics and Chemistry of Material Processing* 6 3-8 (in Russian)

[3] Borcke U V and Cuno H H 1985 *Feldplatten und Hallgeneratoren Eigenschaften und Anwendungen* (Berlin: Siemens)

[4] Del Medico S, Benyattou N and Guillot G 1996 *Semicond. Sc. Technol.* 11 576-581

[5] Yarmoluk N, Vigdorovich V and Kolin N 1980 *Semiconductor Physics and Technology* 17 1311-1314 (translated into English)

[6] Zaitov F I, Isaev F K , Polyakov A Y and Kuzmin A V *1984 Effect of Penetrating Irradiation on Properties of Indium Antimonide and Arsenide* (Baku: Elm) (in Russian)

[7] Zakharenkov L, Masterov V and Khokhlyakova O 1987 *Semiconductor Physics and Technology* 2 347-349 (translated into English)

Chemical Sensors II

ANISOTROPIC ETCHING OF SILICON IN A COMPLEXANT REDOX ALKALINE SYSTEM

Carmen Moldovan, Rodica Iosub, Dan Dascalu
Institute of Microtechnology, P.O.Box.38-160, R72225, Bucharest, Romania, Fax: +40-1-2301553,
email: cmoldovan@imt.ro

Gheorghe Nechifor
Politehnica University of Bucharest, Faculty of Industrial Chemistry, 1 Polizu, Bucharest, Romania,
email: G_Nechifor@chim.upb.ro

Abstract. This paper presents the results from an investigation of the chemical anisotropic etching of silicon in the following solution: KOH 4.5M, $K_3[Fe(CN)_6]$ 0.1M, $K_4[Fe(CN)_6] \cdot 3H_2O$ 0.1M, KNO_3 0.1M and complexant added. The reaction mechanism, the etch rate and the roughness are analysed.

1. Introduction

After many years of research and work in anisotropic etching field, the anisotropic etching of single crystal silicon in alkaline solutions became a key technology for micromachining due to the strong dependence of the etching rate on the crystal direction and the boron concentration.
 Increasing attention of anisotropic etching for crystalline silicon has been according after recognizing its capabilities for micromachining 3D-structures.
We added at KOH 4.5M solution a redox system: $K_3[Fe(CN)_6]$ 0.1M, $K_4[Fe(CN)_6] \cdot 3H_2O$ 0.1M, KNO_3 0.1M and a complexant (ether-crown or calix/4/arene) to improve the quality of the silicon surface and to increase the etch rate.

2. Reaction mechanism

Considering the general mechanism presented in [1], [2] we accept the following reactions:

$$\begin{array}{c} Si \\ \diagup \\ Si \diagdown \end{array} Si: + 2OH^- \rightleftharpoons \begin{array}{c} Si \\ \diagup \\ Si \diagdown \end{array} Si \begin{array}{c} OH \\ \diagdown \\ OH \end{array} + 2e^-_{cond} \qquad (1)$$

$$\begin{array}{c} Si \\ \diagdown \\ Si \diagup \end{array} Si \begin{array}{c} OH \\ \diagdown \\ OH \end{array} \longrightarrow \left[\begin{array}{c} Si \\ \diagdown \\ Si \diagup \end{array} Si \begin{array}{c} OH \\ \diagdown \\ OH \end{array} \right]^{++} + 2\, e^-_{cond} \qquad (2)$$

We supose the following reaction is the most probable to occur:

$$\left[\begin{array}{c} Si \\ \diagdown \\ Si \diagup \end{array} Si \begin{array}{c} OH \\ \diagdown \\ OH \end{array} \right]^{++} + 4OH^- \longrightarrow SiO_2(OH)_2^{--} + Si_{solid} + 2H_2O \qquad (3)$$

We assume the protons from the aqueous solution accept the electrons that appear during the process. The protons exist in aqueous solution due to the equilibrium:

$$HOH + HOH \rightleftharpoons HO^- + H_3O^+$$

This equilibrium, although change of place on the left because the value of pH (pH>12) participate at the medium reactions. The continuum change of place of the equilibrium on the right can be explain by hydroxide ions consumption in the reactions with the silicon surface and the reaction of the protons (H_3O^+) with the free electrons: $H_3O^+ + e^- \rightarrow H_2O + 1/2 H_2$

2.1.The role of the complexants and of the redox system

The complexants can contribute at the etch rate, both, by accepting the electrons from the medium or by consumption of hydroxide ions (4).

$$(4)$$

The ion of phenol type can intervene in the solvability of the silicic acid still at the silicon interface
The reaction (3) becomes (5):

$$(5)$$

Considering the reactions (1)→(3) we concluded that the improvement of the silicon anisotropic etching process can be realised by adding new compounds into the etching system. To choose theses compounds we must consider: a) The acid-basic character; b) the electron transfer capacity (oxido-redox character); c) the complexant character. The new compound must respect the necessity of hydroxide ions generation into the system. OH^- must be generated in important quantities and to be accessible (free).

In our work few aspects are followed: *the process produced electrons capture, the increase of the available quantity of hydroxide ions by complexing of M^+ ions and solvability of $Si(OH)_4$ by complexing.*
During the silicon etching reaction electrons and $Si(OH)_4$ are produced and OH^- ions are consumed. The electrons are captured in aqueous medium the most probable by protons (H_3O^+) and hydrogen atoms are generated. The hydrogen atoms can attack them-self the silicon, or are combined between them and molecular hydrogen is resulting. We propose a schema of the process with one input and two outputs to better explain our point of view (I or II):

Hydrogen atoms generation involve the etching acceleration because H• atoms attack the silicon and form silan bonds at the surface that initiate afterwards another chain of reactions, which realise the solvability of the silicon. To better explain the mechanism, the reaction (*) is detailed as follows (schema III):

If the steps a)÷f) are accepted we can understand the effect of the oxidants (e⁻ acceptors) on the etch rate. In our case the redox buffer (**) fix the electrons e⁻ and the H• atoms quantity is reduced ⇒ the silicon etch rate is stabilised and the roughness is improved.

$$(**) \quad [\,Fe(CN)_6]^{3-} / [Fe(CN)_6]^{4-}$$
$$[Fe(CN)_6]^{3-} + 1e^- \rightarrow [Fe(CN)_6]^{4-}$$

In this way the rate of H_2O/HO^- from the solution is consumed by water hydrated to complexanion $[Fe(CN)_6]^{4-} \cdot 3H_2O$ which is very stable ⇒ H_2O/HO^- decrease very quickly and the reaction is stopped. Addition of KNO_3 allows continuing the reaction with a greater rate compared to the KOH 4.5M solution. It is an autocatalysis reaction. The pH value was found to be the most significant factor affecting the etching characteristics. In comparison to KOH solution, hydrogen evolution is negligible on Si<100> etching fronts with low or no hillocks formation. The schema III shows too, the needs to increase the OH⁻ ions at the silicon surface. We obtained the increase of OH⁻ ions quantity involved in the silicon etching and in [Si(OH)$_4$] solvability by complexing of K$^+$ with: A)ether- crown; B) calix-arene.

A)Ether dibenzo 18 crown 6   B) Calix /4/ arene

Anyone of the complexants (A, B) "immobilize" the ion K$^+$ sufficiently to create an increase of the OH⁻ ion quantity available at the silicon surface.

$$K^+ OH^- + A \text{ (or B)} \rightarrow OH^-_{free} + K^+ A \text{ (or } K^+ B)$$

The effect of very small quantities of A or B complexants is absolutely remarkable on the etch rate and on the surface roughness and can be explain:

The complexant A (ether crown) has a surfactant character $\Rightarrow$ the complexing of K$^+$ ions in the neighborhood of the silicon surface where OH⁻ ions become more accessible.

The complexant B (calix /4/ arene) has a tensioactiv, surfactant character and has phenol groups (similar to pyrocatechol) that participate to remove Si(OH)$_4$ from the surface by complexing in conformity with the general mechanism. The extremely favorable effect of calix /4/ arene on the silicon etch rate is done to its triple effect: K$^+$ ions complexant, tensioactiv character, Si(OH)$_4$ complexant.

3. Experiments and Results

3.1. Determination of the etch rate

Preparation of the samples: 1)Blank wafers: p<100> 14Ωcm, n<100>5-7Ωcm, 2)Cleaning (ultrasound H$_2$SO$_4$:H$_2$O$_2$; BHF 10:1, D.I. water 18MΩ) and patterned wafers: p, n-type (1), Cleaning (2), 3) 2μm BPSG deposition or 3000A SiON deposition, 4) Photolithography, BPSG etching (BHF 10:1) or SiON etching (planar plasma, CF$_4$+O$_2$ t$_c$= 10min), 5) Anisotropic etching of silicon.

The chemical etch rate of the etchant proposed: K$_3$[Fe(CN)$_6$] 0.1M, K$_4$[Fe(CN)$_6$] ·3H$_2$O 0.1M and KNO$_3$ 0.1M and an ether crown complexant is 25% faster compared with KOH 4.5M solution at equal temperatures and 50% faster using calix/4/arene complexant. Both, the etch rate of Si<100> n and the etch rate of Si<100> p [μm/min], function of the temperature [⁰C] are presented in fig.1. The proposed solution is not toxic but it is corrosive.

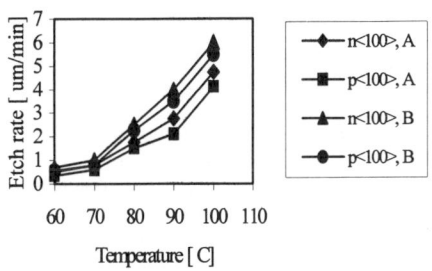

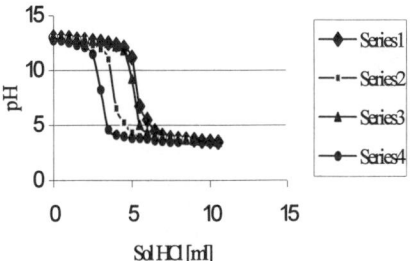

Fig.1. Etch rate of silicon n<100> and p<100> in the redox system + ether-crown (A) and + calix arene(B)

Fig.2. Titration graphics: a) Series1: KOH initial b) Series2: KOH after Si etching; c) Series3: KOH+B initial , d) Series4: KOH+B after etching

3.2. Determination of the OH ions concentration.

Using the "neutralization method" from potentiometric analyse, the "titre" T [3] of the solutions used is calculated. The titration graphics obtained are presented in fig.2. The solutions of KOH and KOH+calix/4/arene are compared before and after 1h silicon etching on blank wafers. The solution quantities used are equals and the silicon samples have equal areas. These determinations are repeated after 2h, 3h, 4h silicon etching.

From [3] the concentration of KOH and the concentration of OH⁻ ions in solutions before and after 1h, 2h, 3h, 4h silicon etching were calculated. The results show that the quantity of KOH consumed in the etching process is increased 50% for the KOH + calix/4/arene solution compared to simple KOH after every hour of etching. That means an increase of the OH^-_{free} ions in the solution, resulting an increase of the etch rate.

For the proposed solution the chemical silicon etch rate is increased, the walls profile for the patterned wafers is better (fig.3a, b), the lateral underetching is negligible (fig.3b) and the roughness (fig.4.a, b) is reduced (one order of magnitude smaller) compared to the usual 4.5M KOH at the same temperature.

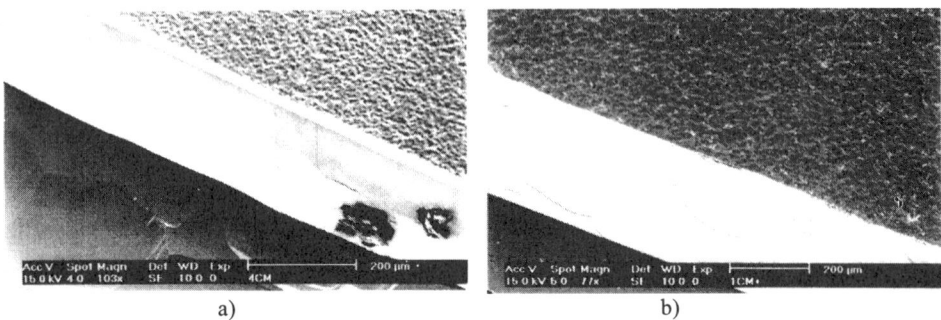

a)   b)

Fig.3. SEM picture of a p<100> silicon etching profile, in the redox system at 80°C; a) ether crown complexant; b) calix/4/arene complexant

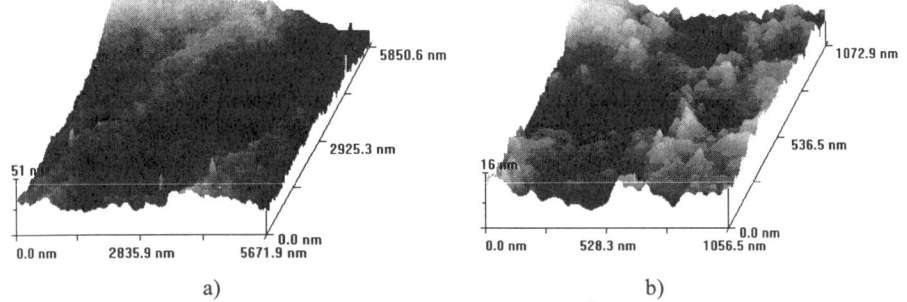

a)   b)

Fig.4. Roughness of a silicon wafer 200μ etched in the redox system at 80°C (3D-AFM investigation) : a) ether crown complexant; b) calix/4/ arene complexant

4. Conclusions:

A new anisotropic etchant of silicon is used. It has the following advantages:

1) Increase the OH^-_{free} ions concentration at the silicon surface by using the complexants of K⁺. The complexants used (A and B) have a complexant and surfactant character. Calix/4/arene has a cleaning Si surface character too (solvability of $Si(OH)_4^-$) and is the best solution of silicon alkaline etching which offer an important increase of the chemical etch rate and an excellent roughness.

2) Utilisation of a redox system that regulate by the capture of e⁻ the increase of the chemical etch rate and the molecular hydrogen output. It is a non-toxic system, but it is corrosive.

3) Minimizing of the roughness. The macroscopic and the microscopic roughness of the silicon etched were very good. The etching samples keep the aspect of a mirror polished wafer independent of the silicon-etched thickness.

References

[1] L. Ristic, Sensor Technology and Devices, Artech House, 1994, pp 67-68

[2] H. Seidel and oth. , J. Electrochem. Soc. Vol.137, No 11, 1990, pp 3612-3632

[3] Linus Pauling, General Chemistry, W.H. Freeman and Company, 1970, Chapter 14-5.-6.-7

Chemical Sensors II
Paper presented at Eurosensors XII, 13–16 September 1998
© *1998 IOP Publishing Ltd*

The influence of surface oxidation on the pH-sensing properties of silicon nitride

T Mikolajick[1,*), R Kühnhold[1), R Schnupp[2), H Ryssel[1,2)

[1) Chair of Electron Devices, University Erlangen, Cauerstrasse 6, 91058 Erlangen, GER
[2) Fraunhofer Institute of Integrated Circuit (IIS-B), Schottkystrasse 10, 91058 Erlangen, GER

Abstract. The pH-sensing properties of silicon nitride were studied with special emphazise on surface oxidation. Silicon nitride was deposited in a LPCVD process with flow ratios of the processing gases NH_3 and $SiCl_2H_2$ ranging from 2 to 20. The samples were characterized by XPS and capacitance-voltage measurements of EIS structures. Samples etched in buffered HF before measurement showed a pH-sensitivity of up to 58 mV/pH, and an increasing hysteresis was found with increasing $NH_3/SiCl_2H_2$-gas flow ratios during deposition. Samples not etched had only a pH-sensitivity between 45 and 52 mV/pH because of their oxygen rich surface layer. The oxygen rich layer was found to be approx. 4 nm deep and soluble in buffer electrolyte. Surface oxidation and dissolution of the oxygen rich layer is claimed to be an important point for silicon nitride pH-sensing layers.

1. Introduction

Ion sensitive field effect transistors (ISFETs) have many advantages over conventional ion selective electrodes. Moreover, the ISFET is a modified standard device of silicon microelectronics which makes this device ideal for a possible integration in silicon micro-systems. On the other hand, there are some serious problems to be solved. Aside from the integration of the reference electrode and the encapsulation, the long term stability is the most serious problem which has to be solved. To improve the long term stability, it is necessary to understand the main features of the ion sensing layer that control the ion sensing properties.

Since silicon nitride is a standard material in microelectronics, it is the most frequently used pH-sensing layer for pH-sensitive ISFETs. On the other hand, the ion sensing properties of silicon nitride are far from being ideal [1,2]. The influence of the fabrication parameters on the pH-sensing properties has been rarely studied. Existing data indicate that adjusting the $NH_3/SiCl_2H_2$-flow ratio during LPCVD may have an influence on the surface oxidation and therefore on the pH-sensitivity and long term stability of silicon nitride [3,4].

2. Experimental

The pH-sensing properties of silicon nitride were investigated by electrolyte/insulator/silicon (EIS) structures. As starting material, p-type <100> silicon wafers with a specific resistivity of 4 - 6 Ωcm were used. A 50 nm thick oxide was grown in dry oxygen. Silicon nitride was

---

*) Current adress: Siemens AG, HL R PZ1, Wernerwerkstr 2, 93049 Regensburg, GER

deposited in an LPCVD process at 770°C. The gas flow ratio $NH_3/SiCl_2H_2$ was set to the values 2, 3.5, 6, 10, 15 and 20 with a total flow of 260 sccm. After removing the nitride and oxide from the back of the wafer, a 0.5 μm thick aluminum layer was evaporated on the back of the wafer. Finally, the wafers were annealed in H_2/N_2 at 450°C for 20 minutes.

One half of the samples were etched in buffered HF (BHF) before measurement in order to remove the oxygen rich surface layer. X-ray photoelectron spectroscopy (XPS) measurements combined with argon sputtering were performed to determine the oxygen content of the surface layer. After each XPS-measurement, an approximately 0.3 nm thick layer of the silicon nitride was removed by 30 seconds of 1 keV argon sputtering.

Measurements on the EIS structures were performed using a computer controlled measurement setup [5]. The voltage required to achieve a preset capacitance in the steep region of the capacitance-voltage curve was used as the sensor signal. After a 12 h stabilization in buffer electrolyte, the pH-value was adjusted by computer controlled titration from pH 7 to pH 3, then to pH 11 and finally back to pH 3 in steps of 0.5 pH. After every change of the pH-value, the sensor signal was monitored for 1 minute before it was defined as the sensor signal for that specific pH-value. The accumulation capacitance, measured at -5 V, was monitored during the titration cycles. The hysteresis of the samples was measured as the voltage difference at pH 7 between titration from pH 3 to pH 11 and pH 11 to pH 3. To determine the long term stability of the samples, silicon nitride deposited with a flow ratio of 3.5 was used. Sensitivity was measured over a period of 3 weeks. Between the measurements, the samples were stored in buffer solution.

3. Results

In Fig. 1, the intensity of the XPS-oxygen(1s)-line is shown in dependence from the number of sputtering steps for silicon nitride deposited at flow ratios of 3.5 and 20. As oxygen can be bound to carbon absorbed at the surface, only measurements from the third sputtering step are shown where the carbon(1s)-line was negligible. The high oxygen content of the as processed silicon nitride is clearly visible. From the measurements, the oxygen rich surface layer can be found to be about 4 nm deep for both samples. The oxygen rich surface layer is removed by etching in BHF in both cases.

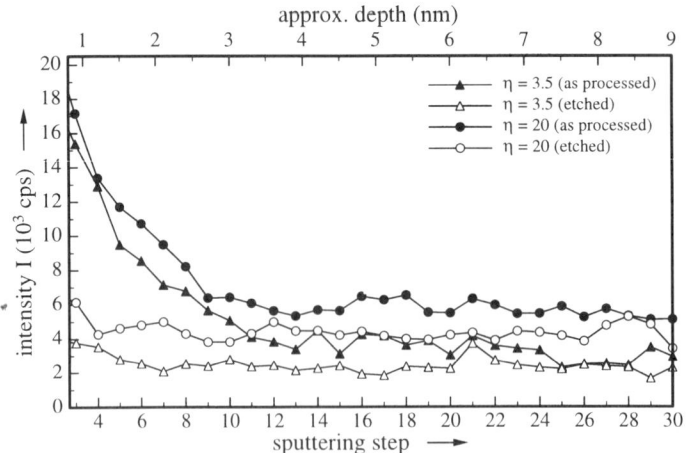

Fig. 1   Oxygen content of as processed and etched silicon nitride deposited with flow ratios of 3.5 and 20 obtained from XPS measurements

The drift rate of the EIS structures after stabilization was typically 2 to 3 mV/h for the as-processed and 0.5 to 1 mV/h for the HF etched samples. Comparing the drift rates, no significant dependence on the flow ratio could be observed. In Fig. 2, the resulting pH-sensitivity and the hysteresis of the EIS structures is shown as a function of the flow ratio of the process gases. The error bars indicate the uncertainty of the sensitivity due to the hysteresis of the samples. A pH-dependent change of accumulation capacitance, resulting from a vertical shift of the C(U) curve, was observed during the titration cycles. Since the change of the accumulation capacitance was independent from the sample under measurement, this effect is likely due to the measurement setup [5]. However, the vertical shift of the C(U) curve leads to errors in calculating pH-sensitivity and hysteresis. The actually measured pH-sensitivity was 3.5 mV/pH higher, and hysteresis 10 mV higher than shown in Fig. 2.

All samples not etched before measurement show poor pH-sensing properties. By etching the surface of the silicon nitride before measurement the pH-sensitivity is increased to 57 - 59 mV/pH. For both as processed and etched samples the sensitivity remains constant for gas flow ratios above 10. For flow ratios between 2 and 6 a strong dependence of the pH-sensitivity on the gas flow ratio can be observed for the as processed samples, with the lowest sensitivity at a flow ratio of 2. After etching the surface, this sample has the highest sensitivity and lowest hysteresis. The hysteresis of the etched samples slightly increases with the gas flow ratio.

The long term stability of as processed and etched silicon nitride films deposited with a flow ratio of 3.5 is show in Fig. 3. After approximatly 12 days continuously in the solution, no difference in pH-sensitivity and hysteresis between the as processed and the etched sample can be observed.

4. Discussion

The reduced pH-sensitivity and the increased hysteresis of the as processed samples compared to the etched samples can be explained by the oxygen rich surface layer confirmed by XPS measurements. No significant influence of the deposition conditions on the depth of the oxygen rich surface layer can be observed. From the pH-sensitivity measurements, it can be seen that the surface oxygen rich layer is effectively removed by one minute etching in BHF.

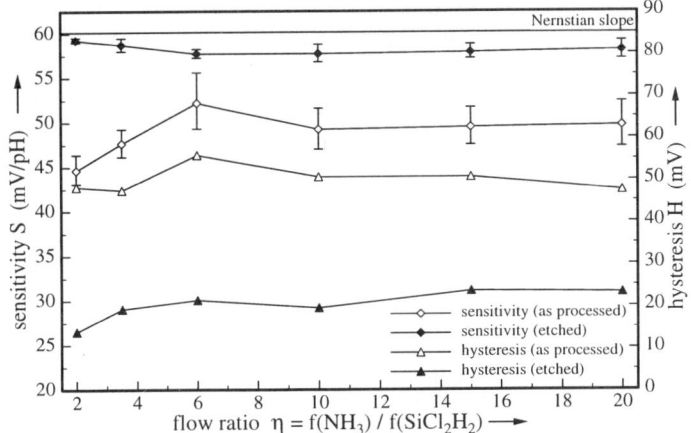

Fig. 2 pH-sensitivity and hysteresis of as processed and etched silicon nitride films depending on the gas flow ratio during deposition

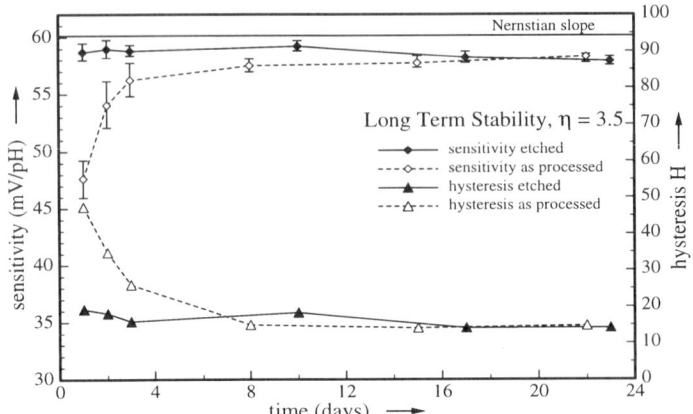

Fig. 3   pH-sensitivity and hysteresis of as processed and etched silicon nitride depending on the time of the sample in buffer solution

The hysteresis of the samples etched in BHF for one minute slightly increases with increasing flow ratio. This may be explained by an increasing hydrogen content of the silicon nitride films found by Pan et al. [6]. Habraken and Kuiper [7] have proposed that hydrogen plays an important role in the oxidation of silicon nitride. As oxidation of the silicon nitride may occure during storage of the sample in electrolyte, the hydrogen content will effect the pH-sensitivity. Combined with the dissolution of the oxide in buffer electrolyte confirmed by long term measurements, the increasing hysteresis of the films can be explained.

5. Conclusions

The pH-sensing properties of LPCVD silicon nitride films were found to be influenced by the gas flow ratio during deposition for as processed and surface-etched samples. However, no sufficient results in pH-sensitivity could be archieved without etching the surface in HF containing etchants. After etching the surface the layers deposited with high flow ratios did not reach the sensitivity of the layers deposited with low flow ratios. The hysteresis was found to be increasing with the gas flow ratio. An increasing hydrogen content which advances the surface oxidation during pH measurment and the dissolution of oxygen rich silicon nitride in the electrolyte is supposed to affect the pH-sensing properties. From our results we claim the best gas flow ratios for producing pH-sensitive silicon nitride is between 2 and 3.5.

References

[1]   Bousse L, Mostarshed S, van der Shoot B and de Rooij N F 1994 *Sensors and Actuators B* **17** 157-164
[2]   Garde A, Alderman J and Lane W 1995 *Sensors and Actuator B* **26-27** 341-344
[3]   Chen K M, Li G H, Chen L X and Zhu Y 1993 *Sensors and Actuators B* **13-14** 209-211
[4]   Chauvet F, Amari A and Martinez A 1984 *Sensors and Actuators* **6** 255-267
[5]   Mikolajick T 1997 *Feldeffekttransistoren zur pH-Wert-Messung und als Transducer für Biosensoren* (Aachen: Shaker); 1997 *Sensors and Actuators B* **44** 262-267
[6]   Pan P, Berry W, Assur I and Burton S 1988 *Proc. Electrochemical Society* **83-8** 88-93
[7]   Habraken F H P M and Kuiper A E T 1990 *Thin Solid Films* **193/194** 665-674

Chemical Sensors II
Paper presented at Eurosensors XII, 13–16 September 1998
© *1998 IOP Publishing Ltd*

Gas sensors for air quality monitoring: realisation and characterisation of SnO$_2$ thin sensing films deposited by the pulsed laser ablation technique

S.Nicoletti[a], I.Elmi[a], L. Dori[a], G.C.Cardinali[a], C. Summonte[a], M.Leoni[b]

(a) CNR-Istituto LAMEL, via P.Gobetti 101, I-40129 Bologna - Italy
(b) Dip. Ingegneria dei Materiali, Univ. di Trento I-38050 Mesiano (TN) - Italy

Abstract. In this work the results of the characterisation carried out on thin SnO$_2$ sensing layers deposited by the pulsed laser ablation technique on a substrate heater element based on 200 nm thick Si$_3$N$_4$ membrane on micromachined Si are presented and discussed. Electron microscopy and XRD analyses have pointed out that the SnO$_2$ thin films are polycrystalline, with equiaxial grains of 40-60 nm as average size; this microstructure exhibits high specific surface area. The functional test has pointed out that 0.5 ppm of benzene added to 30 % RH synthetic air with 20 ppm CO, induced 100% relative conductance variation of the sensing layer. However, prolonged device testing showed that a reduction of the sensor output occurs. Detailed XRD analyses showed a decrease of the residual microstrain after heat treatments, suggesting a reduction of lattice (point and line) defects. Furthermore, photothermal deflection spectroscopy investigations have highlighted a decrease of the SnO$_2$ absorption coefficient, α, at 1 eV, which is related with a reduction of the free-carriers concentration. Taking into account that the structural defects in SnO$_2$ are electrically active, the worsening of the sensor performance with the ageing time might be related with the reduction of free carrier concentration due to the annealing of defects at the sensor operating temperature.

To overcome the decrease of the sensor performance with time, nobles metals were added to SnO$_2$ layers and the sensors were operated in pulsed temperature mode. Preliminar results obtained on Au-doped SnO$_2$ films have shown higher sensitivity and selectivity to benzene and better stability over time.

1. Introduction

In the past few years, the growing interest for the outdoor monitoring of pollutant species coming from automotive exhaust has driven a large research efforts aiming to the development of low-price solid state devices appropriated for the detection of sub-ppm concentrations of such a pollutants [1]. Nowadays, new gas sensor technologies based on free-standing membrane substrates allow the realisation of small-size low-power consumption devices, suitable for distributed sensor networks requested for indoor and outdoor air quality monitoring. As already reported in the literature [2], a heater element device based on suspended membrane exhibits several advantages over a classical alumina substrate. The very low thermal conductivity of the membrane material allows to operate the device, in DC mode and at 400 °C, with a typical power consumption of about 0.1 W. Moreover, the very low thermal mass offers a very interesting possibility to study the sensor performance under a fast pulsed temperature mode. Furthermore, the IC compatible fabrication process of the device allows a low cost mass production and the integration of the electronic circuitry for the sensor signal processing.

On the other hand, the use of a powerful and flexible deposition technique, such as the pulsed laser ablation (PLA), allows an easy investigation of the effects on the performance of the SnO_2 sensing layer by layers with different composition.

In this paper, the gas sensing performance of SnO_2 films deposited by pulsed laser ablation (PLA) technique fabricated on substrate heater element based on micromachined Si and 200 nm Si_3N_4 membrane are presented and discussed.

2. Experimental

A 200 nm thick SnO_2 sensing layers have been deposited using the PLA from a SnO_2 sintered target. To deposit the film only over the heated area of the membrane, a shadow mask has been used. During the sensing layer growth, the substrates were held at 550 °C unbder a 30 Pa O_2 dynamical pressure. To complete the SnO_2 oxidation, the samples were then in-situ annealed for 1 h at 600 °C in pure O_2. A detailed description of the substrate heater element is reported in [2] while a description of the deposition procedure can be found in [3,4]. The sensing layer performance have been evaluated using gas mixtures constituted by benzene (0.5-10 ppm), wet synthetic air, with or without CO as interfering gas. During the test, the SnO_2 layer was heated at 400 °C by a Pt heater resistor integrated on the membrane. At the reported temperature, the device power consumption was ~100 mW. The conductance of the sensing layer was measured biasing the SnO_2 film at $V_{Bias}= 1$ V. The total gas flow, Φ, and the RH values inside the test gas chamber were 300 sccm and 10% or 30%, respectively.

3. Results and Discussion

Figure 1 shows a scanning electron micrograph of the typical surface morphology of the SnO_2 layer. The film grows polycrystalline, with equiaxial grains of an average size of 40-60 nm. This particular microstructure give rise to high specific surface area which is well suitable for the application as gas sensor active component. The crystalline phase, evaluated by the X ray diffraction (XRD) analysis, has shown that the SnO_2 layer has the typical cassiterite structure [3].

Figure 2 shows the response to different benzene and CO concentrations of a 200 nm tick SnO_2 film for an as-deposited (a) and 2 weeks 400 °C operated sample (b).

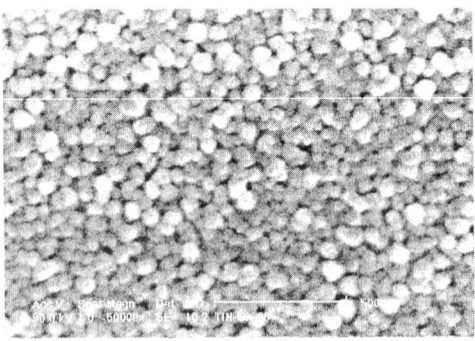

Figure 1: SEM micrograph of the surface morphology of the SnO_2 film.

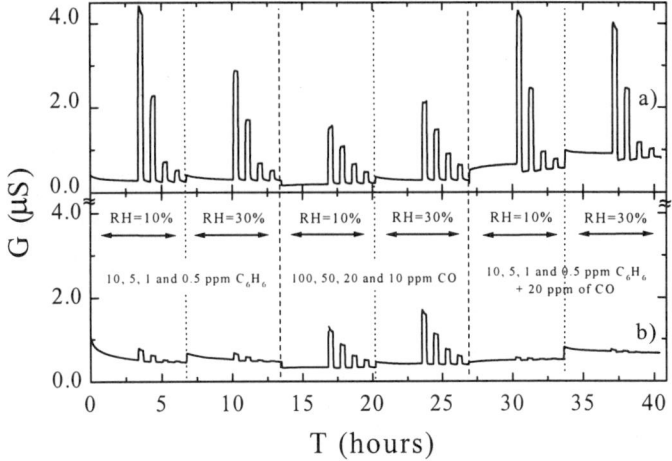

Figure 2: Sensor response to different benzene and CO concentrations 200 nm tick SnO$_2$
film for an as-deposited (a) and 2 weeks operated sample (b).

Under typical working conditions, 0.5 ppm of benzene added to synthetic air at 30 % RH
with 20 ppm CO, induced 100% relative conductance variation of the sensing layer. The rise
and the recovery time were 60 and 120 s, respectively, which allow a fast monitoring of very
low benzene concentrations. However, as showed in Fig. 2b, a prolonged testing has pointed
out a reduction of the sensor output for the same gas mixture composition.

In the attempt to explain the phenomena responsible of the observed signal output
behaviour, deep XRD and photothermal deflection spectroscopy (PDS) characterisations
have been performed on the SnO$_2$ film, before and after ageing. Figure 3a shows the residual
microstrain calculated from the XRD spectra as a function of the ageing. The change of the
strain values for crystallite size L below 20 nm indicates that a significant annealing of point
and line defects is occurring. Figure 3b shows the absorption coefficient α, measured by PDS
technique, as a function of the heat treatment. The noticeable change in the absorption
coefficient at about 1 eV is associated with a decrease in the free-carriers density of SnO$_2$ material
due to a microstructural rearrangement [3].

Taking into account that structural defects in SnO$_2$, such as oxygen vacancies or
dislocations, are electrically active and that variations in the carrier concentration are, at least

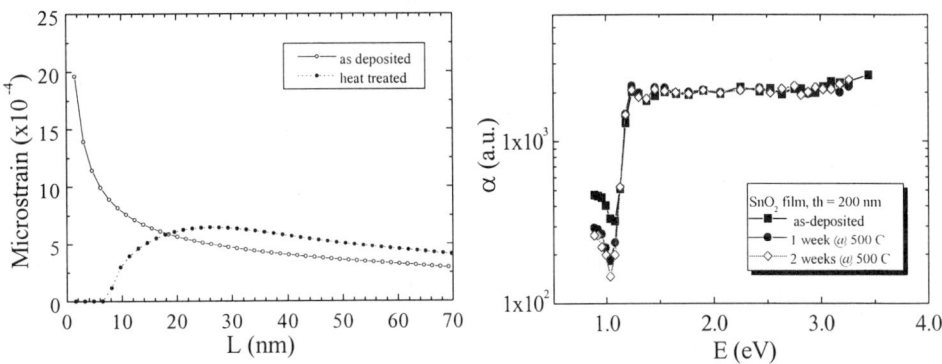

Figure 3: Residual microstrain as a function of the ageing (a) and variation of the absorption
coefficient α for different heat treatments (b) for a 200 nm thick SnO$_2$ layer.

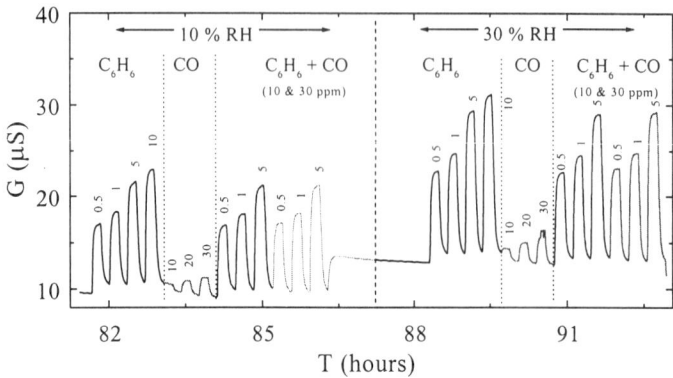

Figure 4: The signal output of a sensor based on Au-doped SnO₂ film
operated in pulsed temperature mode with a duty cycle of 1:6.

in part, related with changes in the defect density a reduction of the sensor sensitivity can be associated with an annealing of those defects.

To overcome the decrease of the sensor performances with time, the doping of the SnO₂ layers with Pt, Pd and Au was investigated. Figure 4 shows the signal output of a sensor based on Au-doped SnO₂ film operated in pulsed temperature mode, with a duty cycle of 1:6. After about 90 hours, the device maintain enough sensitivity and selectivity to benzene with respect to CO. A comparison with pure SnO₂ sensors, where the output signal decrease was consistent already after few hours, points out a significant improvement of the device performance and a better stability with time.

Conclusions

This work report the fabrication and the characterisation of thin SnO₂ sensing layers deposited by the pulsed laser ablation technique on a substrate heater membrane. The SnO₂ films had a polycrystalline structure, with equiaxial grains of average size ~40-60 nm.

The functional test has pointed out that 0.5 ppm of benzene added to synthetic air at 30 % RH with 20 ppm CO, induced 100% relative conductance variation of the sensing layer. However, after prolonged testing, the sensor output significantly reduces. Deep XRD and PDS analyses showed that the worsening of the sensor performance with the ageing time might be related with the reduction of free carrier concentration due to the annealing of defect at the sensor operating temperature.

To overcome the decrease of the sensor performance with time, nobles metals were added to SnO₂ layers and the sensors were operated in pulsed temperature mode. Preliminar results obtained on Au-doped SnO₂ films showed higher sensitivity and selectivity to benzene and better stability with time.

References

[1]   see for example: "Gas Sensors: Principles, Operation and Development" edited by G.Sberviglieri, pp. 1-42, Kluwer Academic Publisher, The Netherland 1992
[2]   Dori L, Maccagnani P, Cardinali G C, Fiorini M, Sayago I, Guerrri S, Rizzoli R and Sberveglieri G, 1997 Proceedings of the EUROSENSORS XI Conference, Warsaw (PL) (to be published)
[3]   Correra L, Nicoletti S, Mater. Sci. Eng. **B32** (1995) 33
[4]   Nicoletti S, Dori L, Corticelli F, Leoni M. and Scardi P. J. Am. Ceram. Soc. (Submitted)

Influence of the Organic Matrix on the Properties of

Membrane Coated ISFETs

Nadejda V. Kolytcheva*,

Department of Analytical Chemistry, D.I. Mendeleev Russian University of Chemical Technology, 125190, Moscow, Miusskaya Sq.,9 (Russia)

Helmut Müller and Jens Marstalerz

Martin-Luther-Universität Halle-Wittenberg, Fachbereich Chemie, Institut für Analytik und Umweltchemie, D-06217 Merseburg,Germany

Key-words: ISFET, nitrate-, potassium-, and calcium -ions, ion selective photopolymerized membranes, polyacrylate polymer matrix

Abstract

Potentiometric sensors with photopolymerized polyacrylate matrix membranes, sensitive to calcium-, potassium and nitrate-ions, were prepared and studied in batch and FIA modes. The encapsulated pH-sensitive field-effect transistors with SiO_2/Si_3N_4 gate insulators were used as transducers. Effects of polymer matrix, ionophore and plasticizer of the membrane on the performance, selectivity, dynamic characteristics and durability of ISFETs were studied. The emphasis was made on the optimisation of the membrane composition of Ca^{2+}-ISFET. A new method of the investigation of a dynamic response of the potentiometric sensor ("a method of the maximal response") was developed.

Introduction

Field-effect transistors (FET) are still widely used as transducers for a preparation of miniaturised ion sensors (ISFETs). Photopolymers are very attractive materials for the preparation of long-lived ISFETs, especially for the production of multisensing commercial devices. These devices can be successfully used in biomedical and clinical investigations. Different kinds of photopolymers have been used for this purpose, see for example [1]. Until now we have used only one type of polyacrylate type polymer as membrane matrix for the preparation of ion selective sensors with photopolymerisable membranes [2-4] (PA-matrix I). This polymer was formed from a mixture of bisphenolA-diglicidyletherdimethacrylate (BIS-GMA) and hexanedioldiacrylate (HDDA)

In the present work calcium, potassium and nitrate ISFETs were prepared by photopolymerization of membranes with the other kind of matrix material: tris-(ethyleneoxide)dimethacrylate ($(EO)_3DMA$) (PA-matrix II). The characteristics of these sensors in batch and FIA mode (performance, selectivity, durability and dynamic response) were compared with those of ISFETs, covered by membranes with PVC-matrix and PA-matrix I. The emphasis was made on the optimisation of the membrane composition of Ca^{2+}-ISFET. Six commercial ionophores (Fluka) and one synthesised ionophore as well as several commercial plasticizers were used in calcium-selective membranes with both polyacrylate matrixes.

Experimental

Transducers

ISFET with SiO$_2$/Si$_3$N$_4$ gate insulator was obtained from Centrum für Intelligente Sensorik, Erfurt, Germany [3,4]. Each chip contained two depletion n-channel mode FETs with independent point contacts; the chip and the gate size were 1.2 x 4.2 mm and 0.016 x 0.4 mm, respectively.

Membrane composition and sensors preparation

Valinomicin (K$^+$), five calcium ionophores: ETH 1001 (I), ETH 129 (II),

Ca-Ionophor III (Calcimicin), ETH 5234 (IV) and Ca-Ionophor V, as well as bis[4-(1,1,3,3-tetramethylbutyl)phenylphosphate calcium salt (VI), di-n-octylphenylphosphonate (DOPP), all plasticizers, potassium tetrakis(4-chlorophenyl)]borate (KpClTPB), Silan A174, PVC were purchased from Fluka, BIS-GMA - from Raducure Specialities, (Drogenbos, Belgium), (EO)$_n$DMA (n=3, 5/6, 23) and lucirine - from Martin-Luther-Universität Halle-Wittenberg, Fachbereich Chemie, Institut für Technishe und Makromolekulare Chemie, HDDA and phenanthrenequinone - from Aldrich. Calcium ionophore, 1,8-bis[2-(diphosphinyl)phenoxy]-3,6-dioxaoctane, was synthesised (see VIII in [5]). Tributyloctadecylphosphonium nitrate was used as electrode active component in NO$_3^-$-ISFET [2].

The ISFETs with PVC- and polyacrylate-matrix membranes were prepared as described [2,4]. Phenanthrenequinone and lucirine were used as photoinitiators for photopolymerization of polyacrylate membranes with BIS GMA and HDDA (PA-matrix I) or (EO)$_n$DMA (n=3, 5/6, 23) (PA-matrix II) as matrix components, correspondingly. The content of components in membranes (mass.%, if not specified) varied as follows: electrode active substances (EAS) - 1- 3 %; plasticizer - 65-67% (PVC-matrix membranes) and 25 45 % (polyacrylate matrix membranes); KpClTPB - 70-200 mol. % in response to ionophore content; photoinitiator - 2%.

The following physical methods were used to characterise the ion-selective polymer membranes: Differential Scanning Calorimetry (DSC), Dynamic-Mechanical Analysis and Pulsed Field Gradient NMR (^{1}H-pfg-NMR). A so-called "method of the maximal response value" was used to determine the dynamic response of the sensors in FIA mode. The electrode potential (E) was measured as a function of time and the maximal value of the slope (E$_{max}$/t) corresponded to dynamic response of the sensor (mV/s). The greater was (E$_{max}$/t), the better was dynamic response.
The usual values t$_{90}$ or t$_{95}$ were also used to characterise sensor response time.

Results and Discussion

K$^+$- and NO$_3^-$-sensors

Three poly(ethyleneoxide) dimethacrylates: **(EO)$_n$DMA** where **n=3, 5/6, 23**, were studied as membrane matrix components, but only one substance, **(EO)$_3$DMA** was suitable. The sensors with other homologues (n=5/6 and 23) had very bad durability (15 and 10 min., correspondingly) and a plasticizer "washing" of the membrane was observed. It could be explained originating the polymer structure and the properties in the membrane. The polymer, based on **(EO)$_3$DMA** had a structure of polycycles, partially hooked into one another (tangled) and partially three dimention crosslinked, it had glass temperature, T_g, higher as room temperature, while the membranes with two other polymers had a microgel structure and $T_g < 25$ °C.

Dibutylsebacate (DBS), o-nitrophenyl-n-ctylether (NPOE) and dioctyphtalate (DOP) were used as membrane plasticizers for polymer, based on **(EO)$_3$DMA**. The best sensor characteristics were obtained with DOP. Its optimal content in K$^+$-selective membrane (Mb) was 45 mas.% and in NO$_3^-$-Mb - 32 mas.% and the content of EAS was 4 and 3 %, correspondingly. The characteristics of the ISFETs with optimal membrane composition are given below:

	K$^+$- ISFET	**NO$_3^-$ -ISFET**
Slope, (mV/decade) (P=0.99)	54.5+1.5	55.5+1.5
Lower detection limit (mol/l) (P=0.99)	$(4.3+0.7) \times 10^{-5}$	$(1.9+0.6) \times 10^{-5}$
Selectivity (log K_{ij}^{pot}, $c_j=10^{-2}$ M)	-1.95 (j=Na$^+$)	-1.5 (j=Cl$^-$)
	-2.50 (j=Li$^+$)	-0.14 (j=I$^-$)
Potential drift (mV/h)	0.05 - 0.30	0.05 - 0.30
Life time	3 months	4 weeks

The characteristics of the sensors with these membranes were compared with those of the corresponding ISFETs with PVC- and PA-membrane matrix I.

The following order of response time and dynamic response improvement was obtained for both sensors with different membrane polymer matrix (EO)$_n$DMA > BIS GMA>PVC.

The optimisation of membrane composition of Ca^{2+}-ISFET

Two membrane matrix materials were studied for these sensors: PA-membrane matrixes I and II. Both kinds of sensors had suitable performance, and for both ones the optimisation of membrane composition was made. For this purpose five commercial ionophores (I-V) and one ion exchanger (VI+DOPP) as well as one synthesised ionophore VIII were tested. The following plasticizers were used: NPOE, DOS, dioctyladipate (DOA), ETH 469, ETH 2041 and ETH 2112. The optimal content of EAS in both kinds of polymer membranes was the same (3.0+0.5 %), but plasticizer content was different: 32+3 % (matrix I) and 54.5+0.5 % (matrix II).

Mechanical properties of membranes, performance and selectivity were better for the sensors with membrane matrix I, but response time and potential reproducibility and stability ws found to be better for the sensors with membrane matrix II.

Both ionophores I and II could be successfully used in both polymer matrixes together with plasticizer ETH 469, but the best selectivity had ISFETs with the membrane composed of

polymer matrix I and ionophors II or IV and ETH 469 or ETH 2112 as plasticizers (see below).

Membrane-active components: Ca^{2+}-ionophore, plasticizer	Ionophore II, ETH 469	Ionophore II, ETH 2112	Ionophore IV, ETH 469
Linear response (mol/l)	10^{-5} - 10^{-1}	10^{-5} - 10^{-1}	10^{-5} - 10^{-1}
Slope, (mV/decade) (P=0.99)	26.5+0.5	27+1	24.5+1.5
Lower detection limit (-log c) (P=0.99)	5.56	5.42	5.12
Selectivity (log K_{ij}^{pot}, j= Na^+,K^+,Mg^{2+}, Fe^{3+}; $c_j=10^{-1}$ M)	-3.35, -3.76, -4.15, -5.23	-3.40, -3.10, -4.15, -3.00	-3.35, -3.76, -4.15, -*
Response time, t_{90}, (s)	2 - 50	2 - 30	30-100
Life time (storage in the solution 0.125 M KCl + 0.1 M NaCl)	> 7 days	> 7 days	> 12 days

Note: * - no responce

Different systems ionophore - plasticizer were studied with this membrane matrix. Short life times had membranes with ionophors V and VIII and with NPOE and ETH 2041 as plasticizers. In the case of very hydrophobic plasticizer ETH 2041 its fast "washing out" and a heterogeneous structure of the membrane were observed due to a bad compatibility of the plasticizer with polymer matrix I, while the "washing out" effect was not observed for also high hydrophobic plasticizer ETH 2112, perhaps due to some chemical interaction (such, as H-bonding of OH- groups) of plasticizer with polymer matrix. Though response time for membranes with ETH 2112 was rather large.

Sensors with membranes, based on ion exchanger showed very bad selectivity to Fe^{3+}- and H^+-ions, large response time and not very good life time.

References

1. *Biosensors and Chemical Sensors. Optimising Performance Through Polymer Materials,* (ed. P.G Edelman and J. Wang), ACS Symposium Series 487, American Chemical Society, Washington, DC 1992.

2. C. Dumschat, R. Frommer, H.Rautschek, H. Müller, and H. -J. Timpe, *Anal. Chim. Acta,* **243** (1991) 179-182.

3. H. Müller and A. Spickermann, *Chem. Anal. (Warsaw),* **40** (1995) 599-608.

4. N.V. Kolytcheva, O.M. Petrukhin, N.V. Filipjeva, and H. Müller *Sensor&Actuators B*, (1998) (in press) (paper of EUROSENSORS XI; 1997, Warsaw).

5. O.M. Petrukhin, E.V.Shipulo et al.*J. Analyt. Chem.* {English translation from *Zh.Analit. Khim.,* (Russ.)}. 49 (1994) N 12 1299-1312.

Chemical Sensors II
Paper presented at Eurosensors XII, 13–16 September 1998
© 1998 IOP Publishing Ltd

Evaluation of detection limit and selectivity coefficient for single ion-selective sensors and an array of non-selective sensors

Legin A.V.*, Rudnitskaya A.M.*, Vlasov Yu.G.*, Di Natale C.**, D'Amico A.**

* Department of Chemistry, St.Petersburg University, 199034, St.Petersburg, Russia
** Department of Electronic Engeneering, University of Rome "Tor Vergata", Roma, Italy

Abstract. A method of estimation of detection limit and selectivity of sensor array has been suggested. Proposed method allows to compare sensor parameters in solutions evaluated for separate sensors (ion-selective electrodes) and multisensor systems. It has been found that detection limit of non-selective sensor array is an order of magnitude lower and that selectivity is 5 times higher in comparison with corresponding parameters of single selective sensors.

1. Introduction.

It is often declared in sensor literature that multisensor systems together with pattern recognition tools for data processing (electronic nose and electronic tongue) can show higher sensitivity and significantly lower detection limit compared to single selective sensors. This conclusion is made by analogy with biological sensor systems whose organization principles multisensor systems pretend to mimic [1,2]. For example human olfactory system is supposed to display three order of magnitude higher sensitivity then a single olfactory receptor. However, clear experimental evidence for this fact neither for gas nor for liquid chemical sensors has not been obtained yet. The aim of the present paper is to compare sensor parameters in solutions obtained using separate sensors (ion-selective electrodes) and conventional least square data processing and a multisensor system together with pattern recognition data processing tools.

2. Experimental.

Sensor parameters have been evaluated in solutions of cadmium and lead ions because there are well-studied electrodes selective to these ions [3]. On the other hand a wide number of materials sensitive to cadmium and lead ions but poorly selective are known. Measurements have been taken using chalcogenide glass cadmium and lead selective sensors and a sensor array. Sensor array included 4 non-specific chalcogenide glass sensors. All sensors have been developed in the Laboratory of Chemical Sensors of St. Petersburg University [4]. For detection limit study selective sensors and sensor array have been calibrated in solutions containing Cd^{2+} or Pb^{2+} in the concentration range $5*10^{-8}$ - $5*10^{-4}$ mol/L. A study of Cd/Pb selectivity has been performed in solutions containing Cd^{2+} in the concentration range $1*10^{-7}$ - $1*10^{-2}$ mol/L on the background of $1*10^{-6}$ mol/L of Pb^{2+}. Measurements of Pb/Cd selectivity have been taken in solutions containing Pb^{2+} in the concentration range $5*10^{-8}$ - $5*10^{-4}$ mol/L on the background of $1*10^{-4}$ mol/L of Cd^{2+}. Experimental data obtained with single sensors have been processed using least squares while data obtained with sensor array have been processed using multivariate methods such as partial least squares (PLS) and back-propagation neural networks (BPNN).

3. Results and discussion.

Statistic approach used for determination of detection limit and selectivity of different analytical methods has not been applied to ion-selective electrodes. Following IUPAC recommendation detection limit of an ion-selective electrode (potentiometric chemical sensor) is defined as concentration where deviation of electrode response from linearity according the Nernst equation at the room temperature is $(59,15/z)*\lg 2$, where z is ion charge [5]. For double-charged ions this deviation is equal to 8,9 mV. A conventional mixed solution method of coefficient selectivity determination require a calibration procedure in solutions of primary ion on the fixed background of interfering ion. Using Nikolsky equation selectivity coefficient is determined as

$$K = C_i/C_j \ (1),$$

where C_j - concentration of interfering ion, C_i - concentration of primary ion [5]. Concentrations are used instead of activities because all measurements have been taken in the presence of supporting electrolyte (10^{-2} mol/L $NaNO_3$). Concentrations C_i and C_j correspond to the point of interception of two linear parts of calibration curve: the first one describing sensitivity to primary ion (A) and the latter one corresponding to fixed content of interfering ion (B) (Fig.1). Thus, for a single sensor such parameters as detection limit and selectivity coefficient are estimated using deviations from theoretical linear response shape.

Such methods can not be applied to data processing from a sensor array because

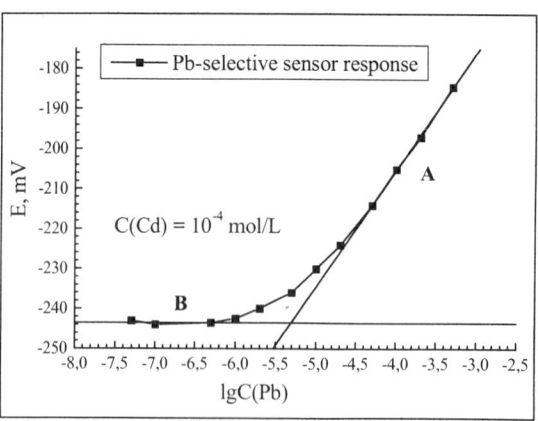

Fig. 1. Determination of selectivity coefficient of lead-selective sensor.

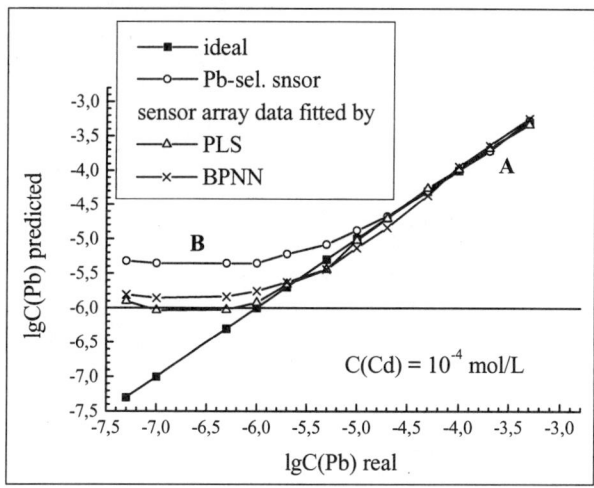

Fig. 2. Determination of selectivity coefficient of sensor array.

there is no theory dealing with sensor array response and non-selective sensor behaviour. Therefore, instead of deviation of measured potential from theoretical one, we tried to use deviation of predicted concentration from real concentration value. Predicted concentrations have been calculated using linear calibration for a separate sensor or multivariate calibration models for a sensor array. Deviation of calculated concentration value from the real one corresponding to the potential deviation of 9 mV makes 0,3 in concentration logarithm. To estimate selectivity coefficient of sensor array we suggest to use predicted versus real concentration plot. Thus the same formula (Eq.1) as for a single sensor can be applied, but here C_i is the concentration of the primary ion at the interception of two linear parts of the plot (Fig.2) and C_j is fixed content of interefering ion. One part of the curve (A) describes sensitivity to primary ion, the other one (B) corresponds to the absence of sensitivity to primary ion. Finally, a value of selectivity coefficient for a sensor array can be obtained.

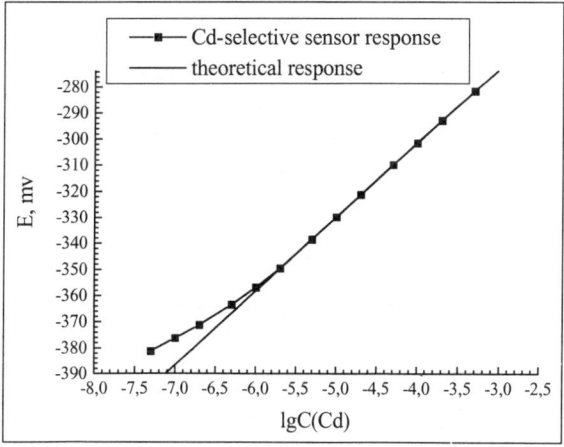

Fig.3. Determination of detection limit of cadmium-selective sensor.

Detection limit of cadmium-selective sensor determined using conventional method is equal to $1,5*10^{-7}$ mol/L (Fig.3). Prediction results obtained by least squares for single sensor and using PLS and BPNN calibration models for sensor array are shown in Fig.4. Detection limit for sensor array is lower than $5*10^{-8}$ mol/L. Similar results have been obtained for lead-selective sensor and sensor array.

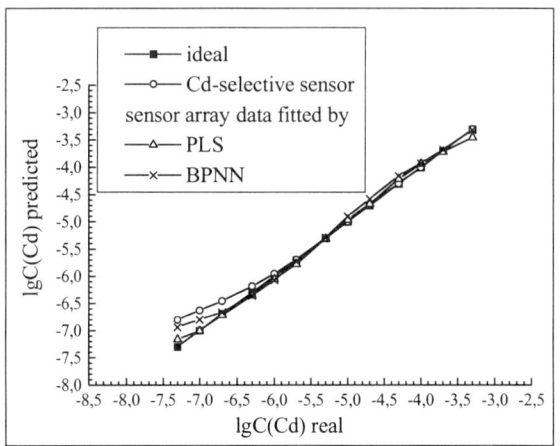

Fig. 4. Determination of detection limit of sensor array.

A comparison of selectivity of a single sensor and a sensor array is illustrated by Fig.1 and 2. Lead-selective sensor response is shown in Fig. 1. Selectivity coefficient of lead-selective sensor estimated using conventional mixed solution method is about 0,05. Prediction results obtained by least squares for single sensor and by PLS and BPNN calibration models for sensor array are shown in Fig. 2. Selectivity coefficient of sensor array is about 0,01. Similar results have been obtained for cadmium selective sensor and non-specific sensor array. Thus, better selectivity has been observed both to Cd^{2+} ions in the presense of lead and to Pb^{2+} ions in the presense of cadmium.

4. Conclusions.

A method of estimation of detection limit and selectivity of sensor array has been suggested. Detection limit of the non-selective sensor array has been found to been an order of magnitude lower and selectivity appeared to be 5 times higher in comparison with corresponding parameters of single selective sensors in solutions of lead and cadmium ions.

References.

[1]   Schild D (ed) 1990 *Chemosensory Information Processing* NATO ASI Series H: Cell Biology **39** (Springer: Berlin)
[2]   Gardner J W and Bartlett PN 1994 *Sensors and Actuators* **18-19** 211-220
[3]   Vlasov Yu G Bychkov E A and Legin A V 1984 *Talanta* **41** 1059-1063
[4]   Vlasov Yu G Legin A V and Rudnitskaya A M 1997 *Sensors and Actuators* **44** 532-537
[5]   Camman K 1979 *Working with ion-selective electrodes* (Springer: Berlin)

Chemical Sensors II
Paper presented at Eurosensors XII, 13–16 September 1998
© *1998 IOP Publishing Ltd*

Adjusting of pMOS dosimeter characteristics by electrical biases under irradiation

V.V.Emelianov, O.V.Meshurov, V.N.Oulimov, V.I.Rogov

RiSI, Turaevo, Lytkarino, Moscow reg., 140061, Russia

Abstract. The goal of this work is the detailed investigation of electrical biases and dose rate influences on standard pMOS transistor characteristics and the subsequent optimisation of a pMOS dosimeter for low dose rate application. Basing on obtained results, we came to the conclusion that the increase in measurement dose range 5-10 times can be obtained by electrical bias optimisation only. The use of dynamical electrical regime allows to decrease radiation induced charge relaxation in the pMOS dosimeter.

1. Introduction

The use of pMOS dosimeters to measure total dose irradiation levels was at first suggested by Holmes-Siedle [1]. Since that time several types of pMOS dosimeters have been proposed [2-5 and another]. The main purposes of developing the new pMOS dosimeter types are the increase in measured dose range and dosimeter sensitivity, constructive optimisation for different applications, the excluding of different additional factors influences (for example, temperature and dose rate).

Most of these dosimeters measures total dose from the threshold voltage shift of pMOS transistor caused by trapping of ionised holes in the gate oxide. Note the most of trapped charge occurs in the oxide near either the top gate-Si/SiO_2 (negative gate bias during irradiation) or bottom Si/SiO_2 (positive gate bias during irradiation) interface. If holes are trapped near the top interface, the measured threshold voltage shift is very small due to a small charge moment arm. For this reason, standard pMOS dosimeters are normally operated with positive applied biases. Unfortunately, for positive applied biases, trapped holes will be neutralised by thermoactivated tunnelling process [6]. This fact will result in fade of dosimeter's output characteristics at the low dose rate or long time applications.

The gate thickness variety is usually used for the dosimeter sensitivity adjusting. However, this way isn't effective for increase in measured dose range which in general determines total number of trap holes at the gate oxide and/or electric field perturbation caused radiation induced oxide charge build-up.

The main goal of this work is the detailed investigation of electrical biases and dose rate influences on standard pMOS transistor characteristics and subsequent optimisation of a pMOS dosimeter. Based on obtained results we came to the conclusion that the increase in measured dose range 5-10 times can be obtained by electrical bias optimisation only. In the same time the using of dynamical electrical regime allows to neglect radiation induced

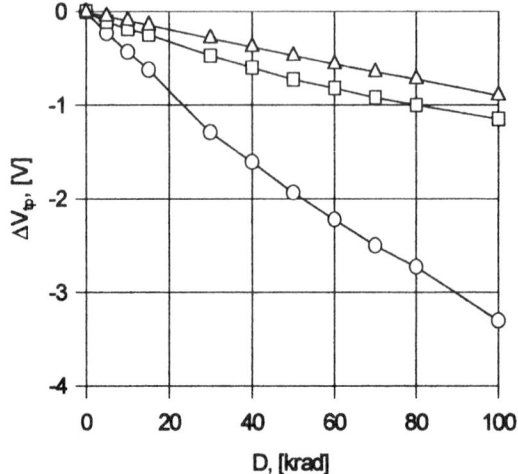

Fig.1. Dose dependencies of p-channel MOSFET threshold voltage at different gate voltage: triangle - -5V; square - 0V; circle - +5V.

charge relaxation in a pMOS dosimeter.

2. Experimental

The p-channel MOSFETs of a specialised test chip with ~60nm gate oxide thickness were used in the experiments. The gate oxide was grown in wet O_2 containing HCl with the subsequent annealing in He.

The irradiation was performed by Co(60) γ -, Sr(90)-Y(90) electron sources, x-ray tubes and an electron accelerator. Devices were irradiated at different static electrical regimes (gate voltages were varied in the range from -10V to+10V, substrate voltage from 0V to +10V) and in dynamical electric regime -10V/+10V gate voltage change with frequency of 10kG.

The characteristics of the MOSFETs were investigated by current-voltage (I-V) measurements.

The typical dose dependencies of p-channel MOSFET threshold voltage at different gate voltages are presented in Fig. 1. Normalised component of radiation-induced positive charge vs. gate voltage after irradiation and annealing is presented in Fig. 2.

3. Radiation calibration

One of main problems of precision of dose measurement by a pMOS dosimeter is the statistical scattering of radiation change of the threshold voltage from chip to chip. In the same time this parameter is changed much smaller from device to device in the range of one chip. For radiation calibration of pMOS dosimeters we suggest to use additional analogous devices in the same chip using "soft" x-ray radiation and masking lid for the shielded real dosimeters. This individual calibration of each dosimeter should be performed before the

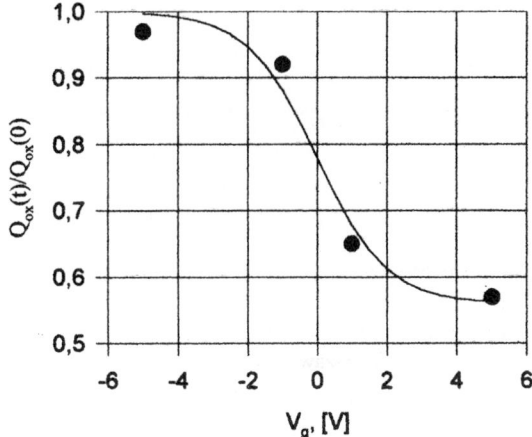

Fig. 2. Normalised component of radiation-induced positive
charge vs. gate voltage after irradiation and annealing.

set-up of packaging lid. The details radiation calibration are presented in this work.

4. Conclusion

Presented experimental results show that sensitivity and dose measurement range for a standard pMOS dosimeter can be increased 5-10 times by electrical bias optimisation only. The use of dynamical electrical regime allows to decrease radiation induced charge relaxation in a pMOS dosimeter.

References

[1] Holmes-Siedle A 1974 *Nuclear Instruments and Methods* **121** 169
[2] Dawes W R Jr. and Schwank J R 1981 *IEEE Trans. Nuclear Sci* **28** 4152
[3] August L S 1984 *IEEE Trans. Nuclear Sci.* **31** 801
[4] Holmes-Siedle A Adams L Marsden S and Pauly B 1985 *IEEE Trans. Nuclear Sci.* **23** 4425
[5] O'Connell Keller A Lane W and Adams L 1995 RADECS 95 Proceedings 481
[6] Emelianov V V .Sogoyan A V Meshurov O V Ulimov V N and .Pershenkov V S 1996 *IEEE Trans. Nuclear Sci.* **43** 2572-2586

Chemical Sensors II
Paper presented at Eurosensors XII, 13–16 September 1998
© *1998 IOP Publishing Ltd*

Determination of sodium and potassium using electrochemical microsensors in flow-cell

M. Chudy, W. Wróblewski, A. Dybko, Z. Brzózka

Department of Analytical Chemistry, Warsaw University of Technology
Noakowskiego 3, 00-664 Warszawa, Poland

Abstract. The paper describes the design of the electrochemical microsensors based on field effect transistors for selective sodium and potassium ions determination. Designed sensors (so-called CHEMFETs) utilized silicon rubber polymer (Siloprene) membrane. The performances of sodium- and potassium-CHEMFETs were determined in the flow-cell system. They showed linear response in the wide range of pMe with a slope of 47-57 mV/pMe and a very short response time not exceeding 50 s. Designed microsensors exhibit good reproducibility and good selectivity towards primary ions required for the determination of blood electrolytes.

1. Introduction

The clinic analysis requires reliable, and real-time measurements of chemical species for accurate diagnostic in health care. Among various analytes the determination of ion concentration (activity) of the electrolytes such as Na^+, K^+, Mg^{2+}, Ca^{2+} and Cl^- in blood plasma is especially important. The determination and the monitoring of the clinically relevant electrolytes can be performed using potentiometric chemical sensors [1-3] which meet the requirements for blood measurements. The silicon technology based ISFET (ion-selective field effect transistor) can be an alternative to the classic potentiometric sensor (ion-selective electrode), exhibiting comparable analytical performances.

Flow-cell systems coupled with chemical sensors detection are especially suitable for clinical analysis [4]. The calibration procedure in such systems is quite simple allowing fast determination and monitoring of the chemical species without necessity of complicated sample preparation and transport. Moreover, flow systems improve the effectiveness and some performances of the sensors used e.g. the response time or sensitivity. The application of microsensor based on field effect transistor (FET) transducer gives the possibility of flow-cell miniaturization.

In this paper the design of sodium- and potassium-sensitive CHEMFETs based on Siloprene membranes is presented. The performance of the microsensors was determined in the flow-cell system. The reproducibility and the selectivity towards primary ions of designed sensors for their further application in blood electrolytes determination were examined.

2. Experimental

2.1. Chemicals

All salts used were of analytical grade and were purchased from POCh Gliwice. The standard stock solutions (0.1M) of metal nitrates were prepared in redistilled water; working solutions were obtained by dilution of the stock solution. The sodium and potassium sensitive ionophores (ETH 2120 and valinomycin, respectively), lipophilic salt potassium tetrakis[3,5-bis(trifluoromethyl)phenyl] borate (KTFPB), Siloprene 1000 and crosslinking agent K-11 were obtained from Fluka. Freshly distilled tetrahydrofuran (THF, from Fluka) was used as a solvent for the membrane components.

2.2. CHEMFET preparation

The membrane solution contained: 2 mg of ionophore, 0.8 mg of KTFPB (only for potassium sensor), 6 mg of K-11 and 65 mg of K1000. The membrane components were dissolved in 0.25 ml of freshly distilled THF. The membrane solution (approximately 8 μl) was deposited on the gate oxide surface of FET covered with poly(2-hydroxyethyl methacrylate) layer. Before the membrane deposition the polyHEMA was conditioned over 3 hours in 0.1M solution of internal electrolyte. After membrane solvent evaporation and crosslinking the sensors were conditioned 2 days in the internal electrolyte.

2.3. Measurements

The performances of the CHEMFETs were examined in the set-up for FET measurements in the flow-through cell in the concentration range of the primary ion 10^{-6} -10^{-1} M. The measuring conditions of FETs: constant drain-current mode $I_D = 0.1$ mA and source-drain potential $V_{DS} = 0.5$ V. The silver chloride electrode with a double junction was used as a reference electrode. Potentiometric selectivity coefficients were determined by the fixed interference method (FIM) [5] by increasing the activity of the primary ion in the solution in steps of 0.5 log a (M).

3. Results and discussion

Chemical sensors designed for clinical analysis must fulfill several requirements which are governed by the composition of the sample. The devices for determination of blood electrolytes should have a good reproducibility, good selectivity towards determined ion over interfering ions (especially sodium) and reasonable sensitivity in physiological concentration range. Additionally the mechanical properties of the sensors must conform with flow-through assemblies which are frequently applied in clinical measurements of small size samples.

In our work back-side contact FET transducers suitable for mass production and especially for flow-cell application were used. Designed sodium- and potassium-sensitive CHEMFETs based on Siloprene membrane containing appropriates ionophores were investigated in the flow-through cell. Na^+-sensitive sensors exhibited a linear response in the physiological sodium range (135-145 mM) with a slope of 47-52 mV/pNa$^+$ in the presence of 0.01 M interfering cation: Li^+, Mg^{2+} (fig.1). Additionally, the responses in 0.01 and 0.001 M K^+ covering potassium concentration in blood plasma (3.5-5 mM) were shown.

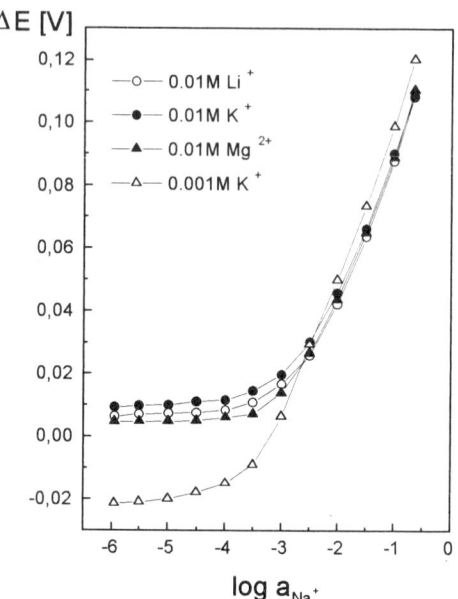

Figure 1. Response curves of Na-sensitive sensors in flow system in the presence of interfering cations.

K[+]-sensitive CHEMFETs showed linear response in the physiological range with a slope of 57 mV/pK[+] in the presence of 0.01 M interfering cations. Figure 2 presents potassium responses of the sensors only in the sodium salts solutions of various concentrations (corresponding to Na[+] concentration in blood plasma).

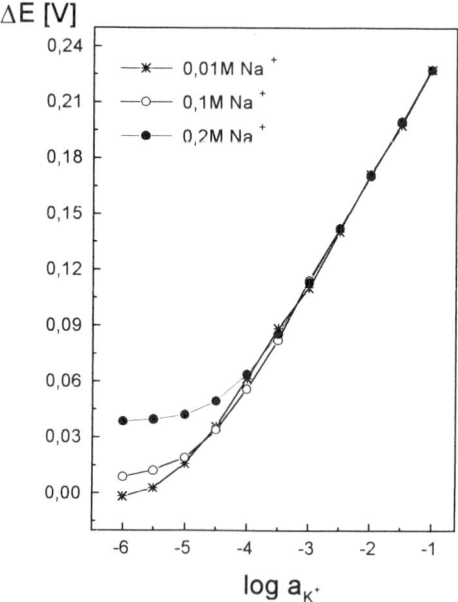

Figure 2. Response curves of K-sensitive sensors in flow system in the presence of 0.01 M, 0.1 M and 0.2 M solutions of $NaNO_3$.

The selectivity of sodium- and potassium sensitive sensors for main interfering cations was determined by the FIM method. Values of selectivity coefficients obtained in the flow-cell are compared in Table 1 with values required for determination of blood electrolytes. Our sensors fulfill the requirements concerning their selectivity against interfering ions occurring in the physiological background..

Table 1. Values of selectivity coefficients required for determination of blood electrolytes and found of sodium and potassium-sensitive CHEMFETs based on Siloprene membranes determined in flow-cell.

Interfering cation M =	$\log K_{Na,M}$		$\log K_{K,M}$	
	required	found	required	found
Li^+	< -0.13	-0.6	< -1.7	-2.6
Na^+	-	-	< -3.6	-3.6
K^+	< -0.6	-0.8	-	-
Mg^{2+}	< -1.2	-1.9	< -2.8	-3.9

The response time of the sensors in the flow-cell system was less than 50 s for small concentration of primary ions (less than 30 s for physiological range). The reproducibility of the CHEMFETs was also tested changing subsequently the concentration of primary ion in the solution from 0.1 M to 0 (0.01 M interfering ion solution). Reproducible steady signals (differences between signals not exceeding 0.2 mV) were achieved.

4. Conclusions

Back-side contact FET transducers were applied to design Na^+ and K^+ electrochemical microsensors suitable for determination of electrolytes in blood plasma. The performances and selectivity of the CHEMFETs were determinated in flow-through system and proved that they meet the requirements for clinical measurements of blood electrolytes.

5. Acknowledgments

The work was supported by the State Committee for Scientific Research, Project No 8 T10C 019 13.

References

[1]   A. Lewenstam, *Anal. Proc.*, **28**, 106 (1991)
[2]   W. Simon, U. Spichiger, *Anal. Sci.*, **7**, 861 (1991)
[3]   J. Wang, *Electroanalytical Techniques in Clinical Chemistry and LaboratoryMedicine*. VCH, New York, 1988.
[4]   M. Valcarcel, M.D. Luque de Castro, *Techniques and instrumentation in Analytical chemistry*. Volume 16, Elsevier Science, 1994.
[5]   Y. Umezawa, K. Umezawa, H. Sato, *Pure & Appl. Chem.*, **67**, 507 (1995)

Chemical Sensors II
Paper presented at Eurosensors XII, 13–16 September 1998
© *1998 IOP Publishing Ltd*

STUDY OF THE INTERFERENCES OF NO$_2$ AND CO IN SOLID STATE COMMERCIAL SENSORS

Miguel A. Martín, J.P. Santos, H. Vásquez*, J.A. Agapito

Dpto. Física Aplicada III, Fac. Ciencias Físicas, Univ. Complutense,
Av. Compluntense s/n, 28040 Madrid, Spain
*Facultad de Ciencias Físicas, Univ. Nacional de Ingeniería, Lima, Perú

Abstract. Commercial solid state gas sensors have been used to measure NO$_2$ and CO concentrations in the range of the first alarm thresholds for urban environments. Their mutual interference has been studied at three operating temperature ranges. The results showed that different sensitivities and selectivities can be obtained with the same type of sensors. This fact is used to successfully train a neural network.

1. Introduction

The simultaneous measurement of the concentrations of NO$_2$ and CO present in dry air atmosphere with discrete solid state gas sensors can not be carried out with a unique sensor, because its response would be the same for a wide range of NO$_2$ and CO concentrations, standing out a near zero sensor signal [1] (due to the concurrent reactions of the oxidizing NO$_2$ and the reducing CO).

We propose the use of a multisensorial system with commercial sensors to solve this problem. These sensors are general purpose (sensitive to a large variety of gases). We use an extensive array, in order to analyze the whole complex response by means of neural network techniques. These techniques are very useful to recognize patterns, that is, NO$_2$ and CO concentrations.

The training of the neural network is based in the back propagation method and will be more successful as the responses of the discrete sensors are more diverse, more numerous and mainly more selective to NO$_2$ and CO (sensors sensitive to CO despite the presence of NO$_2$ and vice versa) [2]. This selective response can is obtained varying the heating temperature of each invidual sensor.

The goal of this work is to obtain an array of different commercial solid state gas sensors whose multiple response, analyzed through a neural network, yield correctly the different NO$_2$ and CO concentrations after its training.

The range tested of NO$_2$ and CO concentrations are around the values of the first alarm threshold limits for NO$_2$ and CO, regulated by the EC ($\approx$ 106 ppb for NO$_2$ and $\approx$ 9 ppm for CO) [3].

2. Experimental

We have used different commercial sensors non specific for the target gases. These sensors are developed using an advanced thick film printing technology on an alumina substrate on which the gold electrodes are printed. A thick film heater of ruthenium-oxide or platinum is deposited on the reverse of the substrate.

Scanning Electron Microscopy has been performed on the sensors in order to check their chemical composition and morphology.

As a measure system we have employed a gas line capable to generate mixtures up to five different gases in dry air, and make the control and the measurement over sixteen different sensors distributed in two chambers. The gas concentration, heating temperature of each individual sensor and resistance measurement are completely automatized.

We have made the following four measurement series, using three different temperature ranges (low, medium and high temperatures) to achieve different selectivities to NO_2 and CO:

a) NO_2 concentration in the range 10 ppb – 250 ppb (ultra low NO_2 concentration) without CO interference.
b) CO concentration in the range 1 ppm – 20 ppm without NO_2 interference.
c) NO_2 concentration in the range before with a CO interference of 5 ppm.
d) CO concentration in the range before with a NO_2 interference of 250 ppb.

The results are given in terms of the sensitivity defined as the percentile ratio of the electrical resistance of the sensor with the NO_2 or CO containing and NO_2 and CO free mixtures.

3. Results and discussion

SEM analysis showed that the sensors are made of polycrystalline SnO_2 with grain sizes in the micron range.

We show clearly the interference of one gas in the presence of the other, and how this interference varies with temperature. Thus, we can use two sensors of the same model in the array operating at different temperatures, as their selectivity and sensivity to NO_2 and CO are different.

3.1 Response to NO₂ and CO

Most of the used sensors were sensitive to the presence of NO_2 and CO at all operating temperatures. Some of the sensors exhibited responses as high as 8000 % to NO_2 and 100 % to CO.

The sensors present their highest response to NO_2 at low and medium temperatures, but this trend is not so clear for CO because the maximum response is obtained at low and high temperatures.

On the other hand, the rate of response to both gases increased with temperature, being very much higher to CO.

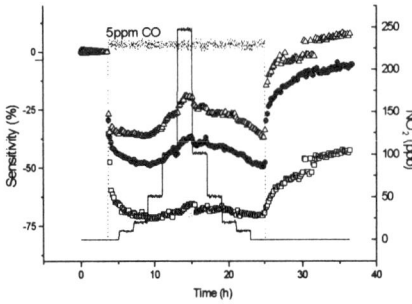

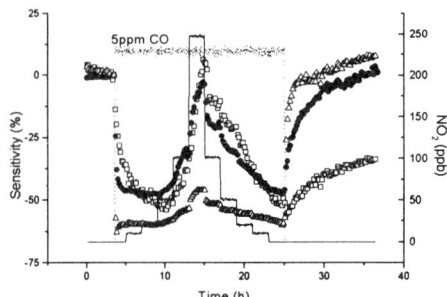

Fig 1: Response of sensor ST11 to NO₂ with an interference of 5 ppm of CO at several temperatures: □ 136°C, • 224°C, Δ 312°C.

Fig 2: Response of sensor ST31 to NO₂ with an interference of 5 ppm of CO at several temperatures: □ 131°C, • 222°C, Δ 313°C.

3.2 Interference

The effects of the presence of CO and NO₂ are opposite on this kind of sensors [4] [5], thus when both gases are present, one effect can predominate over the other, leading to different selectivities.

Figure 1 shows that the reducing effect of CO prevailed over the oxidizing effect of NO₂ at all temperatures. This result is even more remarkable at the lowest temperature.

Different behaviour were found for other sensors. Figure 2 shows that both effects are similar (at the highest concentration of NO₂) except at the highest temperature in which the reducing effect predominates over the other.

Figure 3 shows that the dominant effect is the oxidizing one of NO₂, except at the highest temperature in which the dominant effect is the reducing one of CO.

It can be observed from figures 1-3 that there are some sensors operating at determined temperatures whose responses to CO are almost independent of NO₂ concentration.

On the contrary, a response to NO₂ independent of CO concentration is more difficult to obtain because even in the most favourable case found (figure 4), the CO has a remarkable reducing influence in the response.

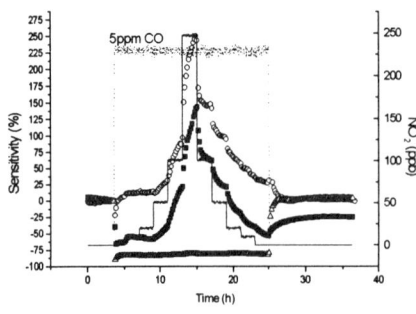

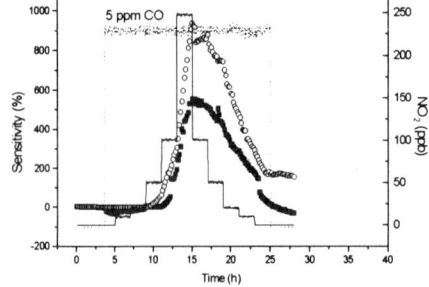

Fig 3: Response of sensor TGS711 to NO₂ with an interference of 5 ppm of CO at several temperatures: ■ 130°C, o 220°C, Δ 310°C.

Fig 4: Response of sensor SP31 to NO₂ with and without the interference of CO at 120°C: o without CO, ■ with CO.

4. Conclusions

We have measured commercial sensors that can be used to monitor simultaneously NO_2 and CO. We have shown that we can obtain different grades of selectivity and sensivity and use it in a neural network to recognize patterns (concentrations of NO_2 and CO in a mixture).

Acknowledgments

The authors would like to thank FIS for the technical information provided and Alejandro Varez for the SEM analysis. One author would like to thank CAJA MADRID for the finnancial support.

References

[1] H.-E. Endres, W.Göttler, R. Hartinger, S. Drost, W. Hellmich, G. Müller, Ch. Bosch-v. Braunmühl, A. Krenkow, C. Perego, G. Sberveglieri, A thin-film SnO_2 sensor system for simultaneous detection of CO and NO_2 with neural signal evaulation, *Sensors and Actuators B,* 35-36 (1996) 353-357.

[2] J. Santos, P. Serrini, B. O'Beirn, L. Manes, A thin film SnO_2 gas sensor selective to ultra-low NO_2 concentrations in air, *Sensors and Actuators B,* 43 (1997) 154-160.

[3] Official Journal of the European Communities, No. L87/5, 1985.

[4] S. C. Chang, Thin film semiconductor NO_x sensor, *IEEE Trans. Electron Devices, Ed*-26, 12 (1979) 1875-1880.

[5] G.Wiegleb and J. Heitbaum, Semiconductor gas sensor for detecting NO and CO traces in ambient air of road traffic, *Sensors and Actuators B,* 17 (1994) 93-99.

Chemical Sensors II
Paper presented at Eurosensors XII, 13–16 September 1998
© *1998 IOP Publishing Ltd*

New DNA based amperometric sensor and its use in immunoassay

S S Babkina, N A Ulakhovich and Yu I Zyavkina

Chemical Faculty, Kazan State University, Kremlevskaya Str.-8, Kazan 420008, Russia

Abstract. New fast way of DNA immobilization on cellulose nitrate membrane which retains maximum of DNA activity has been developed. The amperometric sensor based on the membrane obtained and stationary mercury-film covered silver electrode was used for the determination of the DNA specific antibodies in blood serum. The antibodies content was monitored by measuring the hydrogen catalytic liberation peak current at the potential of -1.2 V which was resulted from Pt(II) complexing with DNA. The scheme of this process was proposed. The peak current value at -1.2 V decreases with an increase of antibodies concentration in blood serum due to partial shielding of the DNA-Pt(II) catalytically active complexes through biospecific interaction of DNA with antibodies. The Ab can be determined in the concentration area of 1.2×10^{-9} - 5.0×10^{-10} M. The sensor can be used within a month without loss of activity.

1. Introduction

The development of new sensitive methods for investigation of the biological macromolecules' (e.g. nucleic acids) participation in immunochemical processes, their reactions with metals and their complexes can be considered as one of the important goals of chemistry.

Since 1969, when Rozenberg et al discovered high antineoplastic activity of cis-dichlorodiamino platinum(II) (DDP) [1], Pt(II) interactions with DNA was studied by various physicochemical methods [2,3]. The use of electrochemical methods has given rise to a large extension of analytical possibilities for nucleic acids and their effectors such as anticancer preparations, metal derivatives, proteins [4,5]. In [5,6] the determination was performed by measuring hydrogen catalytic liberation waves (HLW) which emerge in the presence of some platinum complexes.

In the last few years much attention has been paid to the development of highly-sensitive electrochemical biosensors, based on nucleic acid [7-9]. These sensors not only provide enhanced selectivity and decreased consumption of DNA but can be used in cell phenomena simulation as the nucleic acids adsorbed on different cell's units are known to interact with free molecules including specific proteins and metal ions.

This paper describes the results obtained during the development of the sensor based on the new method of DNA immobilization. This biosensor was used in the assay of DNA-specific antibodies through HLW in the presence of Pt(II) complexes and for diagnostics of autoimmune diseases. The autoimmune diseases of both human and animals are characterised by anomalous increase of DNA-specific autoantibodies concentration in blood-serum , so they can be diagnosed with the sensor developed.

2. Experimental

Voltammetric measurements were performed with a PO-5122/03 oscillopolarograph (Rostov experimental plant, Russia). The biological sensor developed by us and based on stationary mercury-film covered silver electrode (d=0.5 mm) served as a working electrode. . The reference electrode was a saturated calomel electrode - s.c.e. All measurements were performed at 37-40°C in various buffers (pH 4-7).

We used following biological preparations:

- bovine spleen denatured DNA with the concentration of 0.01-0.1 mg ml^{-1};
- immunoglobulin G(IgG) with the concentration of 0.4 mg ml^{-1} from the serum of infected minks, which was salted-out twice with ammonium sulfate;

The reference IgG solutions for graduation were prepared from serum solution obtained from blood of infected minks on the 15th day of disease (IgG concentration was 10mg ml^{-1}). Blood serum activity was determined against Aleutian antigen by the reaction of the immunoelectroosmophoresis (RIEOP) and against DNA by enzyme immunoassay (EIA) on the microtitration plate.

1×10^{-5}-5×10^{-6} mol l^{-1} Pt(II) solutions were used. Highly purified organic solvents (acetone, hexane) and 25% glutaraldehyde solution were used.

For the preparation of biosensitive part of the biochemical sensor 0.1 g of cellulose nitrate (CN) was dissolved in aceton. Solution of a single stranded DNA was added. The mixture was cast onto a glass surface to form a film, then was treated with glutaraldehyde, washed with water and dried. The film was immersed in Bovine serum albumin in PBS-buffer (pH 7.3) to allow free glutaraldehyde active sites to be blocked and thus to prevent non-specific binding.

3. Results and discussion

The use of biochemical sensors gives considerable improvement of the selectivity of determinations. It is particularly important for multicomponent mixtures like biological liquors of an organism. The main step in working out a biochemical sensor is to find a method for immobilization of biomolecules. We proposed a new method of DNA immobilization retaining its maximum biological and immunological activity owing to use of acetone as a CN solvent that provides shorter time of the contact between the biomolecule and organic solvent. The membranes obtained are uniform, do not swell. Immobilised DNA is not washed away from the membrane and remains active within 1 month.

Our investigation is mainly concerned with denatured DNA and its complexing with platinum(II) because toxic influence of metals on human organism may damage DNA structure to the point of denaturation and autoimmune diseases are characterised by formation of the antibodies specific to denatured DNA [10] where antigenic deteminants are not shielded due to the secondary structure unlike native DNA.

The membranes containing immobilised DNA served as biosensitive part of the sensor based on stationary mercury-film covered silver electrode. This biosensor was used for the determination the DNA-specific blood serum antibodies resulted from the damage of the cell units and DNA molecules. Catalytic hydrogen liberation waves obtaining in the presence of Pt(II) complexes with DNA were used as an analytical signal. Pt(II) complexing with denatured DNA seems to occur through Pt(II) binding mostly with N7 atoms of neighbouring guanine moieties of the same spiral i.e. so called intracrosslinking takes place. Unlike native DNA, Pt(II) complexing with N(7) and N(1) atoms of adenine and N(3) of cytosine is possible in our case due to the lack of hydrogen bonds between N-containing heterocycles. This process is also affected by electrostatic interaction of Pt(II) with polynucleotide anion of DNA [2].

The observed cathodic peak current at -1.2 V should be ascribed to catalytic hydrogen liberation process because of the following characteristics of the signal: peak value and wave position depend on pH; the current reaches its maximum at pH 4.5-5.0 the wave shifts anodically with the decrease of pH; the current depends on buffer capacity at given pH, the current is maximum at $\beta=0.5$ mol/l, hydrogen bubbles are liberated under the potentials of peak currents at stationary mercury surface electrode, the current observed is 10-50 times greater than expected diffusion limit, Semerano's factor(ΔlgI/ΔlgV)= 0.2-0.3.

For the determination of the Ab concentration in the blood serum a dependence of Ip vs. cAb was plotted. The Ab presented in the blood serum of minks being sick with Aleutian disease (an autoimmune one) were used as a model. The linear regression was obtained as $y=(-11.82\pm0.28)x+18.96\pm0.27$, $r=0.9983$. The determination was performed on the basis of the decrease of the cathodic peak current due to the presence of the specific Ab in the solution. The decrease of the current is ascribed to the shielding the catalytically active DNA-Pt(II) complexes by antibodies biospecifically bound to DNA.

The results of the determination of the Ab are shown in Table. A detection limit of 5×10-10 M (0.075 μg/ml) was calculated for the Ab using 3s criterion. The specificity of the assay in respect to the autoAb is confirmed by the lack of any response for immunoglobulins from bovine and horse.

Using the new method of DNA immobilisation and the biochemical sensor on its basis together with electrochemical properties of DNA complexes with Pt(II) enables biological molecules to be determined enough selectively and sensitively and in a broad concentration area. This also allows diagnostics of definite diseases. The biosensor can also be used in the assay of different effectors of nucleic acids such as platinum containing anticancer medicines and various environmental contaminants.

Table

Determination of Specific Antibodies(Ab) to DNA
(n=5, P=0.95)

Inserted, 109 mol/l	Found, 109 mol/l	R.S.D.
0.50	0.52±0.02	0.04
0.70	0.71±0.03	0.03
0.90	0.94±0.04	0.02
1.10	1.09±0.05	0.01

References

[1]   Rosenberg B et al 1969 Nature **222** 385-392

[2]   Sherman S E and Lippard S J 1987 Chem.Rev. **87** 1153-81

[3]   Bancroft D P, Lerge C A and Lippard S J 1990 *J.Am.Chem.Soc.* **112**

        6860-71.

[4]   Palecek E 1986 *Bioelectrochem.and Bioenerg.***15** 275-95

[5]   Kim S D *et al* 1990 *Anal.Lett.* **8** 1505-18

[6]   Vrana O and Brabec V 1984 *Anal.Biochem.* **1** 16-23

[7]   Wang J et al 1997 *Anal.Chim. Acta* **1-2** 1-8

[8]   Fojta M and Palecek E 1997 *Anal.Chim. Acta* **1** 1-12

[9]   Babkina S S *et al* 1996 *Anal..Chem.* **21** 3827-31

[10]   Goldfarb D M and Zamchuk L A 1975 Antibodies to nucleic acids(Moscow:

        Nauka)

Influence of tin oxide microstructure on the sensitivity to reductor gases

M.C. Horrillo[a], A. Serventi[b], D.Rickerby[b] and J. Gutiérrez[a]

[a] Laboratorio de Sensores-IFA-CSIC, Serrano, 144, 28006 Madrid, Spain
[b] Institute for Advanced Materials, JRC, 21020 Ispra (VA), Italy

Abstract

The gas response of tin oxide sensors strongly depends on the preparation process and especially on deposition parameters. For this reason, sensor elements have been prepared by depositing nanocrystalline tin oxide at room temperature and at 250°C on polycrystalline alumina substrates by a reactive sputtering technique. It is seen that the response to gases is dependent on the tin oxide deposition temperature due to the structure change undergone by the oxide. The best response to low concentrations of carbon monoxide and propanal is given by tin oxide sensors deposited at 250°C. Microstructural examination of these devices has been carried out using transmission electron microscopy. The main difference observed in the tin oxide microstructure is the porosity. Sensors prepared at 250°C had a more compact microstructure.

1.Introduction

During the last decades, tin oxide has been used in gas sensors for very different applications, obtaining good response to different gases, as reductor or as oxidant. They present the inconvenience of a lack of selectivity in presence of a gas mixture and a degradation with time. Therefore, it is very important to find different approaches to improve the selectivity, such as, dopants, filters, statistical methods and even with the tin oxide acting as a catalyst depending on its morphology and microstructure. It is well-known that the preparation method type of these sensors has a great influence on the response to gases. Their sensitivities, selectivities and degradation with time as sensor-materials depend on the grain size, surface morphology and internal porosity. In general, the development of sensors of the semiconductor oxide type has been based on empirical knowledge and therefore the study of the deposition parameter influence is very important to achieve better devices.

A gas sensor possesses at least two basic functions [1], the function to recognize a particular gas (receptor function) and another to transduce the gas recognition into an electrical signal (transducer function). The first is carried out by the surface chemical process and the adsorption sites (O_2 vacancies) have a great importance, while the second is carried out by physical processes related to the transport of electrons (electrical resistance change). This last one due to the complexity of polycrystalline elements needs studied more, since one of the most important factors affecting the sensing properties of these devices is the microstructure and morphology of polycrystalline sensing element.

In the present work the microstructural and morphological aspects of thin films of tin oxide deposited at room temperature and at 250°C on polycrystalline alumina substrates by reactive rf. sputtering are examined in relation to the performance of this material in gas sensing applications.

2.Experimental

Tin oxide thin films were grown by reactive r.f. magnetron sputtering (Alcatel model SCM 450) using a 99.9% pure SnO_2 target under a 10:90 oxygen-argon mixture at a total pressure of 0.5 Pa. The deposition conditions were the same for all films: forward r.f. power 100 W, substrate to target distance 50 mm The only deposition parameter changed was the deposition temperature: 250°C (S1) and room temperature (S2). Film thickness was 200 nm.
Pt contacts were also deposited by sputtering for specimens that had to be submitted to electrical characterisation. Each specimen was annealed in air at 500°C during 4h, in this way the films are polycrystalline with randomly oriented grains [2].
Conventional TEM observations were made with a JEOL 200CX microscope, while high resolution studies were performed using a JEOL 3010 microscope operated at an accelerating voltage of 200 kV to observe the effects of the substrate temperature on the microstructure of the films. Cross section transmission electron microscopy specimens were prepared by cutting 4 x 5 mm rectangles from the alumina substrate, embedding in a stack of similar sized dummy sheets, drilling out a 2.3 mm diameter cylinder and inserting this into a 3 mm diameter reinforcement tube [3]. Final thinning was carried out by twin ion beam thinning at an impingement angle of 4° using argon ions of 3-5 kV energy .
The response of these specimens used as sensors for CO and propanal were determined using a stainless steel test chamber of 20 cc and the resistance measurements were carried out at a operating temperature range varying from 150 to 350°C with a constant flow rate of 200 ml min^{-1} for low concentrations of CO (from 15 to 150 ppm) and propanal (from 50 to 150 ppm) in air synthetic. Each concentration was prepared by means of a mass-flow controller. The measuring and mixing processes were automatically carried out by means of DMMs and PCs.

3. Results and Discussion

The initial observations carried out with TEM allow some preliminary information to be obtained regarding the film microstructure. It can be stated in general that the films consist of nanocrystalline grains of SnO_2 with a cassiterite structure, the majority of the grains having diameters between 5 and 15 nm. Grains of lengths up to 20-30 nm may be sporadically present. The films replicate the surface morphology of the substrate, though adhesion between the two is not always perfect, which could be a result of debonding during the final stages of the ion beam thinning. There is generally a narrow zone of disordered or amorphous material next to the film/substrate interface. The micrograph shown in Fig.1a shows a bright field image of the film deposited at room temperature with a subsequent thermal treatment at 500 °C. The film thickness varied between 180 and 220 nm depending on location. The grains exhibit a columnar growth structure with homogeneously distributed intergranular porosity. A certain amount of intercolumnar porosity may also occur, particularly in areas of the substrate of small radius of curvature. Figure 1b shows a micrograph of the film deposited at a temperature of 250 °C following the thermal treatment. This film has a compact columnar structure with negligible evidence of intergranular porosity. In regions of pronounced substrate curvature, however, a more open structure may result. The change in microstructure observed in these regions, already noted in the previous specimen, is presumably due to the abrupt change in the angle of incidence at the surface of the substrate [4]. The films have an apparently isotropic texture: the relative intensities of the diffraction rings from the nanocrystalline SnO_2 were not found to vary noticeably with the orientation of the underlying grain of the substrate. Fig 2 shows a typical diffraction pattern showing the SnO_2 polycrystalline ring pattern superimposed on the single crystal spots from an Al_2O_3 grain in near zone axis orientation.

a)   b)

Fig.1 Transmission electron micrograph of an SnO_2 film: a) deposition at room temperature followed by 500 °C thermal treatment. b) deposition at 250 °C followed by 500 °C thermal treatment.

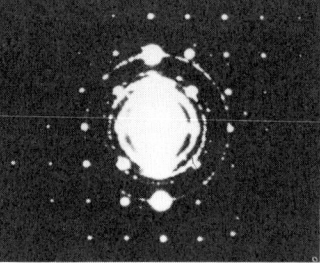

Fig. 2 Selected area diffraction pattern showing diffraction rings from nanocrystalline SnO_2 superimposed on spots pattern from a single Al_2O_3 grain.

With respect to the electrical characterisation, the sensitivity values were calculated from the resistance values measured in each gas. This sensitivity, is defined as the ratio between the resistance change and the semiconductor resistance in the gas target for both gases since they are reductors. Good sensitivities were obtained for tin oxide deposited at room temperature and at 250°C, but were much better at 250°C. This could be due to the fact that the increase of temperature during the deposition makes the film microstructure more compact and that the crystallinity improves, as it has been shown by TEM. Besides it is clear that there is a great ionadsorption of oxygen when the specimen is prepared at 250°C because after annealing, at room temperature (before measuring a gas) these specimens had a resistance value higher than those prepared at room temperature (Fig.3). So, there are evidently available more oxygen species ionosorbed on the semiconductor surface in order to react with CO or propanal. This occurred at every operating temperature but especially at 300°C and 350°C which are the maximum sensitivity temperatures for CO and propanal respectively.

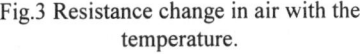

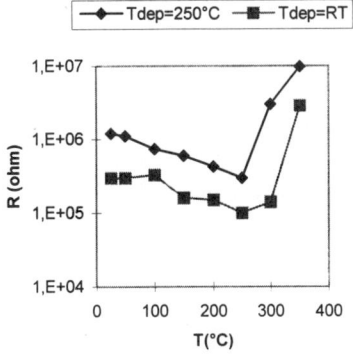

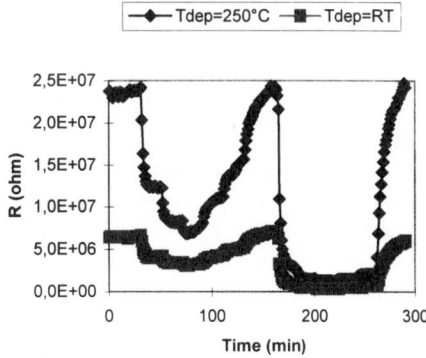

Fig.3 Resistance change in air with the temperature.

Fig.4 Responses to CO and propanal (50,100,150, 100, 50 ppm) in air at 300°C.

In Fig.4, is shown the response to low propanal and CO concentrations. The response to propanal is much higher due to the fact that this gas is much more reactive [5]. In Fig.5 the sensitivity to CO and propanal at the maximum sensitivity temperature is shown, verifying that the best responses are given by sensors prepared at 250°C.

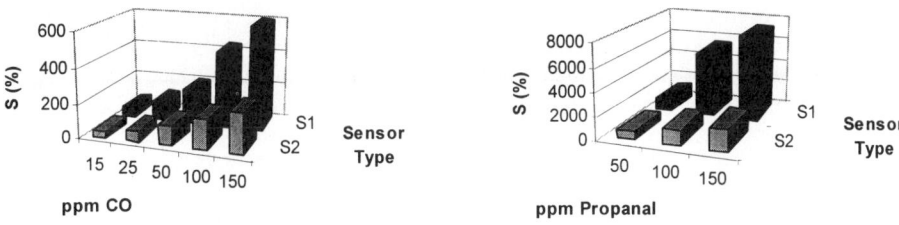

Fig.5 Sensitivities: a) to CO at 300°C; b) to propanal at 350°C.

4.Conclusions

It is possible to conclude that the best sensitivity results were obtained from specimens prepared at 250°C. These presented a more compact structure as the major difference with respect to those prepared at room temperature, for which there is a larger specific surface to react with gases, so that therefore the sensitivity increases. It will be important to analyse the changes in these films with time to see which of them shows more degradation, since it is very interesting to know the life time of these sensors for their commercialization.

References
[1] Chemical Sensor Technology, Vol.4 1992 20
[2] Sanjines R et al. 1990 Sensors and Actuators B1 176-182
[3] Alani R Jones J and Swann P 1990 Mater. Res. Soc. Symp. Proc. 199 85-101.
[4] Smy T Dew S K and Brett 1995 M J MRS Bull. 20 (11) 65-69.
[5] Horrillo M.C et al 1997 Sensors and Actuators B43 193-199

Chemical Sensors II
Paper presented at Eurosensors XII, 13–16 September 1998

CAPACITANCE INVESTIGATIONS OF
Pd/SnO$_2$ SENSOR STRUCTURE

S.V. Ryabtsev, Y.A. Ugai, O.B. Yatsenko, A.N. Lukin
Department of Physics, Voronezh State University, 394693 Voronezh, Russia
e-mail: root@ ftt.vsu.ru

Abstract

In the work the results of investigations for Pd/SnO$_2$ sensor structure by the capacitance methods are presented. We have found the change of capacity of SnO$_2$ thin films with the island-type Pd particles deposited on the film surface after the inlet of H$_2$ into the measuring chamber. C-V characteristic of the system has been studied. Basing on the results of the measurements we have obtained the physical model of the structure and its equivalent electric circuit.

1. Introduction

Sensors based on SnO$_2$ thin films with the island-type metallic particles deposited on their surface have been studied in the number of works [1,2,3 and so on] due to the prospects of their application in a gas analysis. Mechanism of Me/SnO$_2$ sensitivity as well as the physical model of gas interaction with metal particles and semiconductor surface were proposed in [4,5]. Most of the works has been performed using direct current. The aim of our work was to obtain an additional information at the measurements with alternating current.

2. Experimental

SnO$_2$ layers were obtained by magnetron sputtering of Sn onto ceramic substrate (99.7 % Al$_2$O$_3$) with Pt electric contacts and the following oxidation of Sn at 873-973 K. Thickness of SnO$_2$ layers in various experiments was of 30 to 60 nm and their resistance in the air at 423 K exceeded 1 GOhm. After that Pd was thermally evaporated onto the surface of SnO$_2$ in vacuum. The calculated amount of Pd was of 10^{16}-10^{17} at/cm^2. The resistance of the structure just after palladium deposition was of ~ 100 kOhm. Due to the instability of ultrathin metal layer with an increased surface energy shunting the bulk of semiconductor the resistance of sensor was not stable. Before the experiments the structure was stabilized by cyclic ageing in the air and in the mixture of air with 5% of H$_2$ at the temperature of 423 K. Ageing resulted in a sharp increase of resistance in the air and the same increase of sensitivity to H$_2$ (Fig.1). It is connected with the formation of isolated island particles with a lower surface energy than of a thin conductive Pd layer. Microscopic measurements gave the mean size of the island as of ~5 nm. The measurements of capacity of the structures prepared in such a way have been performed at the frequency of 1 MHz with an amplitude of testing signal 250 and 25 mV.

3. Results and discussion

The obtained Pd/SnO$_2$ structures have demonstrated a high sensitivity to H$_2$ at the resistance measurements not only at 423 K but at 293 K as well and demonstrated the characteristic response time of $\tau_{90} \approx 100$ s.

Under inlet of 5% H_2 into measuring cell sensor capacity is increased (Fig.2). Analysis of equivalent circuit of sensor structure (Fig.4) as well as additional experiments has allowed to make a conclusion that the main contribution into the change of sensor layer capacity is provided due to capacitance circuit 3.

Electric circuit 1 (Fig.4) corresponding to contact barriers SnO_2-Pt electrodes (Shottky barrier) does not contribute into the change of capacitance structure. It follows from the fact that the change of Pt ($e\varphi = 5.3$ eV) electrodes by In-Ga eutectics ($e\varphi \approx 4$ eV) did not result in a quantitative change of capacitance dependencies. Moreover, the experiments with the change of symmetric Pt electrodes by antisymmetric ones did not result in a change of C-V curve shape (Fig.3).

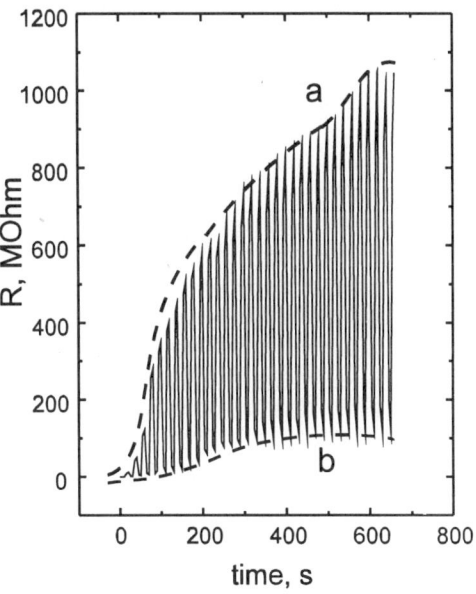

Fig.1 Variation of sensor resistance under cycling ageing: a - sensor resistance in the air;
b - sensor resistance in the mixture of air-5%H_2

Electric circuit 2 which corresponds to intercrystallite barriers also contributes into in the change of structure capacity. It follows from the fact that in the presence of H_2 the height of barrier between the crystallite grains in n-type semiconductors should be reduced [5] and hence the capacity of the structure should be decreased as well that is in a contradiction with our data. Perhaps, the slow processes of recharging of the surface adsorbtion states which are connected with the mechanism 1 considered below are characterized by the time which is insufficient for these processes to proceed at the frequency of 1 MHz.

The increase of a capacity of sensor structure under the inlet of H_2 we explained basing on the known models of sensibilizing effect of additives for catalytic active metals [5]. The role of additives according to one of the two mechanisms is reduced to the activation and spillover of H_2 thus resulting in a change of SnO_2 oxidation state and decrease of a depletion zone which is located normally to the surface. This mechanism does not provide non-uniform charge distribution along the direction of the current flow in a sample and hence should not influence the capacity of a sample(electric circuit 4).

According to the mechanism 2 H_2 causes the change of the work electron escape for metallic island. As a consequence, charge redistribution in a metal and SnO_2 takes place as well as the change of the size for a depletion region located under metal particle. In our mind just this mechanism proves to be the reason for the change of capacity of sensor layer. For the thickness of $SnO_2 < L_D$ sensor layer can be considered as normally closed field-effect transistor (MESFET) with the gate in the form of isolated metal island. This transistor is controlled due to the interaction with the gas ambient. We have used low-signal equivalent scheme of the field-effect transistor for the explanation of C-V curves (Fig.3). Under the change of bias polarity the curves remain symmetric relative to the axis aligning through the point 0 V. The capacity of the structure in this case is determined by a sum of C3 and the less of capacities C1 or C2 depending on the bias polarity. C-V curves were obtained in the presence of 5% H_2 after stabilization of electric parameters of the sensor. The appearence of non-equal capacities C1 and C2 is connected with the disturbance of symmetry for the depletion regions under applying of the voltage along the structure.

According to the represented physics model for Pd/SnO$_2$ structure one can assume that for SnO$_2$ layer thickness > L$_D$ its capacity under the inlet of H$_2$ should not increase but, on the contrary, must be reduced. The experiments with Pd/SnO$_2$ samples having thickness > 100 nm have confirmed this model (Fig.2).

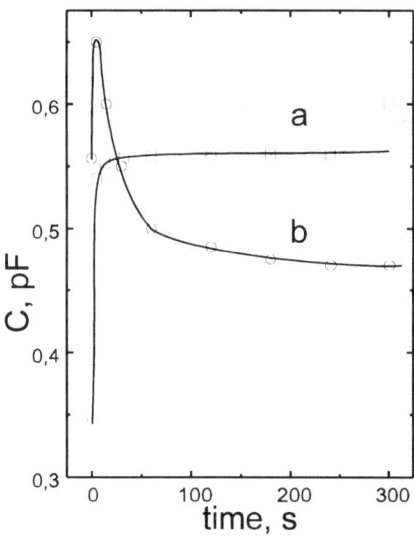

Fig.2 Capacitance of Pd/SnO$_2$ structure under intake of hydrogen: a - thickness of SnO$_2$ layer - 30 nm; b - thickness of SnO$_2$ layer - 110 nm.

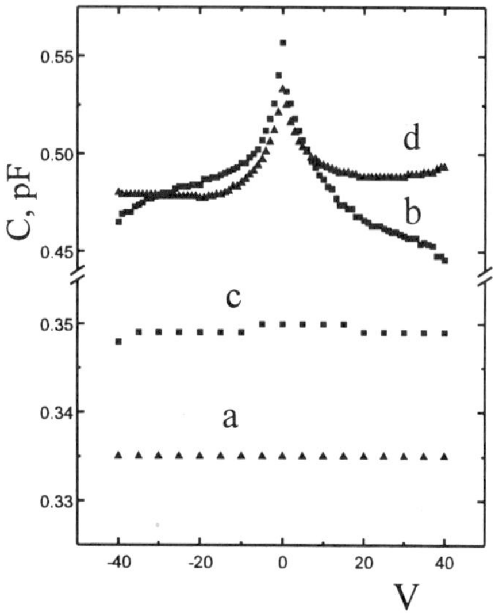

Fig.3 C-V characteristics of Pd/SnO$_2$ structure: a - in the air at 293 K; b - with addition of 5% H$_2$ at 273 K; c - in the air at 423 K; d - with addition of 5% H$_2$ at 423 K

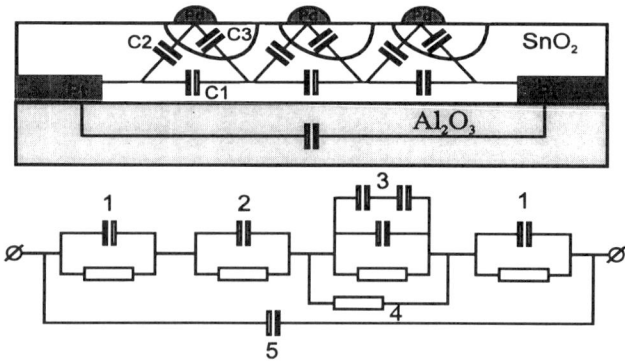

Fig.4 Physical model of Pd/SnO structure (thickness of SnO$_2$ > L$_D$) under bias voltage and its equivalent circuit.

4. Conclusion

High-frequency (1 MHz) measurements of Pd/SnO_2 structure capacity have allowed us to make a conclusion that the mechanism connected with charge redistribution between the metal and semiconductor provides a certain contribution into the sensitivity of the structure to H_2. Capacitance measurements at low frequencies (100-1000 Hz) will possibly allow to make quantitative estimations of the contributions for each of the 2 considered mechanisms.

Acknowledgements

This work was supported by the Grant INTAS-93-0091.

References
[1] K.D.Schirbaum, U.Weimar, W.Göpel, Schottky-barrier and Conductivity Gas Sensors Based upon Pd/SnO_2 and Pt/SnO_2, Sensors and Actuators B, 4 (1991) 87-94.

[2] R.Huck, U.Böttger, D.Kohl, G.Heiland, Spillover Effects in the Detection of H_2 and CH_4 by Sputtered SnO_2 Films with Pd and PdO Deposits, Sensors and Actuators B, 17 (1989) 355-359.

[3] J.Mizseli, V.Lantto, Simultaneous Response of Work Function and Resistivity of some SnO_2-based Samples to H_2 and H_2S, Sensors and Actuators B, 4 (1991) 163-168.

[4] S.R.Morrison, Selectivity in Semiconductor Gas Sensors, Sensors and Actuators , 12, (1987) 425-440.

[5] S.Matsushima, Y.Teraoka, N.Miura, N.Yamazoe, Electronic Interaction between Metal Additives and Tin Dioxide in Tin Dioxide-Based Gas Sensors, Japenese Jornal of Applied Physics, v.27, 10 (1988) 1798-1802.

Evaluation of hydrogen sensitivity and measuring range of integrated MISFET sensors by means of testing the electrical characteristics of sensing elements

<u>S Gumenjuk</u> , B Podlepetsky, M Nikiforova and K Ledovsky

Department of Microelectronics, Moscow Engineering Physics Institute, 31, Kashirskoe sh., 115409, Moscow, RUSSIA

Tel. (095) 323-9190, (095) 323-0184, fax. (095) 324-2111, e-mail: root@d405.micro.mephi.ru

Abstract. Integrated hydrogen sensors based on $Pd-Ta_2O_5-SiO_2-Si$ structure and consisting of MISFET, thermodiode, heater and test elements have been developed and investigated. There has been found the relationships between the electrical characteristics such as the threshold voltage, transconductance etc. of hydrogen sensitive MISFET and the measuring characteristics of integrated sensor based on this MISFET. This allows to except the hydrogen testing stages from the finish testing processes and to use only conventional IC' electrical tests.

1. Introduction

Integrated hydrogen gas sensors based on Pd-gate MISFET have been thoroughly studied during the last years because of their high sensitivity, small size, low cost and so on. It has been demonstrated that the most sensors measuring characteristics depend on the used fabrication processes, sensors constructions [1-2] and electrical modes of employment [3]. All hydrogen sensors fabrication processes include the finish testing stages to examine the electrical parameters of the sensors elements and the measuring capability of integrated sensor, such as the sensitivity, response rates and measuring range. The electrical testing stage is conventional integrated circuits process, but the hydrogen testing is original, difficult and precise process strongly complicating the sensor fabrication cycle.

This paper deals with the experimental investigations of dependencies between the electrical characteristics such as the threshold voltage, transconductance of $Pd-Ta_2O_5-SiO_2-Si$ MISFET as a hydrogen sensitive element and the measuring characteristics of integrated sensor based on this MISFET, which allows to except the hydrogen stages from the finish testing processes and to use only conventional electrical testing.

2. Experimental

The sensors have been fabricated on p-type monolithic silicon wafer of 10 Ω cm relative sensitivity with <100> crystal orientation using the stages of conventional MOS-devices fabrication process. The thin gate insulator have been fabricated by the stages of thin silicon oxidation to thickness about 60 nm, laser deposition of 60 nm thickness Ta layer, photolitography of contact windows and full oxidation of Ta-film in O_2+N_2 at 550^0C. Pd-gate of the sensors has been fabricated by laser deposition in vacuum onto structure heated to 100^0C. The Pd-picture has been formed by lift-off photolitography. To increase the sensors reliability there was used the operation of deposition of Al thin-film to the Pd-gate contact pads of MISFET through a metal mask. The sensors chips have been packaged by using of ultrasonic bonding, and was annealed at 300^0C and trained by gas mixture with high hydrogen concentration. The sensor chip has a size about 1.5 x 1.5 mm^2 and consists of MISFET, thermosensor, heater and test elements.

The simplified circuitry of the sensor testing system is demonstrated in Fig 1. This system provides the constant and manually controlled values of drain current, source-drain voltage and temperature of the chip.

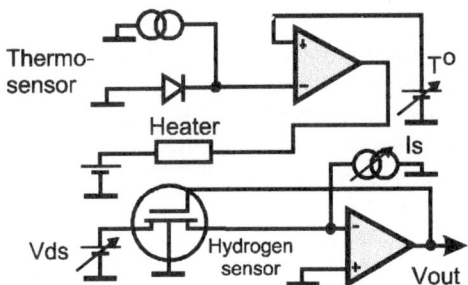

Fig.1. The simplified circuitry of the sensor measuring system

3. Results and discussion

The large quantity of experimental data have been measured and analyzed. Firstly there have been measured the input and output voltage-current characteristics of sensitive MISFETs for different electrical modes and chip temperature. The received experimental dependencies at the temperature typical for the sensor chip (about 100^0 C) are demonstrated in Fig.2.

There was found that all MISFETs have build-in channel and are easily controlled by the gate voltage $U_G>-1$ V, and on the contrary $U_G<-1$ V is not able to control FET drain current. So the next experiments have been made only for $U_G>-1$ V. The values of MISFET's threshold voltages and transconductanses have been defined from the voltage-current characteristics. The set of the sensors with wide variation of these parameters has been sampled to the next experiments.

Then the second part of experiments has been carried out. The integrated sensors with different electrical characteristics have been tested by the influences of the steps of hydrogen with different concentration (from 10 ppm to 2 % H_2 in air) and their measuring characteristics have been defined.

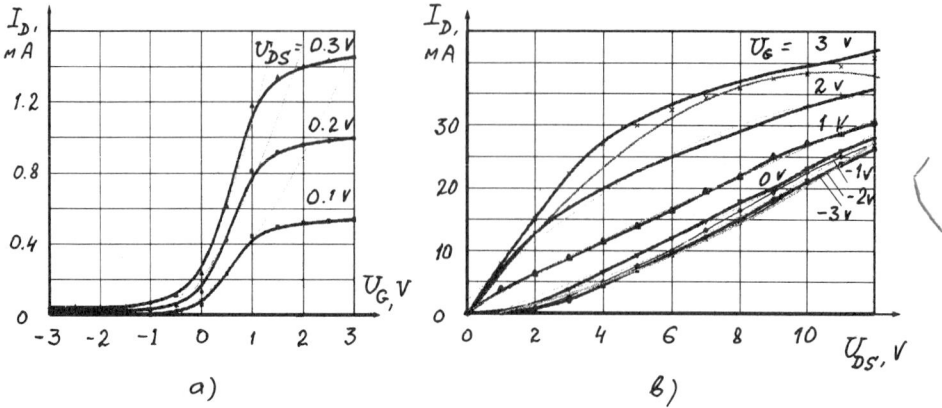

Fig.2 Input (a) and output (b) voltage-current characteristics of MISFET without hydrogen at temperature $100^0 C$

It was found that the sensors response and relaxation times and stability of initial output voltage depend on only a quality of the gate Pd-film, drain current, drain-source voltage and chip temperature and do not vary with the electrical MISFET characteristics alteration. However, the sensors hydrogen sensitivity and measuring range, defined by saturation of transfer characteristic, depend on the MISFET transconductance certainly. The dependencies of these characteristics on FET transconductance and threshold voltage are demonstrated in the Fig.3, 4 respectively.

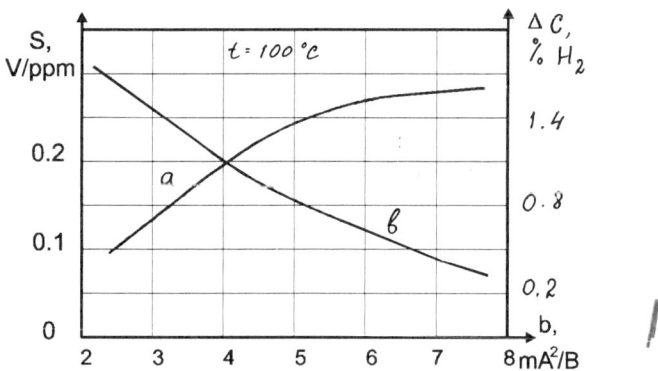

Fig.3. The dependencies of hydrogen sensitivity (a) and measuring range (b) on FET transconductance b

These dependencies are the same to the sensors with different chip temperature. The sensor hydrogen sensitivity increases with rising the MISFET transconductance from 2 to 6 mA/V^2 and rising the threshold voltage from –1.0 to +0.2 V, but the measuring range of the sensor decreases simultaneously more than ten times. The optimal values of threshold voltage for best sensors lie in range from –0.7 to –0.2 V and the optimal values of transconductance belong to the range 4,5 – 6.5 mA/V^2.

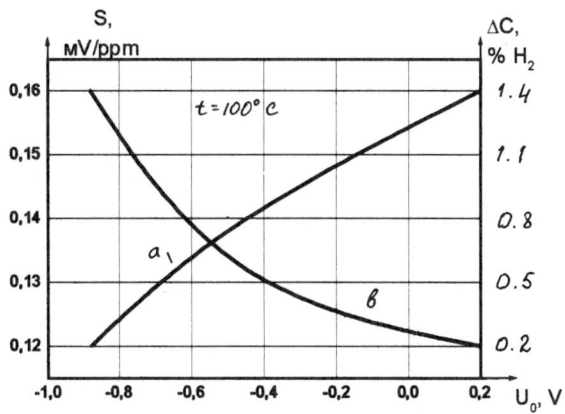

Fig. 4. The dependencies of hydrogen sensitivity (a) and measuring range (b) on FET threshold voltage U_0

So measuring the electrical characteristics of sensing element can evaluate future sensitivity and measuring range of MISFET hydrogen integrated sensor and select necessary sensors for certain areas of employment. For example, the sensors with threshold voltage less than –0.7 V can use in systems with wide hydrogen measuring range and the sensors with threshold voltage more than –0.4 V can use in high sensitive system measuring small hydrogen concentrations.

4. Conclusions

The dependencies of hydrogen sensitivity, response and relaxation times of integrated hydrogen $Pd-Ta_2O_5-SiO_2-Si$ MISFET sensor on threshold voltage and transconductance of the sensing element have been experimentally investigated. It was demonstrated that measuring the electrical characteristics of sensing element can evaluate future sensitivity and measuring range of MISFET hydrogen integrated sensor. This allows to except the hydrogen testing stages from the finish sensors testing processes. The optimal values of threshold voltage and transconductance for best sensors lie in range from –0.7 to –0.2 V and $4,5 – 6.5$ mA/V^2 respectively.

References

[1]. Fomenko S, Gumenjuk S, Podlepetsky B, Chuvashov V and Safronkin G 1992 *Sensors and actuators*, **B 10** 7-10

[2] Podlepetsky B, Gumenjuk S and Fomenko S 1996 *Proceedings of EUROSENSORS X*, (Leuven, Belgium) p 637-640

[3] Gumenjuk S, Podlepetsky B, Ledovsky K and Kozlov I 1997 *Proceedings of EUROSENSORS XI*, (Warsaw, Poland) P1-131

Chemical Sensors II
Paper presented at Eurosensors XII, 13–16 September 1998
© *1998 IOP Publishing Ltd*

Thickness effect on the gas sensitivity in multilayered sensors based on thin metal films

A Galdikas, A Mironas, D Senulienė, A Šetkus

Semiconductor Physics Institute, A.Goštauto 11, Vilnius LT-2600, Lithuania

S Kačiulis, G Mattogno, A Napoli

Institute of Materials Chemistry, CNR, P.O.Box 10, I-00016 Monterotondo Scalo, Italy

Abstract. The multilayered sensors based on the SnO_2 / [metal] / metal thin films are produced and investigated at temperatures from 20 °C to 300 °C in clean and contaminated air. Thickness of the constituent films is varied and the electrical resistance and the XPS depth profiles are measured. It is determined that the resistance response to gases is dependent on the variation of the thickness of the films. The effect is proved to be suitable for optimisation of the characteristics of the multilayered sensors. The metallic character of the conductance in the base film is associated with a unique response to gas in the sensors tested.

1. Introduction

A series of recent investigations has introduced a unique family of the multilayered sensors based on several thin films of different materials one of which is pure metal [1,2,refs. therein]. Being sensitive to gas like the sensors based on metal oxide, the multilayered sensors are more compatible with the silicon planar technology than the other ones. Therefore the multilayered sensors are likely to be easily incorporated into silicon micromachined gas sensing devices. Till now only few microelectronic gas sensors based on layered composition were developed for practical applications [1], however, much larger capabilities of gas detection were demonstrated for this type of the sensors [2]. On the other hand the experimental data are rather mosaic for the multilayered sensors at present. Therefore there is a lack of the knowledge about the gas sensitivity in such multilayered sensors and the mechanism originating the resistance response to gas is still unclear.

In present report we investigated the peculiarities of the sensitivity and the means for optimisation in the multilayered sensors based on the SnO_2/[metal]/metal composition. Thickness of the constituent films was varied for these purposes and the electrical properties were studied. The XPS depth profiles were also analysed. A role of the metallic character of the electrical conductance is discussed on the basis of the experimental results.

2. Experiments and sample preparation

Gas sensitive compositions were grown by a dc-magnetron sputtering from the pure metal targets. Thin films were sputtered one after another like a pile on the SiO_2/Si substrates. The temperature of the substrate was constant and equal to 20 °C during the sputtering. The basis of the layered composition was a thin metal film sputtered in an argon atmosphere. The metals Pt, Au, Mo and Ni were mainly used for the base film. The top film was always obtained by sputtering of pure tin in the mixture of O_2:Ar=6:4. The multilayered sensors produced in this work consisted of two or three thin films.

Thickness of the films was varied by the change of the sputtering time for only one film at a time. The relative thickness was evaluated from the XPS profiles, but the sputtering time and the sheet resistance are the main characteristics indicating the thickness in our present report because these were "easy to get" parameters and more simple for comparison of the multilayered sensors produced.

The dc-electrical resistance was measured for the multilayered sensors in clean and contaminated air. The sheet resistance was calculated from the results after dividing the resistance by the surface area of the sensor. Surface chemical composition and depth profiles were measured for the sensors by using a VG ESCALAB Mk II spectrometer. The details of the experiments were described elsewhere [2].

3. Results and discussions

Except for a Pt film, thin films based on pure metals were insensitive to gas-surface interaction at temperatures specific to each metal when the oxidation was negligible. The metallic character of the films was verified by all the experimental methods used in our work. If the films were covered with the top SnO_2, the resistance response to gas was detected in the layered composition of the SnO_2 / [metal] / metal thin films. Though the response to gas mostly was determined by the metal selected, the response signal was significantly dependent on the thickness of the films in the multilayered sensors.

Depending on the thickness of the bottom film, the electrical conductance in clean air varied from the metallic type to nearly isolating one. If the conductance of the bottom film corresponded with these extremes, the response to gas was small or even negligible. On the other hand a contaminating gas induced considerable change of the resistance if the thickness of the bottom film corresponded with some intermediate conductance which, however, was still nearly metallic. In addition, according to the XPS analysis, only some metals were partially oxidised in the bottom film in the layered composition. The results in Fig.1 illustrate the effect of the bottom-film thickness on the response to gas. In the figure the change of the relative resistance is plotted versus concentration of H_2 gas originating the decrease of the resistance in the two-layer sensors based on the Pt film. In this figure the thickness of the Pt films is defined by the corresponding sheet resistance measured before the SnO_2 coverage. Judging from the temperature dependence measured in clean air, the electrical resistance was similar to the metallic type in all the sensors in which the resistance of the Pt was below the maximum value indicated in Fig.1. It should be noted that the maximum value was individual for the metal selected. In all cases the maximum of the response signal was obtained by diminishing the thickness of the bottom films as it is illustrated in Fig.1.

If the thickness of the bottom film was within optimum interval, the top film of SnO_2 sensitised the layered composition based on any metal selected in the present work. The resistance response to gas increased drastically after the deposition of the SnO_2. An increase

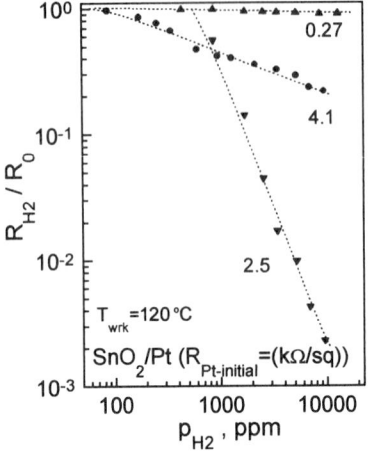

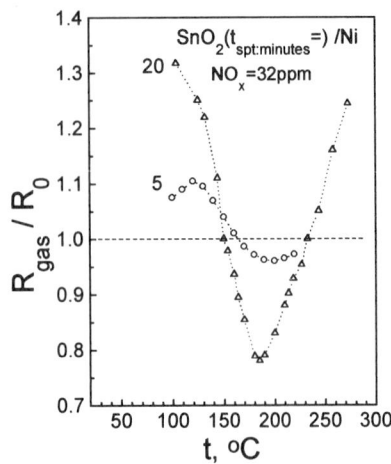

Fig.1. The relative resistance response to H_2-gas in the layered sensors based on the Pt film of different thickness.

Fig.2. The relative resistance response to NO_x in the layered sensor based on the Ni film with different thickness of top SnO_2.

of the response depended on a thickness of the top film. An influence of the top film thickness on the response is illustrated by the results in Fig.2. In the figure the relative change of the resistance is plotted versus operating temperature of the two-layer sensor based on the Ni bottom film. The results in Fig.2 were obtained for the sensor exposed to a fixed amount of NO_x gas in an air. In Fig.2 the thickness is indicated by the SnO_2 sputtering time. As it follows from this illustration, the response signal was increased if the top film was thicker. On the other hand, very thick top film suppressed the response completely.

Judging from the resistance, the intermediate metallic film was much thinner than the bottom film in all multilayered sensors tested. Generally, this fact was verified by the XPS profiles obtained in the three-layer sensors. On the other hand the shape of the inter-layer appeared to be rather spread and, therefore, the boundaries could be indicated only approximately in the XPS profiles. Typical XPS profiles are illustrated in Figs. 3a and 3b. The profiles were obtained in the tree-layer sensors based on the Pt film and with the Ni inter-film

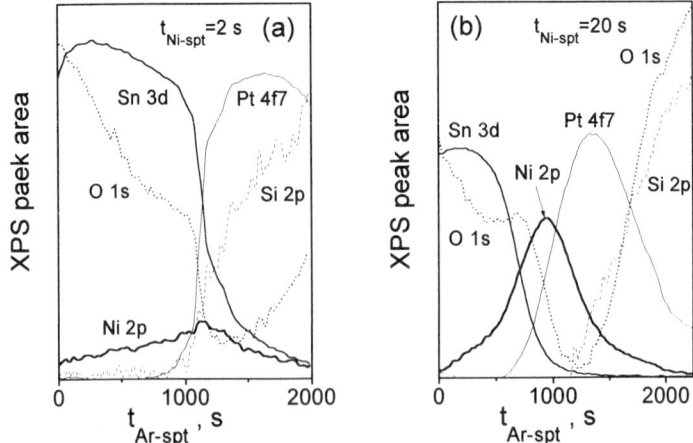

Fig.3. The XPS profiles obtained in the three-layer sensors based on the Pt film with the Ni inter-layer of different thickness.

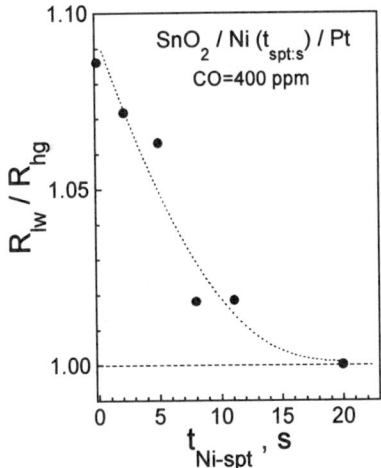

Fig.4. The relative resistance response to CO gas as a function of the thickness of the Ni inter-layer in the Pt-sensor.

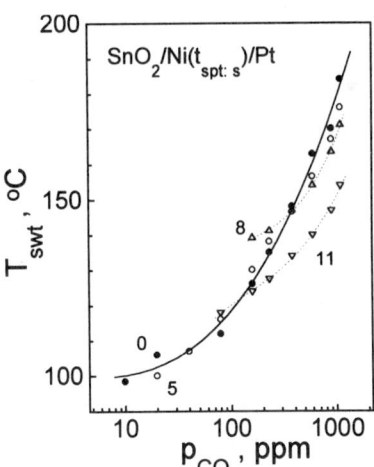

Fig.5. The switching temperature versus CO-amount for the Pt-sensors with the Ni inter-layer of different thickness.

of different thickness included. The thickness of the Ni is indicated by the sputtering time in Fig.3. Fluctuating thickness of the films seems to be the most possible origin of the overlap of the XPS peak areas in Fig.3 representing an amount of the components in the sensors.

The resistance response initially detected in the multilayered sensors was usually suppressed by increasing thickness of the metallic inter-layer. On the other hand the sensitivity to some other gas was revealed instead. Typical suppression of the initial response to CO gas is illustrated in Fig.4. The results were obtained for the three-layer sensor based on the Pt film. The thickness of the Ni inter-layer is defined by the sputtering time. The relative resistance in Fig.4 represents the response to CO gas that could be characterised as a switch in the resistance from the low response state to the high one [2]. As it follows from Fig.4 the switch in the resistance was gradually suppressed by increasing thickness of the Ni. The influence of the Ni could be related with a decrease of the density of the specific defects in the interface between the SnO_2 and the Pt when the Ni-atoms occupy them. The assumption was supported by the results in Fig.5 in which the temperature of the switch effect is plotted versus an amount of CO gas. The dependence in Fig.5 is determined by an interaction between CO and the specific defects in the interface without Ni. The origin of the switch is independent on gradual decrease of the density of the defects due to Ni-occupation, but absence of the defects terminates the switch effect. After the response to CO ceased, the sensitivity to NO_x gas was obtained for the SnO_2/Ni/Pt sensor.

4. Acknowledgement

This work was partially supported by the NATO Linkage Grant HGTECH.LG.960727.

References

[1]   Patel S V, Wise K D, Gland J L, Zanini-Fisher M and Schwank J W 1997 *Sensors and Actuators B* **42** 205-215

[2]   Galdikas A, Kačiulis S, Mironas A and Šetkus A 1997 *Sensors and Actuators B* **43** 186-192

Chemical Sensors II
Paper presented at Eurosensors XII, 13–16 September 1998
© 1998 IOP Publishing Ltd

Field-Effect Semiconductor Sensor for the Detection of Fluorocarbons

W. Moritz*, V. Fillipov**, A. Vasiliev**, A. Terentjev**

*Humboldt-University Berlin, Inst. of Phys. Chemistry, Bunsenstr.1, 10117 Berlin,Germany
**Russian Research Centre „Kurchatov-Institute", 123182, Moscow, Russia

Abstract. The large band gap material SiC was used to develop a high temperature field-effect structure $SiC/epi-SiC/SiO_2/LaF_3/Pt$. The sensitivity for different fluorocarbons as CF_3CH_2F, CF_3CCl_3, $CHClF_2$, CF_3CH_2Cl and CCl_3F was investigated. Beside a more complex behaviour for the temperature range 200-300°C a selective detection of fluorine containing molecules was shown for a temperature of 380°C.

1.Introduction

Fluorocarbon chemistry has importance in a variety of areas including atmospheric chemistry, combustion/flame suppression and plasma etching in the microelectronics industry. On the other hand, the environmental impact of some fluorocarbons is well known especially for the atmospheric ozone depletion. Large amounts of these gases have been used for example in refrigerators in the past. The process of destroying these gases needs sensors for process control. Beside this example there is a demand for sensors for different fluorocarbons.

Sensors for the detection of fluorocarbons have been developed using the metal-oxide-semiconductor type materials as $V-Mo-Al_2O_3/ZnO$ [1] or sulphur doped SnO_2 [2]. Several fluorocarbons as CF_3CH_2F or CCl_2FCClF_2 have been measured at a temperature of 400°C for concentrations down to 5 ppm. It is a disadvantage of these types of sensors that there is sensitivity to other hydrocarbons too and no selectivity to fluorine containing compounds .

Recently we developed a field effect semiconductor sensor for the detection of fluorine and hydrogen fluoride in atmospheric air as in inert gases [3,4]. The active components of this sensor structure $Si/SiO_2/Si_3N_4/LaF_3/Pt$ have been the LaF_3 layer and the Pt gate forming a three phase boundary with the gas under investigation. Already at an initial stage of sensor development it was successfully proved that such a sensor can be used for the detection of fluorocarbons. A signal was found for 1,1,1,2-tetrafluoroethane at a temperature of 180°C [4].

Subsequent investigations showed that this temperature is too small for a stable sensor response and for the detection of other fluorocarbons. Unfortunately, because of the small band gap the silicon based semiconductor field effect sensors can be used only up to a temperature of about 200°C. Therefore we developed a high temperature gas sensor based on the large band gap semiconductor silicon carbide SiC using the same sensitive layers (LaF_3

and Pt). It was demonstrated that for room temperature these sensors have properties comparable to the silicon based devices [5,6]. Using the SiC-based sensors it was shown that fluorine can be detected in the temperature range from room temperature up to 400 °C.

It is the aim of this paper to develop a gas sensor showing selective detection of organic molecules with one or more fluorine atoms and suppressed signal for fluorine free compounds. The sensitivity to CF_3CH_2F, CF_3CCl_3, $CHClF_2$, CF_3CH_2Cl and CCl_3F will be compared to the sensor response to CH_4, CCl_4, HF and F_2.

2. Experimental

The semiconductor substrates (Joffe-Institute, St.-Petersburg) used were 6H-SiC chips (3 x 3 mm^2) with an epitaxial layer of SiC (n-type, carrier concentration $(0.5.\div..8) \cdot 10^{16}$ cm^{-3}). A SiO_2 insulating layer (thickness 28 nm) was grown by thermal oxidation.

These samples were coated with a LaF_3 layer (240nm) using a high vacuum evaporation technique. Platinum gate contacts (thickness 30 nm) were produced by sputtering in argon atmosphere in a way leading to a three phase boundary LaF_3/Pt/gas formation.

The capacitance of the samples was measured using a Hewlett Packard 4284A type LCR meter. A computer controlled system for the determination of the bias voltage shifting at the constant capacitance was used.

The measuring cell, containing the investigated structure, was designed specially for high temperature measurements of the kinetics of gas sensing processes. The nickel cell had an aluminium oxide insulator and a free volume of less than 0.2 cm^3 and was heated by an external heater. The rear and top side contacts of the sample holder were made of platinum and graphite, respectively. This cell assured low noise measurements of the capacitance of the sensor up to a temperature of 400°C in fluorine containing atmosphere.

The gas sensitivity was measured under gas flow conditions. Mixtures of 10000 ppm of fluorocarbons (CF_3CH_2F, CF_3CCl_3, $CHClF_2$, CF_3CH_2Cl and CCl_3F) in synthetic air were diluted with synthetic air using computer operated mass flow controllers to obtain different concentrations of fluorocarbons in the range 200-10000 ppm. Maximum gas flow velocity used was 0.25 l/min.

3. Results and discussion

The electrical properties of the field-effect sensor structure SiC/epi-SiC/SiO_2/LaF_3/Pt were shown to be stable in a temperature range from room temperature up to 400°C. Details of characteristic changes of capacitance-voltage curves with temperature are discussed in ref. [5].

The sensitivity of the sensor to fluorine gas was proved for temperatures up to 400°C[5]. Nevertheless, for the detection of fluorocarbons free fluorine should not be expected as a decomposition product [7]. Therefore, a sensitivity to fluorocarbons can not be explained by detection of the decomposition product fluorine by the sensor.

A catalytic decomposition of hydrofluorocarbons leading to HF as a final or intermediate product is well known (e.g.[7]). Therefore, the sensitivity to hydrogen fluoride was investigated in more detail in the high temperature range. Sensitivity was found to be drastically decreased for temperatures close to 200°C. For the temperature range near to 400°C no sensitivity was found as long as concentrations have been smaller than 2-3 ppm. For more high concentrations a delay in response was observed followed by a potential shift of about 100 mV. This can be explained as an activation of the sensor. The activated sensor is sensitive even for concentrations smaller than 1 ppm of HF, but the potential is still shifted

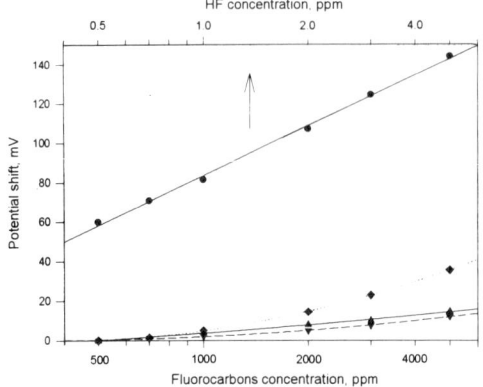

Fig.1 Fluorocarbon and HF sensitivities of the SiC based sensor at 374 ± 3^0C.

● - HF, ▲ - CHClF₂, ▼ - CCl₃F, ◆ - CCl₃CF₃

HF concentration. In contrast to HF the sensitivity was found to be only 14m V/Δlg p(CHClF₂). Similar results were obtained for CF₃CCl₃ and CCl₃F (Fig.1). A value of the shift in potential as observed for low concentrations of HF was not found for all the fluorocarbons investigated up to 5000 ppm.

At room temperature no sensitivity was found for all the gases under investigation. More pronounced differences were observed for the temperature range 200-330 °C. Here the signals were small or sometimes inverted in direction. In some experiments the sign of potential change did even change with time. For the investigation of this opposite in sign signal CH₄ and CCl₄ in air have been involved in the investigation.

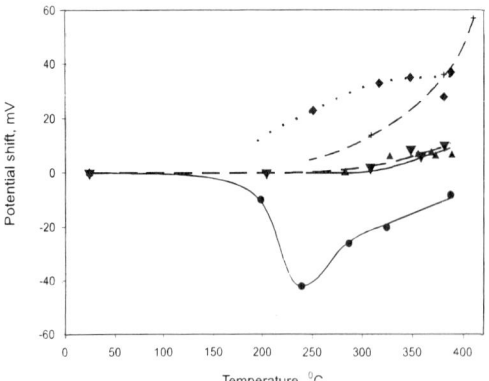

Fig. 2 Sensor response to CF₃CCl₃ (379°C)

compared to the measurements in synthetic air. This behaviour will be very important later on for discussion of sensitivity to fluorocarbons. The sensitivity to HF at 374°C was determined to be 85mV/Δ lg p(HF) (Fig.1).

The sensitivity of the semiconductor sensor to fluorocarbons was investigated in mixtures of the gases under investigation with synthetic air. Gas concentrations have been adjusted to be between 0-5000 ppm. A representative result is given in Fig.2 for the detection of CF₃CCl₃ at a temperature of 379°C. A stable signal was obtained in a time smaller than five minutes. The direction of potential change with increasing concentration corresponds to the direction for increasing

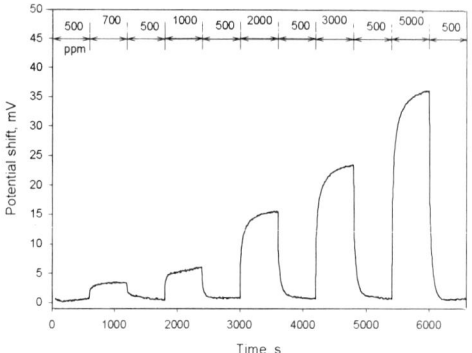

Fig. 3 Temperature dependence of sensitivity; Concentration jump 1000 - 4000 ppm, ● - CF₃CH₂Cl, ▼ - CCl₃F, ▲ - CHClF₂, ✦ - CF₃CH₂F, ◆ - CF₃CCl₃

For these gases we found a sensitivity opposite in sign in the range 300-330 °C. At a temperature of 380 °C this effect is reduced to be smaller than the experimental error. Hence, there is a sensitivity not selective for fluorine containing molecules in the temperature range 200-330°C. Therefore, for the fluorcarbons discussed until now selective detection is achieved for temperatures higher than 330 °C. The influence of temperature on the sensor signal (potential shift) is given in Fig.3. There is one gas (CF₃CH₂Cl) showing an opposite signal for the whole temperature range. Obviously higher temperature is necessary

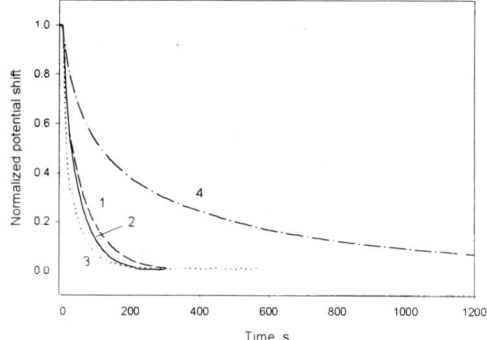

Fig.4 Dynamic response for decrease in
concentration from 5000 to 500 ppm, T=374 ± 3°C
1 - CCl$_3$F, 2 - CHClF$_2$, 3 - CCl$_3$CF$_3$, 4 - CF$_3$CH$_2$F

here to achieve the selective type of detection.

The kinetics of sensor response will be discussed in more detail. In Fig. 4 it is shown that response is slow for CF$_3$CH$_2$F but fast for all other gases. Comparing increasing or decreasing concentrations the sequence of the fluorocarbons is different. CCl$_3$F leads to the fastest response for increasing partial pressure while CF$_3$CCl$_3$ shows the fastest response for changes to smaller concentrations.

For the investigation of the selective behaviour of our sensor a structure SiC/epi-SiC/SiO$_2$/Pt (without LaF$_3$) was investigated. No sensitivity to fluorocarbons was found at 350°C. This is an argument for the selective mechanism of sensor response based on the properties of LaF$_3$.

For the mechanism of sensor response to fluorocarbons a reaction at the three phase boundary LaF$_3$/Pt/gas is to be discussed. A decomposition of the fluorocarbon followed by the detection of HF can be excluded because of the different sensitivity and the absence of the potential shift observed for HF. The high mobility of fluoride ions and the existence of vacancies in the LaF$_3$ is well known. This enables a direct insertion of a fluorine atom of the fluorocarbon into the LaF$_3$ (here fluoride ion) and electron transfer from the platinum for fluorocarbons adsorbed at the three phase boundary.

In conclusion a selective detection of fluorocarbons is achieved for temperatures near to 400°C. Measurements at elevated temperatures are in progress.

Acknowledgments. We acknowledge Volkswagen foundation, INTAS and DFG for the financial support of this work.

References

1   M Shiratori, M. Katsura and T.Tsuchiya, Proc. Int. Meet. Chemical Sensors, Fukuoka. Japan. Sept. 19-22,1983, pp. 119-124

2   T Nomura, T. Amamoto, Y. Matsuura and Y. Kajiyama, Sens. and Act. B, 13-14 (1993) 486-488

3   W. Moritz, S. Krause, A.A. Vasiliev , D.Yu. Godovski and V.V. Malyshev, The Fifth International Meeting on Chemical Sensors, Rome 11-14. July 1994 and Sens. and Act. B, 24-25 (1995) 194

4   W. Moritz, S. Krause, L. Bartholomäus, T. Gabusjan, A. A. Vasiliev, D. YU. Godowski and V.V. Malyshev. Orlando American Chemical Society Meeting. Orlando. USA. August 25-30. 1996, and ACS Series in press

5   A. Vasiliev, W. Moritz, V. Fillipov, L. Bartholomäus, A. Terentjev, T. Gabusjan. Sensors and Actuators, in press

6   Werner Moritz, Vladimir Fillipov, Lars Bartholomäus, Alexander Terentjev, Tigran Gabusjan and Alexei Vasiliev, Proceedings of the 11th European Conference on Solide-State Transducers, Eurosensors XI, Warsaw, Poland, September 21-24,1997 p. 111-114

7   E. Kemnitz, A. Kohne, E. Lieske, J. Fluorine Chem., 81 (1997) 197-204

Chemical Sensors II
Paper presented at Eurosensors XII, 13–16 September 1998
ⓒ *1998 IOP Publishing Ltd*

Chemically Grafted Field Effect Transistors for the detection of potassium ions

Z. Elbhiri[1], J.M. Chovelon[1] N. Jaffrezic-Renault[1], Y. Chevalier[2]

[1] IFoS-PCI, Ecole Centrale de Lyon, BP 163, 69131 ECULLY Cedex (France)

[2] LMOPS-CNRS, BP 24, 69390 VERNAISON (France)

Abstract. K$^+$ ISFETs are prepared by grafting silylated crown-ether molecules which are first synthetized. Three grafting reactions are used. The responses of the grafted ISFETs are sub-nernstian. They are analyzed through the modified site-binding model and the density of grafted sites was found to be equal to 6×10^{12} cm^{-2} with the dimethylamino-silylated molecule, the complexation constant of K$^+$ ions being found to be equal to $10^{2.5}$.

1. Introduction :

Although much research has been done on ISFETs during last decades, several problems still remain unsolved. Among these, the most important is the extension of ISFETs for measuring species other than hydrogen ions. A simple solution which has been previously proposed is to chemically modify the surface sites of the insulator surface by grafting [1]. By this way, the insulator/electrolyte interface remains simple and totally blocking, its potential is determined by surface complexation reaction which can be modelized by the site-binding model as pH sensitivity [2]. This principle insures a low response time and a long life-time for this simple ISFET structure, and moreover, there is no indefinite interfacial potential as it is the case for polymeric membrane/insulator interface. The drawback of the grafting process is that the chemical species are not commercially available and a heavy work of chemical synthesis has to be done. In this study we propose to graft a crown-ether molecule in order to detect potassium ions. The silylated molecule is first synthetized and then grafted on ISFET insulator surface. The responses of the grafted ISFETs are analyzed by the modified site-binding model which allows to determine the density of grafted sites and their complexation constant.

2. Experimental :

Synthesis of the silylated $B_{15}C_5$ molecules

Scheme of the $B_{15}C_5$ silylated molecule [X = Cl, $(CH_3)_2N$ or OH]

The different steps of the synthesis are the following ones :
- preparation of the allyl catechol

A   B

The isomer A was favored by using a copper-base catalysis in the Claisen reaction.
- formation of the allyl benzo-15-crown-5 by condensation of the di-tosylate of tetraethylene glycol on the allyl catechol
- hydrosilylation of the allyl crown-ether

Grafting of the silylated $B_{15}C_5$ molecules on the insulator surface of the ISFETs

The different grafting reactions

The three grafting reactions were performed according to a process already published [1]. For reactions 1 and 2, graftings were carried out directly on silica surface while for reaction 3 the grafting took place on the silica surface firstly aminated by a photochemical vapor deposition [3].

3. Results and discussion :

Response of the grafted ISFETs

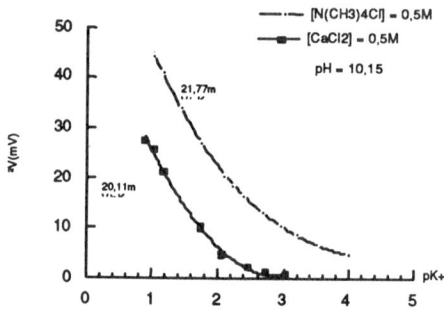

Figure 1 : Response to potassium of the ISFETs grafted with the reaction 1

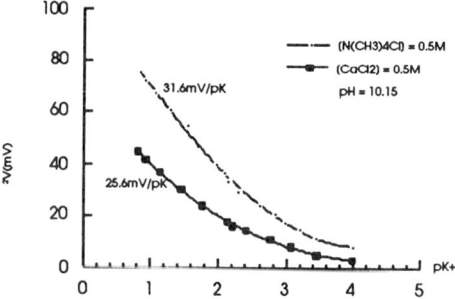

Figure 2 : Response to potassium of the ISFETs grafted with the reaction 2

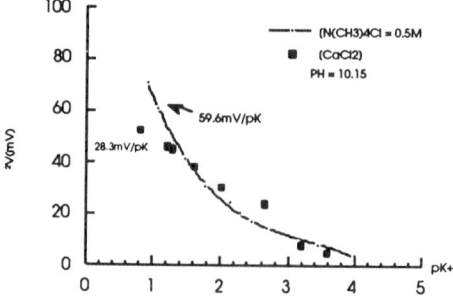

Figure 3 : Response to potassium of the ISFETs grafted with the reaction 3

These curves show that the best responses are obtained with tetramethylammonium chloride solutions, proving the influence of calcium-based electrolytes on the sensitivity. The sensitivities obtained with the three types of grafting reactions are according to the following sequence : reaction 3 > reaction 2 > reaction 1.

Modified site-binding model applied to the ISFET response to potassium [3]

The complexation of the K^+ ions with the grafted sites is schematized by the following reaction :

$$R\text{-}B_{15}C_5 + K_S^+ \rightleftharpoons R\text{-}B_{15}C_5...K_S^+ \tag{1}$$

where R corresponds to $\begin{matrix} & CH_3 \\ \diagdown & | \\ -Si-O-Si-(CH_2)_3 \\ \diagup & | \\ & CH_3 \end{matrix}$ and K_S^+ the potassium ions at the

insulator/electrolyte interface.
The complexation constant is :

$$K_K = \frac{[R\text{-}B_{15}C_5...K_S^+]}{[R\text{-}B_{15}C_5...K_S^+] \quad [K_S^+]} \tag{2}$$

The relation between the surface potential ψ and the concentration of K^+ ions is :

$$pK^+ = -\frac{q\Psi}{2,3kT} + \log\left(\frac{qN_S}{\Psi C_S} - 1\right) - pK_K \tag{3}$$

where q is the electron charge, k the Boltzmann constant, N_S the density of the grafted surface sites and C_S the Stern capacity ($\cong 20\ \mu F.cm^{-2}$)

By minimizing the function $\sum_{i=1}^{i=9}\left(pK_{expérimental} - pK_{calculé}\right)^2$ the ratio N_S/C_S and the

constant pK_K are calculated. The density of the grafted sites through reaction 1 is $N_S = 3\times10^{12}cm^{-2}$, the complexation constant of potassium ions being equal to $10^{+2,5}$ which is close to the value of 10^3 obtained in methanol medium.

For the reaction 2, the density of grafted sites is found to be twice of that obtained with the reaction 1 which explains the higher sensitivity of the ISFET grafted according to reaction 2.

References

[1] Bataillard P, Clechet P, Jaffrezic-Renault N, Kong X G, Martelet C 1987 *Sensors and Actuators* **12** 245-254
[2] Bousse L, De Rooij N F, Bergveld P 1983 *IEEE Trans. Electro Devices,* **ED-30** 1263-1270
[3] Baccar Z, Jaffrezic-Renault N, Lemiti M 1997 *J. Electrochem. Soc.***144(11)** 3989-3992
[4] Perrot H, Jaffrezic-Renault N, De Rooij N F, Van den Vlekkert H H 1989 *Sensors and Actuators* **20** 293-299

Chemical Sensors II
Paper presented at Eurosensors XII, 13–16 September 1998
© *1998 IOP Publishing Ltd*

The influence of ammonia adsorption on stationary photoluminescence of micro porous silicon

V.A. Smyntyna, Yu.A.Vashpanov

Mechnikov University Odessa 270026, Odessa, Dvorjanskja 2, Ukraine

Abstract. Effect of influence of ammonia adsorption on electronic properties and stationary photoluminescence of porous silicon (Por-Si) samples is investigated. The samples of Por-Si have characteristic heterogeneity of porosity and chemical structure along a surface of a material and on its depth. The possible mechanism of the observable phenomena is discussed.

Porous silicon is one of perspective materials of semiconductor electronics, research physical properties of which involves many researchers [1-4]. The interest to this material is supported its unique luminescence properties [5-6]. At the same time a porous silicon has a significant specific surface [7]. It should result in influence of surface processes on electronic and luminescence parameters of this material.

The majority of physical models of luminescence in the visible area connected with availability quantum wires in a structure of measures material and the form of which is supposed similar [8]. It is necessary to notice that in a real structure of a material the thickness filaments and them form essentially depends on a site of a surface of a material. To it given work [9] testify. It is of special interest the study of adsorption of a ammonia, as is established its significant influence to electronic parameters received by structures [10]. The finding out of a physical nature of a luminescence mechanism, and of adsorption sensitivity of Por-Si for sensor's microsystems is represented urgent.

The samples of porous silicon were received by a method anode electrochemical etching of crystalline silicon p-type as the marks < 111> in 48 % water solution of a hydrofluoric acid. Density of a current at etching made 10 mA/cm^2. The anodyne times are 10-30 min. The layers of Por-Si have a thickness 10-20 μm. Porosity of samples made about from 35 to 75 %.

Influence of a ammonia adsorption on electronic and luminescence properties of a material in specially chamber was investigated, in which it was possible to change a structure to gas atmosphere or vacuum processing in the field of temperatures 293- 573 K. We investigated of influence adsorption of a oxygen, oxides of a nitrogen and carbon, hydrogen, methane and ammonia. The concentration NH_3 in a chamber has been changed with the help gas generator GR-645 series. As of a gas of a carrier was chosen very clean nitrogen practically not influencing on electrical properties of material [10].

The received samples have a luminescence in the visible area of radiation at excitation by a laser with a length of a wave 441,2 nm, spectrum dependence of luminescence intensity (I_L) which in atmosphere of nitrogen is submitted on fig. 1, curve 1. The intensity I_L in a maximum is accepted for unit. Admittance in a measuring chamber of gases O_2, NO_2, CO_2, H_2, CH_4 at room temperature of measurements did not practically influence intensity and spectrum of luminescence. However at adsorption of the ammonia with the concentration 490 ppm observed reduction of intensity I_L about in three times in comparison with initial size (fig. 1, curve 2).

If to consider that luminescence recombination in porous silicon is connected to transitions between zones or between the D-A-centres, in this case it should expect wide a spectrum stationary of photoluminescence. However experimentally we observed one maximum in the field of 1,7 eV (fig. 1). Influence of adsorption results in reduction of intensity of this maximum without other sharp of luminescence transitions.

Change of concentration of ammonia **c** in a measuring chamber resulted in reduction of intensity in a maximum of luminescence, the concentration dependence of which is submitted on fig. 2. From this data follows that the threshold of sensitivity of adsorbate effect of NH_3 adsorption makes 7 ppm, and in the field of concentration more it 600 ppm is observed the termination of influence of adsorption on luminescence with growth of concentration of a gas (fig. 2, curve 1). The maximum of adsorption sensitivity ($\beta= I_L^{-1}dI_L/dc$, found on a technique [11]) is in the field of significance's 26 ppm (fig. 2, curve 2).

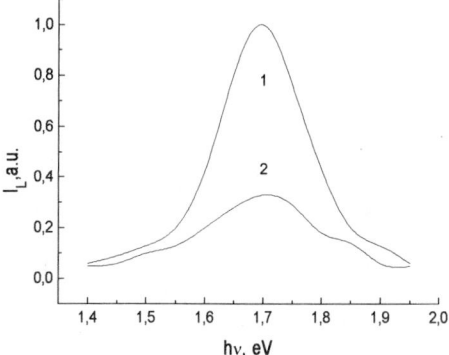

Fig. 1 Spectral dependence of photoluminescence intensity IL measured in very clean nitrogen (1) and in a mix of a ammonia with the concentration 490 ppm in a nitrogen (2).

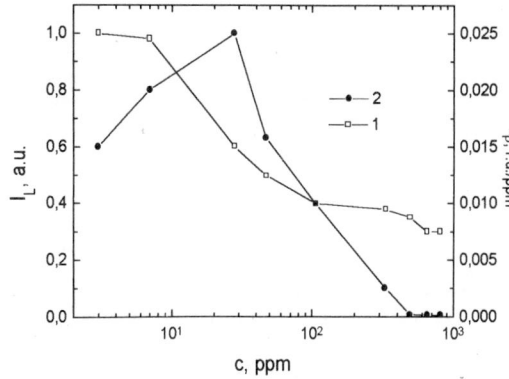

Fig. 2 The dependence of luminescence intensity (1) and of adsorption sensitivity (2) from concentration **c** of a ammonia.

It is necessary to notice that at adsorption of a ammonia the size I_L at room temperature measurements practically but decreased with growth of concentration of a ammonia in a measuring chamber. A time of shift of a gas mix in a measuring chamber with left the order of 1 second. It means that NH_3 adsorption influences intensity of luminescence for times smaller one second. This fact testifies that the ammonia influences a spectrum of fast condition of Por-Si, and the interaction mechanism of NH_3 with a surface Por-Si is connected with physical adsorption. It confirms restoration of intensity luminescence under blowing off of a very clean nitrogen.

Study of morphology of a surface with the help electronic microscopy has shown that the form and sizes of pores and their mutual orientation relative each other depend on a site of a surface of a structure. The research of a chemical structure of micro regions by a method of the energy- dispersion analysis has shown non-uniform distribution of a introduced hydrogen in a structure of a pores surface. The photo of butt-end of a structure of Por-Si also specify reduction of a diameter pores on thickness of samples. Thus in a structure Por-Si have thread-like crystals of silicon with variable thickness.

In considered high porosity non-uniform structures of porous silicon the sizes thread-like crystals of silicon change in a limit 3 up to 1,5 nm. The thread-like the crystals by the size less than 3 nm need to be named quantum wires [12]. According to [1] the effective significance of width of a band gap at reduction of thickness of wires from 3 up to 1,5 nm is increased from 1,2 up to 2,5 eV.

From conducted researches follows that the surface such of thread-like crystals is covered by atoms of a hydrogen in concentration on a surface 46 at%. Besides the surface contains atoms fluorine. A hydrogen and fluorine will be formed as a result of anode electrochemical processing of silicon [13]. Formation of a thin layer of hydrogenated silicon should result in growth of width gap this of very hear surface region. The width of a band gap in a-Si:H makes 2,0 eV at concentration 22 at% of hydrogen [15]. It is possible to consider that the electronic structure of the very hear surface region of wires presented by self graded band-gap semiconductor, width of a band gap of which is maximal on a surface and is minimum on a border of section wires- a hydrogenated layer. The availability of a graded band gap semiconductor is confirmed also by researches of photo-conductivity of porous silicon, observed wide spectral photosensitivity [15].

If to consider that the effective width of band gap of a hydrogenated layer and wires of silicon can change from 1,2 up to 2,5 eV in depended from a hydrogenation degree and diameter of wires, it is possible to speak about formation numerous heterotransitions in a structure of material.

In this case the mechanism of luminescence recombination can be connected to a border of section wires- a hydrogenated layer. For unideal heterotransittions the recombination current on a border of section is described by the formula [16]:

$$j_r = e U_r N_r(E) \sigma_r \int_0^{eU_r} R(E) W(E) N(E) dE$$

Here R- distance between recombination centres, W- probability of transition between them, N- distribution of condition on a border. Conducted size U_r is equal for size of electromotive force, arising on a barrier at the illumination U_o [17].

As well as in work [9] researched samples found out on contacts occurrence of the electromotive force U. The size U was increased at illumination and decreased under adsorption of a ammonia. If to consider a structure of porous silicon, consisting and of huge number heterotransitions, the changes of parameter U should be proportional to changes in each individual barrier U_o.

According to this model the intensity of luminescence through a condition of a barrier should be proportional recombination to a quantum output of luminescence η: $I_L=$ $j_r\eta$. At constant size η the intensity I_L according to the formula for j_r will drop at the expense of reduction of size U_o. The physical reason of change of EMF force on contacts of porous silicon according to [9] is interaction of pole molecules of a ammonia with the micro non-uniform areas possessing by internal micro fields. The size heterogeneity is close on order with dipole by influence of molecules. It is interesting to note that the mechanism of formation of porous silicon connected also with availability micro fields in a material crystal of silicon, which remain and after formation of a porous structure [14].

The adsorption sensitivity to a ammonia grows at additional of surface doping of fluorine [10]. Fluorine on surface research of a physical nature of which it is interest acts by a centre physical adsorption of pole molecules of a ammonia. The absence of influence of gases O_2, NO_2, CO_2 on electrical and luminescence parameters of a material can be explained by availability more active fluorine on a surface of a porous structure, and H_2, CH_4 because of the high contents of introduced atoms of a hydrogen. It is necessary note also that chemical sorption of these gases at room temperatures does not occur because of high heats adsorption. It is thus possible to consider that at the expense of physical adsorption of pole molecules of a ammonia change of size it micro fields in a barrier and according to the formula for j_r reduction of intensity luminescence is observed. Important moment of influence of a ammonia adsorption on intensity I_L is practical absence of shift maximum of intensity. The fact that luminescence recombination in our case occurs in the area of energy 1,7 eV, testifies about formation of local luminescence centres. In view of non-uniform character of a microstructure of a material it means that luminescence transitions take place only in determined points of the space of a porous structure. The nature of luminescence centres should be connected to electronic properties of quantum wires of silicon.

According to considered recombination model through a border of section intensity of luminescence and the form its spectral for depends on character of distribution of condition N on energy and distances centre to recombination centre R. In our case the nature of luminescence centres does not depend on change of sizes micro fields in a barrier. The influence of adsorption is displayed only in restriction of traffic of nonequilibrum carriers to centres of luminescence centres.

References

[1] Bresler M.S., Jassievich I.N.1993 *Physics and techniques of semiconductors* **27**, 871.

[2] Kompan M.E., Shabanov I.Yu. 1994 *Physics of solids* **36**, 2381.

[3] Cullisi A.G.,. Canham L.T. 1994 *J.Appl. Phys.*, **76**, 433.

[4] Ozari T., Araki M., Yuchimura S. 1994 *J.Appl. Phys.*, **76**, 1986.

[5] Astrova E.V., Lebedev A.A. 1995 *Physics and techniques of semiconductors* **29**, 1649.

[6] Chuang A.G., Canham L.T. 1991 *Nature*, **353**, 335.

[7] Labunov V.A., Bondarenko V.P., Borisenko V.G. 1978 *A foreign electronic engineering* (Moscow).

[8] Canham L.T. 1990 *Appl. Phys. Lett.*, **57**, 1046.

[9] Vashpanov Yu.A. 1997 *Letters in J. techniques physics*, **23**, 77.

[10] Vashpanov Yu.A. 1995 *Photoelectronics*, **6**,68.

[11]. Waschpanow Yu. A 1990. *Festkoerperchemie komplexer oxidischer Systeme*, (Greifswald Germany) p.170-180.

[12] Sanders G.D., Chang Y.C. 1992 *Phys. Rev. B*, **45**, 9202.

[13] Kompan M.E., Shabanov I.Yu. 1995 *Physics and techniques of semiconductors* **29** 1959.

[14] Zhu F., Singh J. 1993 *J.Appl. Phys.* **73**, 4709.

[15] Smyntyna V.A., Vashpanov Yu. A. 1997 *SPIE proceedings*, **3359**, 542.

[16] Vasilevski D.L. 1988 Photoelectronics **2**, 12.

Chemical Sensors II
Paper presented at Eurosensors XII, 13–16 September 1998
© *1998 IOP Publishing Ltd*

A new sensor for indoor air quality control

B Hök[1], M Tallfors[2], G Sandberg[1], A Blückert[1]

[1]Hök Instrument AB, Flottiljgatan 55, S-721 31 Västerås, Sweden

[2]Mälardalens Högskola, Box 883, S-721 23 Västerås, Sweden

Abstract. A new sensor for indoor air quality control is presented. It is based on measurements of sound velocity variations related to the molecular mass of the air, and corresponding to CO_2 concentration variations. Temperature and humidity compensation is provided by separate sensor elements. A prototype sensor has been designed, and the expected performance, adequate for the intended application, has been experimentally verified. It is expected to become a cost-effective solution to demand controlled ventilation in the near future.

1.   Introduction

Control of indoor air quality is mostly maintained by non-adaptive passive or active ventilation systems. Adaptation to actual demands would generally result in higher air quality, and improved power efficiency. Carbon dioxide concentration has been suggested as a suitable monitor quantity, being closely related to the number of individuals present in a certain locality, and their level of activity. An upper CO_2 concentration level of 1000 ppm in indoor environments has been recommended by the American Society of Heating, Refrigerating and Air Conditioning Engineers (ASHRAE Standard 62-1989) and other authorities. It should be noted, however, that this upper limit should be related to a varying 'fresh-air' background level. The present atmospheric CO_2 concentration of approximately 360 ppm increases by about 2 ppm per year, with seasonal variations as high as 10 ppm in the Arctic regions [1]. Much larger local background values, up to 600 ppm, can occur due to pollution from traffic, industries etc. Also noteworthy is that the upper concentration limit is not related to direct health effects, but is merely an indirect measure of odours and 'bioeffluents', which are by themselves more difficult to quantify.

Most sensors for measuring CO_2 concentration are based on infrared absorption tuned to the 4.3 μm absorption band of CO_2. Ventostat [2] is a commercially available, non-dispersive ir sensor designed for indoor air quality control. Photo-acoustic devices based on selective ir absorption are commercially available from MSA Instruments, Inc., USA. Although prices are decreasing, these devices are still considered to be too expensive for use in ventilation control. It is the objective of this paper to indicate a possible solution to these and related problems.

2. Theory

2.1. Estimation of possible power savings with demand controlled ventilation

In Fig 1 a simplistic model of a ventilated area is depicted. The volume air flow to and from the area, designated dV/dt undergoes a net change in CO_2 concentration amounting to ΔCO_2 due to respiratory activity, designated dCO_2/dt, of the occupants within the locality. From basic definitions, the following relation can be deduced:

$$dV/dt * \Delta CO_2 = dCO_2/dt = RR * V_T * CO_{2exp} * N \tag{1}$$

where RR is the (mean) respiratory rate, V_T the tidal volume (corrected by dead space), CO_{2exp} the concentration of CO_2 of expired air, N the number of occupants. For estimation purposes, the following values were used: RR=15/min, V_T=0.3 dm³, CO_{2exp}=4%. Then, $dCO_2/dt=3*10^{-6}$ m³ per person and seconds, and corresponds to a moderate level of physical activity [3]. From basic thermal relations, the power Q required to replenish (=heat or cool) fresh air to the locality is given by

$$Q = dV/dt * \rho * C_p * \Delta T = \rho * C_p * \Delta T * dCO_2/dt / \Delta CO2 \tag{2}$$

where ρ and C_p are the density and heat capacity (at constant pressure) of air, and ΔT is the temperature difference between input and output air. Assuming ΔCO_2 =700ppm, and ΔT=20 degC, the power consumption Q extends to more than 100W per person. This estimation clearly indicates that large energy savings are feasible with better indoor air quality control. Adapting the air flow to the actual demands could reduce the energy consumption for air replenishment to half or less than the present level if this is determined by the nominal capacity of localities which are, on average, used to less than half their nominal capacity. This has been experimentally verified [4].

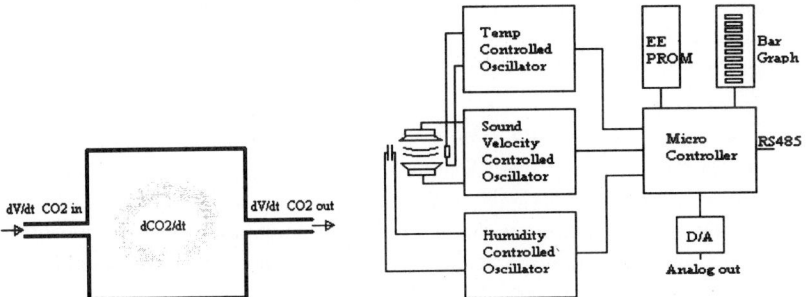

Fig 1 Simple model of ventilated area.   Fig. 2 Block diagram of the new sensor

2.2 Sensor principle

Our new sensor is based on the classical expression for the sound velocity c in a gas [5]:

$$c = (RT\gamma/M)^{1/2} \tag{3}$$

where R=8.314 J/mol K is the general gas constant, T the absolute temperature, γ the ratio between the heat capacities at constant pressure and volume, respectively, and M is the mean molecular mass [kg] of the gas. Eq. (3) has been exploited in a number of gas sensors, mostly in truly binary mixtures [6, 7, 8, 9]. Since the molecular mass of CO_2 differs significantly from that of air (=weighted sum of the main constituting gases, mostly nitrogen and oxygen), c will vary continuously with the CO_2 concentration (as with any gas with molecular mass differing from that of air). Compensation must therefore be provided for other undesired influences, such as that of temperature and humidity.

For small variations of temperature, concentration of CO_2 and H_2O, the accompanying change in sound velocity Δc will be

$$\Delta c/c_0 = \Delta T/2T_0 - (M_{CO2}-M_{air}) / 2M_{air} * \Delta CO_2 + (M_{air} -M_{H2O}) / 2M_{air} * \Delta H_2O \qquad (4)$$

where M_{CO2}, M_{H2O} and M_{air} are the molecular masses of CO_2, H_2O and air, respectively. The quantity $(M_{CO2}-M_{air}) / M_{air}$ has an approximate value of 0.52, and $(M_{air} -M_{H2O}) / M_{air} =0.38$. The undesired dependence on temperature and H_2O are compensated in the new sensor by separate sensor elements. From (4), the required degree of compensation can be calculated. To keep measurement errors below ±25 ppm CO_2, temperature must be measured with a repeatability of ±0.01degC, and relative humidity within ±0.3%.

3. Sensor design, experimental results

In Fig 2, the new sensor design is outlined. It consists of a measuring cell, in which variations of the sound velocity, temperature, and relative humidity, are measured separately. The sound velocity is measured with a transmitter/receiver pair, operating at 40 kHz, and controlling the output frequency of an oscillator circuit (Sound Velocity Controlled Oscillator, SVCO). Similarly, a capacitive humidity sensor and a resistive temperature sensor are controlling the frequencies of two other oscillators (Humidity Controlled Oscillator, HCO, and Temperature Controlled Oscillator, TCO). The instantaneous frequency values are continuously being sampled by the microcontroller, and the corresponding values of CO_2 concentration, temperature and relative humidity are computed according to eq. (4). These values are displayed on a bar-graph display, and can be accessed in analog form via a digital-analog converter, or digitally via a standard RS485 interface.

A prototype of the new sensor has been designed, according to the block diagram of Fig 2, and evaluated. The analog circuitry was designed from low-cost standard operational amplifiers, generating three frequency signals which were directly fed to a microcontroller (MicroChip PIC-17), thereby eliminating the requirement of a high precision A/D converter. To implement the measurement algorithm and support functions, including display and analog outputs, approximately 8k of program memory was occupied.

Fig 3 shows the output frequency of the SVCO as a function of CO_2 concentration, verifying a resolution of ±25 ppm or higher. The observed nonlinearity is probably due to small temperature variations during the measurement.

In Fig 4, the computed sensor outputs of CO_2 concentration, temperature and RH are plotted against time during an experiment, in which the sensor was temporarily exposed to CO_2-rich and humid air. The response time of approximately 120 seconds is determined by the humidity sensor element. While fully adequate for ventilation control, it would require further development in other applications.

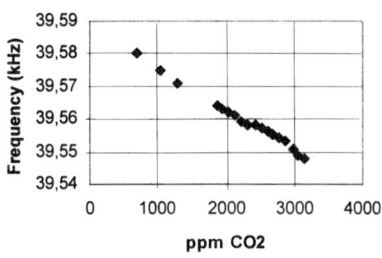

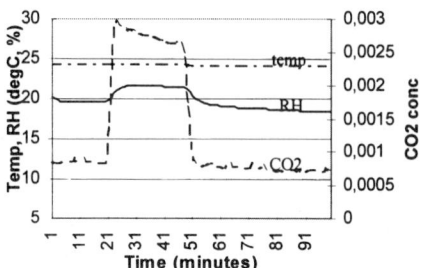

Fig. 3.Output frequency vs CO2 conc.   *Fig. 4. Output signals vs time.*

4. Discussion
The new sensor offers the following possibilities and advantages: The measuring principle is simple, robust and can be implemented with off-the-shelf components, already in high-volume production. Mechanical design and construction is relatively non-critical. The sensor can be readily and effectively screened from undesired environmental influences. This indicates that production cost, maintenance cost and consequently, the total life cycle cost, will be competitive.

Some questions should be the subject of further research: What are the requirements on long term stability, taking the background level variability into account, and the fact that most localities now and then will be 'relaxed' to that level? Are there situations where non-classical intermolecular interactions will make eq. (3) non-valid? Are there other potential applications for this type of sensor?

Acknowledgements. This research is financially supported by NUTEK, the National Swedish Board for Technical and Industrial Development, and Västmanlands FoU-Råd. The sensor will be marketed under the trade name of Q-AIR by Comfort Control AB, Box 15 101, S-750 15 Uppsala, Sweden.

References
[1]   Holmén K *Klimatforskare spår i kristallklar luft*, Forskning & Framsteg 96/1, 4-8.
[2]   Martin H Low cost NDIR gas sensors, Micro Structure Workshop 1996, Uppsala, Sweden, March 26-27, 1996, Abstract No P1.1
[3]   West JB *Respiratory physiology - the essentials*, 3rd Ed., Williams & Wilkins, Baltimore 1985.
[4]   Jansson I, Ahlbeck B, Andersson S, *Behovsstyrd ventilation* Rapport R17:1987, Statens råd för byggforskning.
[5]   Bergmann L, Der Ultraschall und seine Anwendung in Wissenschaft und Technik, S Hirzel Verlag, Stuttgart 1954, p. 501ff.
[6]   Zipser L, Labude J, *Akustische Gasanalyse. Teil 1: Grundlagen* Technisches Messen 58 (1991) 427-432.
[7]   Zipser, L, Labude J, *Akustische Gasanalyse Teil 2: Anwendungen* Technisches Messen 58 (1991) 463-470.
[8]   Zipser L *Fluidic-acoustic gas sensors*, Sensors and Actuators B7 (1992) 592-595.
[9]   Zipser L, Wächter F, *Acoustic sensor for ternary gas analysis*, Sensors and Actuators B26-27 (1995) 195-198.

Sensor Arrays and Multi-Sensor Systems

Sensor Arrays and Multi-Sensor Systems
Paper presented at Eurosensors XII, 13–16 September 1998
© *1998 IOP Publishing Ltd*

The Field-Effect-Addressable Potentiometric Sensor (FAPS) – A new concept for a surface potential detector with spatial resolution

W. J. Parak, M. George, H.E. Gaub, S. Böhm, A. Lorke[*]

Sektion Physik, LMU-München, 80539 München, Germany

Abstract . We propose a new concept for a surface potential detector with high spatial resolution and large number of addressable points. The so-called "Field-Effect-Addressable Potentiometric Sensor" (FAPS) uses the ability to control the resistance of electron channels in semiconductor films using electric potentials. Making use of a grid-structure of perpendicular electron channels and field-effect electrodes, located *below* the electron channels, it is possible to tune a single intersection into maximum sensitivity to electrical potentials acting from *above* the electron channel.

1. Introduction

Within the last decade, many types of chemical and biological sensors have been proposed and developed into applicable technology. At present, for example, a large variety of chemical sensors is available which can quantitatively detect specific ions or molecules in aqueous solution. A number of these devices is based on the measurement of the electric potential at the detector surface. If ions bind specifically to a modified surface, their number can be determined by measuring the change in surface potential induced by the ions' charges. Even though potential sensors without spatial resolution are most commonly used, some sensor also offer the possibility to address specific points on the surface.

As an example, ion selective field-effect transistors (ISFETs) detect the local surface potential at their gate electrode. Optimized setups for a large variety of ions and macromolecules have been reported [1-5] . When ISFET arrays are used, the two-dimensional potential distribution across the sensor surface can be imaged and different ionic species can be detected simultaneously [6, 7]. Then the spatial resolution is determined by the size of the individual electrodes, typically a few micrometer. However, the total sensitive region is restricted by the limited number of fixed gate electrodes. Array sizes containing up to 100 electrodes with a typical spacing of a few ten micrometers are common. Apart from their use as chemical sensors, ISFETs are employed to measure extracellular potentials of electrically excitable cells [8-10]. Changes in membrane potential can be recorded extracellularly by measuring the local surface potential. For such measurements it is important that the spatial resolution is better than the size of a single cell. Here, a conceptual problem is obvious, because for a limited number of addressable points, high spatial resolution and large-area arrays are mutually exclusive. Thus, cells cannot grow freely, but have to undergo a guided growth towards the active regions of the device, which results in further experimental difficulties.

A different kind of surface potential sensor with spatial resolution is the Light-Addressable Potentiometric Sensor (LAPS) [11]. Here, the point of measurement can be selected using a laser pointer. By scanning the laser beam across the detector while recording the local photocurrent, two-dimensional images of the surface potential distribution are obtained. For example, this technique has been employed in spatially resolved measurements of

---

[*] To whom correspondency should be addressed

proton [12] and cyanide [13] concentrations, as well as Fe(II)/Fe(III) redox potentials [14]. Via pH measurements, locally excreted acidic metabolic products from cell colonies could be imaged [15]. Clearly, the conceptional advantage of the LAPS principle is the free selection of the point of measurement. Unfortunately, the spatial resolution is restricted to the diffusion length of electrons and holes in the semiconductor layer of the detector [16]. To our knowledge, no such potentiometric detection schemes are available with a spatial resolution clearly below 100 μm.

As an alternative, purely electrical measurement, action potentials of cells can be recorded using extracellular microelectrodes [17-20]. Beyond the principle of measurement, they have similar geometric properties as ISFETs. The size of their fixed sensitive regions delivers spatial resolution of a few micrometers and there are arrays available consisting of up to 100 single electrodes with a few ten micrometers spacing [21].

Comparing the above described sensors shows that devices with high spatial resolution are restricted with respect to the total addressable area, whereas sensors which offer a large area of freely selectable points of measurement offer only poor resolution. Clearly, a device combining dense packing of active regions and spatial resolution in the micrometer range would be useful for a variety of applications.

2. Principle of the Field-Effect-Addressable Potentiometric Sensor

A general concern regarding sensors with high spatial resolution and large addressable fields is the sheer number of addressable points, which grows quadratically in both size (length × width) and resolution ([points per mm]^2) of the sensor. Electrical measurements with more than a few hundred addressable contacts, however, become cumbersome and prone to faults. Even though schemes like CCDs (charge-coupled devices) can in principle overcome this problem, their fabrication requires sophisticated semiconductor technology and they are difficult to make bio-compatible.

Here, we introduce an electrical potentiometric addressing scheme which is comparably easy to fabricate, offers spatial resolution down to below 1 μm, and requires only a drastically reduced number of electrical contacts. The geometry of the Field-Effect-Addressable Potentiometric Sensor is shown in Fig. 1. It consists of an array of parallel metallic stripes (vertical, dark grey stripes in Fig. 1) which serve as gates for perpendicularly oriented stripes of a field-effect transistor (FET) structure (light grey, comb-like structure in Fig. 1). The FET-structure is designed to be normally "on", i.e. conducting for zero gate bias.

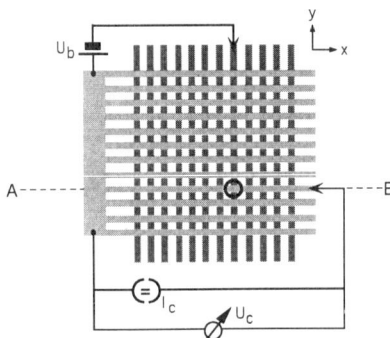

Figure 1: Geometrical setup of the FAPS device. Stripes of field-effect electrodes (drawn in dark grey) are arranged perpendicular to stripes of FET channels (plotted in light grey). A bias voltage U_b is applied to one of the field-effect electrodes and (at constant current I_C) the voltage U_C along one electron channel is recorded. This way, the sensitive region (indicated by a circle) is selected at the intersection of the measured channel and the biased gate.

Let us consider the case shown schematically in Fig. 1, where only one of the gates is biased (U_b) and the resistance of one of the perpendicular FET channels is measured. In the following, we will assume the FET to be n-type. As schematically shown in Fig. 2 and 3, with decreasing gate voltage the channel will become more and more depleted at the point of intersection (circle in Fig. 1) and the resistance will further and further increase. Figure 3 illustrates that near the threshold voltage, U_{th} (where all carriers are depleted and the channel becomes insulating) the

FET is most sensitive to additional potential fluctuations ΔU, which will be provided by the test charge on top of the FET channel. Thus, a change of the test charge located above the intersection between the gate electrode and the measured FET channel will give rise to the output signal measured in the FET circuit.

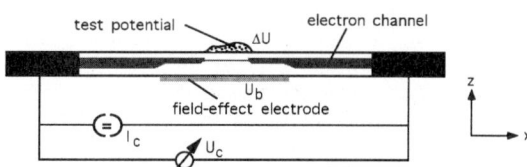

Figure 2: Working principle of the FAPS. The picture shows a section through the device along line A--B in Fig. 1. A suitable gate bias U_b applied to the FET channel from below will locally deplete and tune it to a state of high sensitivity (cf. also Fig. 3). Then a small additional potential ΔU, induced by the test object and acting from above the channel, will lead to a discernible change of the channel's resistance. This change can be detected by driving a constant current I_c through the channel and measuring the corresponding voltage U_c across the channel. Test potentials located away from the point of intersection will only marginally influence the channel's resistance.

A necessary requirement for this addressing scheme is that all surface potentials away from the intersection of interest will not influence the measured signal. This is clearly the case for all points located on unconnected FET channels (above and below line A--B in Fig. 1). For points which lie on the measured channel above unconnected gate electrodes, the sensitivity to test potentials ΔU is negligible, as illustrated in Fig. 3. Thus, indeed, by measuring the resistance of one channel Y and applying a suitable bias to one gate electrode X, a point XY on the surface is uniquely addressed.

    Apart from the simple layout, compared to, e.g., arrays of standard transistors, the main advantage in the FAPS geometry is the great reduction in required electrical leads. Here, the number of contacts only scales linearly with size and resolution or, equivalently, as the square root of the number of addressable points.

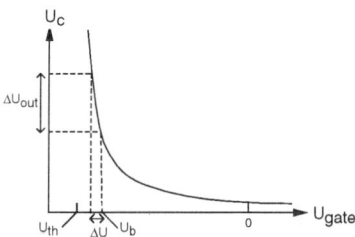

Figure 3: Resistance of a typical n-type FET channel versus gate bias. Here, the resistance is given in terms of the voltage across the channel U_c when a constant current I_c is applied. For a gate bias U_b close to the threshold voltage U_{th}, a small additional potential variation ΔU and thus an effective gate voltage $U_{gate} = U_b + \Delta U$ will lead to a large change in output voltage, ΔU_{out}. Note that at zero gate voltage (unbiased gates in Fig. 1) the same ΔU will lead to a negligible output signal.

3. Concepts for a practical realization

The most challenging problem to overcome in the realization of the FAPS is the fabrication of a thin FET structure which is electrically accessible from both above and below. At present, the most promising and straight forward approach is the so-called "epitaxial lift-off" technique, described by Yablonovitch et al. [22]. Using this technique, it is possible to peel off an entire GaAs/AlGaAs-field-effect transistor structure from its crystal substrate and bond it to almost any smooth surface [23]. Furthermore, it is possible to bond such thin-film transistors onto prepatterned surfaces with structured gates and deplete the FET channel from below [24], as schematically shown in Figs. 1 and 2. For typical semiconductor film thicknesses of ≤ 0.5 µm, resolutions of ≈ 5 µm have been demonstrated and resolutions below 1 µm appear feasible.

    For clarity, an additional insulating protection layer on top of the FET structures has been omitted in the discussion above and in Fig. 1 and 2. This layer has a dual purpose. Firstly, it must protect the tested objects, living cells, e.g., from the volatile As-ions. Secondly, it should prevent ions in the biological or chemical environment from contaminating the FET [25] and from creating electrical contacts between the biased and the unbiased gates. An appropriate layer structure, based on SiO_2 and Si_3N_4, is presently being developed.

It should be pointed out that even though presently an inorganic semiconductor layer seems most promising for the FAPS, the working principle might as well be realized using organic semiconductors, which could even further simplify the fabrication.

For a full working device, an external switching circuit becomes necessary, which will scan the FET channels and gates in a controlled fashion. Furthermore, the discussed detection scheme, namely the measurement of the voltage when a constant current through the FET channel is applied, should be replaced by one which is better suited for electronic signal processing. For example, the FET might serve as the front end of an amplification circuit.

One conceivable application is the detection of extracellular potentials of electrically excitable cells, as already demonstrated using existing sensors. Basically three points have to be considered in the discussion whether the FAPS is an appropriate device for such experiments: electrical sensitivity, time resolution and spatial resolution. Typically, extracellular potentials are three orders of magnitude smaller than intracellular action potentials. Thus, noise in the proposed setup has to be sufficiently suppressed in order to achieve a surface potential sensitivity of better than 100 µV. Since action potentials of neurons only last for a few milliseconds, a time resolution of a few kilohertz would be necessary. The required spatial resolution of better than 10 µm, the size of a single neuron, should not be a limiting factor, since, as described above, electron channels with few microns width and spacing can easily be fabricated.

In summary, the proposed planar multi-site structure promises to combine the advantages of existing surface potential sensitive devices, with improved spatial resolution, a large addressable field and relative ease of fabrication.

References

[1]   Schoot, B.H.v.d. and P. Bergveld, Biosensors, **3**: p. 161 - 186. 1987.
[2]   Haak, J.R., P.D.v.d. Wal, and D.N. Reinhoudt, Sensors and Actuators B, **8**: p. 211-219. 1992.
[3]   Cobben, P.L.H.M., et al., Sensors and Actuators B, **6**: p. 304-307. 1992.
[4]   Reinhoudt, D.N., Sensors and Actuators B, **6**: p. 179-185. 1992.
[5]   Woias, P., et al., Sensors and Actuators B, **15-16**: p. 68-74. 1993.
[6]   Igarashi, I., et al., Sensors and Actuators B Chemical, **1**: p. 8-11. 1990.
[7]   Hanazato, Y., et al., IEEE Transact. ED, **36**(7): p. 1303-1310. 1989.
[8]   Bergveld, P., J. Wiersma, and H. Meertens, IEEE Transact. BME, **23**: p. 136 - 144. 1976.
[9]   Fromherz, P., et al., Science, **252**: p. 1290 - 1293. 1991.
[10]   Offenhäusser, A., et al., Biosensors & Bioelectronics, **12**(8): p. 819-826. 1997.
[11]   Hafeman, D.G., J.W. Parce, and H.M. McConnell, Science, **240**: p. 1182 - 1185. 1988.
[12]   Nakao, M., T. Yoshinobu, and H. Iwasaki, Japanese Journal of Applied Physics, **33**: p. L 394 - L 397. 1994.
[13]   Licht, S., N. Myung, and Y. Sun, Analytical Chemistry, **68**(6): p. 954 - 959. 1996.
[14]   Oba, N., T. Yoshinobu, and H. Iwasaki, Jpn. J. Appl. Phys., Part 2, **4A**: p. L460-L463. 1996.
[15]   Nakao, M., T. Yoshinobu, and H. Iwasaki, Sensors and Actuators B, **20**: p. 119 - 123. 1994.
[16]   Parak, W.J., et al., Sensors and Actuators A, **63**: p. 47-57. 1997.
[17]   Gross, G.W., et al., Neuroscience Letters, **6**: p. 101-105. 1977.
[18]   Regehr, W.G., et al., Journal of Neuroscience Methods, **30**: p. 91 - 106. 1989.
[19]   Jimbo, Y., H.P.C. Robinson, and A. Kawana, IEEE Transact. BME, **40**(8): p. 804 - 810. 1993.
[20]   Gross, G.W., et al., Biosensors & Bioelectronics, **12**(5): p. 373-393. 1997.
[21]   Connolly, P., Trends in Biotechnology, **12**(4): p. 123-127. 1994.
[22]   Yablonovitch, E., et al., Applied Physics Letters, **56**(24): p. 2419-2421. 1990.
[23]   Rotter, M., et al., Applied Physics Letters, **70**(16): p. 2097-2099. 1997.
[24]   Böhm, S., Diplom thesis, 1997, LMU München, unpublished.
[25]   Topkar, A. and R. Lal, Thin Solid Films, **232**: p. 265-270. 1993.

Sensor Arrays and Multi-Sensor Systems
Paper presented at Eurosensors XII, 13–16 September 1998
© 1998 IOP Publishing Ltd

Detection of volatile organic compounds (VOCs) with polymer coated cantilevers

M. Maute†, S. Raible†, H. Ulmer‡, F.E. Prins†, U. Weimar‡, D.P. Kern†, W. Göpel‡

†Institute of Applied Physics, University of Tübingen, Morgenstelle 10, 72076 Tübingen, Germany

‡Institute for Physical and Theoretical Chemistry, University of Tübingen, Morgenstelle 8, 72076 Tübingen, Germany

Abstract. Changes in the resonance frequency of polymer coated cantilevers due to gas absorption can be used as a detection mechanism for gases. An additional mass loading of the cantilever results in a decrease of the resonance frequency. In this approach we prepared cantilevers based on micromachined Si wafers and used polydimethylsiloxane (PDMS) as polymeric prototype coating. PDMS is known to exhibit high selectivity to volatile organic compounds (VOCs). By using the first higher resonance mode we found with our configuration a sensitivity of -0.049Hz/ppm for n-octane.

1. Introduction

Miniaturized sensor devices for the detection of gases and vapors via specific coatings are gaining importance in many fields, especially in the design of so called "electronic noses" [1]. A very promising approach towards this goal is the use of coated single cantilevers [2][3] and cantilever arrays [4]. The resonance frequency changes due to the absorption of analyte vapor in the cantilever coating. This effect can then be exploited to gain a typical response pattern from an array with different coatings which can be used to identify the vapor. This paper deals with our first effort to design and optimize these transducers for VOC detection in the gas phase by using a prototype coating (PDMS) and a typical VOC (n-octane).

2. Theory

The change of the resonance frequency Δf of a cantilever due to additional mass loading Δm is to a first approximation given by

$$\Delta f = -\frac{1}{2}\frac{f}{m}\Delta m$$

where f is the resonance frequency and m the mass of the cantilever. The additional mass loading Δm due to absorption of gas molecules in a polymer layer deposited on the cantilever can be expressed by the change in the gas concentration Δc, yielding

$$\Delta f = -\frac{1}{2}\frac{f}{m}kV_pM\Delta c$$

The additional mass loading Δm is proportional to the partition coefficient k, the volume of the polymer layer V_p, the molecular weight M of the analyte gas and the concentration of the gas molecules. The partition coefficient depends both on the polymer and on the analyte molecules. For PDMS as sensitive layer and n-octane as analyte gas a partition coefficient of 2250 has been reported [5]. It can be seen that large frequency shifts can be achieved by using cantilevers with high resonance frequencies or alternatively by using higher frequency modes. In addition, it is desirable to use a polymer which exhibits high values for the partition coefficient for a specific analyte gas and low partition coefficients for other molecules to ensure high selectivity. A large polymer volume increases the resulting frequency shift since absorption is a volume effect. However, a large polymer volume also decreases the frequency of the cantilever.

3. Cantilever Fabrication

A plasma enhanced chemical vapor deposition (PECVD) process (20sccm SiH_3, 20sccm NH_3, 20Watt, 1Torr, 400°C, 40min) was used to coat a (100) silicon wafer with a 760nm thick silicon nitride layer. After defining the cantilever structures (see figure 1) by means of optical lithography the silicon nitride was dry etched using reactive ion etching (RIE) (27.5sccm CHF_3, 3.3sccm O_2, 100Watt, 100mTorr, 10min). Further anisotropic wet etching in KOH (20%, 80°C, 4h) leads to free standing cantilevers. The clamped ends of the cantilevers are defined by a (111) plane of the underlying silicon carrier. The cantilevers were then coated with a 5nm layer of chrome as an interlayer and 50nm of gold. This metal coating enhances the optical reflectivity needed for the detection mechanism described below (see figure 1). The PDMS was deposited at the square shaped end of the beam to avoid additional frequency shifts due to changes in the spring constant which could arise from swelling stress of the polymer layer during gas absorption.

4. Experimental Setup

The laser beam deflection setup of an atomic force microscope (AFM) was used to measure the resonance frequency of the cantilevers due to thermal excitation. The laser beam (670nm, 3mW) was focused to the clamped end of the cantilever to minimize effects on the absorption mechanism due to laser heating of the cantilever (figure 1). The cantilevers were mounted in a chamber (74ml volume) with a thin glass slide as top window. A pipe connected the chamber with a gas mixing system to expose the coated cantilever to defined gas concentrations. A flow of 100ml/min was chosen to perform the measurements. A PDMS coated quartz crystal microbalance (QCM) was positioned next to the cantilever as a reference [6].

The laser beam deflection signal was investigated with an HP 3561a Signal Analyser and Fourier Transformed (FFT). The FFT's were averaged and the root-mean-square amplitude as a function of frequency was computed.

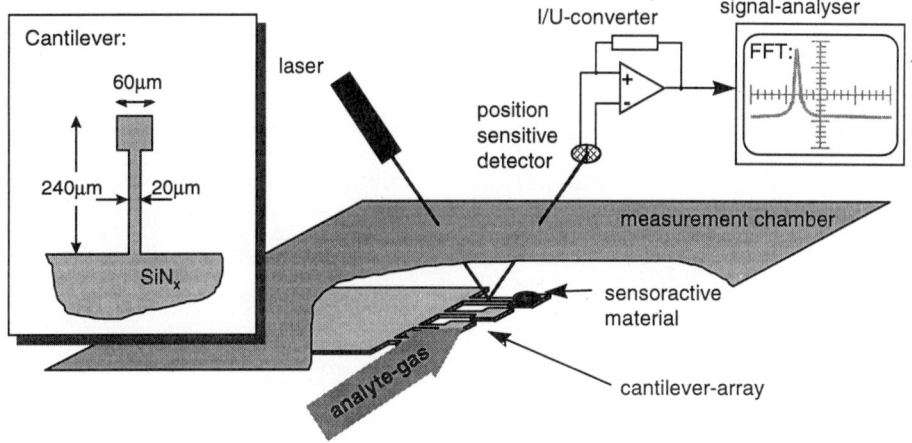

Figure 1. Schematic drawing of the experimental setup.

5. Results and Discussion

The uncoated cantilever had a fundamental resonance frequency of 7093Hz and a first higher resonance mode at 62.83kHz. Coating the end of the cantilever with PDMS resulted in a decrease of the fundamental resonance frequency to 980Hz. The first higher mode shifted to a value of 10.71kHz. Figure 2 shows the frequency response of the coated cantilever to different concentrations of n-octane vapor. It can be seen that the detection mechanism is completely reversible. We found a typical change of the fundamental resonance frequency of 28Hz for a concentration of 12000ppm n-octane. Figure 3 (left) shows the dependence of the fundamental resonance frequency on vapor concentration. The sensitivity of the cantilever response can be calculated as slope of a linear fit resulting in a value of -0.002Hz/ppm for the fundamental resonance frequency. The QCM also showed linear behaviour of the frequency. Hence, the concentration of n-octane vapor in the chamber corresponds to the demanded concentration from the gas flow system. The frequency shift of the first higher resonance mode frequency is much larger. Figure 2 (right) shows the first higher resonance mode for each concentration. A concentration of 12000ppm n-octane vapor results in a frequency shift of 609Hz. The sensitivity for n-octane calculated from the linear fit of figure 3 (right) is -0.049Hz/ppm.

6. Conclusion

We have shown that PDMS can be used as a sensitive layer to transform a gas concentration into an additional mass loading which results in a frequency shift. The shift due to gas absorption shows a linear dependence on the gas concentration. Larger frequency shifts can be achieved using cantilevers which exhibit higher resonance frequencies or alternatively by using higher frequency modes. It is desirable to investigate cantilevers with high Q-factors to increase the accuracy of the frequency measurement.

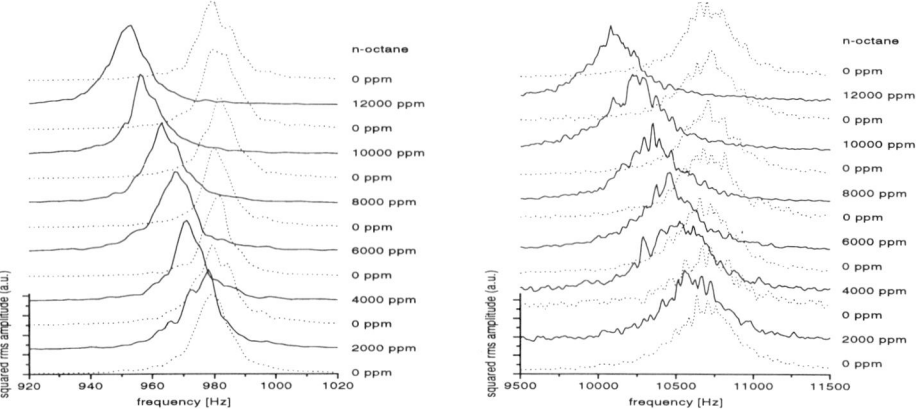

Figure 2. Frequency response of a PDMS coated cantilever for different n-octane concentrations. Left: Fundamental resonance frequency response. Right: First higher resonance mode frequency response.

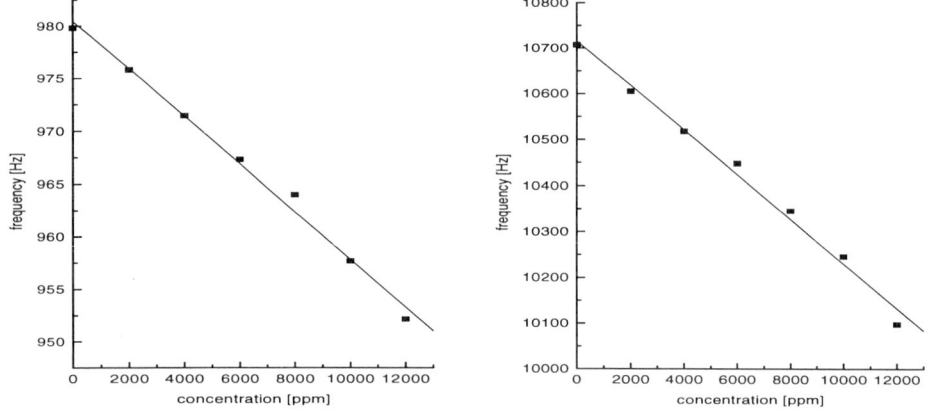

Figure 3. Center frequency dependence on n-octane vapor concentration. Left: Fundamental resonance. Right: First higher resonance mode.

References

[1] H. Baltes, A. Koll, D. Lange, *Proc. IEEE ISIE* (1997) 152-157

[2] H. Baltes, D. Lange, A. Koll, *Proc. SPIE 3224* (1997) 2-13

[3] T. Thundat, G.Y. Chen, R.J. Warmack, D.P. Allison, E.A. Wachter, *Anal. Chem.* **67** (1995) 519-521

[4] H.P. Lang, R. Berger, C. Andreoli, J. Brugger, M. Despont, P. Vettinger, Ch. Gerber, J.K. Gimzewski, J.P. Ramseyer, E. Meyer, H.-J. Güntherodt, *Appl. Phys. Lett.* **72 (3)** (1998) 383-385

[5] K.D. Schierbaum, M. Haug, W. Nahm, G. Gauglitz and W. Göpel, *Conf. Proc. 1st Europ. Conf. Opt. Chem. Sensors and Biosensors, Graz (A)* **(4/1992)**; *Sensors and Actuators B* **11** (1993) 383-391

[6] K. Bodenhöfer, A. Hierlemann, J. Seemann, G. Gauglitz, B. Koppenhöfer and W. Göpel, *Nature* **387** (1997) 577-580

Sensor Arrays and Multi-Sensor Systems
Paper presented at Eurosensors XII, 13–16 September 1998
© *1998 IOP Publishing Ltd*

The combination of an electronic tongue and an electronic nose for improved classification of fruit juices

F Winquist[1], P Wide[2] and I Lundström[1]

[1] The Swedish Sensor Center, Linköping University, Linköping, Sweden
[2] The Department of Physics and Measurement Technology, Linköping University, Linköping, Sweden and
The Department of Technology and Science, University of Örebro, Sweden

Abstract. The combination of an electronic tongue and an electronic nose for classification of fruit juices is described. Using principal component analysis, it is shown, that both the electronic nose and the electronic tongue alone are able to discriminate fairly between the samples. When combining information from both the electronic nose and the electronic tongue, however, the classification properties are clearly improved.

1. Introduction

There is a change in the attitude within measurement technology in how to collect and process information. Instead of measuring single parameters, it has become, in many cases, more desired to get information of *attributes* such as quality, condition, or state of a process. Due to this there is a growing interest for the concept of electronic noses [1]. An electronic nose consists of an array of gas sensors with different selectivity patterns, a signal collecting unit and pattern recognition software applied to a computer. The principle is based on that a large number of different compounds contribute to define a measured smell, the chemical sensor array of the electronic nose then provides a pattern output that represents a combination of all the components. The pattern output is given by the selectivities of the various sensors. The reason for the very large number of odours that can be detected, is that although the specificity of each sensor may be low, the combination of several specificity classes provides a very large information content.

Electronic noses have already proven useful in many applications [2,3]. Recently, similar concepts, but for analysis in liquid, have been described. These systems are in similar ways as for the electronic nose related to the tasting sense, thus, for these concepts the terms "electronic tongue" or "taste sensor" have been coined [4,5]. Recently, an electronic tongue based on pulsed voltammetry was described [6]. It consisted of a number of different types of working electrodes in a standard three electrode configuration. This electronic tongue was able to classify various samples, such as fruit juices, still drinks and milk. It was also used to follow the quality detoriation of milk due to microbial growth when stored at room temperature [7].

By combining information of different origins, an additional dimension of information can be added, since the measurement situation will be enlightened from different positions. This has e.g. been shown in a model experiment for quality studies of crisp bread of different types [8]. It can thus also be expected that the combination of an electronic tongue and an electronic nose is advantageous, especially for measurement situations involving changes in both the aqueous and gas phase.

In the following, it is described how the combination of an electronic tongue and an electronic nose was used for classification of different types of fruit juices.

2. The electronic tongue

The measurement principle of the electronic tongue is based on pulse voltammetry, thus current transients due to onset of a voltage pulse is measured, giving information concerning both amount and type of charged molecules and of redox active species.

The electronic tongue consisted of a six working electrodes of different metals, an auxiliary electrode, and a reference electrode. The electrode configuration was placed in a 150-ml measurement cell, also containing a magnetic stirrer. Current and current transient responses were measured by a potentiostat connected to PC via an A/D converter. A PC was used for onset of pulses and measurement of current transient responses and to store data. To the potentiostat, an electronic filter with a time constant of 0.1 seconds was applied. This will smooth the transient responses.

The voltammograms recorded were based on large amplitude pulse voltammetry (LAPV). A measurement sequence starts by applying a potential of 800 mV during 0.5 sec. The voltage is then set to 0 during the 0.5 sec., whereafter the applied potential is decreased by 100 mV, and the cycle starts again. The sample rate was set to 20 Hz but only the first, second and tenth sample in each 0.5-second interval were used. Each electrode is described by 66 datapoints, hence, a total tongue measurement is given by 396 datapoints.

3. The electronic nose

The sensor array consisted of 10 MOSFET´s with gates of thin catalytically active metals like Pt, Ir and Pd. Four metal oxide (Taguchi) type sensors were also used in the sensor array. The electronic equipment were built at the Laboratory of Applied Physics, Linköping University. The field effect transistors (FET) were placed in a row in a small (volume 1 ml) chamber that was heated by a resistor. The Taguchi type sensors were placed in separate cells after the MOSFET sensor chamber

Gas samples were pumped from the samples by a membrane pump at a flow rate of 100 ml/min and injected in the sensor chamber. The injection of gas samples could be performed at given time intervals by the opening of a valve, operated by a PC. The PC was also used to collect and pre-treat the data from the sensor array. Each measurement contains 14 variables to be further analyzed.

4. Samples

The samples investigated were different types of fruit juices, orange juice (O), apple juice (A) and pineapple juice (P). Two samples of each of the fruit juices were measured three times, making six measurements for each juice.

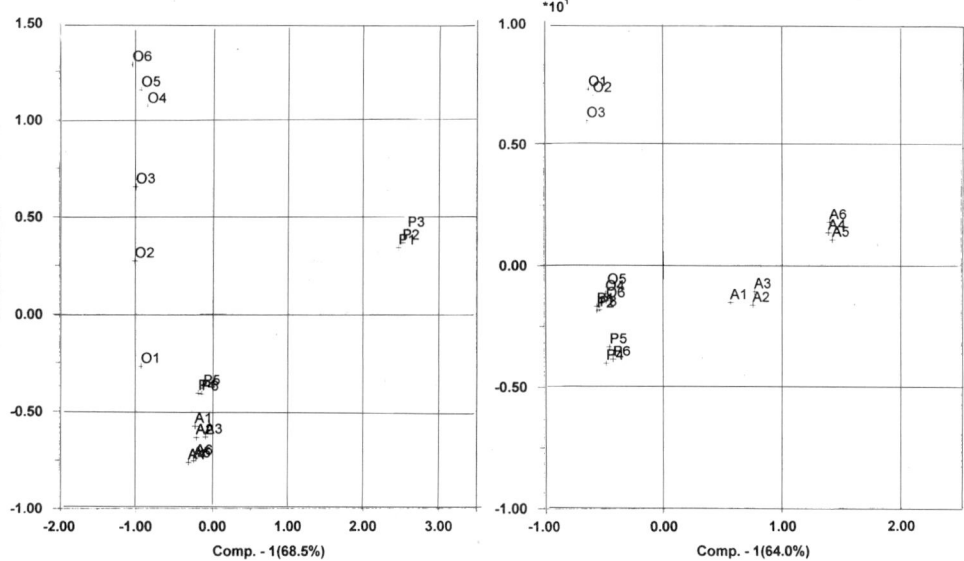

Fig 1. A score plot from values
obtained from the nose.

Fig. 2. A score plot from values
obtained from the tongue.

5. Results and Discussions

Measurements were performed in order to classify the samples, and to follow aging
processes. Principal component analysis (PCA) were performed on data obtained from both
the nose and the tongue. PCA is a mathematical transform which is used to explain variance
in experimental data[9]. A score plot can be made showing the relation between the
observations or experiments, and groupings of observations can be used for classifications. A
corresponding loading plot shows the relationships between the sensor signals and how much
they influence the system. A score plot for data obtained from the nose is shown in figure 1,
and a corresponding score plot for tongue data is shown in figure 2. The classification
properties for both system may be considered to be fair, although, for the tongue, samples P
and O are not separated. For the nose, orange samples are spread and samples A and P are not
separated. A score plot for both tongue and nose data are shown in figure 3, and as can be
seen, the separation has improved. The spread of the points within each class represents the
aging processes. A corresponding loading plot is shown in figure 4, showing the different
areas in which the nose (marked area) or the tongue (the rest of the area) contribute with
information. Since the tongue produce a large amount of data, only each sixth value is
represented in the figure.

It has thus been shown, that both the electronic nose and the electronic tongue alone is able
to discriminate fairly between the samples. It is also shown, that the classification properties
are clearly improved when information from both sources are combined. During these
measurements, it was also observed that aging processes could be followed.

1090

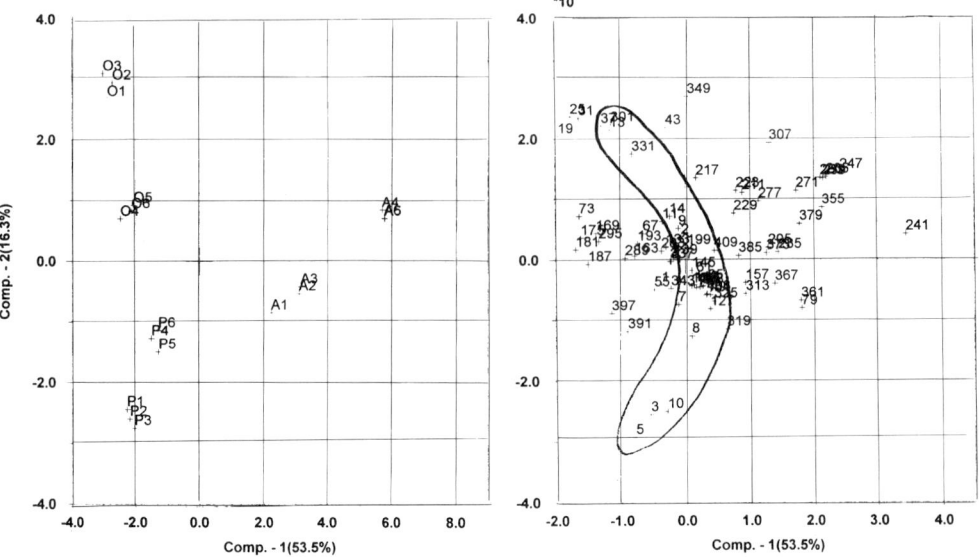

Fig. 3. A score plot from values obtained from both the nose and the tongue

Fig. 4. A loading plot from values obtained from both the nose and the tongue.

References

[1]   J. W. Gardner and P. N. Barlett, "A brief history of electronic noses", Sensors and Actuators B 18-19 (1994) 211-220.

[2]   F. Winquist, E.G. Hörnsten, H. Sundgren and I. Lundström, "Performance of an electronic nose for quality estimation of ground meat", Measurement Science & Technology 4 (1993) 1493-1500

[3]   P. M. Schweizer-Berbereich, S. Vaihinger and W. Göpel (1994). "Characterization of food freshness with sensor arrays", Sensors and Actuators, 18 287-290

[4]   C. Di Natale, F. Davide, A. D'Amico, A. Legin, A. Rudinitskaya, B.L. Selezenev and Y. Vlasov, "Applications of an electronic tongue to the environmental control", Technical digest of Eurosensors X, Leuven, Belgium (1996) 1345-1348.

[5]   K. Toko, "Taste sensor with global selectivity", Materials Science and Engineering C4(1996)69-82.

[6]   F. Winquist, P. Wide and I. Lundström "An electronic tongue based on voltammetry" Analytica Chimica Acta (1997)357,21-31

[7]   F. Winquist, T. Kranz-Rülcker, P. Wide and I. Lundström" Monitoring of milk freshness by an electronic tongue based on voltammetry", submitted for publication.

[8]   F. Winquist, P. Wide and I. Lundström, "Crispbread quality evaluation based on fusion of information from the sensor analogies to the human olfactory, auditory and tactile senses", submitted for publication.

[9]   S. Wold, K. Esbensen, P. Geladi "Principal Component Analysis - A tutorial", Chemometrics and Intelligent Laboratory Systems 2, (1987) 37-52.

Sensor Arrays and Multi-Sensor Systems
Paper presented at Eurosensors XII, 13–16 September 1998

A pyroelectric polymer infra red sensor array with a charge amplifier readout

D Setiadi, H Weller and T D Binnie

Napier University, Department of Electronic and Electrical Engineering, 219 Colinton Road, Edinburgh EH14 1DJ, United Kingdom

Abstract. This paper presents a new pixel architecture for an infra red sensor array based on a pyroelectric polymer. The pyroelectric polymer sensor is integrated with the CMOS charge amplifier. The fill factor of the sensor is optimised by placing the amplifier structure directly below the sensing area. The maximum responsivity is 240 V/W and the specific detectivity is of the order of 10^6 cm.$\sqrt{Hz}$/W.

1. Introduction

The development of low cost pyroelectric sensors has gained a great deal of interest due to the large number of potential applications. A low cost sensor is achieved by integrating the pyroelectric material with their low cost associated electronics. Ferroelectric polymers have been preferred over ceramics and single crystals, due to their fabrication being compatible with semiconductor technology, their low cost, and low thermal conductivity.

Conventionally, the pyroelectrically generated charge is supplied to the gate of a single FET. The impedance of the sensing element is extremely high and consequently it cannot be used for biasing the gate of the FET to supply the gate leakage current. Therefore, a gate bias resistor is necessary. A very high resistance (100MΩ and above) is required to achieve good responsivity. Although on-chip resistors are available, their accuracy, repeatability and linearity in CMOS process technology remain poor and their large values cannot be realistically implemented.

A radiated-damaged diode for biasing the MOSFET is suggested by von Münch[1]. The obtained bias resistor is in the order of 10^{10} Ω with some spreading across a wafer. It requires an extra process from the CMOS standard process. Hammes suggested use of a dc feedback loop[2]. The bias point of the sensor element is measured by a dc amplifier and compared to a reference value. If necessary, the feedback loop can be used to readjusts the biasing. The gate of the readout FET is biased by means of a current source instead of a diode.

In this paper we describe a fully integrated low noise charge amplifier suitable for pyroelectric sensor readout. This charge amplifier is implemented in the standard CMOS process technology with limited chip area. A 2x2 matrix pyroelectric sensor based on PVDF with the charge amplifier has been realized. Responsivity and noise of the sensor will be presented. These values will be compared with those of a conventional single MOSFET pre-amplifier with an external bias resistor.

2. Charge amplifiers

A charge amplifier, as shown in Figure 1, has been realized by a single-ended, folded, cascode amplifier. This structure is extensively employed for the purpose of charge amplification [3-6]. It presents the advantages of having a high dc gain and large frequency bandwidth. The dc operating point of the input transistor is set by means of resistive feedback using a MOS transistor. The feedback capacitor of 1 pF is attained using two polysilicon layers. The charge amplifier is implemented in the double metal, double polysilicon, Mietec 2.4 µm CMOS technology. Specifications of the charge amplifier are given in Table 1. Figure 2 shows a microphotograph of the chip. The chip contains 3 matrices of 2x2 pixel elements; each element with an area of 240 µm x 240 µm. The three matrix configurations are: single MOSFET, voltage-controlled amplifier and charge amplifier. The voltage-controlled amplifiers and charge amplifiers are fabricated underneath the area of the second metal which acts as the lower electrode for the pyroelectric sensor. This can be done by reserving the first metal layer in the process for the amplifier intraconnect, and using the second metal layer to implement the pixels. This allow large area, hence low noise input transistors without sacrificing pixel fill factor.

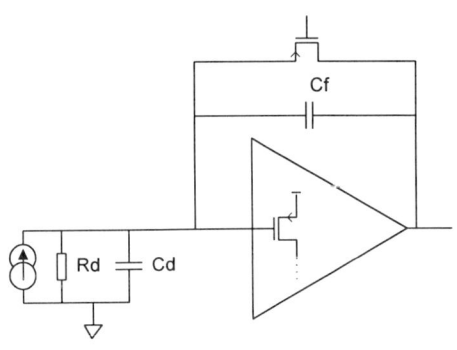

Figure 1 Sensor and charge amplifier equivalent circuit

Table 1 Charge amplifier specification

Properties	Value
DC voltage gain	23
Bandwidth	>1 MHz
Input referred noise (@100 Hz)	57 nV/√Hz
Feedback capacitance	1 pF
Power supply	± 2.5 V
Area	245µm x200µm

3. Technology

A 5 µm-thick capacitive adhesive film (UV curing acrylic) epoxy was applied to the silicon substrate containing readout electronics and 9 µm pre-polled PVDF pressed into place. The PVDF film is metalised only on the top surface, while the bottom electrode was removed after the polarization. The thermal coefficient of the capacitive adhesive film is low and even lower than that of PVDF film. Therefore, this layer acts beneficially as thermal insulation.

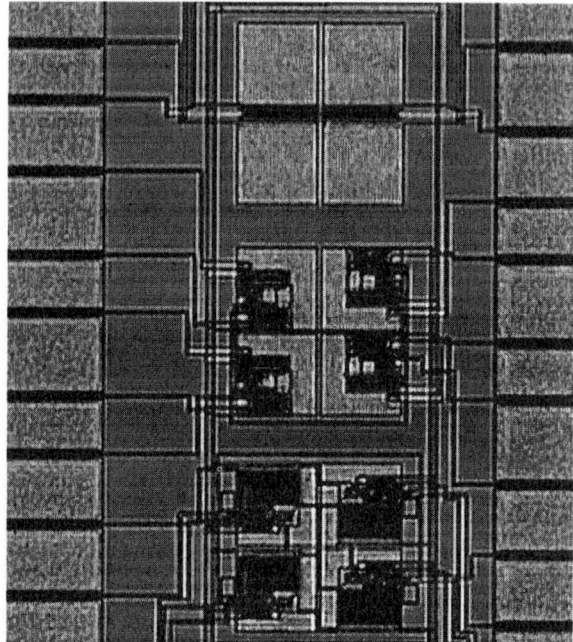

Figure 2 Microphotograph of the three 2x2 pyroelectric sensor with their respective pre-amplifiers

4. Measurement results

A calibrated low temperature black body source of 65 W/m^2 was used to illuminate the sensor. A mechanical chopper modulates the beam to provide an ac signal for the sensor and a reference signal for the lock-in amplifier. A germanium window was used for screening ambient visible light from the sensor. Figure 3 shows the output voltage of the pyroelectric sensor for the charge amplifier and the single MOSFET with 10^9 Ω bias resistor. Table 2 shows the performance of both sensors. The voltage-controlled amplifier had a 2-3 seconds offset drift which could not controlled without resorting to external components.

Table 2 Performance of Pyroelectric sensors with a charge amplifier and a single MOSFET.

No	Parameter (@100 Hz)	Charge amplifier	Single MOSFET (10^9 Ω bias resistor)	Unit
1	Responsivity	240	238	V/W
2	Output noise voltage	1.61	34.2	μV/√Hz
3	Noise equivalent power	6.99	143.6	nW/√Hz
4	Specific detectivity	3.43 * 10^6	1.67 * 10^5	cm.√Hz/W

Output Voltage (V)   D* (cm√Hz/W)

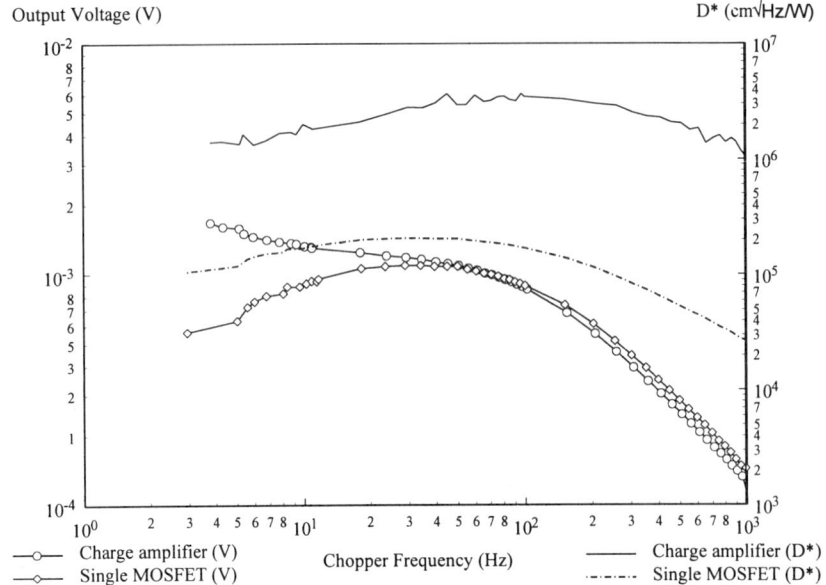

Charge amplifier (V)
Single MOSFET (V)
Chopper Frequency (Hz)
Charge amplifier (D*)
Single MOSFET (D*)

Figure 3 Output voltage and specific detectivity of pyroelectric sensors with the charge amplifier and the single MOSFET with 10^9 Ω external bias resistor

5. Conclusions

An integrated charge amplifier which is implemented in the standard CMOS process for a pyroelectric polymer sensor has been realized. The charge amplifier lies directly under the sensing element. Hence it requires no extra space. The responsivity and Noise Equivalent Power of this sensor are compared with those of the single MOSFET with a 10^9 Ω bias resistor.

Acknowledgment

This work is supported by the UK Engineering and Physical Sciences Research Council (EPSRC grant award GR/K17224).

References

[1]   W von Münch and U Thiemann 1991 *Sensors and actuators* A **25-27** 167-172
[2]   P C A Hammes 1994 *Ph.D Thesis* Delft University of Technology
[3]   A Simoni, et.al 1995 *IEEE Journal of solid state circuits* **30** 800-805
[4]   S Tedja, J van der Spiegel, and H H Williams 1995 *IEEE Journal of solid state circuits* **30** 110-119
[5]   Y Hu, and E Nygard 1995 *Nuclear Instruments and Methods in Physics research* **A365** 193-197
[6]   W M C Sansen and Z Yan Chang 1990 *IEEE transactions on circuits and systems* **37** 1375-1382

Sensor Arrays and Multi-Sensor Systems
Paper presented at Eurosensors XII, 13–16 September 1998
© 1998 IOP Publishing Ltd

Integrated optical sensor utilising a 1D CCD array for multiple output addressing

B. J. Luff*, K. Kawaguchi[†] and J. S. Wilkinson*

*Optoelectronics Research Centre, University of Southampton, Southampton
SO17 1BJ, UK
[†] Kyoto Electronics Manufacturing Co., Ltd., 68 Ninodan-cho, Shinden, Kisshoin
Minami-ku Kyoto 601, Japan

Abstract. An inexpensive and robust method for acquiring multiple outputs from integrated optical sensor devices using a 1D CCD array is described in this paper. An example is given of the development of an instrument based on the use of an integrated optical Mach-Zehnder interferometer (MZI) refractive index transducer. The technique is especially promising for application to multianalyte sensors where several outputs need to be interrogated simultaneously. The high sensitivity and low noise demonstrated by the system will enable the use of cheap, stable LED light sources in practical instruments.

1. Introduction

Integrated optical transducers for the real-time measurement of interactions between biological molecules and for the specific detection of chemical and biochemical species are the subject of growing interest. Applications of this technology include environmental pollution monitoring, industrial process control and medical diagnostics. Integrated optical sensors provide the high detection sensitivity achievable using optical transduction techniques in a compact and robust format. This approach also offers advantages for the fabrication of multianalyte sensors through the integration of multiple transducers on a single chip by straightforward scaling of the photolithographic production process. Several types of integrated optical sensor have been described (e.g. [1-3]), but no commercially viable multianalyte system currently exists and, in order to fully exploit this technology in practical instrumentation, inexpensive and reliable techniques for addressing the multiple outputs of waveguide devices must be found. Fibre-to-chip pigtailing of integrated optical devices formed in 'passive' materials such as glass, where it is difficult to truly integrate monolithic light sources and detectors, is not the best solution when dealing with multiple outputs due to the necessity of producing and pigtailing fibre arrays. For single input devices, however, fibre input coupling is still a viable option as only a single pigtail needs to be made.

In this paper we present measurements on multiple-output integrated optical sensor

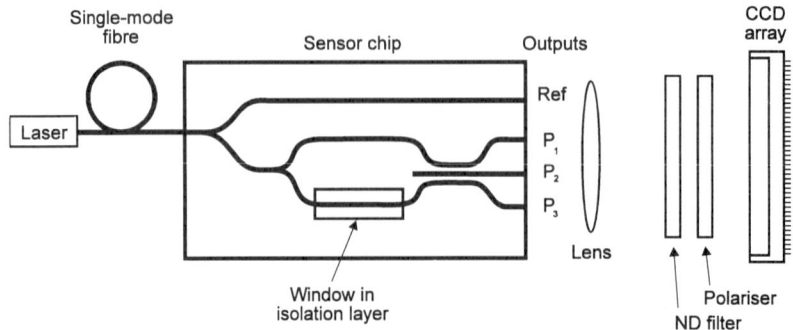

Figure 1 Experimental sensor system

devices using fibre input coupling and a cheap, readily available, 1D CCD array detector to simultaneously address all outputs. A lens is used to focus the waveguide outputs onto the array, resulting in a compact unit that can be housed in a standard instrument package. A further advantage of this arrangement is that other optical elements such as filters and polarisers can readily be inserted into the beam path.

2. Transducer design

The sensor transducer chip is shown as part of the experimental setup in Figure 1: for this work we used an integrated optical Mach-Zehnder interferometer (MZI) device fabricated in BGG36 glass (Schott) by Ag^+-Na^+ ion-exchange [1]. The sensor utilises a three-waveguide coupler to produce three phase-shifted outputs and responds to refractive index changes occurring above the transducer surface within the waveguide evanescent field. Sputtered silica was used as an isolation layer material to isolate the reference arm of the Mach-Zehnder, the input and output transition regions, and the independent reference waveguide, from the analyte. A window was opened in the silica layer using a photolithographic liftoff technique to define the sensitive region of the waveguide. The sum of the intensities of the three outputs of the device potentially yields further information on analyte absorption as the sum reflects the total power loss through the interferometer. The separation between the output waveguides was 250 µm; the chip length was 40 mm. Comparison of the relative intensities of the three outputs enables unambiguous determination of refractive index over one full period of the output interference function.

3. Experimental system

Figure 1 shows the experimental sensor system. Light from a 10 mW 633 nm He-Ne laser was coupled into a single-mode fibre and the fibre was coupled to the sensor chip; the waveguides on the chip were single-mode at 633 nm. A 10X objective lens was used to focus the output signals on a Toshiba 1D CCD array (cost around $40US). The distance from the CCD to the objective lens was 18 cm, giving a total magnification of 14 times. The CCD array has 3648 pixels; the pixel width and height were 8 µm and 200 µm, respectively. A 40 dB neutral density filter was placed between the lens and the detector to reduce the signal so that the CCD array was not saturated and a polariser was used to select TE polarisation. The extremely high

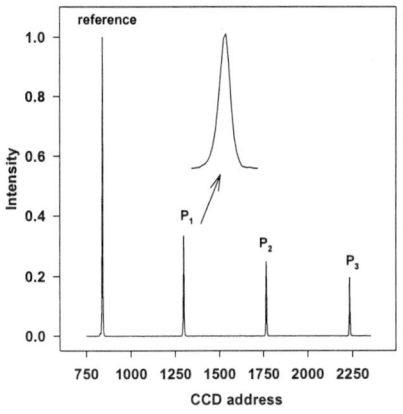

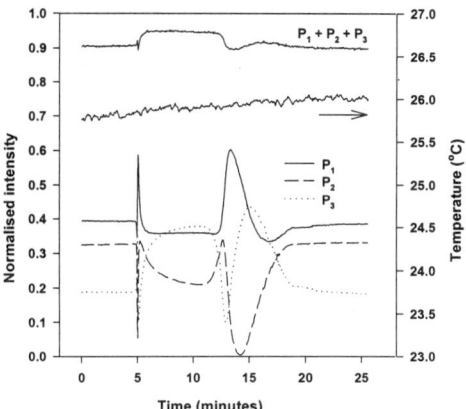

Figure 2 Spatial output on CCD array **Figure 3 Sucrose solution test cycle**

sensitivity of the detector allows less stringent fabrication tolerances as regards waveguide loss when using laser sources; it also opens up the possibility of using other sources that are less intense or which couple poorly to single-mode fibre, such as LEDs. A temperature sensor was placed near the flow cell to monitor ambient temperature during a measurement.

A flow-cell was clamped to the sensor surface and sucrose test solutions of varying refractive indices were applied using a flow-injection system. The refractive indices of these solutions, which are a function of the sucrose concentration, were determined using a KEM refractometer at a wavelength of 589 nm at 20°C.

4. Results and discussion

Figure 2 shows the spatial distribution of the sensor output intensities when water is applied to the chip surface ($n = 1.33299$). The half power full width (HPFW) of the intensity profile of the light from a single waveguide output falling on the array is 34 µm. As the vertical dimensions of the mode profiles are of a similar magnitude, all the light from each output is incident on the 200 µm-high array. The measured signal for each output is taken as the sum of the intensities falling on the pixels within a region surrounding each feature. A background signal measured close to each feature was subtracted for each output. The integration time was 40 ms; for each data point in sucrose solution test cycle measurements several integrations were collected and averaged.

Figure 3 shows the temporal response of the sensor to a sucrose solution of refractive index $n = 1.41045$. Each output P_1, P_2 and P_3 of the three-waveguide coupler is divided by the independent reference output to compensate fluctuations in the input light intensity. Also shown is the ambient temperature and the sum of the three referenced outputs. The sum does not remain constant throughout the measurement due to the redistribution of power taking place between the waveguides of the coupler, which each exhibit somewhat different loss characteristics due to the presence of bends and fabrication errors. Corrections to the measured data to eliminate this effect will be made following further characterisation of the sensor. The sucrose pulse is injected into the flow stream and pumped at a constant rate through the flow cell during the measurement. Equilibrium levels are reached at around $t = 10$ minutes as the pure water is displaced completely from the cell and replaced with the sucrose solution; the sucrose solution is then washed out and replaced with water, and the signals return to their original levels.

The demonstration system shown here is not yet temperature stabilised and therefore there is some drift due to variations in ambient temperatures. However, this setup provides an indication of the limit of detection by calculating the smallest detectable signal change considering the noise in the system: the minimum detectable refractive index change was determined to be 7×10^{-5}. However, the sensor chip employed in this demonstration was not optimised for maximum sensitivity. Work is now in progress to package the transducer and detector in a permanent housing, with temperature control of the active region of the transducer, employing localised temperature sensors, and pigtailed fibre input. The use of more compact and inexpensive light sources is being investigated, particularly 635 nm laser diodes and visible LEDs. Initial experiments indicate that sufficient light is coupled from standard packaged LEDs to single mode fibres to render their use with the current instrumentation a practical possibility. Ultimately, the present test transducer design will be replaced with arrayed MZI devices having more than four outputs and the instrument will be used to probe selective thin films for the detection of specific analytes in solution.

5. Conclusions

An inexpensive and robust technique for the addressing of multiple outputs from integrated optical transducers has been described. The performance of the demonstrator setup indicates that further development of the method and application of low-cost light sources will lead to the first truly practical integrated optical multisensor instrument.

Acknowledgements

This work was funded by Kyoto Electronics Manufacturing Co., Ltd., Japan. The Optoelectronics Research Centre is an Interdisciplinary Research Centre funded by the UK EPSRC.

References

[1] Drapp B, Piehler J, Brecht A, Gauglitz G, Luff B J and Wilkinson J S 1997 *Sensors and Actuators B* **38-39** 277-282

[2] Ch. Fattinger, Koller H, Schlatter D and Wehrli P 1993 *Biosensors and Bioelectronics* **8** 99-107

[3] Mouvet C, Harris R D, Maciag C, Wilkinson J S, Luff B J, Piehler J, Brecht A, Gauglitz G, Abuknesha R and Ismail, G 1997 *Analytica Chimica Acta* **338** 109-117

Sensor Arrays and Multi-Sensor Systems
Paper presented at Eurosensors XII, 13–16 September 1998
© *1998 IOP Publishing Ltd*

Characterisation of a CMOS current drive chip for an array of six polymeric resistive gas sensors

M Cole[1], J W Gardner[1@], A W Y Lim[1], P K Scivier[2] and J E Brignell[2]

[1]Department of Engineering, University of Warwick, Coventry CV4 7AL, UK
[2]Department of Electronics & Computer Science, University of Southampton, Southampton, SO17 1BJ, UK

Abstract. In this paper we describe the characterisation of a full custom analogue current drive chip fabricated using the Alcatel Mietec 2.4 μm CMOS process. The chip has been designed to drive currents in the micro-amp range through an array of six resistive polymer micro-bridge devices. It exhibits good d.c. stability, and a linear temperature coefficient of about $-1.2 \times 10^{-3}/°C$. The chip can be drive in an a.c. as well as in a d.c. and gives a stable pulsed current source at frequencies of up to 10 kHz.

1. Introduction

Arrays of conducting polymer resistors are of increasing interest in gas and odour sensing [1,2]. In order to reduce the effect of operating temperature on the base-line signal of discrete polymer sensors (*ca.* $-10^{-2}/°C$), a silicon micro-bridge device was designed and fabricated at Warwick University with conducting polymer resistive elements on all four arms (two active) in a CMOS-compatible process [3,4]. In order to drive an array of six micro-bridges an ASIC chip was designed at Southampton University which generates six constant reference currents. A constant current source was chosen over a constant voltage source because of the advantage that current driven bridges have less susceptibility to lead resistance, are easier to create, and have better linearity [5]. Preliminary tests have been conducted to investigate the long term stability, load effect and temperature dependence of the current reference. Voltage and current noise measurements were also performed using precision metal film load resistors. Further tests are being carried out with the ASIC chip as a part of a larger circuitry [6], to characterise the response of poly(aniline) micro-bridges to ethanol.

2. Chip design and fabrication

The current drive chip has been fabricated using the Alcatel Mietec 2.4 μm CMOS technology and, prior to fabrication, had been modelled using Cadence Analog Artist HSPICE interface. Figure 1 shows the main part of its schematic diagram where CHCORC

@ To whom correspondence should be sent.

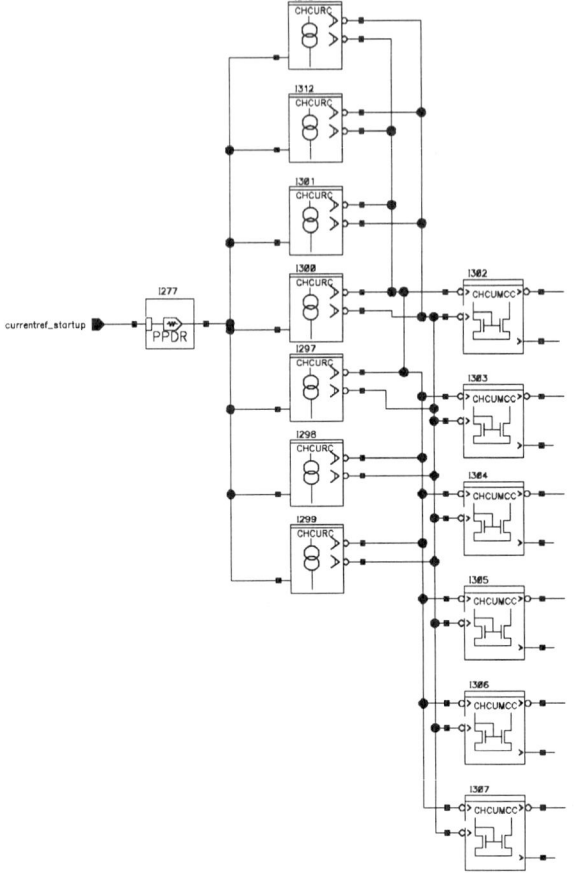

(Core cell High voltage range Current Reference Cascade circuit) represents a full custom Mietec current reference cell with improved power supply rejection; PPDR is the start-up circuit for the current reference and CHCUMCC (Core cell High voltage range Current Mirror Cascade C type circuit) is the corresponding current mirror cell. Each CHCURC circuit delivers a typical current of 2 µA, while each current mirror has a current factor of 2. An input voltage, V_{DD}, for the ASIC chip should be set at 2.5 V. The output from each current mirror drives a polymer micro-bridge with a constant current of 16 µA, thus setting the bridge voltage at the suitable value according to the resistance of the polymer resistors. The chip also comprises of six independent analogue switching circuits, and three operational amplifiers that could be used independently or in conjunction with the constant current reference part.

Figure 1. Schematic of the ASIC current drive chip

3. Tests and Results

The ASIC chip has been designed to work with a voltage power supply V_{DD} optimally around 2.5 V d.c. or pulsed.[1] Initial tests were performed to investigate the stability of the current reference output in relation to the voltage supply of the chip and also to investigate its long term stability. A programmable micropower voltage regulator (MAX666, Maxim) was used to generate the voltage supply for the chip in the range of 2 to 10 V with increments of 0.5V. Each current mirror output of the ASIC chip was connected to a 1 kΩ precision metal film load resistor. Voltages across the resistors were measured using a M810 multimeter calibrated against a Knick 5252 d.c. voltage calibrator. It was apparent that there was an increase in the voltage across, and therefore an increase in current through the resistors, with an increase of V_{DD}. For a given chip supply range of 2 to 10 V, the current output exhibited change from 8.4 µA to 78.5 µA. Although such dependence of the current output on the input of the chip is undesirable, this was not considered a disadvantage because a high voltage regulation for the

---

[1] The ground rail was set to $-V_{DD}$ throughout to obtain suitable currents.

power supply could always be provided. For the long term stability test, the programmable voltage regulator was set at 2.5 V and voltages across six precision load resistors connected to six current outputs of the chip were monitored for 24 h. A virtual instrument (VI) written using Labview software (National Instruments) was used to control and record the data during the test via a PC-LPM-16 12-bit (National Instrument) card. Although the measurements were slightly limited by the accuracy and the noise level of the data acquisition card, current reference outputs have exhibited good stability (< 0.7%/day).

The effect of a load resistor on the current output was also investigated. The programmable voltage regulator was again set at 2.5 V and current from the ASIC current mirror was passed in sequence through the set of precision metal film resistors of 0.75, 1.00, 4.02, 6.81, 10 and 18.2 kΩ. Figure 2 shows a typical effect of the resistive load on the current reference. Voltages across the load resistors and calculated current values were used to determine the Norton short-circuit current and source resistance from the equation

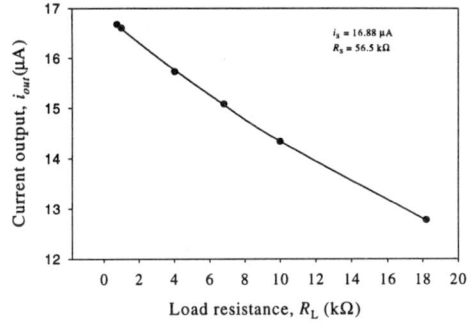

Figure 2. Effect of load on the current mirror output of the ASIC chip

$$i_{\text{out}} = \frac{R_s}{(R_s + R_L)} i_{sc} \qquad (1)$$

where i_{out} is current through the load resistor, R_L is resistance of the load, and i_s and R_s are equivalent source current and resistance, respectively. Values for i_s and R_s were found to be 16.88 μA and 56.5 kΩ by fitting the model to the experimental data (see Figure 2).

In order to investigate the temperature dependence of the current output, the same set-up as before was used: i.e. 2.5 V voltage supply and six current outputs connected to six 1 kΩ precision load resistors. The ASIC chip was isolated and placed in a Dri-Bloc™ heater (Techne Ltd) in order to minimise the influence of other components on its temperature characteristic. Values of current outputs were recorded at seven different temperatures (±0.1°C): 24.3, 30.6, 34.8, 41.5, 46.7, 51.5 and 56.3°C. Figure 3 shows a typical plot of the temperature dependence of the current source. Test results are represented by symbols while the line represents a model based on the assumption that simple polynomial temperature dependence of current output could be employed:

$$\Delta i_{\text{out}} = \Delta i_0 \left(1 + \alpha T + \beta T^2\right)$$

(2)

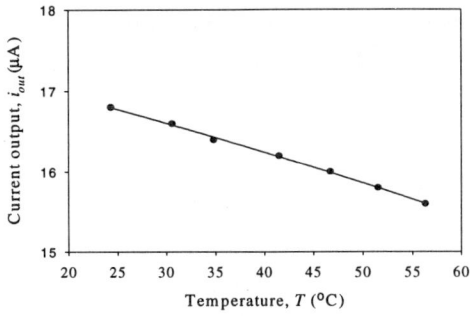

Figure 3. Effect of temperature on the current mirror output of the ASIC chip

where Δi_{out} is change in current output in µA at temperature T, Δi_0 is change at 0°C, and α and β are linear and quadratic temperature coefficients with units of °C^{-1} and °C^{-2}, respectively. The temperature coefficients α and β are calculated for all six current outputs and it was observed that the temperature dependence for all of them is approximately linear with a linear coefficient α of $ca.-1.2\times10^{-3}$/°C. Thus, the temperature coefficient of the current chip is negligible being about $1/20^{th}$ of that for the response from a polymer micro-bridge [4].

4. Noise measurements

The noise measurements were carried out using an Hewlett-Packard HP 35660A dynamic signal analyser with the same circuit as before in which constant current was passed through precision metal film load resistors. Noise spectra have been taken from a number of different precision resistors: 0.75, 1.00, 5.11, 6.81 and 10.0 kΩ. Figure 4 shows

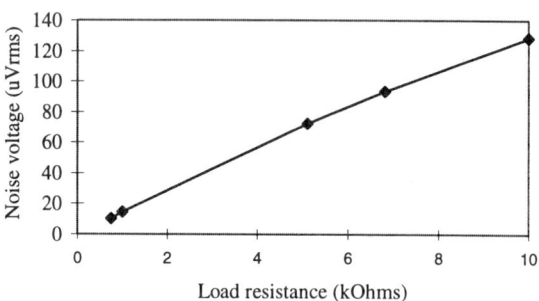

Figure 4. Typical noise voltage measured at 16 Hz for different load resistors

noise voltage measured at a very low frequency of 16 Hz. It can be seen that the noise level increases with an increase of the load. Measurements were also taken from the load resistor of 1 kΩ, changing the voltage supply of the chip in the range of 2 to 10 V. An increase in noise with an increase of supply voltage, and therefore an increase of the current output, was also observed.

5. Conclusion

A current drive CMOS ASIC chip has been designed and fabricated to drive an array of six polymer resistive micro-bridges. The micro-bridges are easy to produce in a CMOS-compatible process and this work is one of the first steps towards the integration of the gas sensors and CMOS circuits to make a smart array device. Characterisation of the chip has shown good stability of the current reference and its suitability for use in future characterisation of polymeric micro-bridges.

References

[1]   Gardner J W and Bartlett P N 1995 *Sensors and Actuators A* **51** 57-66

[2]   Neotronics Scientific Ltd, UK, *Technology Booklet.*

[3]   Pike A C *PhD thesis,* University of Warwick, UK, 1996.

[4]   Gardner J W, Vidic M, Ingleby P, Lloyd C R, Pike A C, Brignell J E, Scivier P, Bartlett P N, Duke A and Elliot J M 1998 *Sensors and Actuators A,* at press.

[5]   Ismail M and Fiez T *Analog VLSI,* Mc-Graw Hill 1994.

[6]   Scivier P K, White N M, Brignell J E, Gardner J W and Vidic M, *Proc. of 11th European Conference on Solid-state Transducers, September 1997, Warsaw, Poland.*

Sensor Arrays and Multi-Sensor Systems
Paper presented at Eurosensors XII, 13–16 September 1998
© 1998 IOP Publishing Ltd

The Application of Genetic Algorithms to Multisensor Array Optimisation

J.Anglesea, P.Corcoran, M.Elshaw.

Sensing and Control Research Group. School Of Engineering, University Of Derby, Derby, DE22
1GB. United Kingdom.
Tel: + 1332 622798 Facsimile + 1332 622739

ABSTRACT

Genetic algorithms have been applied to the task of optimal array configuration and
it is shown that multisensor array performance, in terms of the successful
classification of wines, can be improved through the application of such techniques.
This is achieved by identifying a subset of the available parameters which, when
combined, provide discriminable and repeatable fingerprints of odorous materials in
classification space. In particular, utilising a thermally cycled 8-element array,
which generated 64 parameters, a subset of only 24 parameters was identified which
gave an improvement of 5% above the classification rate derived from the 64-
parameter set. These results, along with previous studies, show that through the
employment of optimisation techniques it is often possible to reduce array
dimensionality by around 50-70% whilst still achieving equal, or better, levels of
system performance.

Introduction

Recent improvements in sensor technology in terms of: new sensor types, the utilisation of dynamic
sensor responses [1] and thermal cycling of sensor operating temperatures combined with multi-
parameter extraction techniques [2,3]. have all conspired to increase the amount of information
available for the description of gases and odours using multisensor arrays. The use of multisensor
systems is common in a number of applications. In particular the use of sensor arrays for the purpose
of gas and odour classification, and sometimes description, has been the subject of much research
activity. It is now accepted that the most promising employment of such technology will most
probably be in application specific instruments, where a limited set of odours would be analysed,
based upon a reference database containing key parameters for each odour.

Ideally, given a database of the responses of n sensors to p species, it should be possible to identify a
subset of n sensor elements, that provide a specified level of performance when applied to the
classification, or description, of a subset of the set of p species. A structured approach to this task
requires the specification of a cost function to quantify the performance of an array of sensors in a
given application, and a search algorithm to identify the optimal or a near-optimal array configuration
[4]. The optimal configuration of existing sensor technology can facilitate improvements in system
performance without the additional cost of developing new sensor technology. Analysing the true
potential of current sensor technology may also guide the development of new sensor technology.
Also since the performance of an electronic nose is based largely on the performance of the sensor
array configuration, it is reasonable to expend effort in trying to improve that selection of elements to
enhance the overall system performance.

With so many different types of sensor and different parameter extraction/generation techniques, it is
necessary to devise some way of measuring the performance of a given set of sensor array responses in
terms of the application at hand. Most applications in the electronic nose domain can be grouped as
either odour detection, classification or occasionally description tasks. Rating the performance of a
given set of input parameters for a given task allows us to configure the parameters or sensors that are
best used for a specific task. This is achieved by rating parameters in terms of their noise,
repeatability, discriminatory contribution, these parameters being indicative of their overall
contribution to the particular discrimination/classification process.

Being able to rate the performance of a set of sensor parameters allows the creation of customised,
application specific sensor arrays, with the most appropriate sensors, perhaps from different

technologies providing the most appropriate combination in order to satisfy the needs of the current application. i.e. application specific multisensor arrays with associated extracted parameters.

Theory

The mechanism by which a genetic algorithm arrives at an optimal solution to a particular problem can be explained by the operation of schemata.[5] A schema is a template that defines a particular pattern of genes within a chromosome. and as well as the usual binary alphabet, (0,1) is made up of 'don't care' symbols that allow a gene to take on any value. A schema is defined in terms of the schema order and length.

The order (o) of a schema is simply the number of fixed positions it contains. For example, **101* is a schema of order three, since it contains three fixed positions. The defining length (δ) of a schema is the distance between the last specific value in the schema and the first. **101* therefore is a schema with order three and defining length 2.

There are 3^l schemata defined over a binary chromosome of length l, since there are three possible values that each gene position can take 0,1 and * (don't care). This means that in a population of n chromosomes. there are a maximum of $n.2^l$ schemata since each string is representative of 2^l schemata. The number of schemata indicates the current state of the population in terms of it's diversity.

Given a schema H within a population $A(t)$ where t is the finite step time, if there are m examples of that schema within the population then m will vary with time and will be defined by $m=m(H,t)$. During reproduction, a string A_i has a probability of being selected for reproduction of:

$$p_i = f_i / \Sigma f_j \tag{1}$$

where f_i is the fitness of A_i and Σf_j is the sum of all the fitnesses in the population. Since $m=m(H,t)$ for $A(t)$, $m=m(H, t+1)$ for $A(t+1)$. The number of this particular schema in the next generation, $A(t+1)$, is therefore the number present in the current population (i.e. $m(H,t)$) multiplied by the ratio of the average fitness of the strings representing H to the average fitness of the population as a whole multiplied across the total number of strings in the population. n. This gives:

$$m(H, t+1) = m(H,t).n.f(H) / \Sigma f_j \tag{2}$$

where $f(H)$ is the average fitness of the strings representing H in $A(t)$. However, the average fitness across the entire population is given by:

$$\bar{f} = \Sigma f_j / n \quad \therefore m(H, t+1) = m(H,t).(f(H) / \bar{f}) \tag{3}$$

which implies that the number of a given schema increases from one generation to the next as a function of the ratio of the average fitness of the schema to the average fitness of the population.

Since this operation is repeated for every schema in the population in parallel, the net effect of reproduction is to increase the number of schemata with above average fitness and to reduce those of lower than average fitness. For a schema with an average fitness $c\bar{f}$, where c is constant:

$$m(H, t+1) = m(H,t).((\bar{f} +c\bar{f}) / \bar{f}) = (1 +c).m(H,t) \tag{4}$$

At $t = 0$ and assuming c is constant:

$$m(H,t) = m(H,0).(1 + c)^t \tag{5}$$

This is a geometric progression showing that reproduction reproduces exponentially increasing (or decreasing) numbers of above (or below) average schemata. Reproduction on its own does not allow new regions of the search space to be investigated. This is the role played by crossover and mutation.

In simple crossover, a random point is chosen along the pair of strings to be crossed and genetic information is swapped after this point. A schema with a high δ is more likely to be destroyed in this case since it is more probable that the crossover point will fall within the defining length of the schema.

The probability that a schema will survive crossover, p_s can be shown to be:

$$p_s \geq 1 - p_c \cdot (\delta(H) / (l - 1)) \tag{6}$$

where p_c is the probability of crossover occurring, $\delta(H)$ is the defining length of schema H, and l is the string length.

The number of schema H present in the next generation now becomes:

$$m(H, t+1) \geq m(H,t).(f(H) / \bar{f}) [1 - p_c \cdot (\delta(H) / (l-1))] \tag{7}$$

The probability of a schema surviving the mutation process is a function of both the order (o) of the schema and the probability of mutation occurring. The probability of mutation not occurring at an individual position within a schema is $(1 - p_m)$ where p_m is the probability of mutation. The probability of the whole schema surviving is therefore the probability of an individual position surviving raised to the order of the schema, i.e. $(1 - p_m)^{\delta(H)}$. Since $p_m \ll 1$, this expression can be approximated to $1 - o(H) . p_m$. Inclusion of the mutation term in the overall expression for $m(H,t+1)$ yields:

$$m(H, t+1) \geq m(H,t).(f(H) / \bar{f}) [1 - (p_c \cdot (\delta(H) / (l-1))) - o(H).p_m] \tag{8}$$

This is known as the *Schema Theorem* or the *Fundamental Theorem of Genetic Algorithms*. The conclusion drawn from this equation is that short, low order, above average fitness schemata will propagate most successfully into the next generation.

A customised software package, GAMOS (Genetic Algorithm Multisensor Optimisation Software), has been developed for the task of optimally configuring sensor arrays used for the classification of measurands. The package uses generic sensor parameters and hence it can be used for a range of sensor array applications. The sensor parameters selected for array configurations used a genetic representation based upon the presence (1) or absence (0) of a sensor parameter. Each parameter is equivalent to a gene, with the actual value at the gene position referred to as the allele. For example, given eight sensors from which an optimal configuration must be identified (S1[msb] to S8[lsb]), the chromosome 01010011 would represent an array configuration S2, S4, S7 and S8

The objective function used to calculate fitness was given by first calculating pairwise normalised mean class differences between classes x and y in the sensor array fingerprint of dimensionality p:

$$D_{xy} = \sum_{n=1}^{p} \frac{(\mu_{nx} - \mu_{ny})^2}{(\sigma_{nx}^2 + \sigma_{ny}^2)} \tag{9}$$

These were then used to calculate the final chromosome fitness which attempts to increase both successful classification rate of all species without bias towards the most easily identifiable. Hence for three classes (1,2,3) with three mean class difference values, D_{12}, D_{13} and D_{23}, the objective function value would be:

$$O_x = (D_{12} + D_{13} + D_{23}) / (max\{D_{12}, D_{13}, D_2\} - min\{D_{12}, D_{13}, D_2\}) \tag{10}$$

The processes of reproduction, crossover and mutation are performed as described earlier.

Experimental Method

An 8-element Taguchi Gas Sensor (TGS) array was thermally cycled in a static rig odour sampling system, as described elsewhere [3]. Two classes of wine, Chardonnay and Muscadet, were each sampled 26 times, by injecting 1μl of wine into a 5 litre flask. The thermal cycle length was 150s, with a variation in temperature range of 175-500°C derived from a triangular heater voltage waveform. The thermal cycling rig utilises *in-situ* temperature measurement to control the sensor element temperature to an accuracy of <1%. Eight parameters were then extracted from each of the sensor elements in line with previous studies [3]. Fractional normalisation was used to extract the fingerprint signatures for each of the wines for each parameter set.

Results

GAMOS was then used to identify a reduced set of three parameters, from each of the 8 sensors, which could be used for the classification of the two wines. The Dynamic Parameter Method data set was used to investigate the application of the GA optimisation and hence the chromosome consisted of 64 genes. The fitness of each chromosome could then be calculated using the objective function (see equations (9) and (10)). Optimisation was undertaken as described earlier. The performance of the complete parameter set was then compared to the performance of the optimised parameter set. Discriminant Function Analysis gave the results shown in Tables 1 and 2.

	Predicted Chardonnay	Predicted Muscadet
Actual Chardonnay	84.6	15.4
Actual Muscadet	42.3	57.7

Table 1. Classification matrix for the full parameter set - 64 parameters (%).

	Predicted Chardonnay	Predicted Muscadet
Actual Chardonnay	73.1	26.9
Actual Muscadet	23.1	76.9

Table 2. Classification matrix for the reduced parameter set - 24 parameters (%).

The results show an overall classification rate of 71.1% for the full parameter set and 75.0% for the reduced parameter set. It should also be noted that GAMOS has identified a classifier which performs equally well for both wines, Muscadet and Chardonnay, whereas the full parameter array configuration shows a bias towards classifying Chardonnay. Hence it is clear that GAMOS is being successful in identifying a subset of parameters that are at least equal, and which may be facilitating a small improvement, in performance when measured in terms of the overall classification rate. Previous studies have shown similar levels of success [3,4].

Conclusions

The use of optimisation techniques for sensor array configuration can solve a number of problems given restricted array sizes, specified confidence levels and the availability of multiple parameters from single sensor elements. Using the GAMOS tool enabled the identification of a subset of parameters that gave an improvement in performance when compared to the full parameter set. The utility of tools such as GAMOS will increase when instrument manufacturers recognise that the benefits of optimal array configuration using sensors from an existing database may outweigh the costs associated with novel sensor development. Furthermore, GAMOS may be used to mitigate against risks in future sensor development programmes.

References
1. D.Wilson and S.DeWeerth, Odour discrimination using steady-state and transient characteristics of tin oxide sensors, *Sensors and Actuators B28* (1995) pp123-128.
2. S. Wlodek, K. Colbow and F. Consadori, Kinetic model of thermally cycled tin oxide gas sensors, *Sensors and Actuators B, 3* (1991) pp123-127.
3. P.Corcoran, P.Lowery and J.Anglesea, Optimal Configuration of a thermally cycled gas sensor array with neural network pattern recognition, *Sensors and Actuators B* (1998) In Press.
4. P.Corcoran, Optimal configuration of sensor arrays for odour classification, *Measurement and Control Vol.30* (1997) pp43-46.
5. D.E. Goldberg, Genetic algorithms in search, optimisation and machine learning, *Addison Wesley* (1989) ISBN 0-201-15767-5.

Sensor Arrays and Multi-Sensor Systems
Paper presented at Eurosensors XII, 13–16 September 1998
© *1998 IOP Publishing Ltd*

Correlation of Human Sniff Tests and of Analytical Data with Results from a Hybrid Modular Sensor System

Heiko Ulmer, Jan Mitrovics, Udo Weimar, and Wolfgang Göpel

Institute of Physical and Theoretical Chemistry; Center of Interface Analysis and Sensors, University of Tübingen, Auf der Morgenstelle 8,
D-72076 Tübingen, Germany, Phone: ++49 7071 297 8768, Fax: ++49 7071 29 5960,
e-mail: hu@ipc.uni-tuebingen.de

Abstract Experimental results of qualitative and quantitative analysis of packaging materials, and pharmaceutical emulsions obtained with a hybrid modular sensor system are presented. Two modules of the hybrid modular sensor system MOSES containing different transduction principles have been applied. In the present study 8 metal oxide-based semiconductor gas sensors (MOX) and 8 polymer coated quartz microbalance sensors (QMB) have been chosen. Pattern recognition was performed by principal component analysis (PCA) and multicomponent analysis by principal component regression (PCR). The obtained results indicate sufficient performance and reproducibility of hybrid modular sensor systems for investigations of a broad spectrum of applications.

1. Introduction

In order to guarantee a certain standard of raw materials as well as final products, quality control in different industries becomes increasingly important. Since many years sensor systems, also known as "Electronic Noses", are gaining interest as tools for the qualitative and quantitative analysis of volatile compounds emitted from materials to be investigated [1, 2]. The sensor systems have the potential as complementary tools to chromatographs, FT-IR-systems or human sensory test panels.

Usually "Electronic Noses" combine several sensing elements but use the same type of transduction principle. Examples of limitations in such an approach have been given earlier [3, 5, 6]. In order to overcome the limitations of every individual tranduction principle, a combination of several of theses principles has been implemented in one instrument. Since Electronic noses are used as laboratory tools at the moment, they are tested for different applications and therefore the detection of the hole spectrum of chemical components is desirable. Consequently a combination of different transduction principles, as realized in the commercially available hybrid modular sensor system MOSES (details of the instrumentation have been published before [3, 4], for further details see http://www.lennatz-electronic.de), increases significantly the performance of gas analysis and the classes of species which may be detected. An overview of investigated applications so far, is given in [5, 6].

2. Experimental

2.1. Packaging materials

Today most food and consumer products are wrapped in packaging materials to avoid spoilage and contamination. In most cases different plastic materials with specific additives like slip agents, cold seals, and glues are used.

The processing of synthetic polymer-based packaging materials like foils, bottles, or boxes is performed by extruding melted pellets. Here the extrusion temperatures vary between 200°C to 350°C. In the presence of oxygen different odorous compounds arise from thermal oxidation of the polymer. The higher the temperature the more odorous aldehydes and ketones are formed. Figure 1 shows the discrimination of three polyethylene foils extruded at different temperatures (low =280°C, medium=305°C, high=330°C). For each sample five headspace vials with 100cm^2 of the material are prepared and sealed with a silicon/PTFE septum. Then the vials are introduced in the headspace sampler and after an equilibration time of 20 min at 80°C the samples are measured.

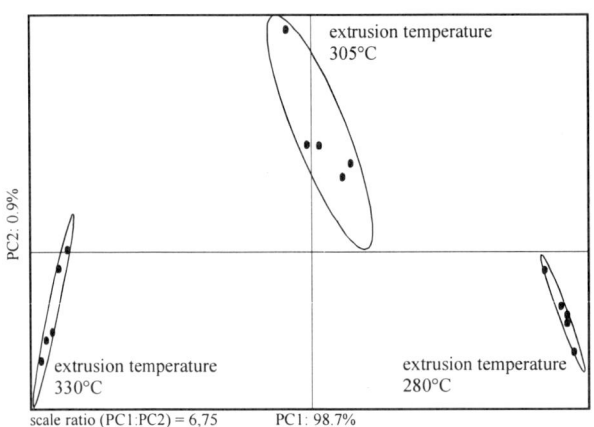

Figure 1: *Discrimination of polyethylene foils with extrusion temperatures between 280°C and 330°C. There is a clear separation along the first principal component (PC1) representing 98.7% of the total information.*

Besides chromatographic measurements the food manufacturers use the Robinson and sniff test procedures [7] for the quality control of packaging materials. Here a defined amount of material is filled into a jar and is then heated to 80°C for 2 hours. Afterwards trained persons classify the intensity of the smell on a scale from 0 (no difference to a given reference) to 4 (big difference to the reference). The same test was performed for the polyethylene foils. Figure 2 shows the correlation of sensor signals and sniff test scores. The straight line represents the ideal correlation between measured and predicted values.

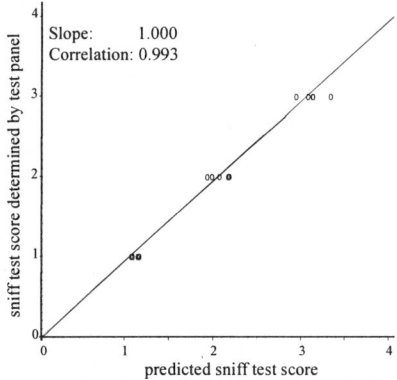

Figure 2: *Linear regression of polyethylene foils extruded at different temperatures. The straight line represents the ideal correlation between values determined by the test panel and predicted by the sensor system. The sniff test score for the sample extruded at low temperature was determined by the test panel as 1 (small difference to the reference). The predicted values vary from 1.0 to 1.15. For the medium temperature extruded sample the determined sniff score was 2 and the variation of the prediction was 1.8 to 2.2. For the high temperature extruded sample the determined sniff scores was 3 (largest difference to the reference) and the predicted values vary in the range of 2.8 to 3.3.*

2.2. *Pharmaceutical emulsions*

For the pharmaceutical and chemical industry quality control of cosmetics, pastes, emulsions, and creams is a very important task. Here the determination of the amount of perfumes, active substances, and additives is of main interest.

Figure 3 shows the qualitative discrimination of pharmaceutical emulsions consisting of the insect repellent DEET (N,N-Diethyl-3-methyl-benzamide) and the carrier MCT (medium chain triglycerides). Six mixtures with an increasing amount of DEET and the pure components are filled in 20ml headspace vials (4 vials for the pure components and 6 vials for the mixtures) and sealed with a silicon/PTFE septum. The vials are introduced in the headspace sampler and after an equilibration time of 30 min at 90°C the samples are measured.

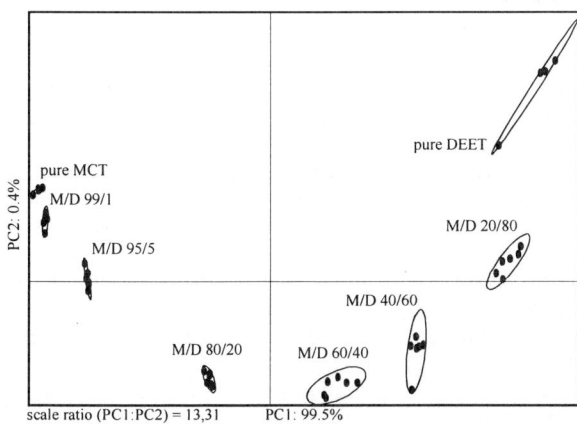

Figure 3: *Discrimination of pharmaceutical emulsion. There is a clear separation between the pure components and the individual mixtures. There is only a small scattering within one cluster which indicates a homogenous distribution of the DEET in MCT.*

In figure 4 the correlation of sensor signals with the ratio of active substance to carrier by means of a principle component regression is shown.

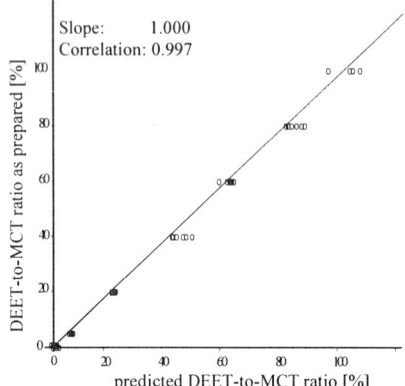

Figure 4: *Linear regression of pharmaceutical emulsions with different amounts of DEET in MCT. The straight line represents the ideal correlation between measured and predicted values. For the different emulsions there is only a small deviation of predicted and measured values.*

3. Conclusions

The different packaging materials and pharmaceutical emulsions could be discriminated successfully. Additionally it was possible to correlate the results obtained by the hybrid modular sensor system with sniff test scores and analytical data.
The result indicates, that hybrid modular sensor systems can be used for both qualitative and quantitative analysis in different applications.

4. Acknowledgement

We gratefully acknowledge Professor Daniels from the Institute of Pharmaceutical Technology at the University of Braunschweig (Germany) for providing us with the pharmaceutical emulsions.

5. References

[1] J.W. Gardner, H.V: Shurmer, T.T. Tan, *Application of an electronic nose to the discrimination of coffees*, Sensors and Actuators B 6 (1992) 71-75.
[2] H. V. Shurmer, J.W. Gardner, *Odour discrimination with an electronic nose*, Sensors and Actuators B 8 (1992) 1-11.
[3] J. Mitrovics, U. Weimar, H. Ulmer, G. Noetzel, W. Göpel, *Hybrid Modular Sensor Systems: A New generation of Electronic Noses*, Proc. of Sensor 97, Nürnberg (1997) B 1.1.
[4] J. Mitrovics, H. Ulmer, G. Noetzel, U. Weimar, W. Göpel, Design of a Hybrid Modular Sensor System for Gas and Odor Analysis, Proc. of Transducers 97, Chicago (1997) 4C2.10P.
[5] H. Ulmer, J. Mitrovics, G. Noetzel, U. Weimar, W. Göpel, Odours and Flavours Identified with Hybrid Modular Sensor Systems, Sensors and Actuators B 43 (1997) 24-33
[6] H. Ulmer, J. Mitrovics, U. Weimar, W. Göpel, Detection of Off-Odors using a Hybrid Modular Sensor System, Proc. of Transducers 97, Chicago (1997) 2C3.03.
[7] L. Robinson, Verpackungs-Rundschau 12 (1961) 3, March, Technical Supplement, p. 17.

Sensor Arrays and Multi-Sensor Systems
Paper presented at Eurosensors XII, 13–16 September 1998
© *1998 IOP Publishing Ltd*

Portable and modular electronic nose for olfactometric measurements

R. Hartinger[1], M. Irsiegler[1], H.-E. Endres[1], S. Drost[1], K. Rieblinger[2], G. Ziegleder[2]

[1]Fraunhofer-Institute for Solid State Technology, Hansastr. 27d, D-80686 Munich
[2]Fraunhofer-Institute for Process Engineering and Packaging, Giggenhauser Strasse 35, D-85354 Freising

Abstract. In this paper we describe an olfactometric measuring system, equipped with a measuring chamber for eight semiconductor gas sensors. The entire system is controlled by a PC/104-system. The measuring software controls the measurement data recording, as well as the process control and the data analysis. As test measurements for the olfactometer nuts with different roasting degrees were selected. A measurement cycle of about two minutes is possible. The main advantages of this device are the modular and portable system concept, the temperature transient operation methods and that most of the different sensortypes (for example semiconductor-, surface acoustic wave-, quartz micro balance- or interdigital capacitor gassensors) can be used.

1. Introduction

The potential application scope of an olfactometric measuring system, also mentioned as electronic nose, is extraordinarily broad: Job security, quality control of food, product check during the production of materials, and environmental protection. On the market there are several of such electronic noses already available. For example Moragas from the Fraunhofer-Institute for Biomedical Engineering, MOSES II from the Institute of Physical and Theoretical Chemistry (University of Tübingen), the Steinbeis Center for Interface Analysis and Sensors and Lennartz Electronic GmbH or QMB 6 of the company HKR sensorsystems [1]. Goal of this paper is the development and testing of an olfactometric microsensor system for online measurements of air quality and for the measurement of smells as a product check.

2. Equipment

An important goal during the development of the olfactometer was to achieve an exchangeability of the individual components, a small size and short gas exchange times. Even the number of individual components can vary, therefore the structure was held open for future extensions.

Such a sensor system can be divided into three levels. The analysis information is passed on thereby from level to level. On each level the information assumes another status. The transformation of one status in the others takes place at the interfaces between the levels.

1. *The physical-chemical level:* The sample handling e.g. with filters or catalysts, the method of analysis can be adapted to the application. The interfaces, which change the chemical information into an electronic signal to the next level, are the sensors.
2. *The electronic level:* Depending upon the transducers (sensors), the different electrical signals of the sensors are the information storage medium. Interfaces to the information level are for example A/D-cards or frequency counters.
3. *The information level:* On the information level the sensor signals are processed by multi-component analysis or neural networks.

The physical-chemical level

Modularity and portability within one level of the system is achieved by standardisation of the interfaces to the adjacent levels, so that an easy exchange of different components is possible.

In this device all capacitive and resistive sensors can be used, which are available on TO-5 bases. This restriction in use of the sensors is given because of the geometry of the measurement chamber. The small volume of the sensor chamber (approx. $1cm^3$) permits small sample quantities. For the measurements semiconductor gas sensors were usually used. These sensors are characterised by high sensitivity, for which they need however a high operating temperature. The sensor can be optimized by selection of the appropriate heater-temperature for a certain gas. Semiconductor gas sensors are operated at temperatures between 200°C and 400°C. The necessary heating is responsible for the relatively high current consumption. Due to the high sensitivity, the commercial availability and the good long-term behaviour, semiconductor gas sensors are well suitably for this range of application.

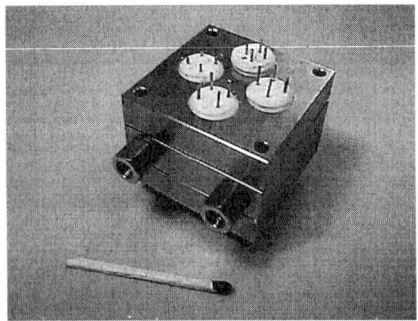 

Figure 1: View of the measurement chamber and the modular sensor system

The electronic level

Because the semiconductor gas sensors change their gas sensitivity dependent on their temperature, an analog controller device was developed. The heating resistor can vary in a wide range from 1 up to 200 ohms. With this electronic the sensor-temperature is stabilized to ± 0.1 °C. This electronic allows temperature transient operations, controlled by an computer.

The other electronic which had to be developed was for the measurement data recording. The developed sensor electronics permits the measuring of capacitive and resistive sensors. The electronics converts an electrical capacity and/or an electrical resistance into a well further-processable electrical signal. The output-signal can either be a voltage- or frequency-proportional signal [2].

The information level

The entire measurements are controlled by a PC/104-system. The program controls the measurement data recording, as well as the process control and the data analysis.

With the available measurement setup the frequency-sensor signal was read in by a 10 channel counter (ComputerBoards PC104-CTR10HD/H50). The valves of the gas mixing system are controlled via a digital input/ output-card (ComputerBoards PC104-DIO48). The data storage can take place either with an externally ZIP-drive, or a remote computer can be attached over a serial link. This remote computer cannot only sample the data, but it can also take over the entire controlling of the measurement equipment.

After the data are sampled, different procedures of mathematical-statistical analysis can be used to optimise the results of the measurement. The method of analysis on the information level can be divided thereby into three functions. First of all the drift elimination or the standardization of the measuring signals. Secondly the suitable values must be extracted from the measuring signals. The third function consists of processing the values by pattern recognition with neural networks or principal component analysis [3].

3. Measurements

This section describes the procedure of measuring the headspace of ground hazelnuts, which were roasted in five different degrees: unroasted, roasted ten minutes at 180 °C and so on, up to roasted 40 minutes at 180 °C.

Eight different metal-oxide semiconductor gas sensors (UST: 1000, 2000, 3000, 6000 and Figaro: 2600, 2610, 2620 and FIS: AQ1) are placed inside the measuring chamber with different sensitivities for various gases. Figure 2 shows a result of a measurement with the hazelnut-samples. The roasting degrees of the samples are identified quite well as shown in figure 3. The different samples are very clearly discriminated. It is obvious from this principal component analysis, that the different samples can be separated quite well.

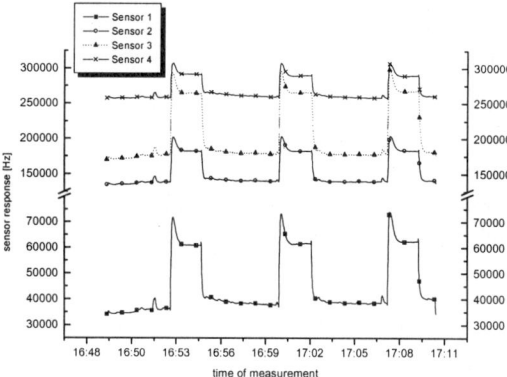

Figure 2 Result of measuring the hazelnut-samples

The procedure of measuring a headspace follows these steps:
The sensors are driven at a higher heater resistance for some time, to get rid of some stains sitting on the surface of the sensor. During this time the stainless steel tubes and the valves are purged for the same reason, while the measuring chamber is still closed. After that begins the purging phase of the measuring chamber, the sensors are now driven at their operating

1114

temperature. The purging time can be set by the software at the beginning of the measurement. The purging medium can either be the surrounding air, synthetic air or any other gas which is connected to the reference gas inlet at the front panel. Then the measuring phase starts, the valve to the measuring chambers opens and the gas connected to the inlet labelled "HS" flows into the chamber. The sensor signals are converted from resistance into frequency which is measured by the counter and written into a datafile. At the end there is another purging of the measuring chamber, the sensor signals come back to equilibrium with the purging medium.

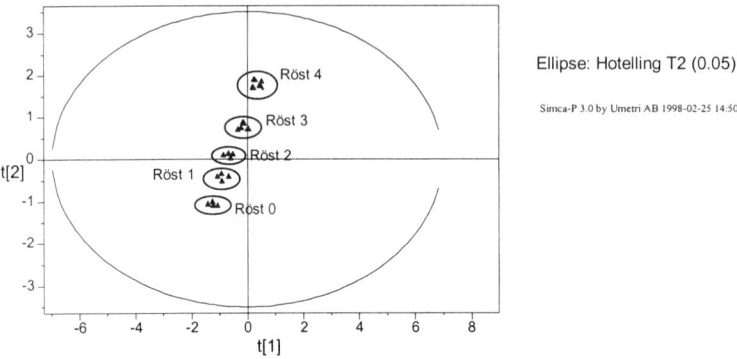

Figure 3 Principal component analysis for hazelnut samples

4. Results and discussion

The prototype model of our process gas measuring instrument is at present equipped with a measuring chamber for an array of eight semiconductor gas sensors, gas feed pump, the appropriate valves and the electronics. The small volume of the sensor chamber (approx. $1 cm^3$) of our process gas measuring system permits small sample quantities. Beside the small size and the open system a further advantage of this device is the transient operation method, which permits a zero point stabilization. The heating of the gas-containing sections should not raise larger constructional problems.

Advantages of this device:

- small structure with minimum gas volume (sensor chamber $1 cm^3$)
- modular concept (simple exchangeability of different sensors because of the standardized interfaces)
- easy adaptation to different measuring problems
- use of commercial sensor technology with low fluctuations of the performance, therefore economical maintenance

References

[1] J. Mitrovics, H. Ulmer, U. Weimar, W. Göpel, Modular Sensor Systems for Gas Sensing and Odor Monitoring: The MOSES Concept, Accounts of Chemical Research, Chemical Sensors and Interfacial design, 1997

[2] M. Irsiegler, Entwicklung und Optimierung eines olfaktometrischen Meßsystems, Diplomarbeit Technische Universität München, 1998

[3] H.-E. Endres, W. Göttler, H. Jander, S. Drost, H. Sandmeier, G. Sberveglieri, G. Faglia, C. Perego, Improvement in signal evaluation methods for semiconductor gas sensors, Sensors and Actuators B, 26-27 (1995) 267-270

Sensor Arrays and Multi-Sensor Systems
Paper presented at Eurosensors XII, 13–16 September 1998
© *1998 IOP Publishing Ltd*

Arylene Alkenylenes as Chemiresistors in an Electronic Nose

E. Vanneste, M. De Wit, K. Eyckmans, H.J. Geise.

University of Antwerpen (UIA), Department of Chemistry. Universiteitsplein 1, B-2610 Wilrijk, Belgium.

Abstract. Five arylenevinylene oligomers, compounds which are easily synthesized and chemically modified, proved useful in conductimetric gas sensors. Three types of sensors were tested: those in which the active oligomer was (i) chemically doped with iodine, (ii) blended in a polycarbonate matrix and subsequently electrochemically doped with lithium perchlorate and (iii) simultaneously electrochemically doped with lithium perchlorate and deposited onto the substrate. All sensors operate at room temperature and have a low power consumption. Responses to organic vapours are fast and high and the life time for type (iii) sensors is over 6 months.

1. Introduction

A trained human nose can detect something is wrong with a load of coffee, when the freshness of fish leaves to be desired whether, or when a flaw in the process of beer brewing occurs. In such cases as well as in others in which human detection is not the obvious solution for the job, an Electronic Nose (EN) may well find good use. Hence, it may become important on the sensor market. An EN is essentially a combination of a sensor array and a pattern recognition system [1,2]. Pattern recognition, ranging from e.g. cluster and factor analysis to neutral networks, will not be addressed here. Instead, we will focus on the search for materials applicable in sensor arrays, and note that sensor materials for the implementation in an EN need a partial selectivity for many vapours. EN sensors can be based on a number of mechanisms [1,3], using e.g. the variation in mass (quartz crystal microbalance, surface acoustic wave sensors [4]), optical (glass fiber sensors [5]), electronic (MOSFET sensors [6]) and electrical (conductimetric sensors [2,7]) properties upon exposure to vapours. For the latter type of sensors, the application of organic semiconductors is advantageous, because in contrast to metal oxide semiconductors [7] they operate around room temperature and have greater possibilities for structural variation. One way to produce organic semiconductors is to blend a neutral polymer with conducting particles [8,9], such as graphite. Another way is in a process called doping to oxidize certain polymers possessing a long conjugated p-system.

Investigating FeCl₃-doped poly(phenylenevinylene) (PPV) and I₂-doped poly(2,5-thienylenevinylene) (PTV), we noted [10-14] that a decreasing conjugation length does not greatly change the sensor responses (resistance changes), but significantly alters the specificity towards certain vapors. These observations led us to test conjugated oligomers in stead of polymers for their sensing properties and also to test electrochemical instead of chemical doping. We report here some results of these investigations on five arylenevinylene oligomers (see Fig 1 for names and structural formulas of the compounds) doped chemically with I_2 or electrochemically with lithium perchlorate.

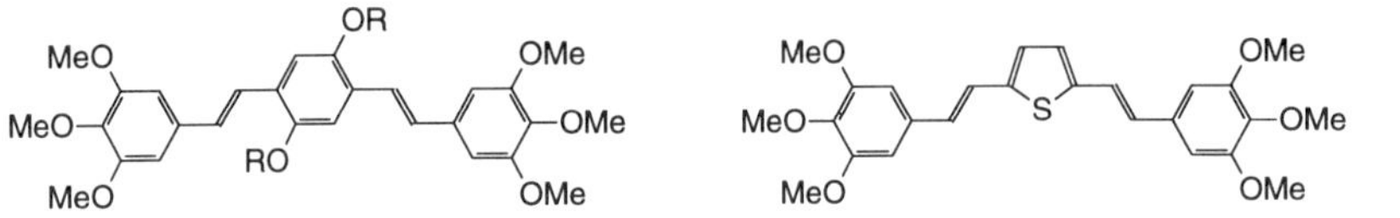

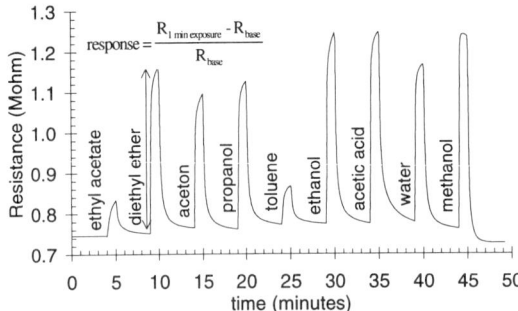

2,5-dialkoxy-1,4-bis(3,4,5-trimethoxystyryl)benzene   1,4-bis(3,4,5-trimethoxystyryl)thiophene = TH-HMT

Figure 1: Structures of 2,5-dialkoxy-1,4-bis(3,4,5-trimethoxystyryl)benzenes (R=methyl, OMT; R=ethyl, DE-HMT; R=propyl, DP-HMT and R=butyl, DB-HMT) and 1,4-bis(3,4,5-trimethoxystyryl)thiophene = TH-HMT).

2. Sensor test procedure

$$response = \frac{R_{1\,min\,exposure} - R_{base}}{R_{base}}$$

Fig. 2: Typical resistance change during consecutive exposures to 9 vapours.

In our standard test [12-14] a sensor is exposed for 1 min to dry air saturated at 15°C with a chemical chosen at random from a list of 9 chemicals (see below). This is immediately followed by a 4 min recovery period using clean, dry air. The sequence is repeated for the remaining eight vapours (see Fig. 2). Four more experimental runs with each time another random sequence of vapours are then performed in a time span of three days. Between the experimental runs clean air is passed over the sensor.

At this point some remarks seem in order. First, the times chosen for exposure (1 min) is a compromise between obtaining high responses and sufficient recovery in a short time (4 min). One may see that for most vapours the maximum attainable response is not reached, nor that the recovery is complete. Furthermore, the sensor is exposed to five sequences of saturated vapours in different orders with only the short 4 min of recovery between the individual vapours. This puts the sensors to a very severe test. Yet, the reproducibility of response values proved to be good. Second, we consider it at this point advantageous to maximize the rates of response and recovery, even when it goes at the expense of the magnitude of the response. A high response rate allows short sampling times which lowers the risks of poisoning or overloading of the sensor. A high recovery rate lowers the risks of memory effects and interferences between measurements. Furthermore, a short cycle time (here 5 min) shows the value of these sensors for on-line monitoring of industrial processes in which odours can be related to the correct progress of the process.

3. Results and Discussion

3.1 The OMT/I₂ and the TH-HMT/I₂ sensors.

3.1 The OMT/I_2 and the TH-HMT/I_2 sensors.

OMT and I_2 were dissolved in dichloromethane in a 1 : 1 molar ratio. Previous experiments had shown that the equimolar ratio leads to the highest conductivity. The solution is then spray-coated onto an electrode substrate (electrode gap 0.2mm) to give an active layer of 20 µm thickness. A 20µm thick TH-HMT/I_2 sensor was made similarly. Table 1 shows the responses of these two oligomeric sensors to the nine test vapours. The responses to water vapour are low (about 3%). Since water is so ubiquitous and often has a poisoning effect on conducting polymer based sensors this may be considered as an advantage. In contrast, the responses to the organic vapours range from 15 to 45% for the OMT/I_2 sensor and from 11 to 43% for the TH-HMT/I_2 sensor, which compares very favorable to polymeric PTV/I_2 sensors [14] ranging from 1 to 16%. Unfortunately, however, both oligomeric sensors showed a drift in the base resistance limiting their life time to about 2-3 weeks.

response %	OMT	TH-HMT
acetic acid	42.2	38.0
acetone	45.1	29.8
diethyl ether	36.6	43.0
ethanol	16.0	12.8
ethyl acetate	36.1	43.3
methanol	15.3	11.5
propanol	18.1	9.8
toluene	36.7	39.4
water	3.1	2.7

Tabel 1: Responses of iodine doped OMTand TH-HMT on saturated vapours

3.2 The four 2,5-dialkoxy-1,4-bis(3,4,5-trimethoxystyryl)benzene perchlorate sensors.

3.2 The four 2,5-dialkoxy-1,4-bis(3,4,5-trimethoxystyryl)benzene perchlorate sensors.

Previous work [17,19] had shown that in case the base line resistance became too high, the sensor could be regenerated by repeating the iodine doping process. This strongly suggested that de-doping possibly by evaporation of iodine causes the instability. To overcome the disadvantage we introduced electrochemically the non-volatile perchlorate dopant ion and encapsulated the oligomer in a polycarbonate matrix. To do so we added a 5% (g/g) polycarbonate solution in CH_2Cl_2 to a solution of the active oligomer to obtain a 50%/50% oligomer/polycarbonate solution. After spray-coating this solution onto the substrate, doping was performed by submerging the substrate into a 0.1 M solution of lithium perchlorate in acetonitrile and applying a +2.7 V potential over the combined sensor electrodes and the central counter electrode. A color change from yellow to black indicates the doping reaction. The thickness of the sensors was 20µm. As we had hoped, the base line resistance became much more stable, increasing the

response %	OMT	DE-HMT	DP-HMT	DB-HMT
acetic acid	47.9	42.5	32.5	28.8
acetone	94.0	24.1	25.3	10.6
diethyl ether	94.0	51.4	82.2	86.9
ethanol	101.6	46.0	32.1	133.9
ethyl acetate	59.9	59.4	98.4	62.1
methanol	73.0	55.2	23.3	94.5
propanol	76.4	43.5	44.7	104.1
toluene	11.2	14.4	33.6	12.8
water	32.8	24.2	31.6	56.2

Tabel 2: responses of perchlorate doped oligomers

life-time of these sensors to 3-4 months. Furthermore, the responses, shown in Table 2, also increased sharply to values being about an order of magnitude larger than those of conducting polymers.

3.3 The OMT/perchlorate/deposite sensor.

response %	
acetic acid	32.9
acetone	32.8
diethyl ether	46.9
ethanol	38.1
ethyl acetate	19.4
methanol	39.1
propanol	36.4
toluene	15.0
water	15.1

Table 3: responses of deposited OMT on 9 vapours.

It is known that the solubility of arylenevinylene oligomers in organic solvents diminishes upon doping [15-17]. Therefore, simultaneous electrochemical doping with lithium perchlorate and deposition of the doped oligomer was considered possible. Indeed, when an alumina substrate was submerged into a solution of lithium perchlorate and OMT (molar ratio 1:2) in acetonitrile and a potential of +1.6 V was applied over the combined electrodes and the central counter electrode, a black film of OMT/perchlorate was obtained. The film had a thickness of 15µm and adhered well to the substrate. Moreover, the stability of the base line resistance had increased further, raising the life-time of the sensor to over 6 months. The responses (see Table 3) to organic vapours are still high (15-47%). The response to water vapour is relatively low, and shows an excellent reproducibility (15.1 ± 0.2%).

References

1. Persaud, K.; Dodd, G. Nature 1982, 299, 352

2. Sensors and Sensory System for an Electronic Nose; Gardner, J.W.; Bartlett, P.N.; NATO ASI Series Vol. 212; Kluwer Academic: Dordrecht, The Netherlands, 1991

3. For a review of types presently available see:
   http://sch-www.uia.ac.be/struct/review/research@sales.html

4. Bodenhöfer, K.; Hierlemann, A.; Noetzel, G.; Weimar, U.; Göpel, W. Analytical Chemistry, 1996, 68, 2210-2218

5. Dickinson, T.; White, J.; Kauer, J.; Walt, D. Nature, 1996, 382, 697

6. Lundström I.; Spetz A.; Winquist F.; Ackelid U.; Sundgren H., Sensors and Actuators B, 1, 1990, 15-20

7. Takahaka, K. Chemical Sensor Technology, 1988, 1, 39

8. Lonergan, K.; Severin, E.; Doleman, B.; Beaber, S.; Grubbs, R.; Lewis, N. Chem. Mater, 1996, 8, 2298-2312

9. Lewis, N.; Freund, M. U.S. Patent 5.571.401; 1996

10. Briers, J. Ph. D. Thesis, University of Antwerpen, Jan. 1994 (in Dutch)

11. Eevers, W.; Cos, P.; Geise, H.J.; Mertens, R.; Nagels, P.; Zhang, X.B.; Van Tendeloo, G.; Herrebout, W.; Van der Veken, B. Polymer, 1994, 35 (21), 4569

12. De Wit, M. Ph. D. Thesis, University of Antwerpen, Sept. 1997 (in Dutch)

13. De Wit, M.; Vanneste, E.; Blockhuys, F.; Geise, H.J.; Mertens, R.; Nagels, P. Synth. Met. 1997, 85, 1303

14. De Wit, M.; Vanneste, E.; Geise, H.J.; Nagels, L.J. submitted to Sensors and Actuators B.

15. Nouwen, J.; Adriaensens, P.; Vanderzande, D.; Gelan, J.; Yang, Z.; Geise, H.J. Synt. Met. 1992, 53, 77

16. Nouwen, J.; Adriaensens, P.; Vanderzande, D.; Gelan, J.; Yang, Z.; Geise, H.J. Synth. Met. 1993, 55-57, 3576

17. Nouwen, J.; Adriaensens, P.; Gelan, J.; Verreyt, G.; Yang, Z.; Geise, H.J. Macromol. Chem. Phys. 1994, 195, 2469

Sensor Arrays and Multi-Sensor Systems
Paper presented at Eurosensors XII, 13–16 September 1998
© 1998 IOP Publishing Ltd

An alternative way to improve the sensitivity of Electronic Olfactometers ('Electronic Noses')

Patrick MIELLE, Florence MARQUIS

I. N. R. A. Laboratoire de Recherches sur les Arômes,
17 rue Sully, F - 21034 DIJON CEDEX
Internet E-mail : Patrick.Mielle @ dijon.inra.fr

Abstract. The quality control, especially for foodstuffs, must include control of the aroma quality of the final products. The new technology of so-called 'Electronic Noses' encounters many problems to come out from research laboratories to plants. The main problem is the limited range of the signal-to-noise ratio. A possible improvement can be done by using the same sensors, but in another way. Up to now, it was difficult to accurately classify samples which were relatively close in terms of flavour or which contain major compounds of a weak flavour impact. This may be achieved by using an alternative way for managing the effluent transfer.

1. Introduction

Now the consumers demand products of quality. A new trend in the food industry is to relate the overall food quality of to the aroma quality. The foodstuffs control must include that of the aroma quality of the final products. From there, everything becomes more complicated... Aromas, flavours and odours are estimated or measured by using reference methods : sensory analysis (human evaluation) or instrumental analysis (separative techniques). Use of gas sensors for global analysis is a promising alternative. This paper deals with the problem of sensitivity exhibited by Electronic Olfactometers (E.O., systems so-called 'Electronic Noses').

2. Development of new methods

The traditional methods used for the characterisation of the food aromas are very accurate, but costly and time-consuming. The new technology of Electronic Olfactometers is evaluated in research laboratories but was not able - until now - to replace reference methods in plants production [1]. The main problem is to improve the systems sensitivity as much as possible. This can be achieved by the classical way, which consists in increasing the sensitivity of the sensing element itself, or by an alternative way, which is increasing the amount of volatiles reaching the sensitive element.

It is possible to point out the sensitivity limitations when studying in details the functioning of such systems, mainly at the system level. It is established that progresses in sensor technology will be very slow [2]. Besides, there is a great need of the industry for global analysis of target foods which can hardly be satisfied by the available commercial systems. A possible improvement can be done by using the same sensors, but in another way. This way is, of course, valid for any type of gas sensors.

2.1 Offer versus industry demand

Progress in gas sensor manufacture in terms of sensitivity and selectivity will be slow and will demand a large amount of investigations including basic research. As a consequence, the limits of capabilities of the E. O. cannot be pushed out in a short delay. There is a great need in the industry for an objective, automated system for analysing the overall aroma of certain foods, but this cannot be satisfied by the current available sensors. This reduces the number of applications. To fulfil all the requirements of the foodstuff quality control, the sensitivity must be improved in the range of 3 orders of magnitude ! And furthermore, the detection limit of E. O. cannot be directly related to those of sensors. We will explain this fact and try to propose solutions. For some systems, a 100 fold improved sensitivity can be obtained.

2.2 Detection threshold, sensitivity and reproducibility

The sensory threshold of the human nose varies according to the individual and the compound. Thresholds are typically located in the range from 1000 ppm to <1 ppt [3]. Usually they are determined in liquid or gas phases. Although the absolute detection threshold of gas sensors should in theory be very low – only a few molecules are required to react with the sensitive elements – in practice, it is closer to the ppm or tenth of a ppm range (measured in the vapour phase). There are three main reasons for this decrease in expected sensitivity:
- the noise level and short-term drift for all type of gas sensors may be high, that drastically decreases the signal-to-noise ratio;
- the sampling method used to generate effluent for transfer to the sensors (e.g. use of static or dynamic headspace (HS) analysis; use of a system that directly samples the vapour, or desorption from a purge-and-trap system) ;
- the effluent-transfer hardware (e.g. choice and control of effluent flow rate; any dilution of the effluent; inadequate temperature control of the effluent-transfer lines, sample-injection ports, and sample measurement cell).

For some types of chemicals, the sensitivity can be excellent -below the ppm range- but at such low concentrations, the scatter of data obtained can be close to 50%, drastically reduces the measurement accuracy [4].

2.3 Relation between sensors and systems sensitivity

First, we must explain why there is a such a difference between the sensitivity of the sensor itself and those of the finished system embedding the same sensor. For this, we will come back to the background of the measurement of food aromas. Gas sensors can only handle volatiles (gas phase). ! This is obvious, but it has a great consequence...

Food samples can be under liquid, solid or powder state (called condensed phase). The volatile (gases) must be extracted from the initial matrix. The efficiency of such an extraction is generally low, and related to many parameters. In addition to external parameters (temperature, equilibration time, volumes, pressure...) there is also important internal parameters (nature of the substrate, interactions between aroma compounds and the matrix or between compounds, oxidation...). The headspace is - on a quantitative point of view - generally poorly representative of the original product sample. In most cases, chemicals having a high volatility are over-represented, in opposite to low volatility compounds. In addition, hydrophilic compounds have a high partition coefficient in aqueous samples while hydrophobic compounds are retented in fat samples (cheese, milk, butter...). That is why the measures on milk samples are very correlated to the fat content [5]. Some authors have investigated this whereas other have ignored this limitation ! At this level, another problem occurs : the major compounds present in the HS have generally a low contribution to the hedonic appreciation while some others present as traces have a high sensory impact (for example the cork odor in wines). That explains why correlations between sensory analysis and instrumental data are so hard to obtain.

These problems are not specific to E-noses, we can find such problems in separative analysis, such as GC or GC/MS. Simply, E-noses add to this some distortion : mainly a poor efficiency and some degradation of the initial chemicals.

For the search of a better efficiency, we must come back to the systems functioning.

3. The basic principle of an Electronic Olfactometer

Before the characterisation of a sample using gas sensors, target or reference products were measured by using the same technique, and profiles were introduced to build the products database. Then, a classification is done by comparison between the sample and the library patterns. Because the sensors can handle only gas-phase samples, it is the vapour above the sample phase (static headspace), rather than the sample itself, that is analyzed. Thus the partial pressure of each component, as well as its molar concentration in the liquid or solid phase of the food sample, has an important effect on the quantitative results. Since the partial pressure is closely related to the temperature of the sample, this must be efficiently controlled.

The Partition Coefficient (or Partial Vapour Pressure in the case of a mixture) represents the ratio of the molar concentration in the vapour phase to that of the original products. It is related to : temperature, equilibration time, nature of the substrate, pressure... When all the parameters are kept constant, it is related to the nature and volatility of the compound. This coefficient may vary in a very large range (approximately from 0.01 to 5,000) [6].

As shown on *Figure 1*, the equilibrium of a static headspace for some pure chemicals (1 µl injected into a 100 ml vial) is quite different in terms of shape and quantitative response, and related to the volatility, expressed as Kovatz indexes (K. I.) on an apolar GC column.

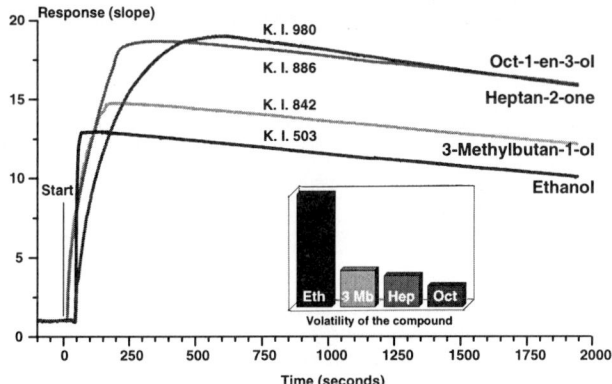

Figure 1 : Influence of the compound on the equilibration kinetics (pure compound)

3.1 About the functioning of an Electronic Olfactometer

Now, we will outline in general terms the stages involved in using an E. O. system in *Figure 2*.

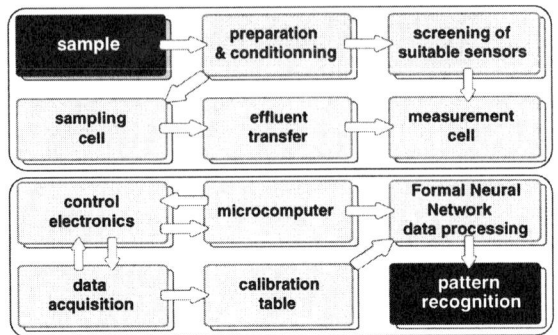

Figure 2 : Simplified synoptic diagram of an Electronic Olfactometer

In this paper, we will only consider the effluent way (upper part of the figure) :

4. Different ways of managing the effluent

Each E. O.-either commercial or prototype- has its proper way of managing the sample HS. *Table 1* summarises the methods which can be used.

TECHNIQUE	METHODE for TRANSFER	ADVANTAGES	DRAWBACKS
Fully Static	Effluent in permanent contact with sensors	Efficiency 100 % Equilibration kinetics	Sensors pollution Long conditioning time
Partial injection of the effluent	Partial injection HS using a gas syringe	Good kinetics Low pollution	Uneasy Dilution of the effluent No automatisation
Displacement of the effluent	*Head-space* displacement using a tight soft pouch	High efficiency Automated	Tight measurement cell Disposable pouch Uneasy, costly
Flow Transfer of the effluent	Effluent transferred into a flow	Low pollution Automated	Huge dead volumes High dilution
Pressurization of the sample	Pressurization of the sample cell using an inert gas	No sample oxidation Good kinetics Automated	Dilution of the HS Unusable with semiconductors problems with powder samples
Autosamplers	Injection into a flow using a gas syringe	Fully automated Continous operation	Repeatability of the HS Very high dilution

Table 1 : Comparison of the effluent managing in an Electronic Olfactometer

5. References

[1] Mielle P 1996 *Trends in Food Science & Technology*, Flavour Perception Special Issue **7**-12 432-438.
[2] Pâtissier B *et al International Meeting on Chemical Sensors,* 20 July 1996, Gaithersburg, USA.
[3] Reineccius G. *Critical Reviews In Food Science and Nutrition* **vol. 29** issue 6 381-402.
[4] Marquis F and Mielle P 1996 *Special Issue of Odours & VOC's Journal* June 96 40-46.
[5] Haugen J E 1998 *Seminars in Food Analysis* in print.
[6] Ettre L S *et al* 1993. *Chromatographia,* **35** 73-84.

Sensor Arrays and Multi-Sensor Systems
Paper presented at Eurosensors XII, 13–16 September 1998

Effect of non-conducting overlayers on the operation of an electronic nose based on conducting polymers

S M Reddy and P A Payne

Department of Instrumentation & Analytical Science, UMIST, PO Box 88, Manchester M60 1QD.

Abstract. This paper demonstrates the use of hydrophobic unplasticised poly(vinyl chloride) as an overlayer to minimise water interference at a pre-formed conducting polymer sensor array. As well as attenuating the signal to water, the response of the modified sensor to polar organic compounds is also reduced with the added effect of a switch in sign of the measured parameter. Photo-dehydrochlorination of the PVC overlayer has little consequent effect on sensor response.

1. Introduction

Electrically conducting organic polymers have been extensively exploited for their gas sensing abilities. The AromaScan 32 element gas sensor employs an array of conducting polymers based on polypyrrole and polyaniline derivatives. In the presence of a volatile organic compound, reversible adsorption results in a dynamic change in the direct current (dc) resistance of the conducting polymer. The applicability of the sensor is widespread including the detection of agricultural malodours [1] and being able to distinguish between fresh and rotting food [2]. Albeit an attractive means of rapid response sensing of volatiles, the sensing element is inherently prone to interference from water due to the charged nature of the polymer in the conducting state.

The aim of the experiment reported here is to reduce the effects of large variations in relative humidity on sensor performance using hydrophobic coatings. This could be accomplished in many ways including the use of lipids. In this study however, a hydrophobic PVC layer has been solvent cast over the conducting polymer array. The response of the modified sensor to a variety of volatiles ranging in polarity is presented.

Photochemical modification of the overlayer is also investigated as a means of introducing an undoped conjugated second layer in place of the original PVC according to the equation:

$$\left(CH_2 - CHCl\right)_n \xrightarrow{\ UV\ } \left(CH = CH\right)_n + HCl \tag{1}$$

2. Experiments

The pre-formed conducting polymer array was based on polypyrrole derivatives and supported on a ceramic substrate [3]. The sensing face was exposed to solvent headspaces of varying polarity including acetone, ethanol and water and the dc resistance at each element was measured.

2.1. Preparation of PVC overlayer

The array was placed on a horizontally flat surface in a Petri dish. A 3% (w/v) solution of high molecular weight PVC was prepared in tetrahydrofuran. One millilitre of the resulting solution was cast uniformly over sensor elements 6 - 26 of the conducting polymer array. The sensor was covered with a glass plate and the membrane allowed to dry over a period of 24 hours. The resulting film-covered array was exposed to the range of solvent vapours. The results were compared with those for the uncoated array.

2.2. Photo-dehydrochlorination of PVC overlayer

Under an argon/water atmosphere, the sensor surface was exposed to UV light at a wavelength of 254 nm. The temperature of the photolysis cell was maintained at 90°C [4]. Exposure over a period of 2 hours resulted in a tanning of the colourless PVC layer due to the formation of polyacetylene according to reaction (1). The modified sensor, when exposed to the solvent headspaces exhibited slower responses probably due to the formation of a more compact (dense) polyacetylene layer in place of the PVC which comprised the bulky chloro-substituent.

3. Data analysis

Figure 1 depicts the base resistance data for the clean array, upon PVC coating and after dehydrochlorination. The base resistances of all the sensors have increased considerably upon PVC coating. The percentage change at each coated sensor is however not consistent. This could be due to non-uniform PVC thicknesses across the array. The fact that adjacent sensors show such large variations disputes this idea. The effect could therefore be explained if the polymer is showing some preference to bind with certain conducting polymers depending upon the side group chemical functionalities of the variously substituted polypyrroles. Upon dehydrochlorination, base resistances have decreased (indicating an increase in the conductivity of the composite). This could possibly be due to inadvertent partial doping of the polyacetylene layer due to UV modified dopant leaching out from the underlying conducting polymer layer [4].

Figures 2(a) - (c) compare the change in sensor responses (dR) relative to the base resistance (R) of both the uncoated (1 - 5 and 27 - 32) and coated (6 - 26) elements when exposed to acetone, ethanol and water respectively. However, there is little effect on response magnitude after dehydrochlorination.

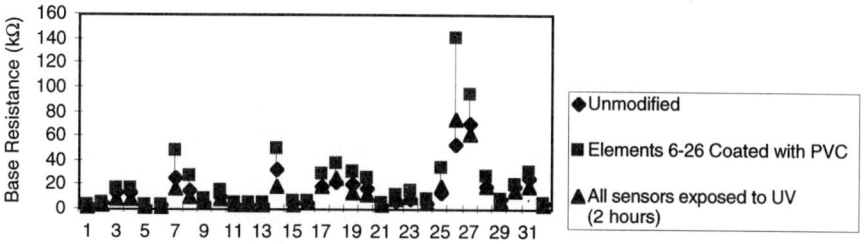

Figure 1 *Effect on sensor resistance upon non-conducting polymer deposition*

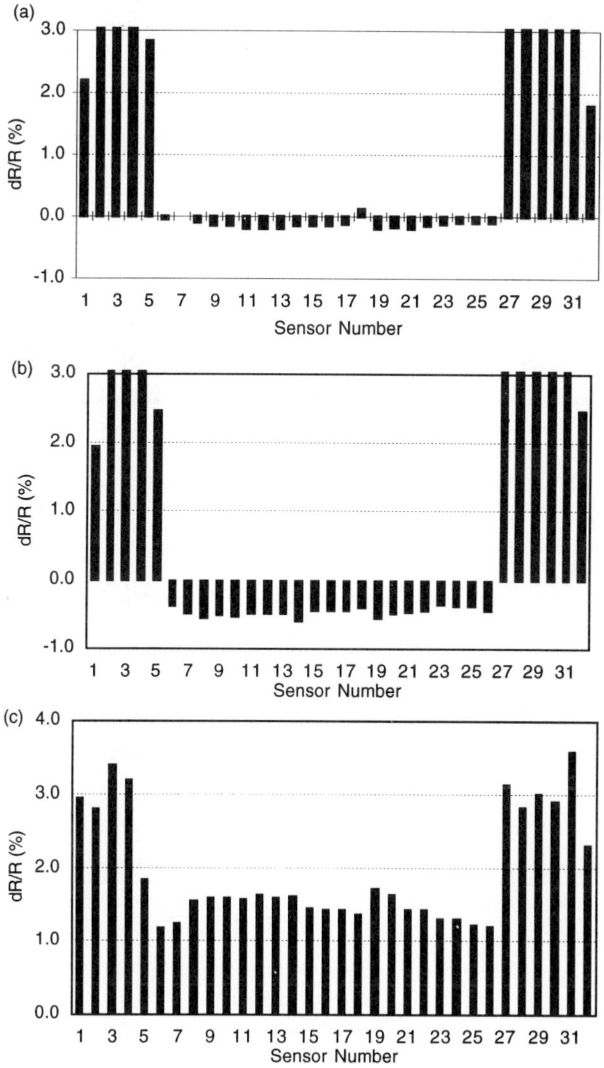

Figure 2 *Response of partially coated sensor (Elements 6 - 26) to:* (a) *acetone;* (b) *ethanol; and* (c) *water*

4. Discussion

The change in sign of response to acetone and ethanol and not water suggests that upon PVC coating, these solvents are altering the resistance of the conducting polymer via a different mechanism to that if the array was uncoated. It is likely that the interaction between solvent and PVC is in turn modulating the coupling between PVC and the underlying conducting polymer. In this respect, the positive resistance change due to water can be attributed to the hydrophobic effect causing the PVC polymer to couple more strongly with the subjacent conducting polymer sensor. Alternatively, the presence of pores in the PVC layer could be allowing a reduced amount of water reaching the underlying conducting polymer layer.

5. Conclusions

By using non-conducting/hydrophobic overlayers, it is possible to isolate a water response from responses to less polar solvents on the basis of the sign change in dc resistance of the underlying conducting polymer sensor.

Acknowledgements

The authors wish to acknowledge financial support from the UK Engineering and Physical Sciences Research Council, the Department of Trade and Industry, AromaScan plc and Terminix Peter Cox.

References

[1]   Misselbrook T H, Hobbs P J and Persaud K C 1997 *J. Agric. Eng. Res.* **66** 213-220

[2]   Persaud K C, Khaffaf S M, Payne J S, Pisanelli A M, Lee D and Byun H 1996 *Sensors and Actuators* B **35-36** 267-273

[3]   AromaScan plc, Crewe, Cheshire CW1 6AZ, UK

[4]   Ogura K, Kisaka K and Furukawa H 1995 *Journal* of *Polymer Science A* **33** 1375-1380

Sensor Arrays and Multi-Sensor Systems
Paper presented at Eurosensors XII, 13–16 September 1998
© *1998 IOP Publishing Ltd*

Lutetium bisphthalocyanine thin films as sensors for organic volatile components of aromas

M L Rodríguez-Méndez[1], J Souto[2], R de Saja[2], J Martínez[2] and J A de Saja[2]

[1] Department of Inorganic Chemistry. University of Valladolid. Pº del Cauce s/n. 47011 Valladolid. Spain.
[2] Department of Condensed Matter Physics. University of Valladolid. Prado de la Magdalena s/n. 47011 Valladolid. Spain.

Abstract. Lutetium Bisphthalocyanine (LuPc$_2$) Langmuir-Blodgett (LB) and evaporated films were investigated as sensitive materials for the detection of organic volatile compounds responsible for the aroma of food products such as olive oil or wine. The sensitive layers were deposited on ITO interdigitated electrodes in order to control the changes in their conductivity. The system was operated at room temperature and a variety of organic vapours with different chemical functionalities were tested. The results demonstrate the viability of the phthalocyanine thin films as the active species for systems specifically designed for the monitoring of aromatic components in food.

1. Introduction

In the last few years, the possible application of phthalocyanine (Pc) thin films as sensors for atmospheric gaseous pollutants has been extensively studied [1]. In these works, the response of thin films towards strong electron donor or electron acceptor compounds such as NOx or NH$_3$ has been evaluated by monitoring the changes in the spectroscopic or the electric properties of the films when exposed to the gases.

Nevertheless, the response of the thin films of Pc towards organic volatile molecules has not been so extensively studied. The main reason is that the exposure of a Pc film to a gas leads to variations in the electronic density in the molecular units of the film. This type of organic compounds generally are not so strong electron donors or acceptors and hence the changes are less intense when compared with those observed with the inorganic gases. Another important reason is that most of the work published in this field is based on thin films of metallated monophthalocyanines, which commonly show a poor reversibility.

Our group has been working for several years with lanthanide bisphthalocyanines, which are sandwich type complexes that can be prepared as thin films using different techniques [2, 3]. These present the advantageous property of a comparatively very low electric resistance. Moreover, these films are electrochromic and when exposed to oxidising or reducing gases, they undergo remarkable colour changes that can be followed visually.

The aim of this work is to quantify the response of LB and evaporated lutetium bisphthalocyanine ($LuPc_2$) films towards a number of organic vapours (esters, alcohols, acids, carbonyls) which are responsible of the aroma of olive oils, wines, etc. For this purpose, the changes in the resistance are measured, and the influence of different parameters such as film thickness and relative concentration of the active species in mixed films prepared with the LB technique using arachidic acid as a transfer promoter are evaluated.

2. Experimental

$LuPc_2$ was synthesised following a previously published procedure. A KSV 5000 Langmuir-Blodgett trough was used for the LB preparation. The phthalocyanine was dissolved in chloroform (5×10^{-4} M) and spread onto ultrapure water (Millipore MilliQ) which was kept at a constant temperature (20°C). The solutions were alternatively mixed with arachidic acid, which was used to reduce the rigidity of the Langmuir film and to facilitate the deposition onto the substrates. After the evaporation of the solvent, the floating film was compressed up to a surface pressure of 22 mN/m. The monolayers were deposited onto glass substrates covered with ITO interdigitated electrodes at a speed of 3 mm/min. Figure 1 shows a schematic diagram of the interdigitated electrode structure used in this work, which allows to prepare three sensors per batch. Films of 10 and 20 layers were built by Z type deposition with a transfer ratio close to 1.

Alternatively, films were prepared by sublimation in a RIAL JEP 300 high vacuum evaporator equipped with an EV1-8 electron beam evaporator from AP&T. The thicknesses of the films were monitored with a S-LDC deposition controller with a quartz crystal oscillator, also supplied by AP&T.

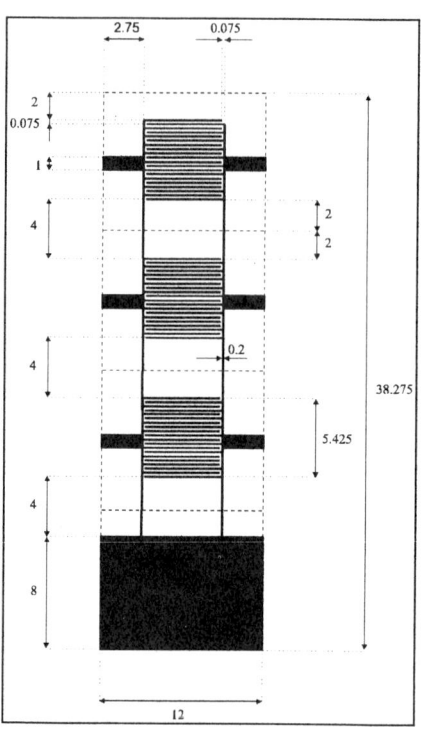

Figure1. Layout of the electrodes used in this work. All dimensions in mm.

The sensors thus prepared were mounted in a test box with a volume of approximately 300 cm^3. Up to ten sensors can be mounted at the same time in the reaction box. A detailed description of the vapour handling system and its operation has been previously reported [3]. A volume of 100 microlitres of the aroma (hexanal, n-butyl acetate, hexanol, or acetic acid) was deposited in a 100 ml flask. An air flow of 300 microlitres per minute was passed trough the chamber until stabilisation. Then, the system was switched and the air was passed trough the aroma and then conducted to the reaction chamber. A Keithley 224 Programmable Current Source was used to provide a direct current of 10^{-6} A to the sensors. The voltage drop across each one of the sensing units was monitored with a computer controlled Keithley 2000 multimeter with a Keithley 2000-SCAN scanner card. The data were collected using TestPoint™ software.

3. Results and discussion

Typical responses of sensors based on LB and sublimed films upon subsequent exposures to hexanal for one and a half minutes at room temperature are shown in Figure 2. As it can be observed, two subsequent exposures give rise to very similar signals for each of the films. Moreover, these are also identical when sensors corresponding to the same batch are compared. Nevertheless, it should be pointed out that the first two exposures give rise to more intense signals. After the third one, the responses are perfectly reproducible.

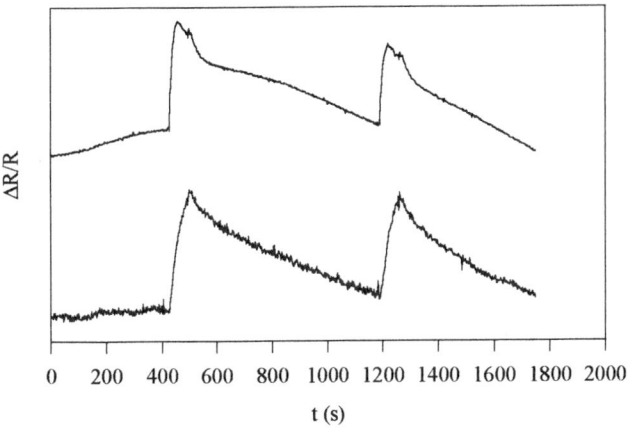

Figure 2. Response of the sensors to pulses of hexanal. The upper trace corresponds to a sublimed film and the lower one to a Langmuir-Blodgett film.

The increase in the resistance could be interpreted bearing in mind that phthalocyanines are p-type semiconductors. Hexanal would act as a weak electron donor, and its interaction with $LuPc_2$ would result in a slight reduction of the number of charge carriers in the organic film. As expected, the kinetics of the adsorption and desorption reactions are different for the LB and sublimed film based sensors. These differences could be attributed to the diverse morphology of the films and to the presence of arachidic acid in the LB films.

A slight drift in the baseline has been observed while monitoring the operation of the sensors. This type of behaviour has been previously reported [4] for other sensors based on phthalocyanine derivatives, and has been attributed to an incomplete recovery of the original properties of the films. This problem could be overcome by heating the samples. Even if strong oxidising materials such as NOx are tested, this proved to be effective for lanthanide bisphthalocyanine based sensors [5].

The exposure of the $LuPc_2$ films to n-butyl acetate also lead an increase in resistivity, which continued to rise until the vapour was turned off. The original conductivity was restored approximately 15 min after the gas flow had been turned off. The relative change of the resistance of the films is higher than that previously observed for hexanal. This shows that the reactivity of the acetate towards phthalocyanine is higher than that of hexanal. This reactivity is even stronger when acetic acid is analysed. This vapour gives rise to the most intense signals, which are depicted in Figure 3.

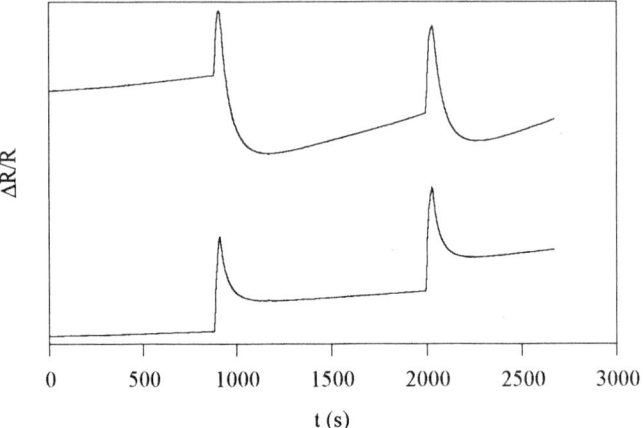

Figure 3. Response of a sublimed film (up) and LB film (down) to acetic acid.

4. Conclusions

LnPc$_2$ is an attractive material for the detection of organic volatile compounds due to its high sensitivity, repeatability and the variety of responses it produces. The sensitive layers are notably stable and have been used for more than one month without significant loses in sensitivity. Although some minor difficulties should be overcome, the application of these materials for the monitoring of the aromatic characteristics of products from the food industry is very promising.

Acknowledgements

We are gratefully acknowledged to CICYT (95-0150-OP and OLI96-2172) for financial support.

References

[1] Snow W and Barger W R 1989 in Leznoff C C and Lever A B P (eds.) *Phthalocyanines. Properties and Applications* (New York; VCH) and references cited therein

[2] Rodríguez-Méndez M L, Khoussed Y, Souto J, Sarabia J, Aroca R and de Saja J A 1994 *Sens. Actuators B* **18-19** 89-92

[3] Álvarez J, Souto J, Rodríguez-Méndez M L, de Saja J A *Sens. Actuators B* accepted

[4] Roisin P, Wright J D, Nolte R J M, Sielcken O E and Thorpe S C 1992 *J. Mater. Chem.* **2** 131-137

[5] Souto J, Rodríguez-Méndez M L, de Saja J A and Aroca R 1994 *Int. J. Electronics* **76** 763-769

Sensor Arrays and Multi-Sensor Systems
Paper presented at Eurosensors XII, 13–16 September 1998
© 1998 IOP Publishing Ltd

From the Laboratory to the Factory: The Use of Electronic Nose Technology in On-line Monitoring

T D Gibson

Biochemistry and Molecular Biology, University of Leeds, Leeds LS2 9JT, UK

O C Prosser, J Peace and J N Hulbert

Bloodhound Sensors Ltd, 175, Woodhouse Lane, Leeds LS2 3AR, UK

Abstract. To date the use of Electronic Nose technology has been restricted to batch sampling in Research, Development and Quality control laboratories. Complex operating procedures, long sample times and unreliable sensor performance are a few of the reasons that have prevented the technology being seen in on-line process monitoring. This paper reports on the progress Bloodhound Sensors has made towards delivering a practical solution to the problems encountered in production line situations.

1. Introduction

The Electronic Nose, based on sensor array technology, has been demonstrated to be capable of distinguishing and identifying complex odours in a number of different applications; micro-organisms [1], beers [2], meats [3] and coffees [4]. Most studies have been carried out under controlled laboratory conditions by highly skilled scientists and the data produced has often been subjected to a high level of processing, PCA, DFA and Neural Networking, in order to produce the desired discrimination. These studies demonstrate the potential power of the technology and have stirred the imagination of many non-scientists in a vast range of areas giving rise to an ever increasing flood of ideas for the use of the Electronic Nose; one of the more interesting possibilities being that of production-line monitoring for the quality control of drinks.

2. Materials and Methods

The Electronic Nose was developed in the Department of Biochemistry and Molecular Biology at Leeds University and has been commercialised by Bloodhound Sensors Ltd. The Instrument used to supply the data presented in this paper was the Bloodhound® BH114 Sensory Array System consisting of a fully integrated sampling system, sensor head and electronic address system controlled by proprietary Windows based software. Data handling and manipulation outside that of the specification of the control software was carried out in Microsoft Excel. The sensor array used consisted of 14 different conductive polymer sensors made up of materials such as polypyrrole, polyaniline and polythiophene. The BH114 operates such that there is a continuous flow of air, at 150ml min^{-1}, through activated carbon and over the sensors until such time as a sample is taken wherein a valve diverts the intake from clean air to the sample for a pre-defined period of time.

Results and Discussion

3.1. Repeatability

Underpinning the notion that the Electronic Nose can be used in real situations with unskilled operators is the assumption that the system is repeatable in its performance; this has not been a proven assumption and the subject of sensors drift is only one complication [5]. For this study the repeatability of the sensors was tested with 40 sequential samples of butanol, each of the measured parameters form the response profiles (figure 1) were monitored (figure 2).

Figure 1   Parameters measured for each sensor response;

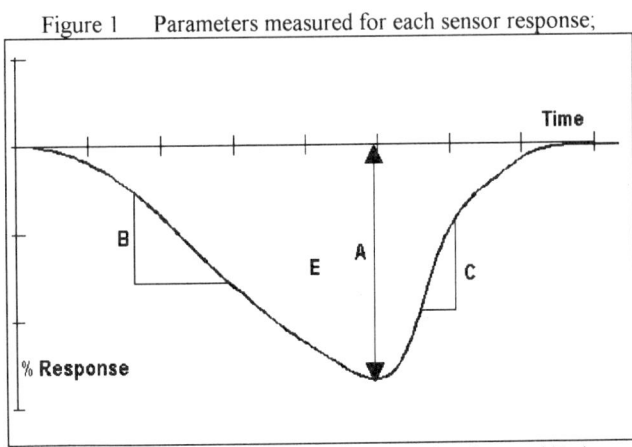

A=Divergence, B=Absorption slope, C=Desorption slope and E=Area.

Figure 2   Repeatability of sampling for 200 samples of Butanol

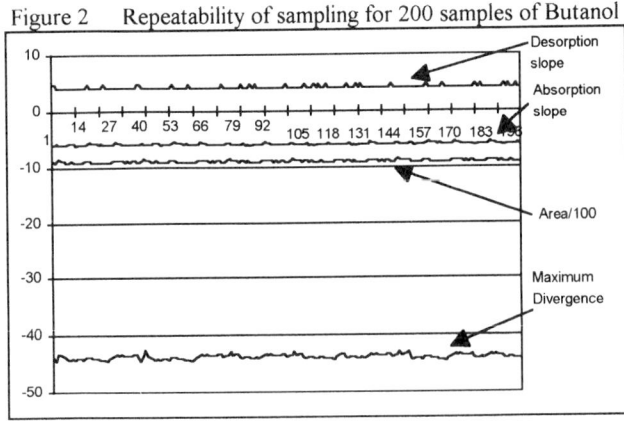

3.2. Principle Data Tracking

Five samples were used to provide an artificial ethanol production-line problem in which methanol was present in increasing proportions, 0%, 2%, 4%, 8% and 20%. Ten samples of the pure ethanol and then 5 samples of each of the remaining solutions were taken in continuous sequence with each sampling cycle taking 1 minute 40 seconds. The sampling cycle consisted of: distilled water control sample (15 seconds), delay (15 seconds), sample (15 seconds) and recovery (60 seconds).

The collected data was subjected to the usual Discriminant Function Analysis to produce clear separation between samples (figure 3). The data from the first Discriminant Function was then used to produce a production chart (figure 4).

Figure 3   Discriminant Analysis of ethanol; legend gives proportion of methanol

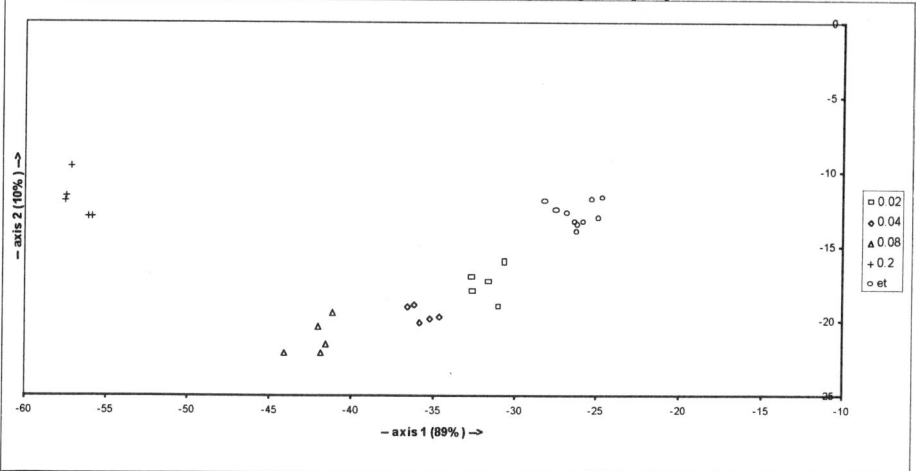

Figure 4   Tracking using First Discriminant Function

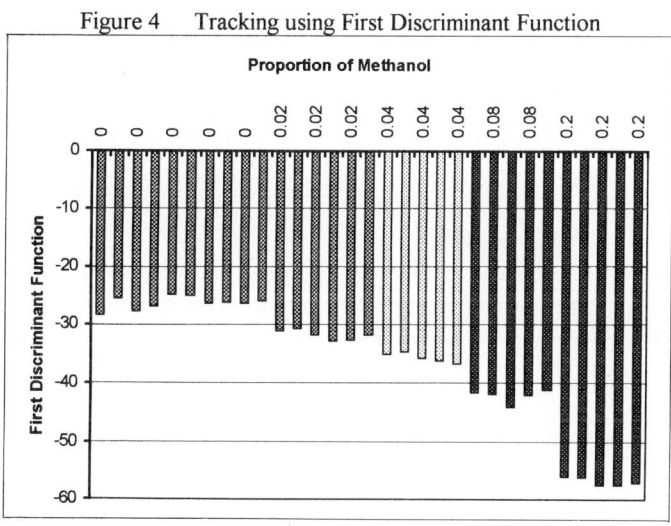

3.3. Algorithm For Data Tracking

Using the Values from the Discriminant Function Analysis provides a means of monitoring a production line but a simpler technique was shown to be feasible. The first step involves calculation of the standard deviation of the standard samples (n) which is used in a sensor performance evaluation and weighting step (equation 1). Step two weights the various parameters; a high standard deviation reduces the effect of the parameter on the final result and *vice versa* (equation 2). Then an overall value for each sample is generated, by a process of addition (equation 3). The average for the group of standard samples is taken in order to produce a reference baseline value (equation 4) and this reference value is compared against the values calculated in equation 3 to give D_b (equation 5) which is then plotted to produce a difference graph (figure 5).

$$S_a = \left[\sqrt{\frac{\sum\limits_{n}^{0} (x_n - \bar{x})}{n-1}} \right]_a^0 \qquad C_{ab} = \left[\left[x_{ab} / S_a \right]_b^0 \right]_a^0 \qquad S_b = \left[\sum\limits_{a}^{0} C_{ab} \right]_b^0 \qquad \bar{x}_n = \frac{\sum\limits_{b}^{0} S_b}{n}$$

Equation 1   Equation 2   Equation 3   Equation 4

$$D_b = \left[S_b - \bar{x}_n \right]_b^0$$

Equation 5

Figure 5   Tracking using custom algorithm

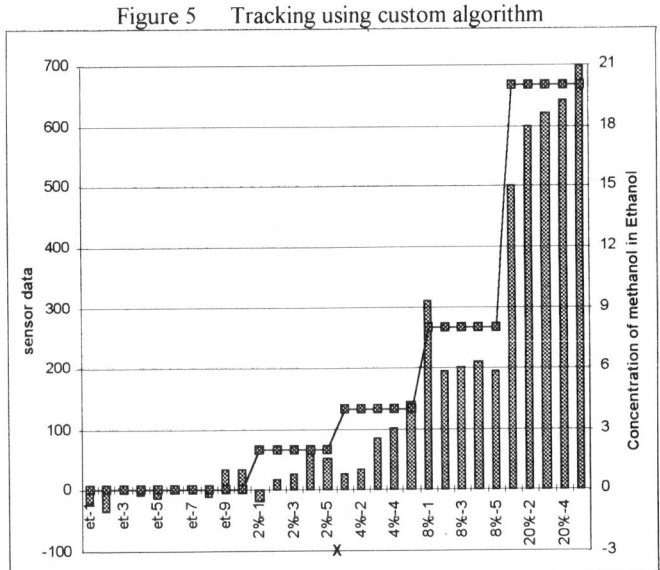

Conclusion

Simple processing combined with repeatability and fast sampling could provide a useful on-line device.

References

[1] Gibson T D, Prosser O P, Hulbert J N, Marshall R W, Corcoran P, Lowery P, Ruck-Keene E A and Heron S 1997 *Sensors and Actuators B* **44** 413-422

[2] Pearce T W, Gardner J W, Friel S, Bartlett P and Blair N 1993 *Analyst* **118** 371-377

[3] Bourrounet B, Talou T and Gaset A 1995 *Sensors and Actuators B* **26-27** 250-254

[4] Gardner J W, Shurmer H V and Tan T T 1992 *Sensors and Actuators B* **6** 71-75

[5] Holmberg M, Winquist F, Lundstrom I, Davide F, DiNatale C and Damico A 1996 *Sensors and Actuators B-Chemical* **36** 528-535

Sensor Arrays and Multi-Sensor Systems
Paper presented at Eurosensors XII, 13–16 September 1998
© 1998 IOP Publishing Ltd

An optimal information fusion framework for multi-sensor object recognition

E.R. van Dop, P.P.L. Regtien, M.J. Korsten
Laboratory for measurement and instrumentation
Faculty of electrical engineering, University of Twente
PO Box 217, 7500 AE Enschede, The Netherlands

1   Introduction

Object recognition is an important issue in computer vision and robotics, where sensory information is processed with the purpose to devide the scene in objects represented by numerical features that discriminate one object class from the other. In many cases one sensory system, which can be a camera, doesn't suffice for object recognition and multiple sensors are required to resolve the classification problem. With the introduction of multiple sensors, information integration or fusion becomes an issue. In this paper the most important frameworks for information fusion are compared for the case of object recognition. Furthermore, general aspects of designing a multi-sensor object recognition system are mentioned by outlining our system developed to recognize electronic components on printed circuit boards.

2   System layout

A general layout of a multi-sensor object recognition system is presented in figure 1. At the bottom of this layout are the physical sensors that measure properties of the scene. The raw data of the sensors is subsequently translated into object features using processes called segmentation and representation.

When the recovered object features are sufficiently discriminative, a classifier can hypothesize the presence of an object with enough evidence to consider the object recognized. Otherwise, new object features have to be determined to verify the hypothesis. When all objects are recognized, objects of interest can be extracted from the scene by an actuator. The system is supported by a set of databases containing information about objects that one wants to recognize and others that contain specifications of hardware and software used in the system.

A succesfull implementation of this layout starts with an appropriate choice of the physical sensors, which depends on the scene to be analysed. In our scene consisting of a set of electronic components, objects have different shapes and colours, are manufactured from different materials and some of them have inscriptions (in particular integrated circuits). Based on this scene analysis an implementation with colour, range and high-resolution imaging sensors has been chosen. Furthermore, an eddy-current sensor that acquires information about the conductivity of objects in the scene can be brought into action when the other features fail to provide sufficient evidence for the classification of an object in the scene.

3   Information fusion frameworks

Uncertainty is an inevitable by-product of sensory information processing systems. As a consequence of uncertainty, the performance of a sensory information processing system regarding its task, for example object recognition, will deteriorate. Reduction of uncertainty is therefore of major

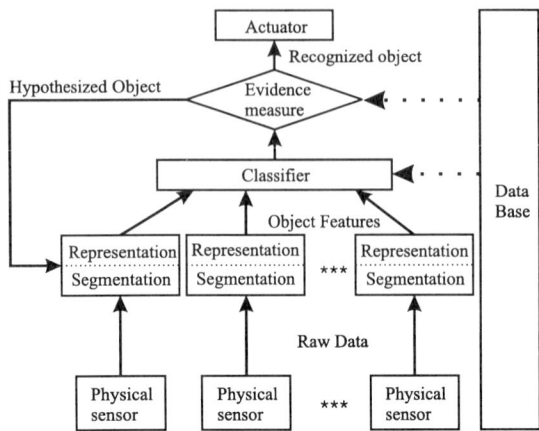

Figure 1: The structure of the multi-sensor object recognition system

interest and can either be pursued by improving the accuracy of the obtained data or by combining several sources of sensory information. Following the latter track, the sensor outputs should be converted to a common representation in order to enable information integration. Furthermore, a framework for reasoning should be selected suited to deal with the available information.

Pure Bayesian probability theory has provided the most influential method for reasoning with uncertainty from multiple sources of information. It represents information by a probability distribution and Bayes' conditioning rule serves as the primary tool for information updating. Although the axioms behind this framework are largely acknowledged, application of the pure (or strict) Bayesian framework in uncertainty management is challenged. Criticism refers to the requirement to specify a complete probabilistic model before reasoning can commence [4]. In other words, information involving ignorance due to incomplete knowledge cannot be represented [6].

To resolve this issue several alternatives have been suggested. First of all, approximate methods are used to complete the probabilistic model. These include the introduction of assumptions like the principle of indifference (for the estimation of prior probabilities) and parameter estimation techniques. Instead of completing probabilistic knowledge, information updating in the presence of incomplete knowledge can be dealt with by adapting or discarding the axioms of probability. This approach resulted in the computation of upper and lower bounds to the probability in the so-called convex Bayesian framework [3, 1] and the invention of the Dempster-Shafer theory of evidence [2, 5]. Still other formalisms, like the certainty factor model, have been developed for certain applications of uncertainty management but their axiomatic foundations are weak as is their theoretical justification. For this reason these methods are not described in this article. Basically non-numerical techniques like fuzzy logic and nonmonotonic logic are not discussed either, since sensory information processing systems generally do provide numerical answers.

In order to compare strict probability theory, convex probability theory and Dempster-Shafer's theory of evidence one has to answer the following questions:

1. What do you want to calculate? (A matter of semantics)

2. What knowledge do you have?

3. What do you need the result for? (The task)

Unfortunately, the answers to these questions will seldom favor one reasoning framework clearly over the other. Most problems need to be restructured to fit the different reasoning frameworks, which often neccecitates the introduction of assumptions and approximations or the acceptance of interval-valued results (which are less expressive than point-values). Choosing a reasoning framework thus becomes a scientific judgement of minimizing the negative consequences of assumptions, approximations and losses of expressiveness.

4   An optimal information integration framework for object recognition

Information fusion for object recognition can be considered as a mapping of a vector of object features to an object identity. In probability theory, this mapping consists of the calculation of probability density functions of the features given the presence of a certain object. Next, when object features are independent, the combined probability density function of the features given an object can be found by multiplication. Analogously, when multiple (independent) sensors are applied, multiplication of the probability density functions of the individual sensors will result in the overall probability density function of the set of features given the object. When this procedure is followed for every object that could possibly be encountered in the scene (the sample space) and when the relative frequencies of occurence of these objects (the prior probabilities) are known, a full probabilistic model is specified.

In this special case few researchers question the superiority of the strict probability approach to information integration. Unfortunately many difficulties obstruct the elaboration of this scheme:

- Independence of object features, and perhaps even sensors, is often not true

- Probability densities of features are only estimates due to approximations in the image formation model

- Defining a complete sample space is often impossible (due to unexpected objects)

- Prior probabilities are seldomly known accurately

Introducing assumptions still enables the implementation of strict probability theory as a tool for reasoning, but the problems circumvented by applying another framework (at the cost of expressiveness of the result) challenge its optimality.

A proper evaluation of the results requires a definition of the task of object recognition: correct classification of as much objects in the scene as possible. From this definition, object recognition can be considered as a decision making problem. A decision pointing to a set of objects is not acceptable in most cases where actions depend on the identity of the recognized object (for example in assembly robots). For these cases of unique identity declarations, the maximum posterior probability (MAP) criterium finds an optimal decision.

Considering Dempster-Shafer's theory of evidence in the light of the above, difficulties arise:

- Belief-plausibiliy intervals that are obtained by this framework have a different meaning than posterior probabilities, so how can the answers provide an optimal decision?

- Even when we find an optimal decision criterium based on belief-plausibility intervals, will it find unique solutions for overlapping or nested intervals?

- If plausibilities are derived from likelihoods [5], assuming conditional independence of features and approximating probability density functions is still required to implement the evidence theory.

- The same goes for lack of knowledge about the sample space: unexpected objects have to be distinguished from modeled objects by introducing the class of unknown objects.

The only problems circumvented by the evidence theory are the need to specify prior probabilities and the frequently hailed property of allowing belief measures for subsets of objects during information updating, instead of a specification of beliefs for all individual objects in the sample space. Since only the information updating rule of convex probability theory and evidence theory differ, the properties of information fusion with evidence theory also apply to the fusion with convex probability theory:

- No prior probabilities need to be specified

- Specification of probabilities to subsets of objects are acceptable

- The class 'unknown object' has to be added to provide a complete sample space

- Approximations and assumtions in likelihoods are still required to evaluaty probability bounds

One very important merit of convex probability theory over evidence theory is its semantics. Probability intervals are calculated, so the maximum posterior probability criterium can be applied. However, another important question remains: How to perform decision making with nested probability intervals? The answer to this question is simple: one has to map the intervals to point values which can be compared easily. One such mapping will be reached by specifying probabilities for all objects individually together with prior probabilities. But in this case one actually applies strict probability theory.

In other words: allowing probability intervals thus shifts, and doesn't resolve, the problem of object recognition from the estimation of prior probabilities and estimation of likelihoods for all individual objects to the handling of overlapping and nested intervals.

5   Conclusions

Application of strict probability theory in information integration requires complete knowledge about sensor models (giving the likelihood $P(E|H)$) and prior probabilities of the possible propositions. Although multiple approximations and assumptions are required for this full specification of the probabilistic model, the approach is still advantageous in object recognition problems over other formalisms. The convex probability theory and the evidence theory need less knowledge during information integration, but the need for assumptions and approximations is only shifted to the decision making stage where lack of expressiveness of the integrated information causes ambiguities.

Optimizing information integration within the strict probabilistic framework asks for a scientific approach to complete the probabilistic model. Those include the training of prior probabilities, careful handling of likelihood approximations close to zero and ensuring a closed world solution by adding the unknown object to the sample space. For a multi-sensor object recognition system, in our case developed to recognize objects on printed circuit boards, this can only be realized by obtaining accurate models of the sensors and the way their data is translated to object features. In other words: optimal information fusion relies heavily on knowledge about the way sensor data is obtained and processed. Designing a multi-sensor object recognition system thus demands an approach that considers the coherence between the various elements of the system under consideration.

References

[1] L. Chrisman, 'Incremental conditioning of lower and upper probabilities', *International journal of approximate reasoning*, 13, 1995, 1-25.

[2] A.P. Dempster, 'Upper and lower probabilities induced by a multivalued mapping', *Annals of Mathematical statistics*, 38, pp. 325-339, 1967.

[3] J.Y. Halpern, R. Fagin, 'Two views of belief: belief as generalized probability and belief as evidence', *Artificial intelligence*, 54, 1992, 275-317.

[4] J. Pearl, *Probabilistic reasoning in intelligent systems: Networks of plausible inference*, Morgan Kaufmann, San Mateo, 1987

[5] G. Shafer, *A mathematical theory of evidence*, Princeton University press, Princeton, 1976.

[6] F. Voorbraak, 'Reasoning with uncertainty in AI', *Lecture notes in Artificial intelligence 1093*, ed. L. Dorst, M. van Lambalgen, F. Voorbraak, 1996, Berlin, Springer-Verlag.

Sensor Arrays and Multi-Sensor Systems
Paper presented at Eurosensors XII, 13–16 September 1998
© *1998 IOP Publishing Ltd*

The approaches to design of multifunctional sensor system for determination of water environment parameters

V. A. Golembo, A. U. Botchkariov

Global Ocean Research Lab, State University "Lviv Polytechnic", Bandery str., 12, 290646, Lviv-13, Ukraine

Abstract. The new approaches to design of intelligent multifunctional sensor system to determine a set of water environment parameters are considered. The system involves three temperature transducers and dot source of programmed temperature disturbances. The distinctive features of the system are multifunctionality, simplicity of constructive implementation and high reliability of functioning.

1. Introduction

Today the global ocean researches are characterized by larger scale, higher precision of measurements and larger number of water environment parameters being measured. One of the most important ways to reduce constructive complexity and cost of measuring devices comprising autonomous buoys is to measure several water environment parameters by one measuring device. The example of such device is the multifunctional sensor system (MSS) offered in the paper.

2. Multifunctional sensor system

2.1. *Structure and capabilities of sensor system*

Considered MSS consists of three identical temperature transducers (TT) and a dot source of programmed temperature disturbances (DSPTD). The transducers are located in the corners of regular triangle and DSPTD is located in geometrical centre of this triangle (fig.1). MSS also contains three converters of measuring signals, a specialized computing device and a source of power supply. The parameters measured by means of MSS are shown in table.

As temperature transducers the piesoelectric temperature transducers are used because of their high precision and large potential of improvement of metrological and operational characteristics [2]. As DSPTD the gas-discharge flashing lamp or other pulse heat source is used in order to provide short radiation of high-strength pulse of temperature in small volume.

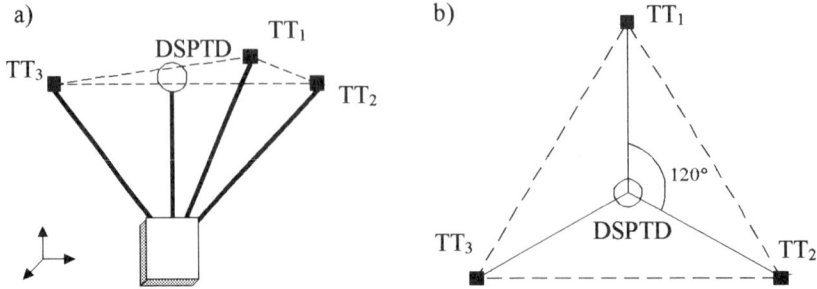

Fig.1. Multifunctional sensor system for determination of water environment parameters: a) general view, b) upper view.

Table.

N°	Water environment parameter being measured
1	dot value of temperature (by indications of any TT)
2	factor of thermoconductivity (on basis of which the preliminary conclusion about concentration of dispersed particles is drawn [1])
3	direction of natural thermal flow*
4	value of natural thermal flow*
5	character of microstructural transference*
6	speed of current*
7	direction of current*

*horizontal or vertical component depending on spatial orientation of MSS.

2.2. Operating modes of sensor system

When measuring water environment parameters listed in item 2.1, three different tasks, which correspond to three operating modes of MSS, are solved:

1. Temperature measurement (parameter 1) realized directly by one of TT. The temperature measurement is characterized by better reliability since in this case the triple reservation takes place.
2. Measurement of natural thermal flow (parameters 2,3,4). To determine these parameters the indications of TT are taken and the problem of stationary heat transfer in semi-infinite area is solved [3, 4].
3. Measurement of water environment transference (parameters 2,5,6,7). For determination of these parameters DSPTD and TT indications are used and the problem of water environment thermosounding (in condition of programmed heating) is solved [3, 4].

The third operating mode is the most complicated and worth of greater interest. In this work we shall consider it briefly.

3. Measurement of water environment transference

The third operating mode has the following stages:

1. DSPTD radiates the high-strength pulse of temperature.
2. After certain time t passed, the temperature values from TT_1, TT_2, TT_3 are measured.
3. Obtained temperature values are sorted in descending order and processed according to the algorithm considered below. The output data of algorithm are parameters 2, 5, 6, 7 (table).

To increase the accuracy of measuring and to adapt MSS for measuring water transference of various intensities, duration of radiation of high-strength pulse of temperature and time t can be adjusted in wide range.

The principle opportunity to determine parameters of water environment transference by means of MSS is shown on a simplified mathematical model of measuring process. According to this model, the duration of radiation of high-strength temperature pulse is negligibly small. After the source radiated, some volume of environment called thermal "cloud" is heated up during the time t due to heat exchange. Expanding thermal "cloud" is carried away by current in some direction. Because of displacement of thermal "cloud" and its expansion, temperature transducers TT_1, TT_2, TT_3 (fig.1) will show different values of temperature T_1, T_2, T_3 (fig.2).

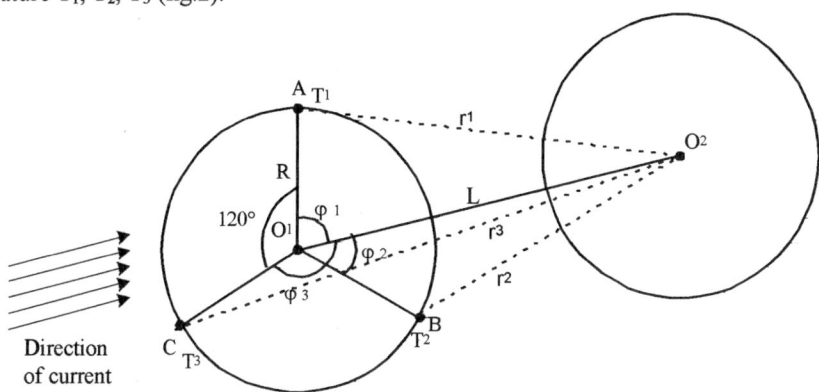

Fig.2. Scheme of measuring process.

Let's assume the following designations: $AO_2=r_1$; $BO_2=r_2$; $CO_2=r_3$; $AO_1=CO_1=BO_1=R$; $O_1O_2=L$. Then the measuring process can be described by the following system of equations

$$\begin{cases} T_1 = f(T_0, t, a, r_1) \\ T_2 = f(T_0, t, a, r_2), \\ T_3 = f(T_0, t, a, r_3) \end{cases} \qquad (1)$$

where t - time from the moment of radiation to the moment of measurement of temperature
        values in points A, B, C (fig. 2);
    T_0 - initial temperature value, to which the point O_2 was heated;
    a - thermoconductivity factor.

The function f describes the heat distribution in conventionally infinite volume from the point heated up to temperature T_0. Thus, function f should be deducted from equation of thermoconductivity for the case of non-stationary heat exchange (equation of non-stationary temperature field) that looks as follows

$$a\nabla^2 T = \frac{\partial T}{\partial t}. \qquad (2)$$

In case of infinite uniform environment with constant temperature and in assumption that initial temperature T (x, y, z) is set in whole space, solution to this equation can be written in closed form

$$T(x,y,z,t) = \frac{1}{(2 \cdot \sqrt{\pi \cdot t \cdot a})^3} \int\int\int_{-\infty}^{+\infty} T_0(x',y',z')e^{-\frac{\Omega^2}{4 \cdot a \cdot t}} dx' dy' dz', \qquad (3)$$

$$\Omega^2 = (x-x')^2 + (y-y')^2 + (z-z')^2.$$

For the case of dot source of initial temperature T_0 the expression (3) is transformed to

$$T(x,y,z,t) = \frac{T_0 \cdot \Delta V}{\left(2 \cdot \sqrt{\pi \cdot t \cdot a}\right)^3} \cdot e^{-\frac{\Omega^2}{4 \cdot a \cdot t}} \qquad \Omega^2 = x^2 + y^2 + z^2, \tag{4}$$

where x, y, z - linear spatial coordinates,

t - time, passed after radiation of temperature pulse,

ΔV - volume of dot source of heating (relatively small),

T, T_0 - relative values of temperature.

Taking into account the expression (4), as well as geometrical dependencies in triangles ΔAO_2O_1, ΔCO_2O_1, ΔBO_2O_1 the system of three equations with three unknown magnitudes is obtained

$$\begin{cases} T_1 = \dfrac{T_0 \cdot \Delta V}{\left(2 \cdot \sqrt{\pi \cdot t \cdot a}\right)^3} \cdot e^{\frac{-R^2 - L^2 + R \cdot L \cdot \cos(\varphi_1)}{4 \cdot a \cdot t}} \\[3mm] T_2 = \dfrac{T_0 \cdot \Delta V}{\left(2 \cdot \sqrt{\pi \cdot t \cdot a}\right)^3} \cdot e^{\frac{-R^2 - L^2 + R \cdot L \cdot \cos(\frac{2 \cdot \pi}{3} - \varphi_1)}{4 \cdot a \cdot t}} \\[3mm] T_3 = \dfrac{T_0 \cdot \Delta V}{\left(2 \cdot \sqrt{\pi \cdot t \cdot a}\right)^3} \cdot e^{\frac{-R^2 - L^2 + R \cdot L \cdot \cos(\frac{4 \cdot \pi}{3} - \varphi_1)}{4 \cdot a \cdot t}} \end{cases} \tag{5}$$

Having solved this system of equations, we find **a**, L, φ_1. The speed of transference of water environment is determined by formula

$$v = L/t. \tag{6}$$

The direction of transference is determined by using the value of angle φ_1 and indications of board navigating subsystem (for example, built-in magnetic compass).

The numerical simulation has shown that it is expedient to use the model considered above for determination of character of microstructural transference when the speeds of water motion are relatively small. If the speed and directions are determined for currents with relatively fast speed of water motion, more complicated model should be built on the solution to the problem of non-stationary heat transfer from sphere moving in viscous liquid [4].

4. Conclusions

The new approach to construction of intelligent MSS for determination of water environment parameters is proposed. This sensor system has the following advantages:
1. multifunctionality;
2. absence of mobile parts and simple constructive implementation;
3. reliable functioning;
4. minimal distortions of physical condition of studied water environment.

References

[1] V.Golembo, A. Botchkariov, A. Roudenko, E. Jorg 1997 *Automatic Determination of Marine Water State Characteristics in Ecological Monitoring System* (Proc. of 4-th International Technical Conference on Ocean Engineering and Marine Technology "Black Sea '97". Varna, Bulgaria) pp.33-35

[2] Golembo V., Botchkariov A. 1996 *Piezoquartz Transducers Parameters Improvement: Theoretical Approach and Practical Realisation Principles* (Proc. of the IEEE Instrumentation and Measurement Technology Conference, IMEKO Committee 7 (IMTC/96). Brussels, Belgium) pp.269-272

[3] Anderson D. et al. 1990 *Computing Hydromechanics and Heat Exchange, Vol. 2* (Moscow: Mir) p 384

[4] Babenko Y.I. 1986 *Heat and Mass Exchange: the Method of Calculation Thermal and Diffusion Flow* (Leningrad: Khimiya) p 144

DEVELOPMENT OF CHEMOMETRIC TOOLS FOR SIGNAL PROCESSING OF SENSOR ARRAYS

G. Barkó, J. Abonyi, J. Hlavay

University of Veszprém, Department of Earth and Environmental Sciences, Veszprém 8201 P.O.B. 158. HUNGARY

Abstract

A piezoelectric chemical sensor array was developed using four quartz crystals. Gas chromatographic stationary phases were used as sensing materials and the signals of the array was processed by different chemometric tools. The application of ANN method proved to be particularly advantageous if the measured property (mass, concentration, etc.) should not be connected exactly to the signal of the transducers of the piezoelectric sensor. After the teaching process the network was used for identification of taught analytes (acetone, benzene, chloroform, pentane). Mixtures of organic compounds were also analysed and fuzzy - clustering proved to be a reliable way to differentiate of the sensing materials and to identify the volatile compounds.

Experimental

In the current contribution, a chemical sensor array consists of four detectors is presented. AT-cut quartz crystals with 9 MHz fundamental frequencies were used. The crystals were arranged in an array and coated by gas chromatographic stationary phases like OV1 (Poly-dimethyl siloxane, SUPELCO), OV275 (Poly-cyanoakryl organosilane, SUPELCO), ASI50 (Poly-methyl-phenil siloxane, Applied Science Laboratories Inc.), and polyphenil-ether (Carlo Erba), respectively. The appropriate stationary phases were found by experimental and theoretical way based upon a principal component analysis. The thin film of the coating was formed by solvent evaporation. The coated surface was 0.2 cm^2 and the frequency of the quartz crystals decreased usually about 8 kHz. Nitrogen was used as carrier gas and 20 L/h mass flow was maintained by a GFM17 3½ digit flow controller. The nitrogen contained 30 ppm water vapor and it was dried by a CRS 202268 packed GC column to remove the traces of water. The analyte was injected by a syringe. A NAFION drying unit was set into the carrier line for declining the interference of the trace amount of moisture. Data handling card was built and a computer program was developed to measure the frequency changes. The computer program compared the measured frequency to that of the clock of the computer.

Results and discussion

Chemical sensor systems and detector arrays are widely applied for determination of organic vapors, gases and odors in the atmosphere. A piezoelectric quartz crystal can be used as a chemical sensor that is able to measure the chemical concentration reversibly. The requirements for a reliable piezoelectric chemical sensor are the sensitivity, selectivity and reversibility towards the analyte to be determined. However, to find an entirely selective material for only one analyte is almost unrealizable. On the other hand, application of the chemical sensor in the environmental analysis is hampered by the fact that several compounds are in the matrix and generally only one has to be detected. Mathematical algorithms have to be developed for the data processing, and chemometric methods are necessary to explain the behavior of non-selective chemical sensors. Principal component analysis (PCA) [1], pattern recognition (PARC) [2], artificial neural network (ANN) [3] and Fuzzy Clustering methods were applied to perform the data evaluation. The artificial neural network based upon the biological neuron model approach and the fuzzy clustering was proved to be the best for classification of the volatile organic compounds.

References

[1] Barkó G and Hlavay J 1998 *Anal. Chim. Acta*, accepted for publication

[2] Barkó G, Papp B and Hlavay J 1995 *Talanta*, **42**. 475.

[3] Barkó G and Hlavay J 1997 **44**. 2237.

Acknowledgements
The authors wish to express their gratitude to the SOROS Foundation for the financial support.

Sensor Arrays and Multi-Sensor Systems
Paper presented at Eurosensors XII, 13–16 September 1998
© 1998 IOP Publishing Ltd

Self Organizing Map analysis of the selectivity properties of QMB sensors coated by porphyrin films

C. Di Natale, R. Paolesse*, A. Macagnano, A. Mantini, A. D'Amico

Department of Electronic Engineering *Dept. of Chemical Science and Technology
University of Rome "Tor Vergata", via di Tor Vergata, 00133 Roma; Italy

Abstract. In recent years metalloporphyrins and their derivatives have been proposed as sensitive layers for mass variation based transducers for the detection of volatile compounds. Despite the positive result obtained so far a systematic study of the kinds of interactions ruling the sensing mechanism has not yet been conducted. This paper reports on a study oriented towards a deeper comprehension of such mechanisms and it introduces a self organizing map based methodology for the analysis of selectivities.

1. Introduction

Metalloporphyrins and related compounds have been introduced as sensing materials in mass transducers for volatile organic compounds detection [1, 2]. They have shown large sensitivities and wide selectivities which are useful features, particularly appealing, for electronic nose applications [3]. Wide selectivities can be related to weak interactions (such as Van der Waals force and hydrogen bonding), but the presence of metal complexes are supposed to induce also mechanisms of coordination of analytes. Past studies [1, 2] shown a certain tendency of QMB sensors coated by these materials to behave according to coordination interactions. On the other hand other researchers in the past [6] have investigated the selectivity of quartzes coated with metallo-phtalocyanines, a class of macrocycles related to porphyrins:, they concluded, using the Linear Solvation Energy Relationship [4] approach, that no evidences of coordination mechanism could be detected and ruled out any role of the coordinated metal in the definition of the sensor selectivities.

In this paper a different analysis of the sensor-analyte interaction is proposed using a data analysis methodology utilized for the electronic nose. In particular the self organizing neural network (SOM) is employed for a deep evaluation of sensors performances. SOM is utilized according to a methodology outlined for electronic nose data analysis and modeling [5]. The analysis takes into consideration a set of QMB sensors coated with various Langmuir Blodgett films of metalloporphyrins. Sensor responses have been measured for different volatile compounds, at different concentrations. According to the chosen methodology sensors have operated in an array configuration. Collective performances have been evaluated and the contribution of each sensors have been then successively extracted from the array. The results indicate the contemporaneous presence of weak interactions and coordination mechanism both cooperating in the definition of the sensors performances.

2. Experimental

Five tetrapyrrolic macrocycles were synthesized and deposited according to the Langmuir-Blodgett techniques. A porphyrin molecule can be considered as composed by three parts: the macrocycle, the metal complexed at the center of the molecule and the eventual lateral substituent around the macrocycle. The used compounds compositions are listed in table 1.

Langmuir-Blodgett films have been deposited onto AT-cut quartzes having a fundamental frequency of 20 MHz. Sensor responses, at different concentrations, have been measured for a number of different volatile compounds chosen as representatives of the following classes: alkanes, aldehydes, alcohols, aromatics and amines.

	macrocycle			lateral group	metal		
	TPP	FP	EMC	HEO	Mn	Co	Cu
H$_2$T(HEO)PP	*			*			
Co- H$_2$T(HEO)PP	*			*		*	
Mn - H$_2$T(HEO)PP	*			*	*		
Mn - EMC			*		*		
Cu - FP		*					*

Table 1: composition of the five selective layers considered in the paper

3. Data Analysis Methodology

Pattern recognition techniques can be applied considering the sensors as elements of an array and studying the interactions between volatile compounds and the array of sensors. In this way there is the possibility to investigate the two main features of the phenomenon: the capability of the array to distinguish each single compounds from the others and to evaluate the contribution that each sensor gives to the whole array; this second feature gives important insights to understand the nature of the interactions occurring at coating-volatile interface.

Self Organizing Map (SOM) is one of the most important neural models. It belongs to the category of competitive learning methods and it is based on unsupervised learning. A comprehensive introduction to the basic principles and many examples of applications can be found in the monograph of T. Kohonen [5]. In the present case SOM gives the opportunity to evaluate the behavior of each sensor by studying the components of the codebook vectors. It is possible to represent graphically this information onto the SOM plane by a 3D surface. This gives the possibility of a visual inspection of the influence of each sensor on the whole data domain.

Before the application of the data analysis method it is necessary to normalize the data in order to eliminate, as much as possible, the influence of the analytes concentration; this because of the interest in studying only the *affinity* between sensor coating and analytes.

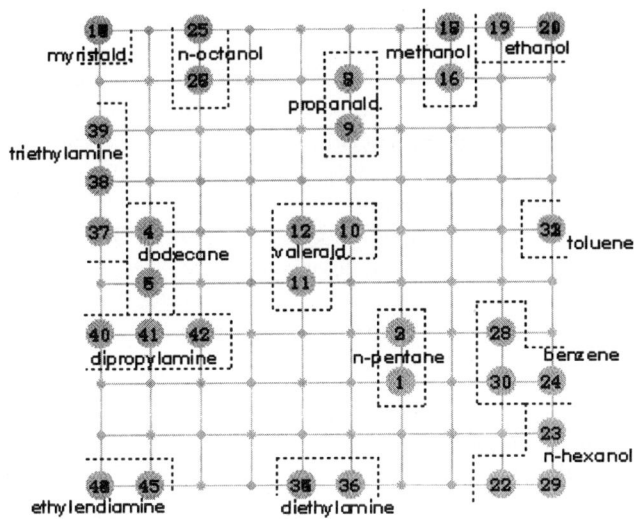

Figure 1: Distribution of the sensors data related to each analyzed compound onto the SOM grid

4. Self Organizing Map Analysis Results

A 10*10 neurons SOM has been trained with the normalized data set. Before the training data have been scaled to zero-mean and unitary variance. Figure 1 shows the arrangement of data onto the SOM grid. A net separation of volatile compounds is obtained showing that the array of five sensors is able to distinguish all the compounds. At the center of the plot those compounds for which are supposed to weak interact with metalloporphyrins are found, while those species for which a coordination is expected tend to lie at the edges of the plot: amines lie leftward and short-chain alcohols lie on the right. This ordering of compounds according is a clue of the influence of the metal part of metallopoprhyrins in defining the performances of the sensors. A more detailed analysis can be obtained studying the component planes of the codebook vectors of the SOM neurons.

Figure 2 shows the component planes related to each porphyrin. Three different behaviors can be recognized. The surfaces related to $H_2T(HEO)PP$ and $Co[T(HEO)PP]$ have basically the same shape, higher influence in amines detection and low influence in short-chain alcohols detection, the surfaces related to $Mn[T(HEO)PP]Cl$ and $Mn(EMC)$ behave in the opposite direction (large response to alcohols and low response to amines, while $Cu(FP)$ has a completely different shape orthogonal to the others surfaces.

From these plots the following conclusions can be drawn:
1. the metal drastically influence the behaviour of the sensor;
2. the addiction of cobalt to porphyrin does not change the sensing properties with respect to the free base macrocycle;
3. manganese influence is stronger than the difference induced by the two different macrocycles {T(HEO)PP and EMC};
4. Cu(FP) shows completely different interactions with the volatile compounds.

5. Conclusions

Pattern recognition has revealed, once again, to be a powerful investigation tool to explore data sets and to infer general behaviour in absence of a certain theoretical model. Among the possible pattern recognition methods Self Organizing Map shows to be suitable to explore multidimensional data, providing both a classification capacity and the ability for a detailed analysis about the behaviour of each sensor.

In this paper these methods have been applied to a set of QMB sensors coated with various metalloporphyrins. Results put in evidence the wealth of possibilities that porphyrins offer to assemble complexes, changing the macrocycle, the metal and the peripheral substituents, in order to get different sensors characteristics.

References
1. Di Natale C.; Macagnano A.; Davide F.; D'Amico A.; Paolesse R.; Boschi T.; Faccio M.; Ferri G.; *Sensors and Actuators B*; 1997, 44, 521
2. Brunink J.; Di Natale C.; Davide F.; Bungaro F.; D'Amico A.; Paolesse R.; Boschi T.; Faccio M.; Ferri G.; *Analytica Chimica Acta*; 1996, 325, 53
3. Schierbaum, K. D.; Zhou, R.; Knecht, S.; Dieing, R.; Hanack, M.; Göpel W. *Sensors and Actuators B* 1995, 24, 69
4. Di Natale C.; Macagnano A.; D'Amico A.; Davide F.; *Meas. Sci. and Tech.*; 1997, 8, 1236
5. Kohonen, T.; *Self Organizing Map*, Springer Verlag, Berlin (Germany), 1995

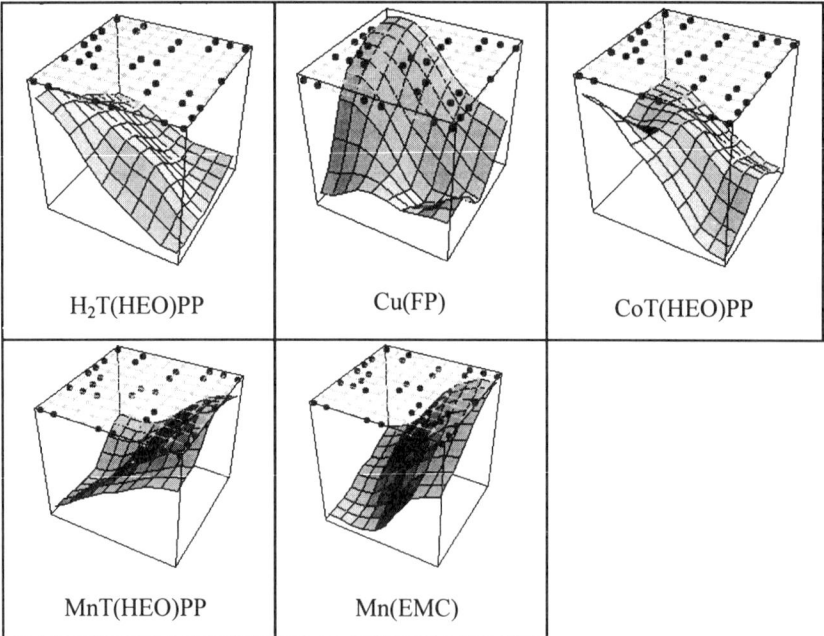

Figure 2: The contribution of each sensor to the whole array is represented by surfaces over the SOM plane. SOM plane has the same orientation of figure 1, so that, as an example, Mn(EMC) surface has its maximum in correspondance of that portion of SOM representing Ethanol data.

THERMOPILE INFRARED SENSOR ARRAYS FOR DETECTION OF POSITION, PRESENCE AND DIRECTION OF MOVEMENT

J. Schieferdecker[1], M. Simon[1], K. Storck[1], R. Jähne[2]

1): EG&G HEIMANN Optoelectronics GmbH, Weher Köppel 3, D-65199 Wiesbaden,
PO Box 3007, D-65020 Wiesbaden, Tel. (++49 611) 492 303, Fax (++49 611) 492 228
2) Fraunhofer Gesellschaft, Institut Mikroelektronische Schaltungen und Systeme IMS2
D-01109 Dresden, Grenzstr. 28 Tel. (++49351) 8823 154 Fax (++49 351) 8823 266

Abstract

Thermopile single element IR-sensors are commonly used as radiation sensors. For some applications like detection of position, size and direction of movement however, the measurement of a single spot is not sufficient. This paper describes the realization of thermopile 3x5 element sensor arrays for low cost applications. Thermopile IR-sensor arrays can be used for several security applications (e.g. person detection in rooms or out-of-position surveillance of the passenger seat in automotive applications, hot spot detection and temperature measurement). Measurement results on sensitivity and detectivity, optical and thermal resolution (with a suitable IR-optics) and crosstalk between neighboring elements will be presented.

Introduction

Heimann is manufacturing single and dual element IR-thermopile sensors with thermocouples made of n-doped poly-silicon and aluminum in a CMOS compatible technology. A membrane is prepared on a silicon-wafer by de-position of a thin SiO_2 / Si_3N_4 layer and anisotropic backside etching of the silicon substrate. The hot junctions of the thermocouples are formed on this thermally insulating membrane and the cold junctions are placed on the rim of the structure which acts as a heat sink. The absorber, which converts the IR-radiation into thermal energy, defines the sensitive area /1,2/. Miniaturized sensormodules including a single element thermopile sensor, mirror optics and an ambient temperature compensation are described in /3/. These single element thermopile sensors were transferred to mass production during the last years. They found a satisfying diversity of applications.

But the detection of position, presence, direction of movement or counting of slowly moving human beings or tools is impossible with only one sensor element. However, large sensor arrays are too expensive for mass production applications. The solution can be a thermopile linear- or two-dimensional sensor array with only a few sensor elements. As a first approach, thermopile linear arrays with 8 elements were described in /4,5/. This paper introduces 3x5 element 2D arrays in a similar technology.

Sensor design and Fabrication

Single element sensors and array infrared sensors have some different criteria for figures of merit and optimization of sensor performance. Single element sensors are optimized towards high sensitivity, high detectivity and good linearity. Sensor arrays have additional optimization parameters, e.g. a homogenous sensitivity, high fill factor and small thermal and electrical crosstalk. An efficient signal processing close to the sensor chip becomes very important. Table 1 gives some of the design parameters for our 3x5 element array sensor:

substrate	(100) silicon
chip size	3,5 mm x 3,7 mm
number of elements	15
element pitch	500 μm (vertical) x 1100 μm (horiz.)
element size (absorber)	375 μm (vertical) x 425 μm (horiz.)

Table 1: Design parameters for the array sensor

The overall size of the sensor chip is kept small to allow its integration together with the ASIC-Chip in a small TO5-type housing.

Each thermopile element consists of 26 thermocouples made of n-doped poly-silicon and aluminum as thermoelectric materials connected in series. The pixel size is defined by the IR-radiation absorbing layer which covers and heats the „hot" contacts. The adequate „cold" contacts are located above the silicon rim surrounding the membrane. The silicon rim acts as heat sink.

The sensor array chip is fabricated in a CMOS compatible wafer batch process. At the end of the process the silicon below the membrane area is removed by anisotropic backside Si-etching.

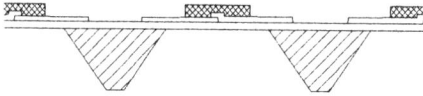

Figure 1: Schematic cross-section

Figure 1 shows the schematic cross section of a thermopile sensor on a (100) oriented silicon substrate and figure 2 displays the layout of the 3x5 element array sensor.

In order to meet security and automotive requirements the sensor chip contains a self-test-circuit. Externally applied digital pulses at the selftest pins generate a thermal excitation for each individual sensor element of the array. As a result the sensor provides an electrical output sequence to allow the required self test function during the measurement.

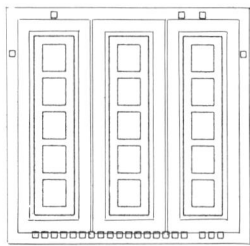

Figure 2: Layout 3x5-element sensor chip

Signal condtioning chip (ASIC)

Since the thermopile sensor elements create only rather small dc signals it is recommended to place the preamplifier very close to the sensor chip. Therefore we designed a small signal conditioning chip of about 2.5x1.2 mm². This chip contains Analog-Multiplexer (MUX), preamplifier, voltage reference and temperature reference.

The ASIC is designed in a CMOS technology, which is characterized by a CMOS chopper stabilized amplifier with low power and very low noise voltage. The amplification factor of about 3000 raises the low thermopile signal voltage before leaving the metal package. The effect of electromagnetic interference to external signal lines outside the package is considerably reduced.

For the sensor operation using signal conditioning circuitry only five connections are necessary: Analog output, Ground, power supply (5V) and two connections for clock (SAMPLE) and synchronisation (RESET). Figure 3 shows the time sequence for the clock and synchronization (Reset) as well as the output signal.

The serial output provides after reset the temperature and voltage reference followed by the 15 signal voltages of the individual elements. Typical clock frequencies for the output signal are about 3 kHz. The frame rate for a complete image is approximately 200 Hz. External averaging can be done to adapt frame rate and noise for the specific application. The block diagram is shown in Figure 4.

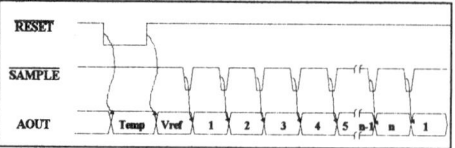

Figure 3: Block diagram for the ASIC

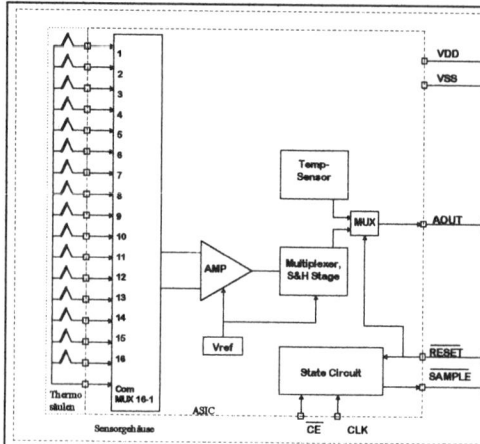

Figure 4: Serial output signal sequence

Packaging and Assembly

The thermopile array and the signal conditioning chip fit onto a 7 pin TO5 baseplate. Five pins are required for the ASIC and two for the self test feature. Figure 5 shows both chips mounted on the baseplate. The sensor elements are symmetrically in the center to ease optical adjustment (e.g. outside lenses).

A transistor cap with an infrared window is then hermetically sealed in a dry nitrogen atmosphere. The size of the window is 5,2 mm x 4,2 mm so that none of the elements is blocked off. Transmission range is typically 6..14 µm but other filter ranges can be chosen in accordance with the application.

Figure 5: TO5 baseplate with sensor chip and ASIC

Optics

For sensor arrays with only a few elements a single lens optics is sufficient. Depending on the field of view requirements external lenses with focal length of 3..15 mm can be mounted above the sensor package. The total field of view can be established from about 10..50 deg. Figure 6 shows the measured output signal of the array in a 3D-graphics. A human object is crossing the optical field of view from left to right. An infrared Si-lens (focal length 3.8 mm) was established in front of the package. It is demonstrated, that the sensor can detect position/presence of objects and measure their surface temperature.

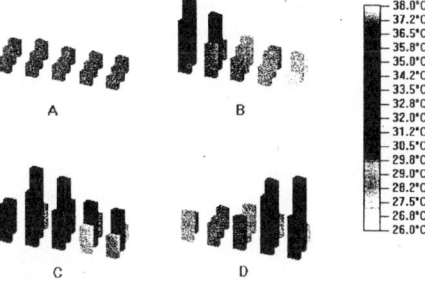

Figure 6: 3-D picture of the measured output signal of the sensor array. A human object is crossing the optical field of view from A to D

Results

Some basic figures of merit of thermopile sensor arrays are sensitivity, noise voltage, noise equivalent power (NEP) and specific detectivity (D^*). More detailed descriptions of their calculation can be found in /2,5/. To evaluate the thermal resolution of the sensor array the noise equivalent temperature difference (NETD) is important. To compare NETD of different infrared array sensors the f-figure of the optics must be mentioned (we refer to a f/1 optics, i.e. focal length and lens diameter are equal). Measured data of the sensor array (including ASIC) are depicted in table 2:

parameter	typical values	unit	conditions
sensitivity	43	V / W	1 Hz, 500 K, without filter
sensitivity	23	V / W	1 Hz, 500 K, 6..14 µm filter
thermopile resistance	25	kΩ	
noise voltage	48	$nV / \sqrt{Hz}$	300 K, rms
NEP	1.2	$nW / \sqrt{Hz}$	1 Hz, 500 K, 6..14 µm filter
detectivity	$0.35 \, 10^8$	$cm\sqrt{Hz} / W$	1 Hz, 500 K, 6..14 µm filter
NETD	0.5	K	f/1 optics, frame rate 200 Hz
	0.1	K	f/1 optics, frame rate 8 Hz [1]
	0.035	K	f/1 optics, frame rate 1 Hz [1]
time constant	20	ms	
temperature range	-20..85	°C	operation

Table 2: Parameters of thermopile 3x5 array sensors; [1] frame rate reduced by external averaging

Applications

Major applications for the thermopile sensor arrays are detection of position, presence and direction of movement. The sensor application for simple imaging purposes is very easy, since neither cooling nor mechanical choppering is necessary. The total size of the sensor system (containing lens optics and sensor package) can be less than 1 cm³ . That small size combined with low costs allow many new applications in security, automotive, industrial and even medical market areas. The automotive passenger occupation detection (smart airbag) using infrared thermopile arrays is decribed in /6/.
Another application is contactless temperature measurement. Therefor the internal temperature reference can be used for ambient temperature compensation of the output signals.

Summary and Conclusions

The step from single element thermopile sensors with known high reliability and temperature stability to thermopile array sensors opens new fields of applications. An IR-array sensor with an additional temperature reference and a suitable optics can be used to survey simultaneously the temperature of different spots in a chosen area.
A two dimensional low cost sensor array with signal processing integrated in a TO5-package allows low cost mass production due to an excellent compatibility to CMOS processes.

Acknowledgments

This project is funded by the German BMBF under contract number 16SV286/1.

References:

/1/   J. Schieferdecker, R. Quad, E. Holzenkämpfer, F. Plotz:
Congress „Sensor 93", Nürnberg, 11.-14. Okt. 1993, Proceedings Band 3, p. 171-178
/2/   J. Schieferdecker, R. Quad, E. Holzenkämpfer and M. Schulze:
Sensors and Actuators A 46-47 (1995) 422-427
/3/   J. Schieferdecker, M. Schulze R. Quad, A. Beudt
Congress „Sensor 95", Nürnberg, 9.-11. Mai 1995, Proceedings p. 613-618
/4/   M. Simon, J. Schieferdecker, M. Schulze, R. Gottfried-Gottfried, M. Müller, R. Jähne
Congress „Sensor 97", Nürnberg, 13.-15. Mai 1997, Proceedings Vol. 2, p. 83-88
/5/   J. Schieferdecker, M. Simon, K. Storck
Congress Sensors Expo, Detroit (MI), Oct. 21-23 1997, Proceedings, p. 33-38
/6/   M. Simon, J. Schieferdecker, M. Schulze, K. Storck, M. Rothley, E. Zabler:
4. Fachtagung Infrarotsensoren und Systeme, Dresden, Sept. 1997, Dresdner Beiträge zur Angewandten Sensorik, p. *tbd*

Simulation and Modelling

Simulation and Modelling
Paper presented at Eurosensors XII, 13–16 September 1998
ⓒ *1998 IOP Publishing Ltd*

Numerical and Experimental Analysis of Distributed Electromechanical Parasitics in the Calibration of a Fully BiCMOS-Integrated Capacitive Pressure Sensor

G. Schrag, G. Zelder, H. Kapels* and G. Wachutka

Institute for Physics of Electrotechnology, Technical University of Munich, Arcisstr. 21, D-80290 Munich, Germany

* Corporate Technology, SIEMENS AG, Otto-Hahn-Ring 6, D-81730 Munich, Germany

Abstract. The experimental characterization of integrated microsensors strongly benefits from the accurate modeling of its operation. This is demonstrated for a fully BiCMOS-integrated pressure sensor, where we studied the influence of distributed parasitic cross-coupling effects on the signal read-out by coupled-field electro-mechanical simulations. In addition, the combination of measurement and simulation enabled the correct identification of all parameters required for the sensor calibration.

1. Motivation

As the operation of any microsensor exploits one or more coupling effects between different energy and signal domains, it is an inherent problem that the signal extraction and thus the device calibration is quite often strongly affected by undesired cross-coupling effects. Additional complications arise from the use of industrial fabrication processes which allow the integration of the sensor elements in the microelectronic circuitry, as it is the case with the fully BiCMOS-integrated capacitive pressure sensors investigated in this work. This is because the restriction to materials and process steps which are fully compatible with a given integrated circuit technology inevitably leads to complex sandwich-like sensor structures which consist of a sequence of thin layers with almost unknown process-dependent material properties such as, e.g., the electromechanical parameters and built-in prestresses. Hence, the device behavior is governed by an intricate interplay of the desired sensor effect and perturbing parasitics which can hardly be separated by measurements alone. In this situation the combination of numerical simulation and inverse modeling techniques with experimental characterization methods proves to be an efficient and practical way to extract the required material parameters, to identify the relevant physical effects and eventually to understand all the details of the device operation.

In the following we demonstrate this approach considering a fully BiCMOS-integrated micromechanical pressure sensor with capacitive read-out. We perform a comprehensive continuous-field analysis of the underlying physical effects using mechanical, electrical and coupled electro-mechanical FEM simulation combined with static and dynamic measurements.

2. Micromechanical Capacitive Pressure Sensor

The micromechanical pressure sensor investigated has been fabricated using an industrial standard BiCMOS process which includes all surface micromachinig steps as integrated part of the process sequence [1]. The polysilicon membrane of the sensor structure is released by sacrifial layer etching of the field oxide [2] , while the boss attached at the surface of the membrane to increase its stiffness consists of a sandwich of several oxide and nitride layers as obtained by the BiCMOS process (see Fig. 1). The membrane structure can be displaced under the action of electrical attraction by a counterelectrode, which is formed by highly doped n$^+$ implantation layers located at the bottom of the cavity. The structure is electrically screened by a guard ring formed by a surrounding p$^+$ implantation. The sensor geometry is basically quadratic, with the option of varying the lateral extension according to the needs of model verification and parameter extraction.

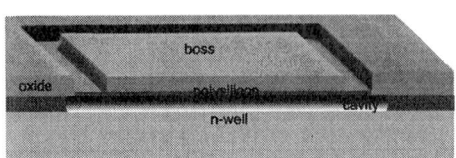 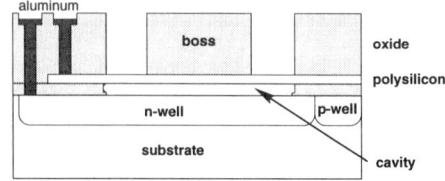

Figure 1. SEM image (left) and cross sectional view (right) of the BiCMOS-integrated pressure sensor.

3. Experimental Characterization

The static and dynamic properties of the device structure were characterized using several complementary measurement techniques. For the static characterization the change of the capacitance between the polysilicon membrane and the counterelectrode caused either by external pressure or by an applied voltage is determined by an LCZ-meter. In order to eliminate the parasitic capacitances originating from other BiCMOS-substructures in the vicinity of the membrane, the signal obtained from a reference structure with a non-deflectable rigid membrane is subtracted from the sensor signal, yielding a quantity which reflects the displacement of the membrane alone and, hence, determines the sensitivity.

The dynamic behavior of the sensor was studied by applying an AC voltage between membrane and counterelectrode. The sensor response (i.e., the frequency-dependent amplitude and phase shift of the membrane displacement) was monitored by an impedance analyzer and exploited to determine the resonance frequency and Q-factor.

4. Numerical Simulation

The complete numerical analysis of the sensor operation comprehends merely mechanical simulations (capacitance change vs. applied pressure C(P), modal analysis), coupled electro-mechanical analysis (capacitance change effected by the applied bias voltage C(V)), and merely electrical simulations of the parasitic small-signal MIS- and pn-junction capacitances, which contribute to the observed C(V) characteristics in both the sensor and the reference structure.

The mechanical part of the sensor is modelled by means of finite elements using a commercially available general purpose simulator [3]. Non-linear behavior as well as built-in prestresses are taken into account. For the calculation of the parasitic C(V) characteristics, first the semiconductor transport equations together with Poisson's equation have to be solved at a given static operating point with bias voltage U_{dc}. Then, in a second step, a small signal analysis is performed by superimposing a small AC signal $U_{ac}(t)$, from which the voltage- and frequency-dependent admittances can be extracted [5]. Advantages of this method are that it simulates the measurement process itself and that, meanwhile, it is implemented in most general purpose device simulators (e.g. [6]). The unperturbed C(V) characteristics of the sensor alone has to be determined from the balance of electrical and mechanical forces along the interface between cavity and self-consistently displaced membrane. This implies a strongly coupled electromechanical problem, as the mechanical displacement of the sensor membrane influences the attractive electrical force and vice versa. Our simulations follow an iterative (Gauss-Seidel) solution scheme, where the mechanical displacement effected by the external pressure is passed to the electric field simulator which, in turn, computes the resulting electric force on the membrane and sends it back to the mechanical FEM solver, which recalculates the mechanical displacement, and so on, until convergence has been obtained.

Hereafter, the influence of the membrane displacement on the C(V) characteristics of the parasitic capacitances is included by repeating the small-signal analysis with the self-consistent membrane location.

5. Results

One major aim of our investigations was to examine whether the mechanical and the electrical characterization methods give equivalent results. To this end we proceeded as follows:

Mechanical and coupled-field electro-mechanical simulations (FEM): In a first step we attempted to accurately reproduce the merely mechanical behavior, allowing us to extract a correct parameter set for the further simulations.

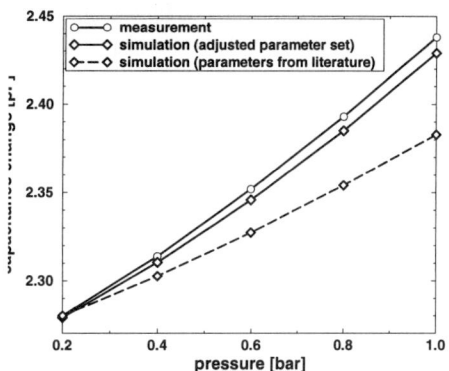 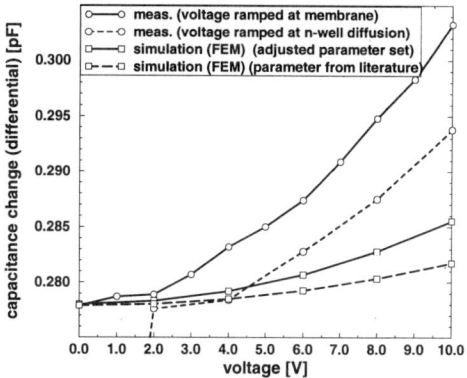

Figure 2. Capacitance change due to an applied external pressure (left) and to an applied voltage (right) for one single sensor geometry. Comparison between measurement and 3D FEM model, with parameters taken from literature and extracted from process-specific data. The signal of the reference structure is subtracted from the measured data.

Starting with material parameters taken from literature (e.g. [4]) and geometrical data as specified by the design we arrived at a model with too large stiffness (see Fig. 2 and Table 1). But after a careful analysis of experimental data obtained from different sensor geometries we could identify

	sensor 1	sensor 2	sensor 3	sensor 4
$f_{meas.}$	1.9 MHz	2.0 MHz	2.1 MHz	1.4 MHz
$f_{lit.}$	2.3 MHz	2.4 MHz	2.7 MHz	1.6 MHz
$f_{adj.}$	1.9 MHz	2.0 MHz	2.2 MHz	1.3 MHz

Table 1: Fundamental frequency of different sensors. Comparison between measurement $f_{meas.}$ and FEM (parameters from literature $f_{lit.}$, adjusted parameters $f_{adj.}$)

the relevant parameters and influences and extract a new parameter set which allowed the reproduction of the mechanical behavior (eigenfrequencies and C(P) relation) within acceptable tolerances, but failed with respect to the voltage-dependent measurements (Fig. 2, right). Moreover, this figure reveals the peculiarity that we obtain a different C(V) characteristics if the voltage is not ramped at the membrane but at the n-well diffusion. This is a strong indication that the voltage-dependent parasitic capacitances represent a considerable contribution to the sensor signal.

Small signal analysis (electric simulation): The simulated high-frequency C(V) curves of the parasitic capacitance of the reference structure (Fig. 3, left) agree fairly well with the measured characteristics, though no interface and surface charges have been taken into account, and are highly sensitive to the magnitude and the sign of the applied voltage.

If the voltage ramp is applied at the membrane, the C(V) curve is composed of several vertical MIS-capacitances in parallel (cf. steps at -20 V and 10 V) combined with distributed lateral capacitances associated with the n-well and p-well diffusions (cf. slope of the capacitance increase). If, on the other hand, the voltage is ramped at the n-well diffusion, the characteristics is dominated by the presence of the n-well/p-well junction in the substrate, while the typical steps originating from the MIS-structures vanish.

Influence of membrane displacement on parasitic capacitances: We studied this effect by passing the self-consistent membrane displacement obtained by the coupled-field electromechanical FEM analysis to the small-signal C(V) simulator. It shows that the parasitic C(V) curve is notably shifted by the membrane displacement. Thus the parasitic capacitances of sensor and reference structure are not identical for the same applied voltage and, therefore, the difference signal of sensor and reference structure as shown in Fig. 3 (right) is an intricate non-linear combination of the mechanical and the parasitic capacitances, which does not represent the capacitance change caused by the mechanical deformation alone. This finding qualitatively explains the discrepancy between measurement and FEM simulation as displayed in Fig. 2 (right).

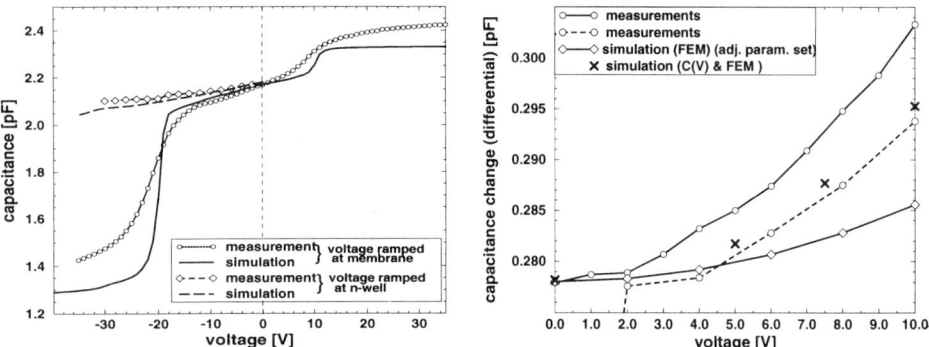

Figure 3. <u>Left:</u> Results of the small signal analysis at 100 kHz (measurement and simulation) with voltage ramped at the membrane and at the n-well diffusion. <u>Right:</u> Capacitance change versus applied voltage. The crosses indicate the result of combined electric and FEM simulation leading to a much better agreement with the measured characteristics than the merely electro-mechanical FEM simulation.

6. Conclusions

We demonstrated that the experimental characterization of integrated sensors must be accompanied by accurate numerical simulation to enable the correct extraction of model parameters and to understand the details of the signal read-out. Since the latter is strongly affected by distributed parasitic cross-coupling effects, coupled-field numerical simulation is required to identify and assess the influence of these effects. It seems that inverse modeling is becoming indispensable for the interpretation of the sensor output and the calibration of the sensor signal.

References

[1] Scheiter T., Oppermann K.-G., Steger M., Hierold C., Werner W. M., Timme H.-J., Proc. of EUROSENSORS XI, Warsaw, 1997, pp. 1595-1598.

[2] Scheiter T., Näher U., Hierold C., Proc. of EUROSENSORS XI, Warsaw, 1997, pp. 1369-1372.

[3] Swanson Analysis Systems: ANSYS User's manual, Rev. 5.3, Houston, PA, 1997.

[4] Beadle W.E., Quick Reference Manual for Silicon IC Technology, 1985, Wiley & Sons, N.Y.

[5] Wachutka G., Fahrner W.R., J. Appl. Phys. **71**, 1,1992.

[6] Silvaco International, Atlas User's Manual, Santa Clara, CA, 1997.

Modelling Of An Air Cone-Jet Sensor Using Finite Element Method

T Q Xie, Q Yang, Y J Au, B E Jones

Department of Manufacturing & Engineering Systems, Brunel University, Uxbridge, Middlesex, UB8 3PH, UK

Abstract. This paper presents some results obtained from the modelling of an air cone-jet sensor using finite element method. The results have indicated that a cone-jet sensor has much greater stand-off distance, wider sensing range and better lineairy than a back-pressure air gauge. The sensor design may also be optimized using finite element method.

1. Introduction

It is often desirable to implement the automated dimensional inspection and the tool condition monitoring in-process (i.e. during machining) [1]. The information from this instant feedback may be used to minimise any possible costs due to significant changes in tool conditions or process. There have been online in-cycle probing systems installed on CNC machining centres since the 1970's. But they are not regularly used because they generally lack the necessary measuring accuracy and reliability.

Pneumatic sensing, in particular, air gauging based upon back-pressure principle has been well established in manufacturing metrology as a means of measuring small dimensional variations. The air flow at speed acts as both the measuring medium and the cleaning jet, the latter is very important in in-process inspection [2]. Despite its success in post-process inspection, however, its use as an in-process measuring device is somewhat restricted because of its rather limited operating range.

A novel probe has been proposed and it consists of a primary air cone-jet sensor and a secondary optical fibre sensor. The cone-jet functions as a buffer for converting a stand-off distance to a pressure that pushes a diaphragm, the optical fibre sensor then detects this displacement. However, since it is difficult to establish the theoretical relationship between the pressure and the measured distance for cone-jet sensor, finite element method has been used to model the sensor and further optimize its configuration, which is the main purpose of this paper.

2. Cone-jet sensor and simple back-pressure sensor

For a simple back-pressure sensor [3], fluid at a regulated pressure is supplied to a fixed-flow restriction and varied by moving the measured object to change the stand-off distance x_i. This causes a change in output pressure p_0, which may be expressed as

$$p_0 = \frac{p_s}{1+16(d_n^2 x_i^2 / d_s^4)}$$

where p_s is the supply pressure; d_n the nozzle diameter; d_s supply-orifice diameter. Thus, a pressure-measuring device connected to p_0 can be calibrated to read x_i.

Shown in Fig. 1 is an air cone-jet sensor. It can be used with a stand-off distance some ten times greater than the simple back-pressure sensor for an equivalent flow consumption, with the measurement range typically 2-4 mm, and it also has a relatively fast response time [4]. Compared with back-pressure sensor, however, it is more difficult to derive the analytical solution and optimize the design for a cone-jet sensor. Finite element analysis has been used for the sensor modelling.

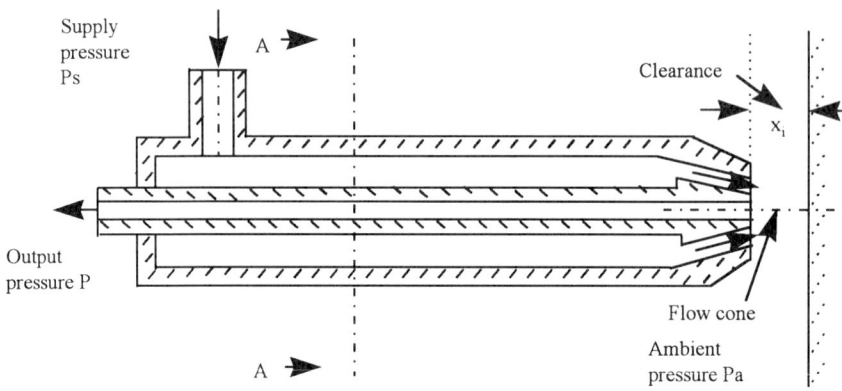

Fig. 1 Cone-jet sensor

3. Modelling

Since finite element analysis (FEA) was used in aircraft design in 1950's [5], this method has been used increasingly in other branches of engineering. A finite element model is a mathematical simulation of the actual physical structure of an object. The model is divided up into a number of regions known as elements, each of them is defined by a pattern of nodes. The method defines an approximation within each element, appropriate continuity conditions being imposed on the inter-element boundaries. Improvement in accuracy is achieved either by decreasing the size of elements, or by increasing the number of terms in the approximations within the elements.

The geometry of a simple back-pressure sensor with FE mesh is shown in Fig. 2 and the meshed cone-jet sensor with loading and boundary conditions in Fig. 3.

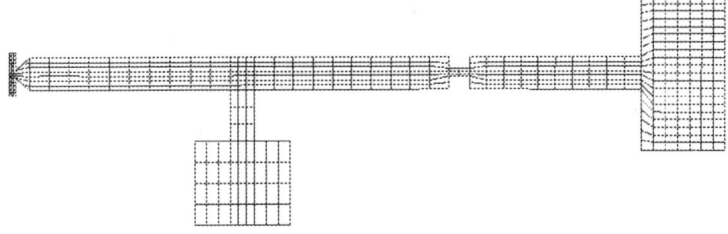

Fig. 2 The meshed back-pressure sensor

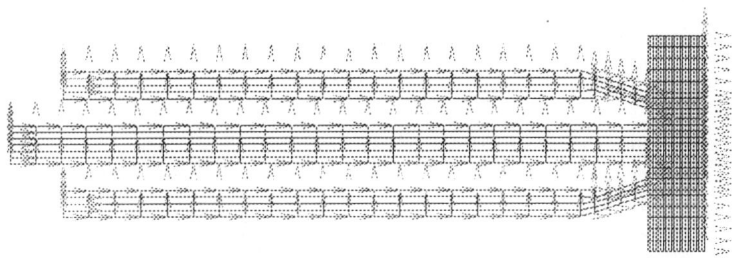

Fig. 3 The meshed cone-jet sensor with loading and boundary conditions

A gaseous medium was assumed since a majority of practical applications utilize low-pressure ($p_s = 1.37$ to 2 bar) air as the working fluid [3]. The input pressure p_s is constant, the output pressures p_o will be changed by moving object to change the distance x_i. The gaseous medium is air, its density ρ 1.293 kgm^{-3}, viscosity η 18.325Nsm^{-2}. Given the fixed boundary conditions and distance x_i, the output pressure p_o can be calculated using finite-element method in fluid flow field. Thus, different sensor designs may be analyzed and compared, resulting an optimised sensor design.

4. Results and discussions

As shown in Fig. 2, where both the supply-orifice and nozzle diameter are 0.794 mm, the results obtained using finite-element method are shown in Fig. 4. It is obvious that the measuring ranges are about 0.2 to 0.6 mm, which typically increases with nozzle diameter. The measuring range is too small for most in-process applications.

The cone-jet sensor as shown in Fig. 3 has good linearity and measuring range. The results in Fig. 5 have confirmed this. It can be seen that the relationship is approximately linear over a range of clearances from 1.6 mm to 4.5 mm. The line contour map of the pressure field is shown in Fig. 6.

The above analysis can be easily extended to analyze the effects of the cone-jet shape and parameters on the sensor characteristics.

Pressure (bar)
4.5
4.1
3.7
3.3
2.9
2.5
2.1
1.7
1.3
0.9
0.5
0.2
0.36
0.52
0.68
0.84
1
Displacement (mm)

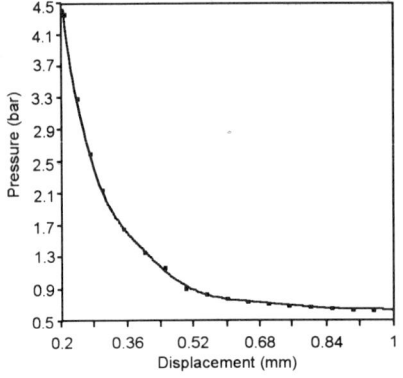

Fig. 4 Nozzle-flapper characteristics

Pressure (Pa)
3500
3057
2614
2171
1729
1286
843
400
1.5
2.5
3.5
4.5
Displacement (mm)
y=a+bx+cx²+d/x+e/x²
a=6697.0733, b=-1495355.4
c=90126885, d=-8.4420209
e=0.010676919, r²=0.999978716

Fig. 5 Cone-jet sensor characteristics

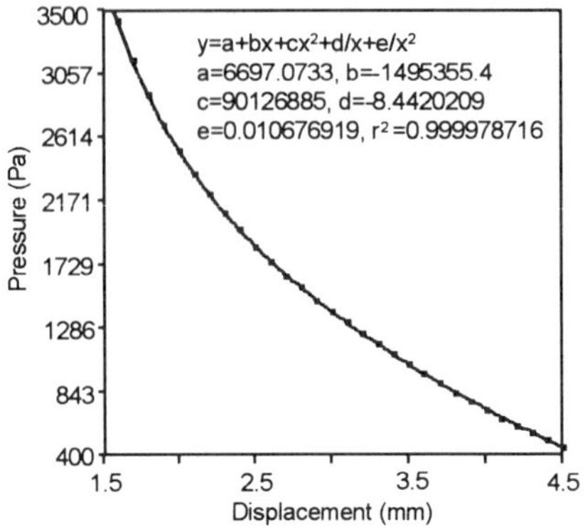

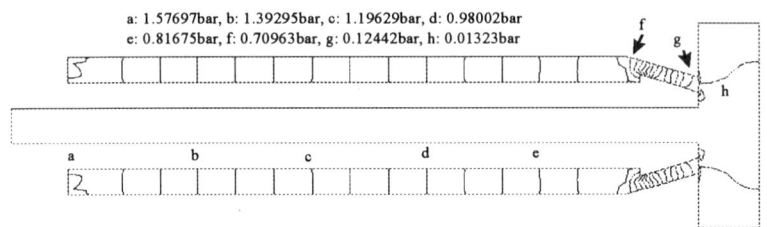

a: 1.57697bar, b: 1.39295bar, c: 1.19629bar, d: 0.98002bar
e: 0.81675bar, f: 0.70963bar, g: 0.12442bar, h: 0.01323bar
f
g
h
a
b
c
d
e

Fig. 6 Line contour of the pressure field

5. Conclusions

The finite element modelling has provided a powerful tool for the study of a cone-jet sensor performance. The preliminary modelling results have indicated that the cone-jet sensor has a wider working range than back-pressure sensing, although when the clearance becomes very small, the behaviour of the device reverts to that of a back-pressure sensor. The cone-jet sensor has also proved to have a better linearity than a back-pressure sensor.

The optimisation of the air cone-jet sensor will be reported in the future.

References

[1] Murphy S D 1990 *In-process Measurement and Control* (New York: Marcel Dekker)

[2] Rutelli G, Cuppini D 1988 *J. Eng. Materials & Technology* Vol. **110** 59-62

[3] Doebelin E O 1990 *Measurement Systems Application and Design* (New York: McGraw-Hill Publlishing Company) 292-300

[4] Hall L C, Jones B E 1976 *Proc Instn Mech Engrs* Vol **190** 23-30

[5] Carey G F, Oden J T 1986 *Finite elements: Fluid Mechanics* (Englewood Cliffs NJ: Prentice-Hall)

Simulation and Modelling
Paper presented at Eurosensors XII, 13–16 September 1998
© 1998 IOP Publishing Ltd

Stress analysis on the open window plastic package for sensors

C.V.B. Cotofana, A. Bossche, P. Kaldenberg* and J. Mollinger

Electronic Instrumentation Laboratory, Delft Institute of Microelectronics and Submicron Technology (DIMES), Delft University of Technology, Mekelweg 4, 2628 CD Delft The Netherlands
* Eurasem B.V. PO-box 566, 6500 AN Nijmegen, The Netherlands

ABSTRACT

Sensor packaging has become a critical element of a competitive sensor industry, being determinant of sensor performance and cost. Our recent work was focused on a new packaging concept, the open window concept that allows the phenomenon to be measured to reach freely the sensor active area, without impeding on the sealing of the circuit. The open window concept provides low cost packaging solutions by simple customization of standard plastic packages that can consequently be used for a wide range of sensors. A first product was already developed for an optical sensor, using the single access window package derived from the open window concept. This paper presents the results of stress analysis performed on this package in order to solve the problems related to high stress levels during soldering. The finite element modeling enabled us to analyze the stress levels in critical points within the package and decide on the most appropriate solution for optimization. From stress analysis results we were able to decide on the optimal choice for a specific application.

1. Introduction

A large number of chip-based sensors have been developed in the last years but their packaging however received attention only in individual cases, custom solutions being used. Standard plastic packages have been applied to sensors that do not require open access paths to the environment, e.g., magnetic sensors, thermal sensors, and closed cavity accelerometers. For other types of sensors other solutions have to be found. We focused on finding such solutions, by developing the open window packaging concept that led to the first application: a single access window plastic package [1]. The open window concept provides the possibility to develop different packages that can be used for a wide range of sensors, e.g., optical, pressure, flow or chemical sensors, showing great potential as low cost sensor packaging solutions [1].

2. Open window package concept

In [1] we presented a low-cost reliable package concept, the open window concept, that allows the active area of the device to be exposed to the outside world, while the rest of the chip and lead-frame are covered by plastic as in a standard plastic package. It can be used to build packages with one or more environmental access windows at costs comparable with those of the standard plastic package. A first application was developed for an optical sensor at Eurasem BV, a Dutch packaging company. Figure 1 shows a picture of this application that uses a single access window package we called open quad flat pack (OQFP). A characteristic of this application is that in order to protect the active area of the optical sensor, a glass lid is mounted on top of the open window by means of gluing. The glue used to mount the protection glass lid has to ensure that no direct path exists between the cavity and the outside world. Also the adhesion of the glass lid and the glue has to be comparable to the adhesion between glue and plastic.

Figure 1: OQFP with a single access window

The first concept of the package used a glass lid 5mm wide and 0.5 mm thick. The attachment of the glass lid was done with an SMD glue (Permacol 2035Z), an epoxy-based material that cures in 3 minutes at 125 °C. Lid attachment solution used for this product is presented in Figure 2. The major problem we encountered during the product reliability testing was the poor behaviour at Solder Dip Reflow test, while all the other tests were successfully passed. In order to find the cause of this problem and be able to solve it we decided to make a model of the package and study the stresses arisen due to the high temperature and pressure that the package withstands during the Solder Dip Reflow test.

3. Finite element modelling (FEM)

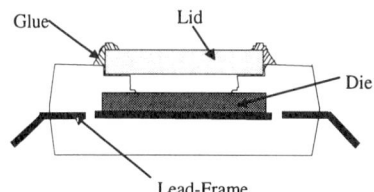

Figure 2: OQFP Structure

3.1 Modeling technique

A schematic drawing of the open window package is depicted in Figure 2. We developed a three dimensional model for this package, using brick elements. Since we are more concerned with the behavior of the die, glass and glass/epoxy glue, the meshing is refined in the region of interest. Due to the four-fold symmetry, only a quarter of the package is used for building the model. Lead-frame fingers have been omitted in the model for simplicity [2]. OQFP specific dimensions are presented in Table 1. The figures for the new approaches we analyzed are depicted in italic characters.

Table 1: OQFP Package Dimensions

Item	Outer Dimensions [mm]	Inner Dimensions [mm]	Thickness [mm]
Die	2*3.5	-	0.45
Die attach Glue	2*3.5	-	0.05
Cu lead-frame	2*5	-	0.2
Glass	2*2.5, 2*3	-	0.5
Low-stress Epoxy	5	2*2.75, 2*3.25	0.9
Glue glass/epoxy	2*2.75, 2*3.25	2*3, 2*3.5	0.1^2

Table 2: Material Properties used in the Finite Element Modeling (FEM)

Material	Young's Modulus (10^3 MPa)	Poisson's Ratio	Thermal Expansion Coefficient (10-6/K)
Silicon	169	0.26	2.4
Die attach Glue	0.2	0.3	56
Cu lead-frame	1.21	0.33	16.2
Glass	56	0.31	8
Low-stress Epoxy	10.8	0.27	12
Glue glass/epoxy	1.5 10^{-3}, 2.6,	0.31, 0.35	250, 65

Material properties are listed in Table 2. We assumed that material properties are homogenous on all directions and independent of temperature. The boundary conditions are zero displacements in the symmetry directions and a zero displacement for the ground plane node situated at (0,0,0) coordinates of the Cartesian system (in order to prevent rigid-body motion).

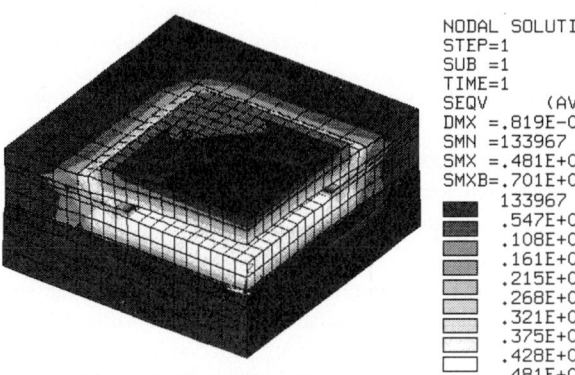

```
NODAL SOLUTION
STEP=1
SUB =1
TIME=1
SEQV      (AVG)
DMX =.819E-05
SMN =133967
SMX =.481E+08
SMXB=.701E+08
    133967
    .547E+07
    .108E+08
    .161E+08
    .215E+08
    .268E+08
    .321E+08
    .375E+08
    .428E+08
    .481E+08
```

3.2 Linear Elastic Finite Element Analysis (FEA)

Due to the fact that in any package we deal with a number of materials with different characteristics, e.g., thermal expansion coefficient, Young's modulus, Poisson ratio, the packaging process can induce significant thermal stresses. There is also stress induced by the air pressure built inside the cavity when the package is prone to high temperature. Consequently, we used two load sets:

Figure 3: Von Mises Stress distribution for the OQFP

temperature body load and elemental pressure load. The analysis was performed with Ansys 5.3. The first objective was to evaluate stress magnitude and distribution inside OQFP, particularly what happens in the glass and glass/epoxy glue layer, as it is in this region that the package had problems (glass popping out at Solder Dip Reflow test).

Figure 3 depicts the stress distribution inside OQFP as Von Mises contour stress. We also analyzed the in plane shear stress (Sxy) for the glass/epoxy glue, the glass lid, die and plastic smear on the top of the die. In this model perfect adhesion was assumed at all interfaces. As it can be observed in Figure 3, stress has increased values in the middle of the package, concentrated on the die, glass/epoxy glue, and glass lid. As our main problem concerns the glass/epoxy glue stress, we were interested to see it in more detail. We can see in Figure 4 that the stress inside the glue layer is larger under the glass and next to the glass lid and epoxy.

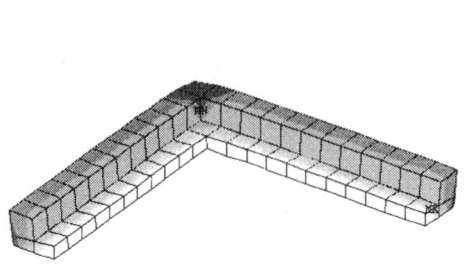

```
NODAL SOLUTION
STEP=1
SUB =1
TIME=1
SEQV      (AVG)
DMX =.819E-05
SMN =.335E+07
SMX =.329E+08
SMXB=.439E+08
    .335E+07
    .664E+07
    .993E+07
    .132E+08
    .165E+08
    .198E+08
    .231E+08
    .264E+08
    .297E+08
    .329E+08
```

Calculating the yield stress in the glass/epoxy glue layer we were able to appreciate that the von Mises stress given by the temperature load set was up to two times bigger than the yield stress. Analyzing different SMD glues we realized that using this assembly configuration, normal SMD glues are not fit and that we have to find

Figure 4: Von Mises Stress distribution for the glass/epoxy glue

either another glue geometry, change dimensions in this critical area, or use other glue families. So, our analysis focused on these three possible choices. Consequently, we built models for new approaches having different glue and glass lid geometry or using other glues, as follows:

- A model using a glue layer only under the glass lid was simulated and the stress levels showed that the maximum stress level was 26% lower than in the model built for the initial application. However the stress levels were still 53% higher than allowed. Another possible inconvenience of this approach may be poor package sealing.

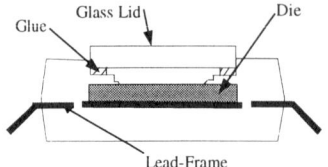

Figure 5: OQFP modified glue geometry

- Another approach was to modify the open window geometry (glass lid and package inner dimensions) in order to analyse the effect on stress magnitude and distribution. The glass lid and package inner dimensions were increased by approximately 20%. The model showed that the changes in stress configuration were insignificant. Consequently, this approach was also left aside.

- Our next approach was to use silicone-based glues that cure at normal temperature, like UV or condensation curing silicones. These materials have a different range of properties that can be successfully used for our development [5]. We tried different silicones that are considered to have good adhesion qualities on many substrates and can withstand the soldering temperature. Our first choice was a condensation cure silicone from Shin-Etsu Chemical Co. The stress levels for this model were much lower, in the range of 0.04 to 1.38 MPa, considering both temperature and pressure load sets. Taking into account that the yield stress for this material is 2MPa, this glue could very well suit our purpose. Furthermore, the maximum stresses in the die were 30% smaller in this approach than in our first development. We have however to test the quality of the adhesion between glass and epoxy. Considering the stress analysis results we can conclude that we should focus for further developments of the OQFP construction to silicone based glues from the condensation or UV cure class.

4. Conclusions

The goal of the present work was to find a way to reduce the high stress levels encountered during the OQFP packages soldering. Stress analysis permitted us to choose from a range of different approaches the right one for a reliable product, without loosing time and money with tests performed on non-viable solutions. A thorough analysis showed that our first development was not going to provide a solution to high stress. Our attempts to reduce stress levels by using other OQFP geometrical structures could not provide significant improvements. However, the OQFP construction using room temperature curing silicone-based glues instead of SMD glues is the appropriate solution, providing significant improvement in stress lowering in the die and in the glue layer. Consequently our work will further focus on this new development.

References

[1]   Cotofana C, Bossche A, Kaldenberg P and Mollinger J, " Low-cost Plastic Sensor Packaging using the Open Window Package Concept", to appear, Sensors and Actuators vol. A 1998.

[2]   Pendse R and Demmin J, "Test Structure and Finite Element Models for Chip Stress and Plastic Package Reliability", IEEE 1990 Vol.3, p 155-156.

[3]   Simon B R, Yuan Y and Umaretiya J R, Parametric Study of a VLSI Plastic Package Using a Locally Refined Finite Element Models", 5th IEEE Semi-Therm Symposium 1989, p 53-54.

[4]   Condra L, Nguyen L T and Hakim E B, "Comparison of Plastic and Hermetic Microcircuits under Temperature Cycling and Temperature Humidity Bias", IEEE Trans. CHMT, vol. 15, p 640-650, 1992.

[5]   Davis J H "Silicone Protective Encapsulants and Coatings for Electronic Components and Circuits", Dow Corning Ltd., Barry, Wales, UK, 1994.

Simulation and Modelling
Paper presented at Eurosensors XII, 13–16 September 1998
© *1998 IOP Publishing Ltd*

New Coupled-Field Device Simulation Tool for MEMS Based on the TP2000 CAD Platform

Eva-Renate König [1] **, Peter Groth** [*] **and Gerhard Wachutka**

Institute for Physics of Electrotechnology, Technical University of Munich, Arcisstr. 21, D-80290 Munich, Germany

[*] IGF Ingenieursozietät Groth – Faiss, Pfullingen, Germany

Abstract. We describe and demonstrate the capabilities of a new coupled-field simulator based on the industrial CAD platform TP2000 for the multidimensional numerical analysis of the operation of electro-mechanical microdevices. The coupling of the finite element and the boundary element methods makes the simulation tool particularly suited for microdevices where movable parts are deflected, displaced, or rotated by electrostatic forces. As an illustrative example, we study the fully coupled electro-mechanical behavior of a deflectable micromirror.

1. Motivation

A variety of electro-mechanical microdevices (MEMS) have now made their way into practical industrial application. Typical examples include micropumps which constitute key components of microfluidic systems used for chemical microanalysis or medical applications such as microdosed drug infusion. Deflectable micromirrors arranged on a chip in large arrays are encountered in scanners, copy machines or large screen projectors. For optical eye surgery, micromirrors are used for the exact positioning of the laser beam. Interdigitated comb structures are applied as sensors in accelerometers and gyroscopes or used as driving unit in microresonators and other actuators.

With a view to cost-effective and time-economizing design optimization of MEMS, computer-aided engineering methods have to be employed, leading to a demand for simulation tools which are especially dedicated to the accurate analysis and optimization of the device and system performance. In this work, we describe a new electro-mechanical device simulation tool based on the coupling of finite elements for the thermomechanical analysis and boundary elements for the electrostatic part of the problem. Among others, it is particularly suited for the analysis of microdevices where movable parts are deflected, displaced, or rotated by the action of electrostatic attraction, one of the most commonly used actuation principles in MEMS. Here, the coupled electro-mechanical simulation problem consists in the self-consistent calculation of the balance of the elastomechanical, electrostatic, and inertial forces acting on the device.

The paper is organized as follows: After a short problem description, we discuss the solution strategies and computational methods. Then the special functionality of our implementation especially dedicated to the treatment of MEMS is demonstrated with reference to a micromirror as typical practical application.

[1] E-mail: koenig@tep.e-technik.tu-muenchen.de

2. Physical Model and Problem Definition

On the continous field level, the device operation of MEMS is modelled by a system of coupled partial differential equations, comprising the structural equations (stating the balance of forces) and Poisson's equation (governing the electric field). Each of these equations applies only in its respective space domain; the coupling takes place along the common interfaces through electrostatic surface forces and the momentary geometrical location of these interfaces. In consequence of the displacement of the movable parts of the device, one or both domains are strongly deformed, which must be accounted for by an especially adapted space discretization used in the numerical simulation.

Each of the governing equations is quasi-linear, but the coupling term is highly non-linear, and therefore we arrive at an all-in-all non-linear system, which is capable of describing well-known instabilities of high practical interest such as, for instance, the so-called snap-down effect arising in membrane drives. This phenomenon occurs when, at sufficiently small distance between neighboring electrodes, the electric attraction can nolonger be balanced by the mechanical forces, with the consequence that the electrodes snap together until they touch each other. Tackling this unstable solution requires the application of special numerical methods like homotopy algorithms.

The situation, when a movable electrode touches its counterelectrode, implies a contact problem which concerns the mechanical problem only and can adequately be treated by introducing additional compulsory mechanical forces (penalty forces).

3. Solution Strategies for Coupled-Field Problems

For surface-coupled problems, efficient global solution methods based on partitioning and domain decomposition are available, but rarely implemented in existing simulation environments. Typically we face a situation where we have well-performing device simulators adapted to one of the single physical effects, suggesting the use of iterative interfacing techniques for the external coupling of the existing tools.

Basically three approaches have proven to be successful. In the semianalytical approach, one or several of the coupled parts of the electro-mechanical device are described by simplified analytical models which interact through interface conditions with the remaining parts described by a numerical model(e.g., FEM model). The overall accuracy of this method strongly depends on the idealizations made, restricting it to simple geometries and material laws. In the second approach, existing simulators calculate in their specific space and energy domain(s) and communicate with each other along the connecting interfaces according to an iterative scheme (typically a relaxation algorithm). However, this method may suffer from convergence problems under strong coupling conditions. The third approach avoids convergence problems by the simultaneous solution of all governing equations using special, problem-specific numerical methods (e.g., domain decomposition, non-conforming grids etc.). As this approach cannot be based on available commercial tools, it requires new software implementations. In addition, the computational expense may easily exceed the hardware limitations.

4. Computational Methods

Our approach follows the idea of using finite element codes for the mechanical and the electrical subproblems and coupling them through the common domain interfaces. Concerning the electrical subproblem, it is important to note that, if finite element methods (FEM) were used, the strong deformation of the electrical domain would require frequent remeshing to avoid ill-conditioning of the discretized Poisson equation. Furthermore, regarding the coupling to the mechanical subproblem, it is only the boundary value (trace) of the potential along the common interface which is needed for the data interchange, but not the values inside the electrostatic volume.

Both problems are properly solved by the use of boundary element methods (BEM) for the electrostatic subdomain. So we decided to choose a combination of FEM and BEM, using the mechanical solver TP2000 with a new BEM extension and providing two alternative implementations for the coupling scheme, an iterative method and a simultaneous method. The iterative method follows a Gauss-Seidel-like relaxation scheme. The simultaneous method collects the linearized equations at a given operating point from the FEM and the BEM modules, which then are augmented by the nonlinear coupling equations and solved all together as nonlinear equation system. This establishes a predictor-corrector method which finds the solution along a chain of operating points. The numerical results presented below refer to the iterative method, as it is computationally much cheaper (provided convergence can be obtained!). The full analysis program has been linked to the widely used commercial pre- and postprocessor FEMAP, thus offering the microdevice designers a comfortable CAD platform that allows easy access to the above-described methods in combination with very comfortable options such as automated and semi-automated grid generation.

Functionality particularly useful for tackling typical MEMS problems is provided by three additional program modules. The first one allows for the treatment of unstable operating points such as the above-mentioned snap-together effect, where none of the usual solution schemes will converge. In this situation homotopy methods are the appropriate means to handle the problem. To this end, a properly chosen homotopy parameter is introduced, which allows to externally control the state variables. Starting with a parameter value where the solution can easily be computed, the desired operating point is attained by path continuation [1]. This algorithm is applicable as exterior loop in any of the coupling schemes mentioned in section 3.

The second module refers to the contact problem, which is a typical feature of many actuator structures, and can be used for each of the two possible cases occuring as contact problem: movable structure against rigid contour, as it is the case with the micromirror studied below, and movable structure against movable structure. As soon as the presence of a contact is detected, the direction and strength of the friction forces and the conditions for gliding and traction friction are automatically set up and evaluated.

The third module addresses the case of structures which exhibit high or extreme aspect ratios, a situation which occurs quite often in micromachined devices. Here, a fully FEM-based method would be advantageous, because it is well known that in such situations BEM tends to converge very slowly or even fails. On the other hand, the usual FEM procedure would require a frequent remeshing of the FEM grid in the electrostatic domain, leading to the problem of prolonging the solution vector onto the updated mesh. To cope with this problem a novel "node deactivation and condensation"algorithm is being implemented in the present version of the TP2000 code.

5. An Illustrative Example

The new implementations in TP2000 were demonstrated with reference to selected microstructures such as, e.g., the micromirror displayed in Fig. 1. The numerical results were obtained by employing the iterative coupling scheme as discussed above.

Fig. 1 shows a schematic view of a two-dimensional cross section of the micromirror structure investigated. The dimensions of the plate are $30\mu m \times 30\mu m \times 2\mu m$. The black rectangular region represents the cross section through the torsion beams which act as hinges for suspending the plate. Their size amounts to $0, 2\mu m \times 2\mu m \times 15\mu m$. The nominal distance between plate and substrate is $2\mu m$, whereas the parameter d in Fig. 1 is $3\mu m$.

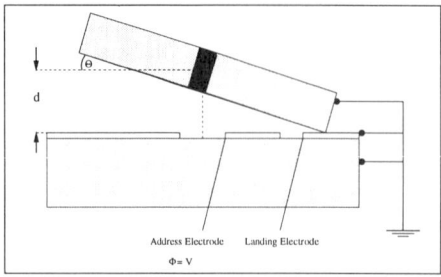 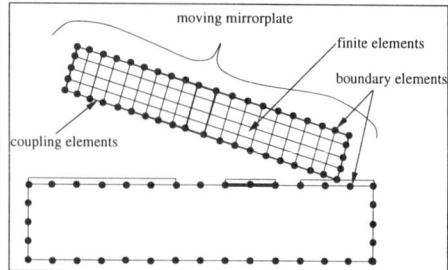

Fig.1: Schematic view of a micromirror structure

Fig.2: Discretization grids with FEM nodes and BEM nodes

In Fig. 2 the FEM and BEM discretization nodes are illustrated for the 2D case. Pair coupling of nodes occurs along the bottom side of the deflectable mirror plate and the top side of the substrate. At its right edge the mirror plate gets contact with the rigid contour of the landing electrode. At this node pair the reaction forces calculated in the BEM-simulation have to be transferred to the FEM nodes on the mirror plate.

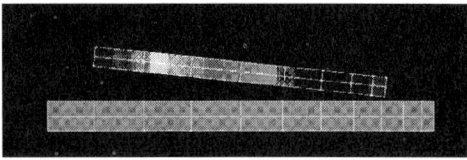

Fig.3: Equilibrium deflection of the micromirror after applying a voltage between mirror plate and address electrode.

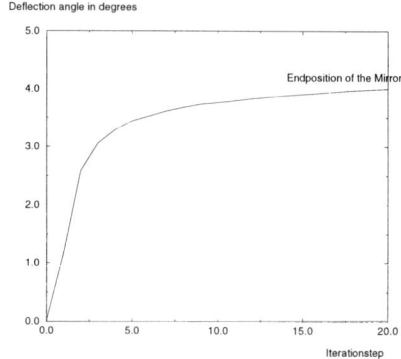

Fig.4: Convergence history expressed by the deflection angle vs. iteration step.

Fig. 3 shows the equilibrium deflection of the micromirror after an iterative simulation of the coupled problem. Because of the large aspect ratio the deflection is stretched by a factor of 4 for the sake of better visibility. Finally, Fig. 4 displays the convergence history (i.e., the deflection angle versus the iteration steps) reflecting the iterative rotation of the mirror plate until the equilibrium position is reached.

References

[1] R.H.W. Hoppe, E.-R. Sieber, G. Wachutka, U. Wiest, *"Mathematical Modelling and Numerical Simulation of a Free Boundary Problem for an Electromechanical Micropump"*, Rep. No. 358, Institut für Mathematik, Universität Augsburg, Germany, (1996)

[2] G. Löbel, W. Sichert, *"BETTI-Boundary Element Code for Heat Conduction"*, Forschungskuratorium Maschinenbau e.V., Frankfurt, (1987)

[3] R. Bausinger, H. Faiss, P. Groth, *"TPS10 Benutzerhandbuch"*, TSE GmbH, Reutlingen, (1993)

[4] V.P. Jaecklin, C. Linder, N.F. de Rooij, J.-M. Moret, R. Vuilleumier, *"Line-addressable torsional micromirrors for light modulator arrays"*, Sensors and Actuators, A41-42, pp.324-329, (1994)

Simulation and Modelling
Paper presented at Eurosensors XII, 13–16 September 1998
© *1998 IOP Publishing Ltd*

Investigation of Leakage Flux in a Capacitive Angular Displacement Sensor Used in Torque Motors by 3D Finite Element Field Modelling

S H Khan   K T V Grattan   L Finkelstein

Measurement and Instrumentation Centre, Department of Electrical, Electronic and Information Engineering, City University, Northampton Square, London EC1V, 0HB, UK

Abstract–This paper concerns the finite element (FE) investigation of leakage flux in a capacitive angular displacement sensor used in specialised limited-angle torque motors for rotor position sensing. This is done by 3D FE modelling and computation of electrostatic field distribution in the complex geometry of the cylindrical sensor. The effects of the axial leakage flux are investigated by comparing the results of 3D and 2D electric field distribution in this small length-to-diameter ratio sensor. Results are presented in terms of 3D potential plots and the effects of leakage flux are assessed by comparing the output characteristics, $V = f(\theta)$ of the sensor obtained by modelling and experiment.

1.   Introduction and Constructive Features of the Capacitive Sensor

The sensor under investigation is an all-metal cylindrical displacement sensor designed for sensing the angular position of rotor of limited angle torque motors used for frame scanning in thermal imaging devices. Fig. 1 shows the longitudinal (a) and cross (b) sections of the sensor. It consists of four outer electrodes, one centre electrode and two rotating electrodes. The four segmented outer electrodes, insulated from one another by thin insulation strips (Fig. 1b) are

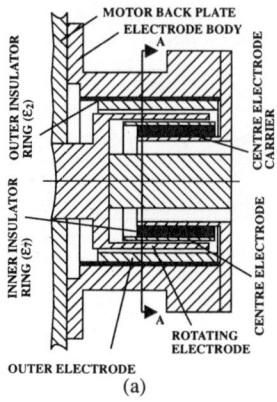
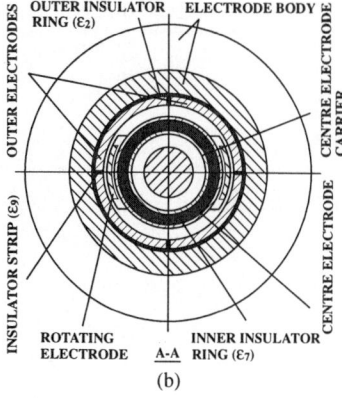

Fig. 1. Longitudinal (a) and cross (b) sections of the capacitive angular displacement sensor under investigation showing various constructive features (not drawn to scale).

mounted symmetrically on the inner surface of the electrode body. The cylindrical centre electrode is positioned concentrically with the outer electrodes by the centre electrode carrier. The rotating electrodes, attached coaxially to motor rotor, rotate in the airgap between the inner and outer electrodes. Potentials of the same magnitude but opposite polarities are given to the outer electrodes, while the rotating electrodes are kept at zero potential. The electrode body, the centre electrode carrier and the motor body are earthed and act as a shield against stray fields. For precise and effective position sensing the variation of the sensor output voltage V, measured between the centre electrode and virtual earth, with angular position θ needs to be highly linear. The design target for the output characteristic $V = f(\theta)$ is to achieve a linearity of 0.02% for a limited angle frame scan motor that has maximum end to end deflection of ±8°.

As can be seen from Fig. 1, the sensor under investigation possesses small length/diameter ratio which tends to increase the proportion of axial leakage flux in the total flux between the outer and centre electrodes. The magnitude and distribution of this leakage flux change with rotor position. Hence a qualitative and quantitative estimate of the effects (if any) of this flux on the output voltage is necessary in order to be able to design sensors with desired output characteristic $V = f(\theta)$. This is especially true when comparison of the results of 2D FE modelling of the sensor with experimental data have shown some differences [1, 2] attributable to several factors, one of which may be the leakage flux which could not have been considered in 2D models. Hence the necessity for 3D modelling described in this paper.

2. Realisation of the 3D Finite Element Models of the Sensor

The 3D FE modelling of the sensor shown in Fig. 1 is carried out by setting up appropriate FE models of the sensor and solving the following Laplace's equation [3, 4]:

$$\nabla \cdot [\varepsilon(x, y, z)\nabla\Phi(x, y, z)] = 0 \tag{1}$$

which gives the electric field distribution in the sensor. Under given boundary conditions in terms of specified potentials, the above equation is solved for unknown potential distribution $\Phi(x, y, z)$ by finite element method (FEM) [4, 5] using commercial FE package TOSCA and OPERA-3D [6]. Fig. 2 shows the typical FE model used for simulation purposes. It contains about 75000 8-noded hexahedral elements made up of the same number of nodes. For the sensor

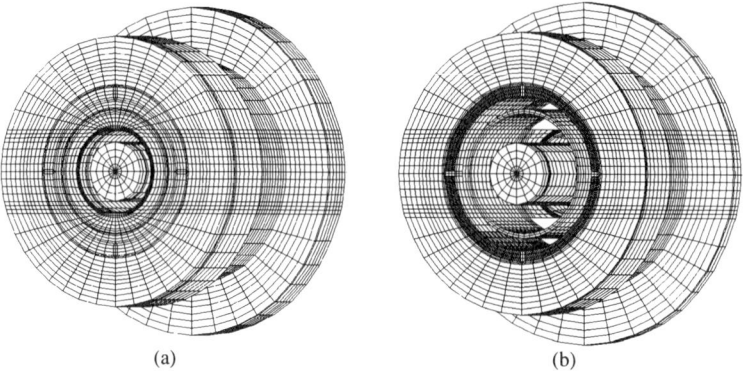

(a)   (b)

Fig. 2. 3D finite element model of the capacitive angular displacement sensor under investigation (a) with and (b) without the centre electrode carrier showing electrode body, centre electrode, motor back plate etc.

with approximate outer diameter of 25 mm and axial length of 17 mm, the resulting mesh density provides modelling accuracy comparable to that obtained from accurate 2D models [2].

3.   Results and Discussion

Figs. 3-4 show some of the results obtained by 3D FE modelling of the sensor. Fig. 3 shows the equipotential contours distributed on the radial plane of the sensor, the detailed study of which may be carried out adequately by 2D modelling [1, 2]. Fig. 4 shows the distribution of equipotential contours on the axial plane of the sensor obtainable only by 3D modelling. It clearly shows the distortion of potential distribution at the ends (marked **A** and **B**) of the outer electrodes. A close look at these two ends (Figs. 4b and 4c) reveals that at **A**, there is less fringing of field in the axial direction than that at **B**. This is due to the close proximity of the earthed central electrode carrier (Fig. 1a) at **A** which acts as a more effective electrostatic screen than the motor back plate, placed further from the electrodes at the other end **B**. However the axial potential distribution is more distorted radially at **A** than that at **B**. As can be seen from Fig. 4b, this is attributed to the differences in axial lengths and alignments of the outer, rotating and centre electrodes. The effect of these factors on the output characteristic of the sensor may be seen from curve 3 in Fig. 4d, which is calculated approximately by introducing correction factors to curve 2 obtained by 2D modelling. For various angular positions θ, these correction factors are calculated as the ratio of charges on the centre electrode determined by 3D and 2D modelling. Although this gives a preliminary indication of the effects of leakage field on the output characteristic, further modelling investigations are needed before any definitive conclusions can be made.

4.   Conclusions

Preliminary results based on 3D modelling show that the axial leakage field does not seem to have much effect on the linearity of the output characteristic of the sensor investigated. From the results and discussions in Section 3, it seems it should be relatively simple to reduce the extent of fringing and distortion of this field by redesigning the rectangular base of the rotating electrodes (for example, to a circular disc that would more effectively screen the field at **B** (Fig.

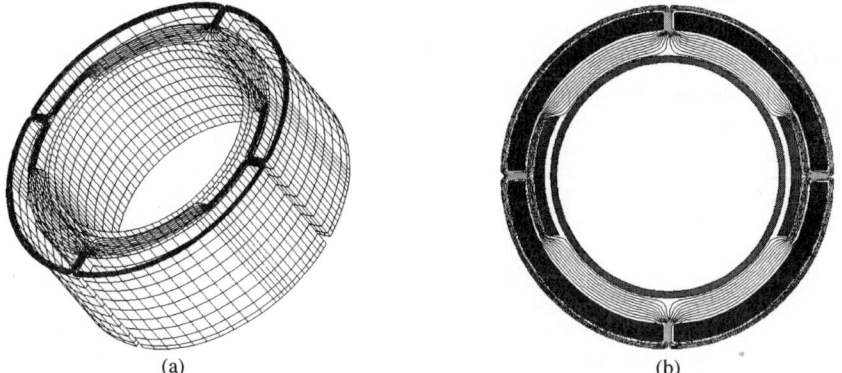

(a)   (b)

Fig. 3. Results of 3D field modelling: (a) radial equipotential contours in the capacitive sensor for angular position θ = 0, and its (b) 2D equivalent, the detailed study of which is done by 2D finite element modelling [1, 2].

1174

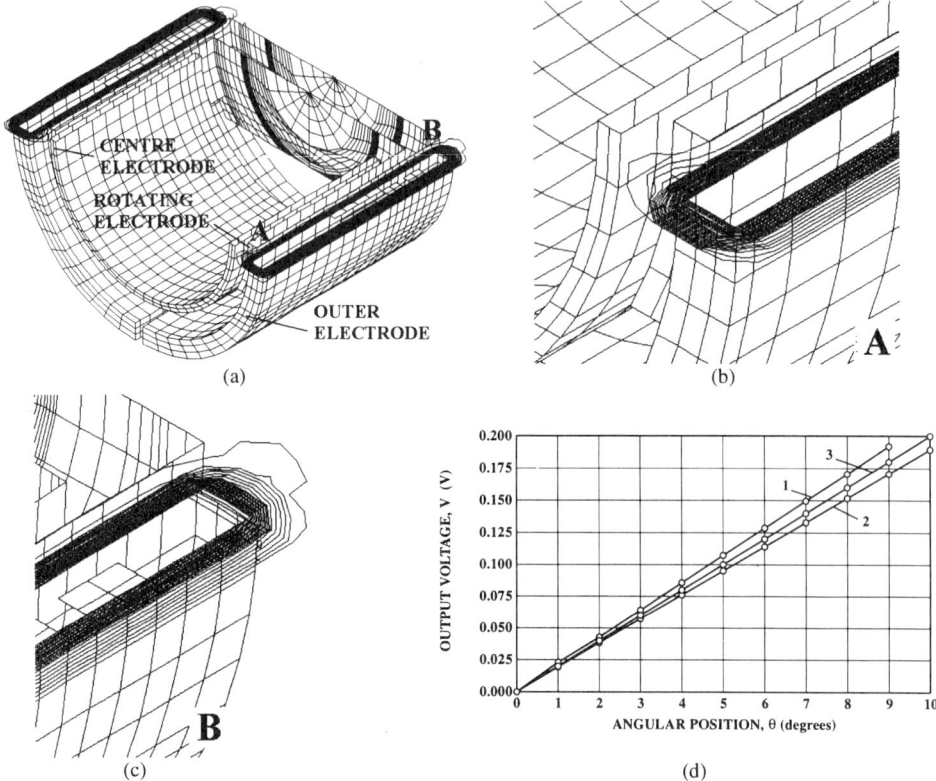

(a)

(b)

(c)

(d)

Fig. 4. Results of 3D field modelling: (a) axial equipotential contours around the outer electrodes showing distortion in potential distrubution at the ends (marked **A** and **B**); (b), (c) detailed distribution of potential at these two ends; (d) output characteristics of the sensor obtained by 2D (curve 2) and 3D (curve 3) modelling and their comparison with the experimental characteristic (curve 1).

4c)) and by aligning the ends of the outer, rotating and centre electrodes along the same plane at **A** (Fig. 4b). Further 3D modelling investigations are needed to address these design concepts.

5. Acknowledgement

The authors would like to thank the Engineering and Physical Sciences Research Council (EPSRC), UK for financing the work under an EPSRC Grant.

References

[1] S H Khan and F Abdullah 1996 *Studies in Applied Electromagnetics and Mechanics* 10, eds. A. J. Moses and A. Basak, (The Netherlands: IOS Press) 704-707.

[2] S H Khan, L Finkelstein, and F Abdullah 1997 *IEEE Trans. on Magnetics* **33** 2081-2084.

[3] J D Kraus 1992 *Electromagnetics* (London: McGraw-Hill Inc.).

[4] K J Binns, P J Lawrenson, and C W Trowbridge 1992 *The Analytical and Numerical Solution of Electric and Magnetic Fields* (Chichester: John Wiley & Sons).

[5] P P Silvester and R L Ferrari 1990 *Finite Elements for Electrical Engineers* (Cambridge: Cambridge University Press).

[6] TOSCA, Version 6.6, OPERA-3d, Version 2.609,1997, Vector Fields Limited, Oxford, UK.

Simulation and Modelling
Paper presented at Eurosensors XII, 13–16 September 1998
© 1998 IOP Publishing Ltd

A new fitting method for data analysis in impedance spectroscopy

A Schwake, H Geuking and K Cammann

Institut für Chemo- und Biosensorik e.V. (ICB), Mendelstraße 7, D-48149 Münster, Germany, tel.: +49-251 980-2895, fax: +49-251 980-2890, e-mail: schwaka@uni-muenster.de

Abstract. A new method for analyzing impedance data presented in the complex plane plot is introduced. The advantage of our approach is that it can be easily employed with commonly available spreadsheet calculation software like Microsoft Excel™. It is possible to define a semicircle by three points of the circumference. Knowing the trigonometric function, the radius, e.g. representing the bulk resistance of ion selective membranes, can easily be calculated. The error of the data fit is determined by considering the standard deviation of the data points refered to the fitted semi circle. The geometric capacitance of ion selective membranes can be calculated by plotting the real part of the impedance against the angular frequency. If the equivalent circuit of the system is given it is possible to obtain the geometric capacitance by employing the non linear least square method. The efficency of our method has been verified by analyzing different networks.

1. Introduction

The impedance method is a technique for complex plane analysis of networks. Reduction to equivalent, lumped R, C, L circuits is an essential tool in understanding microscopic processes occuring in real systems. In the case of ion selective electrodes it is possible to determine important parameters, e.g. salt mobilities in the membrane phase, ion pairing constants and surface rates [1-2]. To obtain the values of these physical quantities from the measured impedance spectra, data analysis is necessary. Nowadays, the preferred method of analyzing impedance data is the application of the complex nonlinear least squares fitting (CNLS) [3-6]. In such CNLS fitting the entire impedance function (real and imaginary parts or modulus and phase) is fitted to a model or equivalent circuit thought to represent the measured response. The best model for data fitting can be found by the recently described method of Zoltowski [7-8]. However, for performing the CNLS method on a computer a special data evaluation software is necessary. In contradistinction to this, our approach described in this paper can be applied by the use of commonly available spreadsheet calculation software like Microsoft Excel™.

2. Analyzing the real part of the impedance

Typical impedance spectra of simple RC circuits in the impedance complex plane have the shape of semi circles. The diameter of the semi circle represents the value of the resistor. Hence, the trigonometric function of the semi circle has to be fitted from the data points. This

Figure 1. Analyzing the ohmic resistor in the impedance complex plane plot by calculating the trigonometric function of the semi circle.

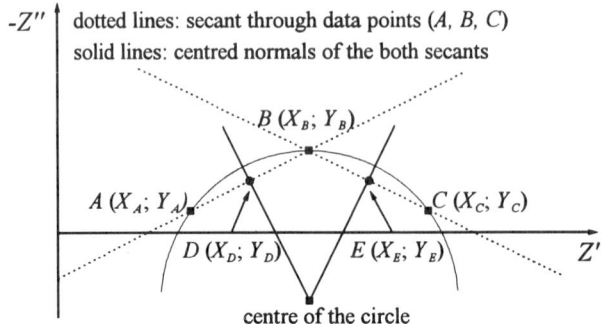

centre of the circle

can easily be done by choosing three data points (A, B and C in fig. 1). These define two secants (dotted lines in fig. 1). The points D and E in figure 1 are given to be the centres of the both secants. Using the coordinates of D (X_D; Y_D) and E (X_E; Y_E) the function of both centred normals (solid lines in fig. 1) can be determined. The intersection of the centered normals is defined to be the centre of the circle. Equation 1 gives the coordinates X_M and Y_M of the centre:

$$X_M = \frac{(Y_B - Y_A)\cdot(Y_C^2 - Y_B^2 + X_C^2 - X_B^2) - (Y_C - Y_B)\cdot(Y_B^2 - Y_A^2 + X_B^2 - X_A^2)}{2\cdot[(X_C - X_B)\cdot(Y_B - Y_A) - (X_B - X_A)\cdot(Y_C - Y_B)]}$$

(eq. 1)

$$Y_M = \frac{1}{2}\cdot(Y_A + Y_B) + \frac{X_B - X_A}{Y_B - Y_A}\cdot\left[\frac{1}{2}\cdot(X_A + X_B) - X_M\right]$$

With equation 2 it is possible to calculate the radius r of the semi circle, representing the half of the value of the ohmic resistor in ohms using the coordinates of one data point on the circumference (e.g. X_C; Y_C) and the coordinates of the centre (X_M; Y_M):

$$r = \sqrt{(Y_C - Y_M)^2 + (X_C - X_M)^2}$$

(eq. 2)

The whole trigonometric function of the semi circle is given by equation 3:

$$r^2 = (X - X_M)^2 + (Y - Y_M)^2$$
$$\Leftrightarrow (Y - Y_M)^2 = r^2 - (X - X_M)^2$$
$$\Leftrightarrow Y - Y_M = \pm\sqrt{r^2 - (X - X_M)^2}$$
$$\Leftrightarrow Y = Y_M \pm\sqrt{r^2 - (X - X_M)^2}\;;\; +\Rightarrow \text{semicircle up}\;;\; -\Rightarrow \text{semicircle down}$$

(eq. 3)

Additionally a complete data analysis requires an error specification of the calculated parameter as well. In our case this is done by determing the deviation d form the fitted trigonometric function of every data point in radial direction (fig. 2, next page). The values of d can be calculated using equation 4:

Figure 2. Calculating the deviation d of the data points from the fit function.

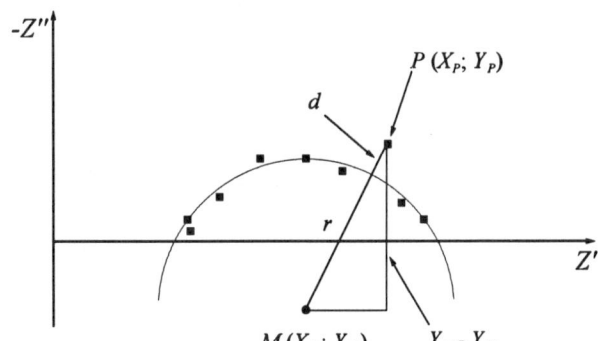

$$d = \sqrt{(P_y - M_y)^2 + (P_x - M_x)^2} - r \qquad \text{(eq. 4)}$$

Terming the number of acquired data points with N and the deviation of any data point with d_N the standard deviation σ_r of the radius r is given considering equation 5:

$$\sigma_r = \sqrt{\frac{\sum_1^N d_N^2}{N}} \qquad \text{(eq. 5)}$$

The confidence interval is then set to be $2 * \sigma_r$. This ensures 96% probability for accuracy of the calculated error. For obtaining the optimal data fit it is nescessary to choose the best of all $z = \binom{N}{3} = \frac{N!}{3! \cdot (N-3)!}$ possible data triples (N= number of data points). Each of those z triples forms a semi circle. The best of them is the one with the least standard deviation σ_r.

3. Analyzing the imaginary part of the impedance

For analyzing the imaginary part of the complex plane plot it is necessary to know the equivalent circuit of the experimental system. Here the easiest case, a simple RC parallel circuit is assumed. The impedance Z for this network is given by equation 6:

$$Z = \frac{1}{R^{-1} + iC\omega} = \frac{R}{1 + iRC\omega} \qquad \text{(eq. 6)}$$

Multiplication with the conjugated complex form of the denominator results in equation 7:

$$Z = \frac{R(1 - iRC\omega)}{1 + R^2 C^2 \omega^2} = \frac{R}{1 + R^2 C^2 \omega^2} - i \frac{R^2 C\omega}{1 + R^2 C^2 \omega^2} \qquad \text{(eq. 7)}$$

Equation 7 is a description of a circle in the parameter form since both the real and imaginary

Figure 3. Verification of our method by analyzing a parallel RC circuit.
(denoted values: *R= 130 kΩ; C= 150 pF*)

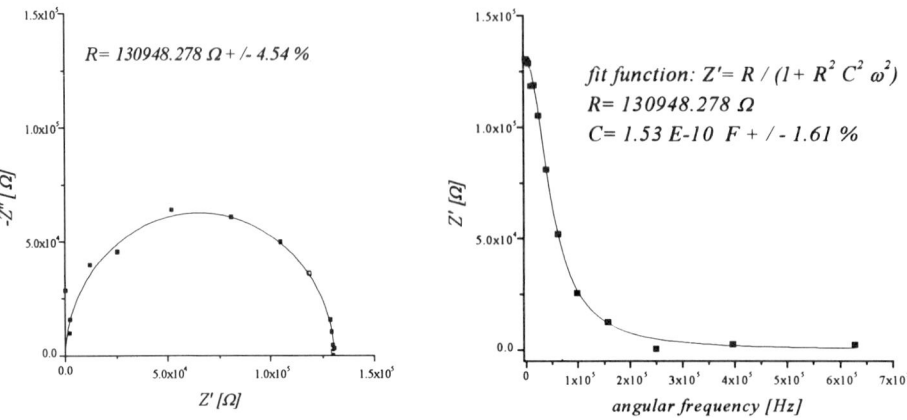

$R= 130948.278\ \Omega + /- 4.54\ \%$

fit function: $Z'= R\ /\ (1+\ R^2\ C^2\ \omega^2)$
$R= 130948.278\ \Omega$
$C= 1.53\ E\text{-}10\ F\ +\ /\ -\ 1.61\ \%$

part of the impedance Z depends on a third variable, the angular frequency ω, which represents the third dimension in the impedance spectroscopic plot. The calculation of the capacity C is done in the Z' - ω - projection. Based on equation 7 the fit function for the capacity is found to be:

$$Z'=\frac{R}{1+R^2C^2\omega^2}\Leftrightarrow Z'\left(1+R^2C^2\omega^2\right)=R\Leftrightarrow 1+R^2C^2\omega^2=\frac{R}{Z'}\Leftrightarrow R^2C^2\omega^2=\frac{R}{Z'}-1$$

$$\Leftrightarrow C^2=\frac{\dfrac{R}{Z'}-1}{R^2\omega^2}\Rightarrow C=\sqrt{\frac{\dfrac{R}{Z'}-1}{R^2\omega^2}} \qquad\qquad (eq.\ 8)$$

4. Application of the new method

The efficency of our method was shown by analyzing a parallel RC network (*R= 130 kΩ; C= 150 pF*). The fit functions for the Z' - Z'' and the Z' - ω planes calculated with the approach presented in this paper are shown in figure 3. The fitted values (see fig. 3) conformed very well with the given values keeping in mind that the error of both circuit elements were denoted by the manufacturer with 1 %.

5. References

[1]   Buck R P 1980 *Hungarian Scientific Instruments* **49** 7-23
[2]   Buck R P 1982 *Ion-Selective Electrode Review* **4** 3-74
[3]   Macdonald J R 1992 *Solid State Ionics* **58** 97-107
[4]   Macdonald J R 1993 *Electrochimica Acta* **38** 1883-90
[5]   Macdonald J R, Schoonman J, Lehnen A P 1982 *J. Electroanal. Chem.* **131** 77-95
[6]   Macdonald J R *Impedance Spectroscopy* (New York: Wiley Interscience)
[7]   Zoltowski P 1997 *J. Electroanal. Chem.* **424** 173-8
[8]   Zoltowski P 1994 *J. Electroanal. Chem.* **375** 45-57

Simulation and Modelling
Paper presented at Eurosensors XII, 13–16 September 1998
© *1998 IOP Publishing Ltd*

Genetic Learning of Neural Networks for Calibration in Temperature and Humidity Measurement

G.Kleymenov[1], S.Krutovertsev[2], A.Piskounov[1] and S.Vdovichev[1]

1) Moscow Institute of Electronic Technology (Technical University), Moscow, 103498, Russia
Phone: 7-(095)-534-0264   Fax: 7-(095)-530-2233   E-mail: irmiet@orgland.ru

2) Joint Stock Company "Practic-NC", Moscow, 103498, Russia
Phone/Fax: 7-(095)-531-9751   E-mail: pnc@orgland.ru

Abstract: Monitoring the temperature and humidity in an isolated chamber is looked upon as an approximation task. The article is devoted to the usage of neural networks for solving the task. A method of tuning the neural network parameters is described, based on genetic algorithms. The approximation of a two-variable function is considered.

1. Introduction

In the process of humidity measurement an obstacle concerning selectivity difficulties is occurred. The reason is that a capacitor commonly used as a humidity sensor, has also a temperature-dependant parasitic resistance that influences the sensor reading [1]. It results in a sensor depending not only on humidity but temperature as well. A properly developed approximation method can help to solve the problem. Moreover, in calibration one cannot avoid methodological errors, so, the approximation method must perform a smooth approximation. Besides, when implementing one or the other method in a device, both the computational complexity and memory required for storing and transmitting the calibration data, become important as well.

2. Existing Approaches

At present there are two most popular approaches to performing the approximation. *The tabular method* is the most widely used one, since it eliminates sophisticated computation. However, this method requires a large memory for storing and transmitting the calibration characteristics. As for approximation of functions of two or more variables, the amount of memory becomes unacceptably high. Also, an auxiliary smoothing is often needed to reduce the noise effect. *The polynomial method* consists in finding a polynomial $Q(x)$ (usually obtained by the least square method) with the power of n (or less), which gives output values close enough to $\{y_i\}$ for the input $\{x_i\}$. The resulting curve is to be smooth and reduce noise effect.

Compared to the tabular method, the polynomial one has obvious advantages. Its implementation imposes less strict memory requirements as well as a better smoothing capability, which decreases the noise effect. On the other hand, this method presumes sufficient accuracy of mathematical operations (mainly, multiplication). Its application to approximating a function of two or more variables faces considerable difficulties, since there is neither an standard form of approximating polynomials, nor effective algorithms for finding their coefficients. Also, the polynomial method demonstrates lack of flexibility (see fig.1). While the polynomials with the power of 1 or 2 do not ensure the accuracy required, the polynomials with power of 3, 4 or higher do not provide adequate approximation: although giving good values for the calibration points, they do not match the dynamic behaviour of the function being approximated (see fig.2).

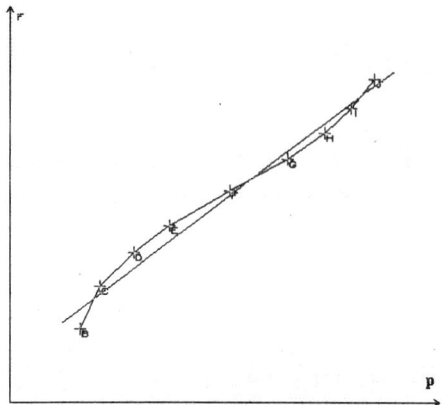

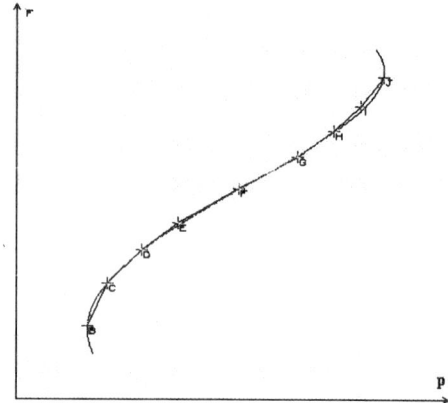

Fig.1. *Polynomials with the power of 1 or 2*   **Fig.2**. *Polynomials with the power of 3 or 4*

3. Neural Network Approximation With Genetic Tuning

Some successful solutions of the approximation task using neural networks have made us believe in the efficiency of this approach, although it is not an easy one to implement. For instance, the commonly used methods of learning neural networks are gradient oriented, which makes it difficult to exploit them when:

1) the network structure is not fully interconnected;
2) the output function carried out by neurones, are not differentiable;
3) weights and biases conform some restrictions, e.g. in contrast to a generally used universe of real numbers, the learning algorithm should allow integer values of weights and biases;
4) in computation of the neurone outputs, such mathematical operations as multiplication of two real figures, are to be avoided.

To avoid limitations of that kind, one needs non-analytic methods for learning neural networks. Genetic Algorithms method is one of the appropriate ones.

Any application of genetic algorithms implies determining the form of chromosome representation, the way of generating the initial population, developing genetic operators (selection, crossover, mutation and replacement) and an evaluation function. In respect with the approximation of a real variable function $y=f_i(x)$, defined on a set of calibration points $\{x_q,y_q\}$, with no limitations for the operation of neurones, genetic algorithms can be applied as follows (taking a feedforward network as an example).

A chromosome consists of weights $W=\{w_{ij},w_{jk}\}$ and biases $B=\{b_j,b_k\}$ (here the indexes i correspond to the input layer, j - the hidden one, and k - to the output layer), representing a sample of tuning of the network. Since a genetic algorithm starts searching from an arbitrary point in the search space, the initial population of chromosomes is formed randomly.

The training set includes the calibration points $\{x_q,y_q\}$, determining the function $y=f(x)$. For each x_q arriving at the input of the neural network, a reaction y'_q is computed. The closer y'_q is to the correct values y_q of the function $f(x)$, the more feasible is the combination of weights and biases represented by a chromosome. So, its fitness is determined as follows:

$$fitness = \sqrt{\frac{1}{K}\sum_{q=1}^{K}(y_q - y'_q)^2}$$

The search of weights $W_p=\{^pw_{ij},^pw_{jk}\}$ and biases $B_p=\{^pb_j,^pb_k\}$ is carried out until a termination condition of the genetic algorithm is reached. The coefficients W_N and B_N are the desirable quasi-optimal result for approximation.

4. Experimental results

As mentioned above, the calibration procedure can be considered as a task of approximating a function $y=f_1(x)$, which provides an output value y for one input x. The simplest feedforward neural network needs to have one neurone at the input layer as well as one at the output layer. For the hidden layer two neurones may be used, each of them performing the following function:

$$h_j = \frac{1}{1+e^{-s_j}}, \quad \text{where } s_j = \sum_{i=1}^{L} w_{ij} \cdot x_i + b_j, \quad j=1...M.$$

For the purpose of this task it is reasonable to use a simpler expression - first three members of a relevant series:

$$h_j = \begin{cases} 1 - \dfrac{2}{1+(1-s_j+\dfrac{s_j^2}{2})}, & s_j \geq 0 \\[4mm] \dfrac{2}{1+(1+s_j+\dfrac{s_j^2}{2})} - 1, & s_j < 0 \end{cases}$$

The output neurone sums up the weighed values arriving from the hidden level, and adds a bias:

$$y_k' = \sum_{j=1}^{M} w_{jk} \cdot h_j + b_k, \quad k=1...N.$$

The example below represents the operation of such network. Suppose that it is necessary to calibrate a unit converting frequencies x_q of the incoming signal into the temperature values y_q which is represented in the following table:

Q	1	2	3	4	5	6	7	8	9
x_q (kHz)	8.8	22.8	33.8	42.6	54.0	64.3	72.8	80.9	90.4
y_q (°C)	7.9	14.0	23.8	34.3	51.4	68.5	79.0	86.3	93.3

These are the parameters of the genetic algorithm used when learning the neural network: number of iterations - 3000; population size - 50 chromosomes; mutation rate - 1.0; crossover rate - 1.0; selection operator - the best chromosome and the one chosen at random; crossover operator - real-coded crossover [2,3]; mutation operator - real-coded mutation [2,3]; replacement - of the least feasible chromosomes. Learning of this neural network resulted in the following weights and biases:

$$w_{11} = 0.0283 \quad b_{11} = -2.2314 \quad w_{21} = 14.103 \quad b_{21} = 55.345$$

$$w_{12} = 0.0625 \quad b_{12} = -3.3345 \quad w_{22} = 49.826$$

The fig.3 depicts both the initial calibration points and the approximating curve. The mean square error in the calibration points is $\sigma = 0.229$, which is as good as a value that can be obtained via the traditional learning procedure with the gradient descend method.

Reduction of the temperature dependence of a humidity sensor can be considered as approximation of a two-variable function. This may be implemented as a neural network with a similar structure as above, with two neurones at the input layer (each corresponds to one of the variables). Having tuned the network by a genetic algorithm, the results of calibration obtained are presented in the fig.4. The calibration points B, C, F, G, H, J and L correspond to the mean temperature (+20°C). The other four points were measured at the border of the changes of both temperature and humidity (points D and I – at +40°C whereas E and K – at +0°C). The calibration points are denoted by 'x', and the curves represent the functions being approximated.

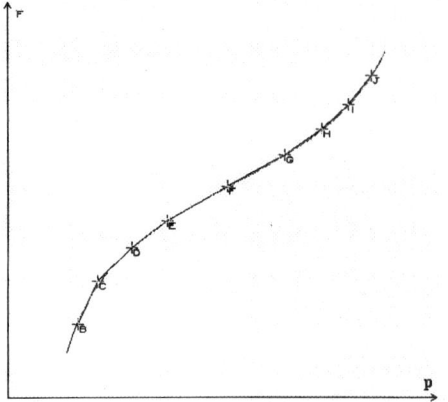

Fig.3. *Approximation with Neural Networks*

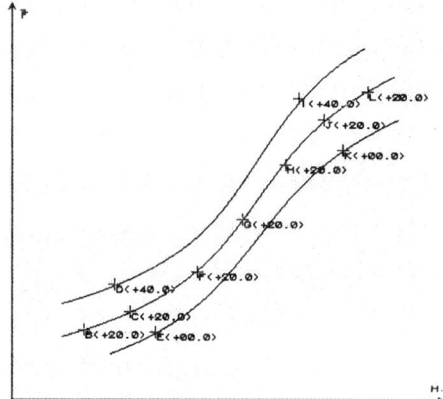

Fig.4. *Approximation of function of 2 variables*

5. More opportunities

The results obtained make it possible to reduce the effect of the temperature dependence of humidity sensors. Besides, the approach suggested in this article has advantages when some auxiliary restrictions are introduced, e.g. approximation of an integer variable function $y=f(x)$, determined on a calibration set $\{x_q, y_q\}$, with a limited set of operations used for computing the neurone's output. These restrictions arise from an applied task where the core of a control system is a plain microcontroller processing 8-bit (or 16-bit) integers and performing simple arithmetic operations such as summing, shifting to the left and to the right, etc.

6. Conclusion

The approach suggested here provides a new basis for the implementation of the calibration procedure, reducing the effect of the temperature dependence and integrating two modern intellectual technologies (neural networks for approximation and genetic algorithms for their learning). Its practical realisation needs a moderate amount of memory and computing resource, which eventually result in reduction of costs.

References

[1] S.Krutovertsev, A.Tarasova, L.Krutovertseva, A.Zorin *Integrated multifunctional humidity sensor*, Sensors and Actuators A, vol.62, 1997, pp.582-585.

[2] Alden Wright H. *Genetic Algorithms for Real Parameter Optimization* ed. by Rawlins Gregory J.E. Foundation of GAs, 1991

[3] Piskunov A., Puchinin D., Vdovichev S. *Genetic Algorithm Using for Control System Identification* Proceedings of the 5-th European Congress on Intelligent Techniques and Soft Computing. Aachen, Germany, 8-11 September, 1997, vol.2, pp.1456-1460

Simulation and Modelling
Paper presented at Eurosensors XII, 13–16 September 1998
© *1998 IOP Publishing Ltd*

Modeling of microsystems with analog hardware description languages

S. Marco, M. Carmona, J. Sieiro, J. Samitier

Enginyeria i Materials Electrònics (associated unit to CNM-CSIC),
Departament d'Electrònica, Universitat de Barcelona
Av. Diagonal 645-647, 08028-Barcelona, Spain

Abstract: A methodology for the modellization of microsystems with analog hardware description languages is presented. This procedure implies several steps with emphasis in model complexity reduction, model adaptation, identification of critical parameters and partial validation of the system sub-components followed by full verification of the complete model. This procedure will be applied to the modellization of a themo-pneumatic micropump

1. Introduction

The complexity of microsystems makes modeling an essential need for successful development. While up to now, modeling has been mainly focused on the device level, typically using FEM, the maturity of the microsystems technology needs more and more simulation to be carried out at the system level. It is our purpose to present a methodology for microsystems modeling at high levels of abstraction. We understand by modeling the rigorous process to be follow to ensure a fair transition from the 'real' system to a mathematical 'object', necessary for simulation and control. This approach will be illustrated with the modeling of a thermopneumatic micropump. The main problems found will be briefly discussed in the final part of this extended abstract. As a tool for the implementation of system level models we will use analog hardware description languages[1]. Among their advantages we should remark the possibility to carry out simulations mixing different levels of abstraction and the possibility to couple signals from different physical domains[2].

2. Model Building Procedure

The simulaion of a microsystem at the system level needs a reduction in the number of DOF to be calculated. This abstraction implies a loss of information about the system. For reliable model use, with only minimum information loss, a rigorous procedure for the abstraction is going to be outlined.Our strategy to build such physical models is as follows (Figure 1):

    step 1) First an examination of the system architecture allows to define an adequate lumped representation of the system which retains an adequate topology and the couplings of different physical phenomena.

    step 2) Then it is necessary to write down the model describing the behavior of every individual element.

    step 3) Except for very simple elements, this stage requires a previous analysis at the

1184

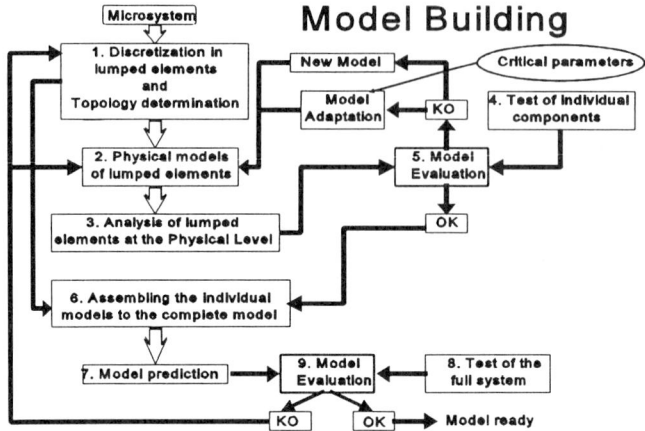

Fig.1 Model building procedure

physical level using Finite Element Analysis. From this physical level, some integrated output can be extracted and the equation of the component can be written down. Some times the inherent distributed nature of the problem makes difficult the extraction of a lumped representation. In this case, a discrete approximation to the continuous model should be introduced. One typical example are thermal problems: except in very simple geometries where the heat flow paths are clear, the problem has to be analyzed at the physical level, or alternatively, the geometry can be substituted by a macromodel consisting of a network of resistors and capacitors.

Verification of the models through the comparison between experimental data and model predictions is a keypoint in the model building procedure. Moreover, we recommend verification to be carried out at two levels. One the one hand, individual component models have to be tested On the other hand, the whole assembled model at the system level has to be tested again.

Step 4) For model component verification it is necessary to plan a set of tests for every individual component.

Step 5) Despite the development of a physically correct model, it is usual that the predicted behavior exhibits *significant departures* from the real behavior. This is mainly due to the lack of well determined values of the parameters describing the model. A typical example are residual stress values of thin-films which are very sensitive to processing conditions. It is then necessary to *adapt the model* to the experimental data to extract the values of these parameters. Here a double-check is necessary: first the extracted values have to fall within the expected range according previous works, and secondly, it is always useful to check the dependencies among the different parameters. It is usual that these parameters appear to be correlated. (When modelling thermal problems, we found such problem related to the values of the heat exchange coefficients due to natural convection and conduction and emissivities) Then it is difficult to restrict them to a small region of the parameters space. When this occurs, a different set of parameters has to be defined or more experimental data has to be collected.

Step 6) Once the quality of every individual component model has been assesed the complete model can be assembled.

Step 7) This model will output some system behaviour predictions which have to be compared again to the experimental behaviour of the microsystem.

Step 8) For such purpose it is necessary to test the complete demonstrator. We would like to remark that verification of the model reliability has to be tested in a different set of experimental data than those used for model adaptation. This would serve as a check for evaluating the predictive capabilities of the component models.

Step 9) The verification of the complete model is quite important especially when the coupling of different physical domains has been done at the system level. Because this coupling is usually inherent to the operation of transducers and actuators it can not be neglected at all. This coupling usually can change the behavior of the individual components. From our experience we recommend at this stage to simulate the whole test system, not only the microsystem under test. This procedure will show perhaps unexpected influences of the measuring set-up in the microsystem behavior.

3. Modeling Examples

Because of space reasons, it is not possible to describe in detail several examples of this modeling procedure, so we will present here just an example: a brief description of the development of an analog behavioral model of a thermo-pneumatic micropump[4].

The problem is quite complex due to 3D effects and the simultaneous presence of thermal, structural and fluidic problems. In this case, our choice has been to use the fluidic network as the main criteria to define the topological structure of the model (step 1). Then we have chosen a lumped representation where the pumping membrane, the channels and the microvalves are the constituting elements (Figure 1) Every individual element has been analyzed at the physical level to extract the corresponding models (step 2-3). The pumping membrane has been analyzed by FEM to find the displaced volume-pressure characteristic of the element. In a similar way, channels have been analyzed by FEM to solve the Navier-Stokes equation. A series of analysis for a range of pressures has permitted to extract the flow-pressure relationship as an empirical law. Modeling the microvalves is a complex problem because of the fluid-structure interaction. In this case we chose to study by FEM the structural part, while the fluidic part was represented by a set of algebraic equations. The fluidic-structure interaction was also handled at the behavioral level (Figure 2). It remains to decide how to model the thermo-pneumatic actuation. This is a

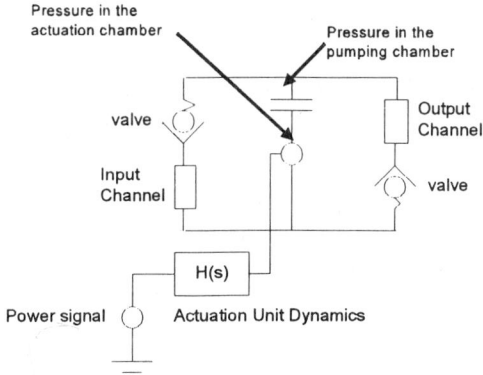

Fig 2: Schematic representation of the behavioral model.

1186

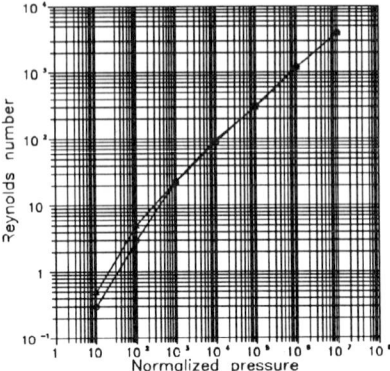

Fig. 3 Flow-pressure characteristic for microholes: comparison FEM simulations with
experimental measurements

coupled thermo-mechanical problem where we chose as input the power to the heater and as
output the pressure in the pumping cavity. In this case, we analyzed the dynamic problem at the
physical level as a way to effectively tackle the distributed characteristic of the problem. The
dynamics of this system has been identified as a functional transfer-function. The black-box
approach has the disadvantage that the parameters are not easily related to the geometry[3].

Every model of a system subcomponent has been tested: temperature distribution on the
chip, thermo-mechanic dynamic behaviour, fluidic pressure losses in the inlet and outlet. *Model
adaptation* was required in the thermal model, where the results, although qualitatively correct,
showed non-negligible differences against the experimental values. The critical paramenter in this
case was the thermal exhange coefficient whose value was a rough estimation in the 1st evaluation
of the model. The correctness of the individual models give more reliability to the full model,
although model faults concerning ignored cross-talks can still persist. The evaluation of the
correctness of the topology can only be done with the complete model.

4. Summary

System level simulation using Analog Hardware Description Languages is a powerful tool.
However, simulation results are only reliable if simulation is supported by high quality models.
A procedure for model definition has been proposed and some critical points have been discussed.
An brief example has been given concerning the simulation of a thermo-pneumatic micropump.

References

[1]   Folkmer B. and Sandmaier H. 1995, in *Simulation and Design of Microsystems and
        Microstructures*, (Southampton, Computational Mechanics), p. 321.
[2]   Shi CJR and Vachoux A, 1995 *Modeling in Analog Design, Current Issues in Electronic
        Modeling*, al., (Dordrecht: Kluwer Academic), pp. 1.
[3]   Marco S, Lundquist U, Carmona M, Pardo A, and Samitier J, 1996 *Micromechanics
        Europe 1996, Barcelona*, p. 200.
[4]   Carmona M., Marco S, Nguyen H, Lundquist U, Pardo A, and Samitier J, 1996 *Design
        Conference on Integrated Systems, Sitges*, p. 271.

New model for piezoresistive properties of highly doped p-type polysilicon

S V Spoutai*'**, H G Chun**

*Novosibirsk State Technical University, 20, K.Marx prospect, Novosibirsk, 630092, Russia e-mail:ssput@hotmail.com; **University of Ulsan, P.O. Box18, Ulsan, 649-780, Rep. of Korea.

Abstract. This paper describes new approach to modellingpiezoresistive properties of polycrystalline semiconductors. The model combines the electro-physical properties of microcrystalline silicon films with that of monocrystalline silicon in the way that the former properties are attributed to the grain boundaries regions whereas the latter – to intragrain regions. The model is used to describe the dependence of gauge factor (GF) and its temperature coefficient (TCGF) of the highly boron doped polycrystalline silicon, $4 \cdot 10^{25} \dots 10^{27}$ m^{-3}, on the dopant concentration for different grain sizes, $10^{-8} \dots 10^{-6}$ m. It is concluded that for very high doping level the zero or even positive TCGF can be obtained.

1. Introduction

The possibility of application of thin polycrystalline silicon films for measurement of various mechanical quantities have been demonstrated and discussed by many researchers [1-4]. Several approaches were introduced for the modelling of the electro-physical properties of the polysilicon. The streamline of the modelling efforts was to consider grains as the small perfect crystals having the properties of monocrystalline silicon and being differently oriented with respect to some specific directions (in-plane orientation, texture). The properties of grain boundaries were calculated on the basis of theoretical models [4] or contribution of the grain boundaries to piezoresistance effect was "ignored" [1-3] mainly due to the lack of any experimental data and, hence, validated theoretical approach.

In our model we relied only on the experimentally obtained data and experimentally proven intermediate models.

2. Model

The model combines the electro-physical properties of microcrystalline silicon films with that of monocrystalline silicon in the way that the former properties are attributed to the grain boundary regions whereas the latter – to intragrain regions.

The properties of the regions of grains and grain boundaries in the case of highly doped polysilicon film may be obtained by techniques developed in [5]. Some results were presented recently [5, 6].

On the basis of the close agreement of the results [5] for the grain boundary regions at the paricular doping level and results [7] for the μc-Si (10 nm grain size) for the same doping level, we assumed that the resistive and piezoresistive properties of the grain boundaries in the highly boron doped polysilicon can be described by properties of μc-Si from [7].

The properties of the grains were assumed to be described by the properties of monocrystalline silicon with the same doping level. The gauge factors for the grains were taken from [2] as the different orientations and concentration dependence were already taken into account there. In addition, the gauge factors obtained in [5,6] for the intragrain regions were found close to data in [2].

As it was found in [5] the share of the total resistance of the highly boron doped polysilicon attributed to the grain boundaries is constant for the particular grain size. The dependence of the share on the grain size was taken from [8].

Thus, the proposed model uses data which are directly related to the experimentally observed entities.

3. Results

The results of the modelling are represented on the fig.1 and fig.2. The appearance of the graphs look similar to that in [1,3,4] but our results are extended to higher doping concentrations.

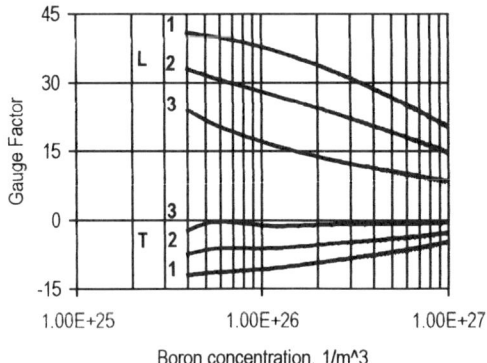

Fig.1 Calculated dependence of longitudinal (L) and transverse (T) gauge factor on the boron concentration for polysilicon with different grain size: 1, 0.1, 0.01 μm (1,2,3, respectively).

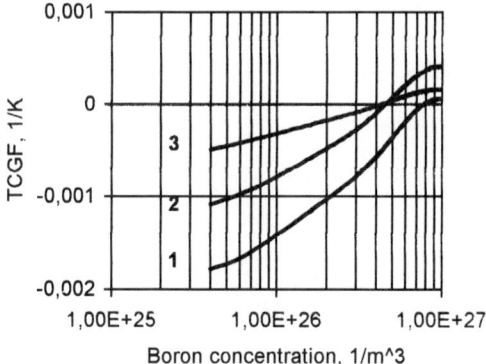

Boron concentration, 1/m^3

Fig.2 Calculated dependence of temperature coefficient of the longitudinal gauge factor on the boron concentration for polysilicon with different grain size: 1, 0.1, 0.01μm (1,2,3, respectively).

For the very high doping levels our model predicts the TCGF equal to zero or even positive that is not the case for monosilicon. The previous models [1-4] fail to explain the experimentally observed positive TCGF for p-type polysilicon with concentration of free carriers of $2 \cdot 10^{26} m^{-3}$.

4. Summary

The proposed model sinthesized on the basis of experimentally observed entities describes piezoresistive properties of highly boron doped polysilicon films in the range of boron concentrations $4 \cdot 10^{25} \dots 10^{27}$ m^{-3} and polysilicon grain sizes $10^{-8} \dots 10^{-6}$ m.

References

[1]   Mosser V, Suski J, Goss J and Obermeier E 1991 *Sensors and Actuators A.* **28** 113-132
[2]   Gridchin V A, Lubimsky V M and Sarina M P 1995 *Sensors and Actuators A* **49** 67-72
[3]   Shubert D, Jenchke W, Uhlig T and Schmidt F M 1987 *Sensors and Actuators* **11** 145-155
[4]   French P J and Evans A G R 1989 *Solid-State Electronics* **32** 1-10
[5]   Spoutai S V 1995 *Cand. of Sci.Dissertation* NSTU Novosibirsk Russia p 255 (in Russian)
[6]   Spoutai S V and Gridchin V A 1997 *Proc. Int. Symp. on Science and Technology, Ulsan,* (University of Ulsan, Rep. of Korea)
[7]   Germer W 1985 *Sensors and Actuators* **7** 135-142
[8]   Sarina M P 1992 *Cand. of Sci.Dissertation* NSTU Novosibirsk Russia p 196 (in Russian)
[9]   Golovko V, Gridchin V and Berenshtein O 1983 *Poluprov. Tenzometriya* 29-34 (in Russian)

Study of the mass sensitivity of SH-APM sensors with Maple V

Irène ESTEBAN, Corinne DÉJOUS, Dominique REBIÈRE, Jacques PISTRÉ and Roger PLANADE *

IXL - Université Bordeaux I (UMR CNRS 5818)
351, Cours de la Libération - F-33405 TALENCE Cedex
Phone : (33) 05 56 84 65 40 - Fax : (33) 05 56 37 15 45
email : esteban@ixl.u-bordeaux.fr

* Centre d'Études du Bouchet (DGA-DCE)
Le Bouchet BP n°3, F-91710 Vert le Petit

Abstract. Shear horizontal acoustic plate mode (SH-APM) devices offer interesting possibilities to probe liquid or polymer physico-chemical properties. As many parameters can influence their sensitivity, our research on these sensors involves modeling. This was based on a "matrix approach", and was carried out with MapleV. In this paper, we study the mass sensitivity of SH-APModes, in the case of a quartz plate covered by a copper phthalocyanine layer. The influences of geometrical parameters of the device and of the mode number are discussed. Results are compared to the mass sensitivity of SAW devices.

1. Introduction

SH-APM sensors have demonstrated the ability to measure modification of the physical or chemical properties of liquids [1]. The acoustic wave is generated and received by interdigital transducers deposited on a thin piezoelectric plate.

A matrix method was developed by J.C Andle [2] in FORTRAN language to model the propagation of acoustic waves. The SH-APM sensor can be considered as a multilayer waveguide displayed in figure-1. Our work consists in implementing a similar model using MapleV software, as it was depicted previously [3]. In this paper we recall briefly in section 2 the modeling of an SH-APM waveguide. Section 3 gives the results obtained by the software for the mass sensitivity of SH-APM devices.

2. Matrix approach

The matrix approach consists firstly in determining the acoustic displacements and electric potential of the waves that can propagate in an infinite medium [3]. Secondly, the modeling of the multilayer waveguide also requires boundary conditions on both plate faces : the continuity of U_i (i=1..4), T_{3j} (j=1,2,3), D_3. The complete system of boundary conditions was theoretically developed by J.C Andle [2].

With this model, the presence of a sensitive coating deposited on the piezoelectric plate only modifies the boundary equations.

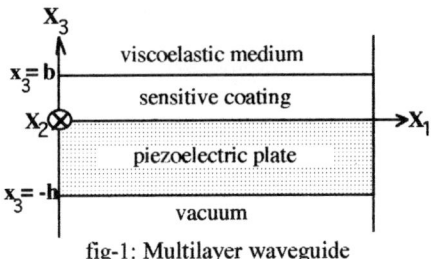

fig-1: Multilayer waveguide

3. Results

We consider the multilayer waveguide [vacuum / quartz / sensitive coating / vacuum]. The quartz plate cristallographic cut is defined by $(0,\theta,90°)$ Euler angles, for which the modes have a pure shear horizontal polarisation. The sensitive coating is a copper phthalocyanine (CuPC) layer considered as an isotropic medium, of density $\rho=1620$ kg/m^3. Its elastic stiffness coefficients were previously determined [4] : $\lambda=2.10^9$ N/m^2, $\mu=5.10^9$ N/m^2.

3.1. Influence of the thickness of the sensitive coating

Figure 2 gives the frequency variation absolute value, Δf_m, of the six first modes (m = 0 to 5), as a function of the CuPC thickness (b). For thin CuPC layer, Δf_m varies linearly with b. This linear domain is greater for higher order modes. Saturation occurs for thick CuPC layer : this can be attributed to the fact that the acoustic wave is only sensitive to perturbations close to the quartz/CuPC interface. Furthermore, one can see that the frequency shift is also greater for higher order modes.

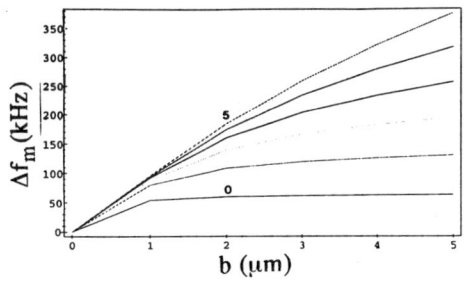

fig-2 : Influence of CuPC thickness b (μm)
(device: quartz-h=0,5mm-λ=60μm)

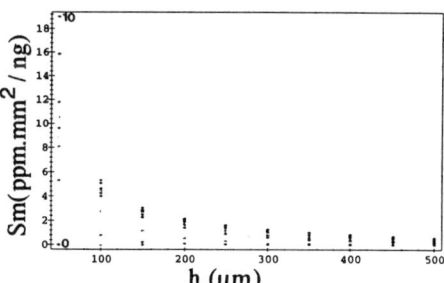

fig-3 : Influence of quartz thickness h (μm)
(device: quartz-λ=32μm-b=1μm)

3.2. *Mass sensitivity*

Adsorption of chemical species by the sensitive coating can modify its density and its elastic or electric properties. In the case of a density variation with b constant, this effect is called "mass loading". In order to study SH-APM sensors, the influence of various waveguide parameters on the relative mass sensitivity (S_m) for a relative density variation $\Delta\rho/\rho-5\%$ and a CuPC layer of constant thickness b=1 μm, was modeled:

$$S_m = \frac{\Delta f_m}{f_m} \times \frac{1}{\Delta m} \quad \text{where}: \Delta m = b \times \Delta\rho \quad (\Delta m \text{ is the mass variation per unit area (kg/m}^2\text{)}.$$

The simulations have been made on quartz plates with ST and BT cuts defined respectively by (0,132.75°,90°) and (0,32°,90°) Euler angles. Since the results were very similar for both cuts, only the results of the ST cut are presented.

3.2.1. Influence of the quartz plate thickness h
Figure 3 shows the influence of the quartz plate thickness, h, on the relative mass sensitivity, S_m, for modes m=0 to 10, for a quartz plate with $\lambda-32$ μm. We note that, S_m increases as h decreases : this can be due to a higher acoustic energy density at the quartz/CuPC interface. For h $\leq$ 100 μm, higher order modes are more sensitive, and for h~50 μm, $S_{10} \approx 19$ ppm.mm^2/ng. However, for thicknesses less than 300 μm plates become too fragile and are prohibited for SH-APM sensors.

3.2.2. Influence of the electrode periodicity of IDTs, λ
Figure 4 shows the IDTs electrode periodicity influence on the relative mass sensitivity S_m for modes m=0 to 10, for a quartz plate of thickness h=300 μm.
- For $\lambda \geq 60$ μm, all the modes of the device have a relative mass sensitivity S_m quasi independent of λ and approximately equal to 1,3 ppm.mm^2/ng.
- For $\lambda \leq 60$ μm, and for the first modes, the relative mass sensitivity S_m depends strongly on λ, but for higher order modes, the sensitivity tends towards the limit given before.

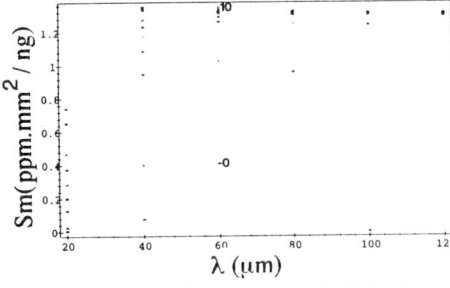
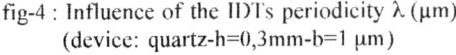
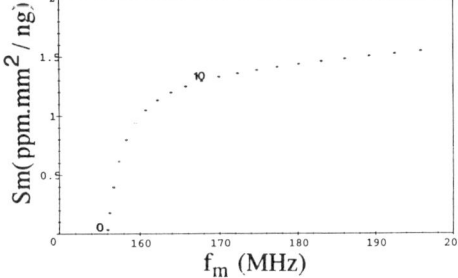

fig-4 : Influence of the IDTs periodicity λ (μm)
(device: quartz-h=0,3mm-b=1 μm)

fig-5 : Influence of the mode frequency
(device: quartz-h=0,3mm-λ=32μm)

3.2.3. Influence of the mode number, m
Figure 5 gives S_m for each mode as a function of the mode frequency f_m, for a device with h=300 μm and λ=32 μm. The curve confirms that, for small λ, the sensitivity increases

linearly with the mode order, up to m≈5. For λ=120 μm (curve not shown), all modes have quite the same sensitivity as that obtained in section 3.2.2.

3.2.5 Comparison with the mass sensitivity of SAW sensors

For a SH-APM device with h=300 μm and b=1 μm, the estimated mass sensitivity limit is about 1.5 ppm.mm^2/ng, while for a typical SAW device (f=158MHz) the sensitivity has been evaluated at S_{SAW}≈20 ppm.mm^2/ng [5].

Moreover, as it is shown in figure 4, the relative mass sensitivity S_m of higher order modes of an SH-APM device is quasi independent of the mode frequency f_m, while for SAW sensors it is proportional to the operating frequency f_{SAW} [5].

4. Conclusion

The mass sensitivity of feasible SH-APM sensors appears to be ten times lower than for SAW, probably due to the less acoustic energy density on the detection surface. The quartz plate thickness should be h=50 μm, to obtain the same mass sensitivity. However, SH waves that propagate in contact with liquids suffer low attenuation, and SH-APM devices offer an advantageous packaging in which the IDTs can be protected from the test medium. So, for biochemical applications or for the detection in an agressive medium, improving SH-APM sensors are still a deal.

Moreover, the mass sensitivity of an acoustic wave sensor depends not only on the wave sensitivity to mechanical or electrical perturbations, but also on the elastic properties of the sensitive coating. Preliminary simulations were undertaken, considering a quartz plate covered by a fluoropolyol (FPOL) sensitive coating of elastic constants : λ=4.7.10^9 N/m^2, μ=1.10^9 N/m^2 [6]. The study of the influence of λ and μ on the mass sensitivity is in progress. The aim remains to improve the mass sensitivity of SH-APM sensors.

In addition, experimental studies will be investigated in order to validate the theoretical results given in this paper.

Acknowledgements
This work was sponsored by the French Ministry of Defence, through a specific program from the Direction Générale de l'Armement (DGA-DCE) and especially from the Centre d'Etudes du Bouchet (CEB).

References
[1] Martin S J, Ricco A J, Niemczyk T M and Frye G C 1989 *Sensors and Actuators*, **20** 253-268
[2] Andle J C 1993 PhD dissertation University of Maine.
[3] Esteban I, Déjous C, Rebière D, Pistré J and Planade R, to be published in *Sensors and Actuators 1998*
[4] Rebière D 1992 PhD dissertation University of Bordeaux I (FRANCE).
[5] Grate J W, Martin S J and White R M 1993 *Analytical chemistry, vol* **65**, *NO* 21 941-948
[6] Ballantine D S, Wohltjen H 1989 *America Chemical Society*

MODELING OF THIN FILM GAS SENSORS. KINETICS OF GAS SENSITIVITY

V. Brynzari, G. Korotchenkov, S. Dmitriev

Microelectronics Laboratory, Technical University of Moldova,
Bld. Stefan cel Mare., 168, Chisinau, 2004, Moldova,
Fax: (373-2)-237509, E-mail: optolab@ch.moldpac.md

ABSTRACT

We develop our computation approach to the gas sensitivity of tin dioxide thin film in transient case. It is shown the possibility of quantitative interpretation of observed experimental facts.

INTRODUCTION

In previous report [1] we had developed steady-state model of thin film gas sensor (TFGS) based on Volkenstein' chemosorptional theory. This model allowed to described and calculate the main characteristics of gas sensors (GS).

In present report we examine the applicability of proposal model for the analysis in transient case, i.e. for kinetics of GS. At this stage we put the task of investigation of gas sensors behavior in oxygen atmosphere in the case of absence active gas. Without study of this mode it is difficult to interpret subsequently the kinetic of gas sensors in active regime (in the presence of active gas).

THEORY

Expanded equations of gas particles balance on semiconductor surface for the neutral and charge forms of chemosorbed oxygen [1] are the basis of our theoretical study

$$dN_O^o/dt = \alpha_o P_{O_2}(N^* - N_O)^2 - \beta_o(N_O^o)^2 - \beta_1 N_O^o + \beta_2 N_O^-, \tag{1}$$

$$dN_O^-/dt = \beta_1 N_O^o - \beta_2 N_O^-, \tag{2}$$

where, N^* - the number of adsorption centers; $N_O^o; N_O^-; N_O$ - the number of oxygen atoms in neutral and single charged form, and the total number of oxygen atoms, respectively; P_{O_2} - partial pressure of oxygen in gas phase; $\alpha_o; \beta_o$ - coefficients of adsorption and desorption; $\beta_1; \beta_2$ - coefficients of charging and neutralization of chemosorbed oxygen.

As in our last works, we considered the case of quasi-continuous distribution Hauss type of chemosorbed states in the bad gap. This significant feature has the principal importance for understanding of gas sensors characteristics. This fact can be illustrated, for example, for the dependence of surface potential on partial pressure in steady-state case. We got such relationship between neutral and charged forms of oxygen

$$N_O^o = N_O^- \exp((U_S + E_V - U_O^M)/E^*) , \qquad (3)$$

where E_V - bulk Fermi level position; U_O^M - position of maximum of oxygen chemosorbed level distribution; at this $E^* = \varphi_T + 1/2\sigma_o$, where φ_T - temperature potential; σ_o - dispersion of oxygen level distribution.

It is easy to set expression for the surface potential takes into account both (1) and (3) in steady-state case (i.e. $dN_O^o/dt = 0; dN_O^-/dt = 0$) and Schottky equation

$$U_S = U_O^M + E_V + 1/2(E^* \ln\tau_d/\tau_a) + C \cong const + E^* \lg P_{O_2} , \qquad (4)$$

where $\tau_a = 1/(\alpha_o P_{O_2} N^*)$ $\tau_d = 1/(\beta_o N^*)$; and C - some term independent on P_{O_2}.

As further will be showned τ_a and τ_d are the time constants, connected with adsorption and desorption processes. Thus, if the change of P_{O_2} is one decade $\left(P_{O_2}^{(1)}/P_{O_2}^{(2)} = 10\right)$, then $\Delta U_S = E^*$. In the case of discrete distribution we have $\Delta U_S = \varphi_T$.

Mathematical difficulty of the solution of equations is conditioned by the fact, that (1) and (2) are nonlinear and $\beta_1; \beta_2$ depends on surface potential. Moreover, in general case with the absense of equilibrium it is necessary to introduce Fermi quasilevel for the oxygen chemosorbed states, because there isn't equilibrium between charging and neutralization of surface oxygen.

Analysis shows that in this case initial equations have such forms:

$$dN_O^o/dt = \alpha_o P_{O_2}(N^* - N_O)^2 - \beta_o(N_O^o)^2 - dN_O^-/dt \qquad (5)$$

$$dN_O^-/dt = (1/2)\beta_1 N_O^o\left[1 - \exp(-\Delta F/\varphi_T)\right] - (1/2)\beta_2 N_O^-\left[1 - \exp(\Delta F/\varphi_T)\right] \qquad (6)$$

$$\Delta F = E^* \ln(N_O^o/N_O^-) + U_O^M - U_S - E_V , \qquad (7)$$

where ΔF - the difference between Fermi level position for the electrons and Fermi quasilevel position for the chemosorbed states; and $\beta_1 = a_1 n_S$, where a_1 - velocity constant of electron capture at oxygen level; n_S - electron surface concentration.

Numerical modeling shows that essential deviation from quasiequilibrium is possible when $\tau^o \gg \tau^*$, $\tau^- \gg \tau^*$, where $\tau^o = 1/\beta_1, \tau^- = 1/\beta_2; \tau^* = \sqrt{\tau_a \tau_d}$. In this case at the beginning of recovery, the changing of oxygen neutral states number takes place ($dN_O^o/dt \gg dN_O^-/dt$), and only then in condition $N_O^o \approx const$ one can observe the changing of oxygen states number in charged form ($dN_O^o/dt \ll dN_O^-/dt$). For response mode we have reverse process. However, for chosen changing of oxygen partial pressure the calculation and comparison with experiment shows that $\tau^o, \tau^- < \tau^*$. So, quasiequilibrium between neutral and charge forms of oxygen is observed. It means that $\beta_1 N_O^o \approx \beta_2 N_O$ and $dN_O^o/dt \gg dN_O^-/dt$ takes place in entire time interval of observation. Proceed from this facts one can see that equation (1) and (2) becomes independent and one can solve it analytically. We got such expression in case $N^* \ll N_O$

$$N_O^o(t) = N_O^o(0)\left(1 + \frac{N^*}{N_O^o(0)}\sqrt{\frac{\tau_d}{\tau_a}} \cdot th\frac{t}{\tau^*}\right)\bigg/\left(1 + \frac{N_O^o(0)}{N^*}\sqrt{\frac{\tau_a}{\tau_d}} \cdot th\frac{t}{\tau^*}\right) \qquad (8)$$

$$N_O^-(t) = (\beta_1/\beta_2)N_O^o(t), \quad where \quad N_O^o(0) = N_O^o \quad at \ t=0. \qquad (9)$$

Equation (7) is nonlinear relative to N_O^- because $\beta_1/\beta_2 = f(U_S)$. We found that this relationship is

$$\beta_1/\beta_2 = \exp\left(-(U_S + E_V - U_O^M)/E^*\right) \qquad (10)$$

Taking into account, that conductance of SnO_2 thin films at large surface band bending is determined approximately by such relationship

$$\sigma = \sigma_O \exp\left(-(U_S / \phi_T)\cdot(L_D / D)\right), \quad \text{where } L_D \text{ - Debye length; and } D \text{ - film thickness,} \quad (11)$$

one can show that

$$\sigma \cong \sigma_0 \left[N_O^o(t)\right]^{-\eta}, \quad \text{where} \quad \eta = (E^* / \phi_T)\cdot(L_D / D). \quad (12)$$

At that, changing of conductance on the time $\Delta\sigma(t)$ is near to exponent dependence at $t>0.3\ \tau^*$ for the charging curve (recovery) and at $t>\tau_d$ for the discharging curve (response). Precise calculation of conductance were carried out according to algorithm, described by us in [3].

For both response and recovery characteristics it is distinctive that there is two regions on dependence of $N_O^o(t)$. In the region of small times ($t< 0.3\ \tau^*$) $N_O^o(t)$ has such forms

$$N_O^o(t)(\text{re cov ery}) = N_O^o(0)\left(1 + (N^* / N_O^o(0))\cdot(t/\tau_a)\right) \quad (13)$$

$$N_O^o(t)(\text{response}) = N_O^o(0)\Big/\left(1 + \left(N_O^o(0)/N_V\right)\cdot(t/\tau_d)\right). \quad (14)$$

For these conditions kinetics is determined by only adsorption or only desorption processes with the time constant τ_a and τ_d, respectively. In the region of large times there is combine mechanism with interaction both processes ,which determined by equation (8) and time constant $\tau^* = \sqrt{\tau_a\tau_d}$.

We can note such significant features of kinetics:

1. time constant of gas sensors and gas sensitivity is determined by kinetics of oxygen neutral form, at that, total time of transient mode (at level=0.9) $\sim \tau^*$;

2. $\tau_a \sim 1/P_{O_2}$, whereas $\tau^* \sim 1/\sqrt{P_{O_2}}$ (we have to note that P_{O_2} is the pressure at $t >0$, at $t=0$ there is stepped changing of oxygen partial pressure);

3. ratio $\tau_{res}^* / \tau_{recov}^* = 10$ for chosen changing of pressure ($P_{O_2} =20\%$ and $P_{O_2} =0.2\%$);

4. activation energy of time constant τ^* is equal $1/2(E_a + q_{O_2})$, because $\tau^* \sim \sqrt{\alpha_o\beta_o}$, where $\alpha_o \sim \exp(-E_a/kT)$; $\beta_o = \exp\left(-q_{O_2}/kT\right)$ and E_a, q_{O_2} - activation energy of adsorption and desorption, respectively;

5. from equation (8) one can see such inequality $\tau_d < \tau_a$, which takes place for all cases , in opposite case $(\tau_d > \tau_a)$ $N_O^o(\infty) > N^*$. This relationship is fundamental for the chemisorptional processes;

6. in response mode significant decreasing of P_{O_2} can lead to such case, that dependence $N_O^o(t)$ will be circumscribe by expression (4) in entire time interval of observation. One can see that total time of response (at the level 0.9) $t_{des} \approx 10(N^* / N_O^o)\cdot\tau_d$.

At the region of small pressure $t_{res} \to t_{des}$. This is in agreement of general reasoning because only desorption process must determine the kinetic in vacuum.

COMPARISON WITH EXPERIMENT.

In this section we present kinetic dependence obtained by simulation and measured in experiment. Kinetic characteristics were measured at various temperatures. Undoped SnO_2 films were obtained on Al_2O_3 substrates by spray pyrolysis method with thickness ~ 20-30 nm [2].. Estimations of electron concentration were gave $n\sim(5-10)10^{18}$ cm^{-3}.

One can to note some significant features of experimental dependencies (fig. 1(a)):

1. response dependencies are near to exponent in entire time interval whereas for the recovery dependencies there are the deviation from this law at the beginning of transient process;

2. $\tau(\text{response}) \gg \tau(\text{recovery})$, their rations is approximately equal ($P_{O_2} =20\%$ / P_{O_2} $0.2\%)^{1/2}$;

3. activation energy of time constants ~ 1,3 eV at the entire time for response and at middle and large times for recovery. At small times activation energy for recovery ~ 0.4-0.5 eV.

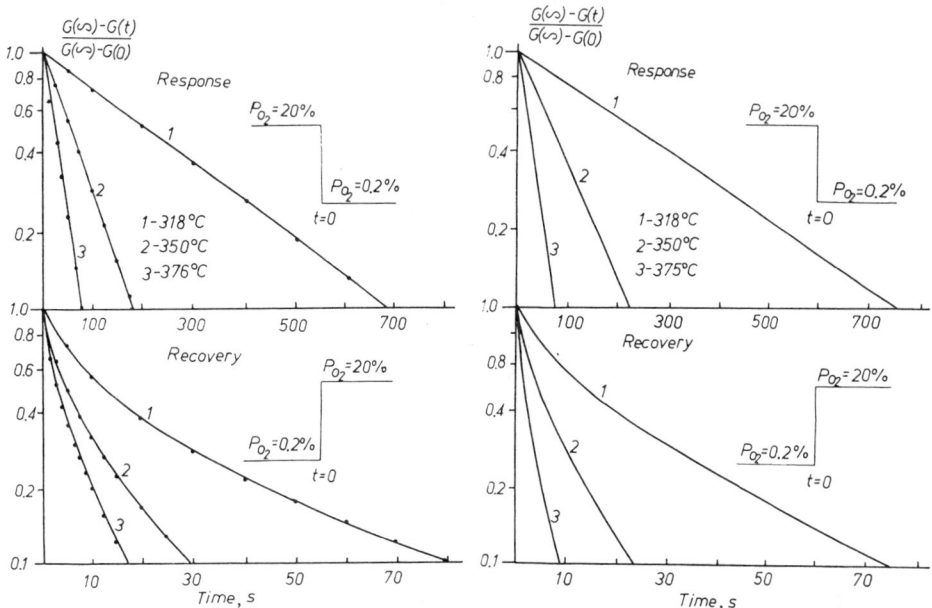

Fig.1. Experimental (a) and theoretical (b) kinetic characteristics of gas sensitivity.

Calculation dependencies are shown in Fig.1(b). The values of parameters, used in calculations were the next: $N_C=2.10^{20}$ cm^{-3}; $N_d=10^{19}$ cm^{-3}; $\mu=10$ cm^2/V.s; $D=20$ nm; $\varepsilon=12$; $N^*=10^{14}$ cm^{-2}; $\sigma_0=0.3$ eV; $U_O^M=1.3$ eV; $\alpha_0=(0.01/N^*)\exp(-E_a/kT)$; $\beta_0=(10^{16}/N^*)\exp(-q_{O2}/kT)$; $E_a=0.5$ eV; $q_{O2}=2.2$ eV; $a_1=10^{-9}$ cm^3/s. The pri-exponent multiplies at the α_0 and β_0 were estimated according to [3] in assumption immobile adsorption layer.

We can note the quite well accordance between results of calculations and experiment. Within the framework of this model the values of activation energy of τ^* in the region of middle and large times are found its explanation.

In the region of small times the activation energy for $\tau_{responsw}$ ($t<\tau_d$) is q_{O2}, and for $\tau_{recovery}$ ($t<0.3\tau^*$) is E_a. However, since $\tau_d<<\tau^*$ this region practically is not observed in response characteristics, whereas it has pronounced form in recovery characteristics.

Also we present in our report results of modeling thermal desorption spectra, which closed connected with this problem.

REFERENCES

1. V.Brynzari, G.Korotchenkov, S.Dmitriev. Proc. of Intern. Conf. EUROSENSOR XI, Warsaw, Poland (21-24 Sept., 1997), ISBN, 1997, V.1, p.91-94.
2. S.Dmitriev, G.Korotchenkov, V.Brynzari, Proc. of Intern. Conf. on Microelectronics, Nis, Serbia (12-14 Sept.,1995), IEEE,1995, V.2. p.585-588 and 589-592.
3. Trepnel. Hemosorbtsiya. Moscova, 1958.

Coherent light detector on a HTSC film

V M Aroutiounian and V V Buniatyan

Yerevan State University, 375049, Yerevan, Armenia

Abstract. Using the phenomenon of the change of the surface microwave impedance of the HTSC films at its photoexcitation, we analyse the work of the HTSC photodetector, considering it as the photon counter. The results obtained suggest the possibility of the detection of optical coherent signals by such a detector at the non-bolometric absorption of optical radiation. The detectivity, signal / noise ratio and response of photosensor are estimated.

1. Introduction

Let an optical signal modulated on the intensity $\Phi = \Phi_0 (1+sin\omega_s t)/2$ incidents on a HTSC film. Here Φ_0 is the intensity of the incident radiation, ω_s is the frequency of modulation, t is time. It is known [1-2] that the active and reactive components of the surface microwave impedance can be modulated under a pumping law. High sensitivity of superconductors and HTSC films to optical exposure revealed in a number of works opens a possibility of realisation optically controlled sensors, switches as well as mixers, delay lines and other microwave devices [3-4].

Setting that at the absorption of the optical radiation by a HTSC film non-bolometric transfer of electrons through the superconducting gap (the Cooper pairs breaking or destroying) takes place, but the general concentration of electrons within the framework of the two-liquid model [2] is kept ($n = n_s + n_N = n_{s\Phi} + n_{N\Phi}$), the concentration of electrons in a superconduction condition n_s decreases, and the concentration of "normal" electrons n_N increases by the same magnitude ΔN. Then we can write down:

$$n_{S\Phi} = n_S - \Delta N , \quad n_{N\Phi} = n_N + \Delta N . \tag{1}$$

As the dark surface active resistance R_S of the HTSC film [5] is proportional to λ_L^4 σ_N, where λ_L is the London penetration depth, σ_N is the conductivity of the film in the normal condition and $\lambda_L \sim n_S^{-1/2}$, $\sigma_N \sim n_N$, it is easy to show that at the absorption of light by the film the light surface active resistance $R_{s\Phi}$ is proportional to ΔN^m , where m is a parameter depending on the absorbing and reflecting ability of the film and the recombination and electron - phonon interaction processes. As the consequence, it is natural to expect that at the absorption of optical energy a change (reduction) of the surface superconducting current will be observed also. Researches of the excitation (laser radiation) influence carried out in Refs. [6-9] have shown that the formation of additional non-paired electrons reduces the gap, but the superconduction condition is not destroyed down up to determined concentration of non-equilibrium (additional) electrons. And, as it is shown in these papers, in result of the interaction of the laser radiation with the superconductor an arising in non-thermal mode the intermediate condition (where the superconducting phase and normal one co-exist) is accompanied by the emit of phonons.

For the concentration of non-equilibrium (additional) quasi-particles the following expression is received in Ref. [6,7]:

$$\Delta N \cong \eta P \tau_e / d \Delta(T, \Delta N) , \qquad (2)$$

where P is the power of the incident radiation per unit of the area (J/cm^2), τ_e is the effective lifetime of additional quasi-particles, $\Delta(T, \Delta N)$ is the magnitude of the energy gap, η is an effective quantum exit indicating which part of power (energy) absorbed by the film is spent for the creation of additional quasi-particles, d is the thickness of the film.

Using Eq. (2), it is possible to show that at the absorption of light by the film we have [2]:

$$R_{s\Phi} \cong R_{s0} + \Delta R_s \sin\omega_s t ,$$

$$\qquad (3)$$

$$R_{s0} \cong R_{sd} + \Delta R_s , \quad R_{sd} \cong \frac{a_1 n_N}{n_s^{3/2}} , \quad a_1 = \frac{\omega^2 \sqrt{\pi m_s \tau_e}}{qc} , \quad \Delta R_s \cong \frac{a_1 n_N}{2 n_s^{3/2}} a_2 , \quad a_2 = \frac{\eta \Phi_0 \tau_e}{n_N V} ,$$

where ω is the frequency of a microwave field, m_S is the effective mass of electron in the superconduction condition, V is the volume of the HTSC film, c is the velocity of light.

In the present work possibilities of the detection (mixing) of optical coherent signals by the HTSC films and using this phenomenon for the creation of promising photosensor are analysed.

2.Theoretical analysis

Assume that the density of the incident photon flow Φ modulated on the intensity depends on time and can be presented in the form:

$$\Phi = \Phi_0 + \Phi_1 exp(i\omega_s t) \qquad (4)$$

Consider two beams of the coherent light with an identical polarisation with frequencies $\omega_m/2\pi$ and $\omega_n/2\pi$ (let their difference is equal $\omega_s/2\pi$). Let these beams superimposed on one another on a photosensitive surface of the HTSC film having the length W. If it is possible to neglect by signals of all combinational frequencies, except $\omega_s/2\pi$, we receive the following expression for the ratio of the amplitude Φ_1 of the instant flow, appropriate to the differential frequency, to the steady component of the flow Φ_0:

$$\Phi_1 / \Phi_0 = 2\sqrt{\Phi_m \Phi_n} / \left(\Phi_m + \Phi_n\right), \qquad (5)$$

where Φ_m, Φ_n are the densities of the photon flows for two optical (mixed) signals.

The starting position for an analysis spent below just as it carried out in Ref. [10], is that fact that the photosensor made of the HTSC film is determined as a photon counter. It is possible to judge the amplitude and frequent characteristics of such photomixer by determining a pulse of the current on an output at the absorption of one photon. It is assumed that, at the photon absorption and the occurrence of additional normal quasi-particles, the surface microwave impedance of the film increases which results in a reduction of the surface superconducting photocurrent (SSPC). For determination of the alternating SSPC j_{sl}, caused by the alternative component of the density of the photon flow $\Phi_l \sin\omega_s t$, we proceed as follows. Let photon $h\nu$ of the incident radiation is absorbed in a point x inside the HTSC film. Electrons after the breaking of the Cooper pairs passed in the normal condition. It should result in a reduction of the current. If at single photoexcitation a pulse (change) of the current is known, the amplitude and frequency characteristics can be determined by the Fourier transform of the pulse of the current. The spectral decomposition of an "elementary" alternating current per unit of the length is possible to calculate proceeding from the multiplying together the Fourier transform of the transfer function of the system with the velocity of the Cooper pairs breaking on the input. Let t_r is the transfer time of electrons through the HTSC film, $g(x)$ is the rate of "phogeneration" of electrons from the superconduction condition in normal one per unit of the length due to the flow incidenting on the law $\Phi_l \sin\omega_s t$, $t(x)$ is time on which electrons remain in the normal condition, i.e. $t(x) \equiv \tau_e$ At uniform illumination of the HTSC film $g(x) \cong \Phi_1 / SW$, where S is the surface of the film. Let $P(t_1-t_2)=P(t)$ is the probability of the transition of electrons in the superconduction condition or the probability of the recombination (drift) during the time $d(t_1-t_0)=dt$ after the start of the photoexcitation. If the probability $P(t)$ has the exponential character [6-9] $P(t) \cong exp(-t / \tau_e) / \tau_e$,using the Fourier analysis, we receive for the alternating surface "super-photocurrent " the following expression:

$$\tilde{J}_s(\omega_s) \cong \frac{2\eta q \, \Phi_1 S \tau_e}{t_r} \cdot \frac{1}{1 + j\omega_s \tau_e} . \qquad (6)$$

For the high frequency ("optical") equivalent circuit, if loads are matched, we have the following expression for the nominal power

$$\tilde{P}(\omega_s) \cong \frac{1}{8}\left(2\eta q \, \Phi_1 S\right)^2 \frac{\tau_e^2}{t_r^2}\left(1 + \omega_s^2 \tau_{\ni\varphi}^2\right)^{-1} G_p^{-1}. \qquad (7)$$

where G_p is the microwave conductivity, which is a function of the intensity of the light flow [7-12].

If we neglect the $1/f$ noise, amplifier noise [8,9] for the case of the non-bolometric response, we obtain the following expressions for the NEP and signal/noise ratio, respectively:

$$NEP \cong \frac{2\sqrt{kTG_p}}{\Re}\left\{1 + \frac{\eta q \Re h\nu S \Phi_1}{kTG_p}\right\}^{1/2} \quad, \quad \Re = \frac{2q\tau_e}{h\nu t_r \sqrt{1 + \omega_s^2 \tau_e^2}}$$

$$\left.\frac{S}{N}\right)_p \cong \frac{\left(\eta \Phi_1 Sh\nu\right)^2 \Re^3}{16 G_p}\left\{kTG_p \Delta f + \frac{\eta q \, \Phi_1 Sh\nu \Delta f \Re}{2}\right\}^{-1/2}$$

(8)

where $\Re$ (A/W) is the current responsivity, Δf is the frequency band.

3. Discussion

If $\Phi_1 = 0,1 \times \Phi_0$, $\Phi_0 \cong 1,5 \times 10^{11}$ photon s^{-1} cm^{-2}, $\tau_e \cong 10$ μs, $T=64K$, $\Delta f=30$ Hz, $\lambda=12,5$ μm, $h\nu \cong 0,1$ eV [8], $S \cong 50 \times 100$ μm^2 [9], $G_p \cong 0,5$ sm [11] and $t_r \cong 10^{-7}$ s, $\eta \cong 0,1$, $\omega_s = 10^6$ Hz, from Eq. (8) we obtained $\Re \cong 0,2 \times 10^3$ (A/W), NEP$\cong 2 \times 10^{-13}$ WHz$^{-1/2}$ and for the detectivity $D^* = (S\Delta f)^{1/2}/NEP \cong 3,5 \times 10^{10}$ cmHz$^{1/2}$/W.

Above-mentioned parameters of the photosensor working in the non-bolometric mode are better or comparable with the bolometric and other type photosensors and can find some inportant applications in low temperature sensoric and electronics.

References

1. Carlsson E, Gevorgian S, Kolberg E et al 1994 *IEEE MIT Topical Meeting on Optical Microwave Interaction.* 195-197
2. Aroutiounian V M and Buniatian V V 1997 *J. Contemporary Physics* **32** 310-312; Aroutiounian V Buniatian V and Gevorgian S 1997 *Proc. of the 97 GigaHertz Symp., Kista, Sweden,* 118-119
3. Gupta D, Donalson W R et al 1993 *IEEE Trans. Appl. Supercond.* **3** 2895-2898
4. Track E K, Drake R E and Hobenwarterg G K 1993 *IEEE Trans. Appl. Supercond.* **3** 2899-2902
5. Vendik O, Galchenko S, Kaparkov D et al 1994 *Models of HTSC Transmission Lines as Appl. for CAD of Microwave Integrated Circuits. -Rep. N9, ISSN 1103 - 4599, ISRN CTH - MVT - R - 9 - SE*
6. Parker W H and Williams W D 1972 *Phys. Rev. Lett.* **29** 924-927
7. Owen C S and Scalapino D J 1972 *Phys. Rev.Lett.* **28** 1559-1561
8. Bluser N 1995 *J.Appl.Phys.* **78** 7340-7351
9. Moix D B, Scherrer D P and Kneubuhl F K 1996 *Infrared Physics and Technology.* **37** 403-426
10. Domenico M Di and Svelto O 1964 *Proc. IEEE.* **52** 142-148
11. Kleinhammes A, Chang C L, Moulton WG et al LR 1991 *Phys. Rev. B.* **44** 2313-2319
12. Mai Z, Zhao X, Zhou F et al 1997 *Infrared Physics and Technology.* **38** 13-16

Simulation and Modelling
Paper presented at Eurosensors XII, 13–16 September 1998
© *1998 IOP Publishing Ltd*

Thermal simulation of surface micromachined polysilicon hot plates of low power consumption

Marius Dumitrescu, Cornel Cobianu, Dan Lungu, Dan Dascalu,
Adrian Pascu*, Spas Kolev** and Albert van den Berg***

National Institute of Microtechnology, P.O. Box 27-17,
77550 Bucharest, Romania. E-mail: ccobianu@pcnet.pcnet.ro,
* "Politehnica" University Bucharest, Mechanics Faculty, Romania
** University of Sofia, Faculty of Chemistry, Sofia, Bulgaria
***MESA Research Institute, University of Twente, Faculty of Electrical
Engineering, P.O. Box 217, 7500 AE Enschede, The Netherlands.

Abstract. A simple, IC compatible, surface micromachined polysilicon membrane was technologically designed and thermally simulated by 3D finite element "COSMOS" program in order to investigate its capability to work as a micro hot plate for a gas sensing test structure of low power consumption. For an optimised lay-out based on four "poly" suspended bridges and a central "poly" pillar supporting the 110x110 µm "poly" membrane separated from the silicon substrate by 1 µm of air gap, temperatures as high as 673 K were obtained for an input power of 100 mW.

1. Introduction

Recently, there is a strong trend to the integration of gas sensors due to the advantages offered by the silicon microtechnology. Hot substrates (aimed at low input power) necessary for SnO_2 -based gas sensors and micropellistors were microfabricated by releasing a dielectric membrane through back side [1] or front side bulk micromachining [2]. Earlier, we have made a thermal simulation of the temperature profile on a dielectric membrane [3]. Pioneering work on low power surface micromachined hot plates for micropellistors was reported recently [4].

It is the purpose of this paper to investigate by means of 3-D COSMOS simulator the capabilities of polysilicon-based surface micromachining technology to provide an alternative route for making hot plates acting as substrates for gas sensing applications at a relative low input power.

2. Description of sensor structure and thermal simulation

For the thermal simulation we have considered a simple, IC compatible, test structure consisting of doped polysilicon suspended membrane (acting as heater and temperature sensor) released by the etch of sacrificial PSG layer (1 µm in thickness) and which is isolated from SnO_2 sensing layer by a SiO_2 CVD layer. A W/Au metallization for both heater and sensor electrodes is used as an example. In order to avoid polysilicon membrane sticking to the substrate, a thin polysilicon pillar was preserved in the centre of the membrane, in addition to the four poly feet. We have considered two cases. In the first case, the four "poly" bridges are suspended close to membrane and make the contact to silicon at about 60 μm far from the poly membrane. We call this lay-out "bridge" type. When the four "poly" feet make the contact to the silicon exactly at the corners of the squared membrane, we call this lay-out "table" type. The sensor lay out (top view) and cross sections through two axes are presented in Fig. 1, for the "bridge" type.

The sensor geometry including the bulk silicon and the packaging support of the chip is meshed in 4709 nodes and 3992 elements. The last ones can be volume elements (SOLID), when the three dimensions are comparable (in our case bulk silicon, packaging support and air gap) or surface elements (SHELL) when one element dimension is much less than the others (thin films of polysilicon, SiO_2, W/Au and SnO_2). For each type of element the thermal and mechanical constants are loaded.

The boundary conditions for the steady state thermal analysis considered that room temperature is obtained at the bottom of silicon chip. The heat losses by convection and radiation are introduced for all the external surfaces in contact with air (excepting the air gap region where the conduction through air is considered). The radiation to the cold walls of the test structure package was also considered by means of different view factors. The heat flux applied to the heater is obtained from the conversion of the input power.

3. Results and Discussion

The real advantage of integrated hot plates for gas sensing applications appears when membrane temperatures as high as 500-700 K are obtained for input power less than 100 mW. Taking into account the concern about the heat loss by conduction through air due to the small air gap between the silicon surface and the "poly" membrane (this is the price paid by the surface micromachining), and by conduction through polysilicon "feet" and "pillar" keeping the suspended membrane, preliminary simulations were performed on different lay-out configurations to find an optimum sensing area (for the minimisation of surface area dependent heat losses, convection and radiation) and optimum dimensions of the feet.

In Fig. 2 we show the simulation results of maximum temperature on the membrane as a function of heat flux for the two geometrical configurations of the membrane. At the same input power and membrane surface area, the temperature can be increased by about 200 K for the "bridge" type lay-out with respect to "table" type as a result of introducing long and thin suspended feet and thus minimising the heat losses by conduction through the solid suspended bridges. On the other hands, it is proved that for the "bridge" type it is possible to

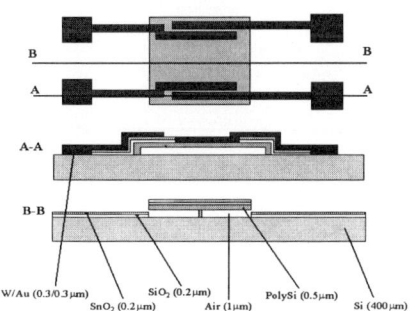

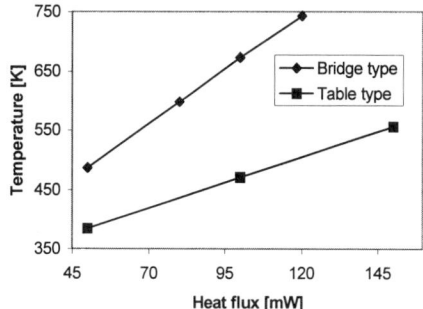

Fig. 1. Top view and two cross sections through "bridge" gas sensing microstructure.

Fig. 2. Temperature vs. input power for "bridge" and "table" type lay-out.

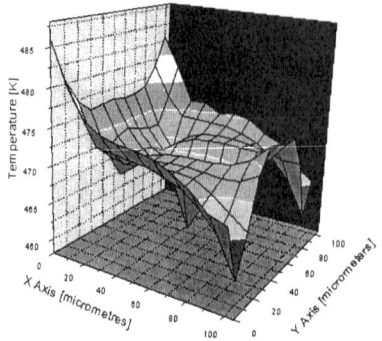

Fig.3. Temperature mapping on the membrane surface for an input power of 50 mW.

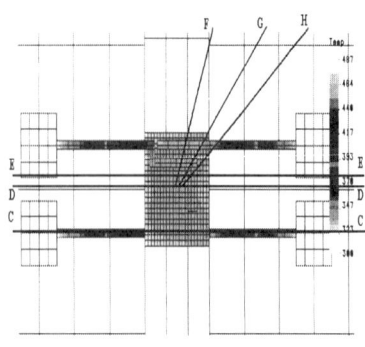

Fig.4. Mesh of the "bridge" type sensing structure.

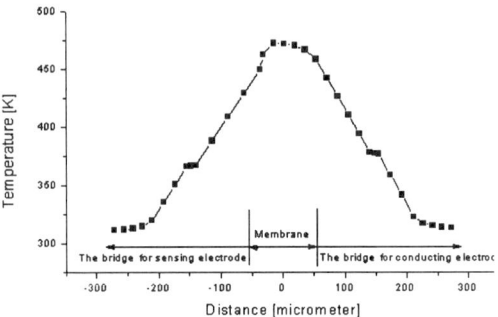

Fig. 5. Temperature profile along the C-C axis from Fig. 4.

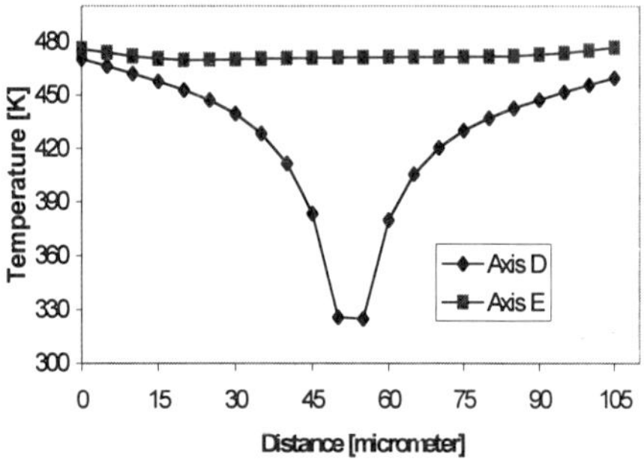

Fig. 6. Temperature profile along the D-D and E-E axis from Fig. 4.

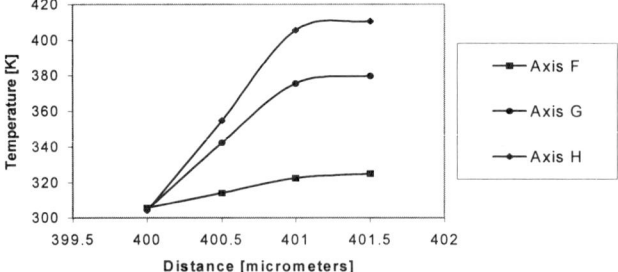

Fig. 7. Temperature profile through the air gap on the vertical direction recorded perpendicular on the membrane in the point F, G, H from the Fig.4.

get a temperature of 673 K (400°C) for an input power of 100 mW.

In Fig. 3 we present the 2-D temperature profile on the membrane surface (110x110 μm) in the presence of the central pillar for an input power of 50 mW and air gap of 1μm. As expected, in the immediate vicinity of the central pillar, on a small area the temperature is deeply decreasing.

In Fig. 4 we show the top view of the meshed test structure and the temperature mapping along the whole sensor structure including the four suspended bridges and metallic pads (two for sensor electrodes (on the left side) and the other two for heating resistor). In order to offer a detailed information about the temperature profile on different directions, we have defined on this figure the three axes (C-C, D-D and E-E) and the three points F, G, H, the later used for the temperature profiling in the z direction to the bulk of the silicon.

In Fig. 5 we present the temperature profile along the axis C-C of the Fig. 4, for the input power of 50 mW. One can see that close to the pads the room temperature is obtained.

In Fig. 6, the temperature profiles along axes D-D and E-E of Fig. 4 are shown for an input power of 50 mW. It is demonstrated that the temperature on the pillar is almost equal to room temperature but increases rapidly outside of it. At only 10 μm far from the pillar in all directions on the membrane, the temperature comes close to the high value.

In Fig. 7, the temperature profiles of the air gap in the "z" direction starting from the point F, G, H of the membrane (marked on the Fig. 4) to the surface of the bulk silicon are presented (the origin of axis is at the bottom side of the silicon wafer) for an input power of 50 mW. From this figure it is obtained that the (substrate) silicon surface and central "poly" pillar are practically at the room temperature, while at distance of 5-10 μm far from pillar, the temperature on the membrane is close to maximum value.

4. Conclusions

Steady state thermal simulations by 3D, FE "COSMOS" program have shown that temperatures as high as 673 K can be obtained on the surface of the surface micromachined polysilicon membrane with an air gap of 1 μm for an input power of 100 mW and a membrane surface area of 110 x 110 μm in the case of four suspended bridges and a central "poly" pillar (to avoid "poly" sticking to the substrate) supporting the membrane. The presence of central thin pillar alters the membrane temperature on an area of about 10x10 μm around it. The temperature at the surface of silicon substrate is close to room temperature.

Acknowledgements
EU COPERNICUS project CP 940963 "PORSIS" and the Romanian Ministry of Research and Technology are kindly acknowledged for the financial support of this paper.

5. References

1. N. Najafi, K. Wise and J.W. Schwank, 1994 *IEEE-ED*, **vol. 41**, 1770-1777.
2. D-D. Lee, W-Y Chung, T-H Kim and J-M Baek, 1995 *TRANSDUCERS '95 EUROSENSORS IX*, Stockholm, paper 210-D5, 827-830.
3. M. Dumitrescu, C. Cobianu, D. Lungu, D. Dascalu, A. Pascu, A. van den Berg, J.G.E. Gardeniers, S. Kolev, D. Csaba, and I. Barsony, 1997, *Proceeding of EUROSENSORS XI*, Warsaw, September 21-24, 743.
4. A.J. Ricco, J.H. Smith, R.P. Manginell, R.C. Hughes, D.J. Moreno, R.J. Huber and S.d. Senturia, 1996 May 5-10 Los Angeles *ECS Meeting Abstracts* **vol. 96-1,** 1506.

Author Index